Urban Sustainability

The Urban Sustainability Book Series is a valuable resource for sustainability and urban-related education and research. It offers an inter-disciplinary platform covering all four areas of practice, policy, education, research, and their nexus. The publications in this series are related to critical areas of sustainability, urban studies, planning, and urban geography.

This book series aims to put together cutting-edge research findings linked to the overarching field of urban sustainability. The scope and nature of the topic are broad and interdisciplinary and bring together various associated disciplines from sustainable development, environmental sciences, urbanism, etc. With many advanced research findings in the field, there is a need to put together various discussions and contributions on specific sustainability fields, covering a good range of topics on sustainable development, sustainable urbanism, and urban sustainability. Despite the broad range of issues, we note the importance of practical and policy-oriented directions, extending the literature and directions and pathways towards achieving urban sustainability.

The series will appeal to urbanists, geographers, planners, engineers, architects, governmental authorities, policymakers, researchers of all levels, and to all of those interested in a wide-ranging overview of urban sustainability and its associated fields. The series includes monographs and edited volumes, covering a range of topics under the urban sustainability topic, which can also be used for teaching materials.

Jian Feng · Xueming Li · Yungang Liu

Urban Living Space in China

Science Press
Beijing

Springer

Jian Feng
College of Urban and Environmental Sciences
Peking University
Beijing, China

Xueming Li
College of Geographical Sciences
Liaoning Normal University
Dalian, Liaoning, China

Yungang Liu
College of Geographical Sciences
South China Normal University
Guangzhou, Guangdong, China

ISSN 2731-6483 ISSN 2731-6491 (electronic)
Urban Sustainability
ISBN 978-981-95-0594-4 ISBN 978-981-95-0595-1 (eBook)
https://doi.org/10.1007/978-981-95-0595-1

Jointly published with China Science and Technology Press Co., Ltd.

Translation from the Chinese Simplified language edition: "《中国城市生活空间》" by Jian Feng et al.,

This Springer imprint is published by the registered company Springer Nature Singapore Pte Ltd.
The registered company address is: 152 Beach Road, #21-01/04 Gateway East, Singapore 189721, Singapore

Preface

Over the past 40 years of reform and opening up, with the development of economic growth and rapid urbanization, the lifestyles and living spaces of Chinese urban residents have undergone radical changes, and the new phenomena and problems related to urban residents' lives that have emerged along with this process have attracted extensive attention from academics at home and abroad. In the new period, human beings, human activities and human life are the most important elements to be concerned about in the research of new urbanization in China. Against this background, urban living space has undoubtedly become an important hand in the research of new urbanization and urban spatial transformation in China, and is a hot area in the research of urban geography and urban-rural planning in China.

The general idea of this book is to start with an overview of the overall development and evolution of urban living space in China and a summary of the state of research to build up a general impression; then move on to the basic content sections of "urban lifestyles" and "evaluation of the quality of urban life and quality of the human habitat", which are the more obvious intersections between geography and sociology, and the science of the human habitat to help understand urban living space in China from a broader disciplinary context; then it enters into the interpretation of different types of urban living space, including residential space, interaction space, activity space and leisure space, which is the main content of the book, reflecting the main framework of the composition of urban living space under the perspective of geography; on this basis, it explores the living space of different groups and new forms of living space, which expands the scope of attention of the traditional research on urban living space, and in particular highlights the impact of the new technology development and the emergence of new living concepts on urban living space, which not only increases the interest of the book, but also reflects the contemporary nature of the book; finally, the whole book is summarized and the future research direction is prospected.

The book is divided into 11 chapters. The main contents of each chapter are as follows.

The first chapter is an introduction, which mainly introduces the research background of urban living space in China, the framework structure of this book, and identifies the relevant important concepts. Chapter 2 provides an overview of urban living space in China, the traditional living space of Chinese urban residents, and the development and evolution of the living space of Chinese urban residents on the basis of a large number of Chinese and Western research literatures, and sorts out the relevant research results about urban living space in China, so that readers can build up a general impression of the urban living space in China and grasp the context of the era in which it has evolved, and pave the way for the following discussion.

Chapter 3 discusses urban lifestyles and their evolution, mainly analyzing the traditional lifestyles of Chinese urban residents, the impact of economic and social changes on lifestyles, the characteristics of the evolution of Chinese urban residents' lifestyles, and the formation of new lifestyles. Chapter 4 discusses the quality of urban life and the evaluation of the quality of the human environment, focusing on analyzing the types and characteristics of the differentiation of the quality of life in Chinese cities, the types of urban human settlements and the factors affecting them, the evaluation of the quality of urban human settlements and its spatio-temporal differentiation, as well as the outlook of the reconstruction of the quality of life in Chinese cities based on the improvement of the human settlements. These two chapters are not the research scope of pure geography in the traditional sense, but rather emphasize the intersection of sociology and human settlement science with geography, which are important foundations for understanding urban living space in China, providing a broader disciplinary support for the systematic interpretation of the various types of urban living space in China below, and forming an important feature of the framework system of this book.

Chapters 5–8 are sub-types of urban living space in China. That is, the authors systematically explore urban living space in China by subtypes from four aspects, namely residential space, interaction space, activity space and leisure space respectively. Chapter 5 discusses residential space, mainly analyzing the urban residential spatial structure and the pattern of residential areas, the differentiation of urban residential space and the differentiation of groups of residents, the habitability of urban communities and their differentiation, migration and the change of residents' living space, and the interaction between residence, work and commuting, and so on. Chapter 6 explores communication space, and based on the analysis of the "network of relationships" as a way of carrying out interpersonal relationships and social interaction among urban residents in China, the authors study the role and characteristics of "social networks" in the daily social interactions of urban residents at a smaller spatial scale, as well as the "virtual space" dominated by the Internet-based community. Chapter 7 explores activity space, and the orientation of this chapter is not to expound all types of activity space, but to grasp several important clues to interpret the activity space of residents, such as the activity space based on life paths and life course in the long time scale, the activity space based on life paths and life logs in the short time scale, as well as the activity space based on the nesting of the family and the local order, which is at the forefront of the international research field. activity space. By interpreting the characteristics of the activity space of Chinese urban residents from these clues, we can avoid the tediousness that may be generated by interpreting each type one by one. Chapter 8 explores leisure space, combining numerous research literatures on Chinese urban leisure space, mainly analyzing the formation and evolution process of Chinese urban leisure space, the structural pattern and spatio-temporal differences, as well as the interaction between leisure space and leisure behavior in urban China. It is worth pointing out that, in the process of researching different types of living space, we focus on the combination of theoretical generalization and empirical research, the combination of quantitative analysis and qualitative description, the combination of textual analysis and spatial visualization, and the combination of academic discussion and storytelling, and the use of multiple research methods and paradigms, so as to strengthen the research norms based on a certain degree of in-depth empirical research, and then to summarize the general characteristics of urban living space in China, in order to contribute to the relevant academic research in this field.

Chapter 9 explores the living space of different groups, which is an interesting and popular topic. It focuses on the more representative and diverse groups within Chinese cities, especially by analyzing the living spaces of transnational ethnic groups, disadvantaged groups and minorities with their collage and mosaic characteristics, fully demonstrating the plurality and complexity of urban living space in China.

The theme of Chapter 10 is the exploration of "new" living spaces. In order to reflect the impact of Internet technology, information and communication technology and new life conceptual changes on urban living space, it is devoted to exploring the impact of "new" living spaces, especially paravirtual communities, micro-life, fashionable living spaces and so on, on the formation of urban living space for urban residents, as an extension and supplement to urban living space in China.

Chapter 11 summarizes the whole book, and at the same time looks forward to further research directions and research topics on urban living space in China. In short, through the study of this book, we hope to have a more comprehensive and era-specific grasp of the formation background, development characteristics, evolution mechanism and future research direction of urban living space in China, and also hope to further enhance readers' understanding of the theoretical framework and empirical research norms of urban living space research in China.

This book is designed to introduce the theoretical framework, basic concepts, cutting-edge methodologies, and important findings and conclusions of urban living space in China to a wider readership, based on the concepts of "cutting-edge", "scholarly", and "interesting". This book brings some research inspirations to readers who are interested in this field by looking into the future research direction of urban living space in China.

The specific division of the chapters of this book is as follows:

Chapter 1 was written by Jian Feng; Chap. 2 was written by Jian Feng, Zhu Ye, and Ying Yang; Chap. 3 was written by Jian Feng, Xin Shen, Qinan Zhang, Xiaoxue Chen, and Shiqi Li; Chap. 4 was written by Xue Ming Li, Liga Tong, Xu Zhang, Zhizhen Bai, Da Chuan Chen, and Miao Lin; Chapter 5, Section 1 was written by Jian Feng and Xin Shen, Sections 2 and 3 were written by Jian Feng and Yichun Zhong, Section 4 was written by Xue Ming Li, Jianjun Liu, and Bosi Yang, and Section 5 by Xuming Li and Xueping Cong, Section 6 by Jian Feng and Ying Yang, and Section 7 by Jian Feng; Chapter 6 by Jian Feng and Fangfang Wu; Chapter 7, Section 1 by

Jian Feng and Ying Yang, and Sections 2, 3, 4 and 5 by Jian Feng and Hongbo Chai; Chap. 8 by Xuming Li and Huanhuan Li; Chap. 9 by Yungang Liu, Wenting Zhou, Lulu Hou, Yue Zhang, Minying Wei, Boya Wang; Chap. 10 by Jian Feng, Xin Shen, Qinan Zhang, Shiqi Li, and Xiaoxue Chen; Chap. 11 and the preface were written by Feng Jian. Jian Feng was responsible for the outline and the final draft of the book. Each chapter is responsible for its own text.

The writing of this book was entrusted by Prof. Chaolin Gu, the editor-in-chief of "China Urban Geography Series", and the editorial committee of the series, and it took several years and many working meetings to complete. In the process of writing, the three editors-in-chief of the fascicle (including Jian Feng, Xueming Li and Yungang Liu) and their team worked closely together, united as one, absorbed the opinions of experts, leaders and publishers in time, and made several drafts, and did their best to make the figures and tables as exquisite as possible, in order to meet the requirements for the publication. We would like to thank the National Publication Fund and the National Natural Science Foundation of China (42171194, 42230705, 41671157, 41671158, 41571130) for their financial support of this book. We thank Prof. Chaolin Gu, Prof. Yanwei Chai, Mr. Feng Zhao, Mr. Chuanbiao Ding, and Ms. Lei Wang, for their contributions to the publication of this book. In preparation for the English version of this book, prof. Jian Feng and his team at Peking University were responsible for the translation of this book. The translators of this book include: Jian Feng, Hongbo Chai, Huali Hou, Qiqi Chen, Yuni Zhong, Yuhan Yuan, Hyeonuk Kang, Haitian Lu, Wenchu Fang, Meixi Wang, Ningjing Lu and Yexun Meng. We would like to thank them for their hard work in getting this book published in English.

We hope that this book can realize the goals of "cutting-edge", "academic" and "interesting" as expected, and can bring some help and inspiration to readers who are interested in the research field of urban living space in China. If so, the authors will be very pleased.

Of course, this book inevitably has some shortcomings. For example, in some areas or directions where the authors have not carried out empirical research, they can only rely on the research results of domestic and foreign academics to sort out and summarize them, which will result in the difference in the depth of the research between the chapters. For another example, the authors come from different colleges and universities and different groups of subjects, and there are big differences in the way of research thinking and the way of recognizing the problems, which will inevitably be reflected in the style of their writing. For the third example, urban living space in China is a larger subject, and many aspects, especially the emergence of new forms and characteristics of living space due to the influence of technological innovations and changes in the concept of living, have not yet been ready for investigation and research. Improvement of these shortcomings awaits further research in the future and the joint efforts of all colleagues in the academic community.

In the process of writing this book, a large amount of research literature in English and Chinese has been consulted and referred to, and all of them have been labeled as far as possible, so please forgive us if there is any omission! We also welcome readers and friends to make pertinent suggestions and comments on this book with a view to improving it.

Beijing, China Jian Feng
Dalian, China Xueming Li
Guangzhou, China Yungang Liu
April 2025

Contents

Chapter 1
Introduction

1.1 Background

1.1.1 China's Rapid Urbanization Has Triggered a Drastic Restructuring of Urban Space

After China's urbanization entered the mid-accelerated stage, the characteristics of rapid urbanization became more and more obvious (Zhou and Cao 1999), which was more prominent in the rapid expansion of urban space and the rapid changes in urban population, land and economic composition caused by it. From the perspective of spatial structure, rapid urbanization has accelerated the speed of urban spatial transformation, and heterogeneity, polycentricization, and fragmentation have become the most important characteristics of the spatial evolution of the large cities in China in the transition period. From the perspective of evolutionary dynamics, the space of large cities in China has become a spatial carrier where centrifugal diffusion forces and centripetal accumulation forces converge (Zhou and Meng 1998, 2000; Gu and Sun 1998; Feng and Ye 2013). On the one hand, since the 1990s, the suburbanization of large cities in China has become more and more obvious, and suburbanization has led to the discretization of urban population and economic and social activities (Zhou 2010, 2013), and its important spatial destinations are various suburbs and development zones that are far or near from the "urban area". On the other hand, the agglomeration effect and agglomeration economy of large cities continue to develop, and the floating population and many economic factors continue to concentrate in large cities, but the spatial distribution pattern within cities has changed.

From the perspective of the composition of population and social elements, the composition of residents on the carrier of urban space is becoming more and more complex, and these complex social compositions of residents and their changes have left traces or trajectories in urban space. In particular, the composition of residents' social life and social activities is often intertwined with urban spatial elements to

J. Feng et al., *Urban Living Space in China*, Urban Sustainability,
https://doi.org/10.1007/978-981-95-0595-1_1

form more complex characteristics of urban living space. Moreover, the rapid transformation of space and land use in China's large cities is accompanied by economic, social, governmental, and institutional transformations, which are intertwined with the reproduction of space. In short, the process of rapid urbanization in China has brought about a drastic restructuring of urban space, and living space, as the most important component of urban space, is undoubtedly an important research field that can arouse the research enthusiasm of geographers.

1.1.2 Urban Living Space Has Become an Important Vehicle for Interpreting the Current Rapid Urbanization in China

In a sense, urban living space is an important window to show the characteristics of China's rapid urbanization and reflect the problems of rapid urbanization. The study of urban space in the traditional sense is more concerned with the development of urban physical and economic space. Since the 1990s, China's market economy has become increasingly mature. The growth of economy, the rising of income, and the differentiation of occupational have had the most direct impact on urban residents in the form of changes in their social lives, which have left traces in space. In the past 20 years, the development of China's Real estate market has been booming, from the early housing system reform and the disintegration of the Danweilization system to the later stage when residents' home purchasing behavior has basically become a completely market-oriented behavior, and the development and change of residential spaces and living environments are an important aspect of China's economic marketization and urbanization. Improvement of economic level and increase of income have also led to the increase of residents' leisure demand, and accordingly, urban planning and construction have also met the growing leisure demand of residents in terms of hardware and leisure facilities.

Activity space is an important part of the living space for urban residents. In the study of urban activity space, the most widely used research method is the spatio-temporal activity path and spatio-temporal constraints in time geography, which can be used to explore human spatial behavior in the spatio-temporal context of large-scale environments as well as to study the behaviors and activities of the sample individuals and households under analysis (Golledge and Stimson 1997). It can be said that human beings are active every day, and all activities require a spatial vehicle, which can be either the "home" or a space outside the "home". At the same time, all activities have time continuity. This time continuity can be a long-time scale, such as a person's life course, can also be a medium time scale, such as a year, a month or a week, can also be a short time scale, such as a day. As far as residents' activities are concerned, time and space exist in an organic combination. Therefore, activity space records the spatio-temporal activity trajectories of residents' activities, including both the spatio-temporal activity trajectories of urban natives and the spatio-temporal activity trajectories of urban floating population, and is thus an important window reflecting China's rapid urbanization development.

In addition, the space for social interaction has become an important part of the living space of modern urban residents, and as the standard of living in society has risen, the ways and places of social interaction have changed considerably. With the deepening of cultural and social research, urban subcultures and special social groups have attracted more and more attention from academics. On the one hand, to study them is a manifestation of the concept of "people-oriented" in the field of urban development and urban planning, and a reflection of respect for "human nature"; on the other hand, to study them is a demand for the production of urban culture and social knowledge.

1.1.3 Technological Change Has a Profound Impact on Urban Living Spaces

In the past 20 years, the development of information technology and Internet technology has been rapid and has brought great changes to the way of life of the people. As early as in the twentieth century, Castells suggested that in the information society, the type of work, the nature of work, the place of work, and the way of work will change, and that informationization will lead to a new production system and the creation of a flexible workforce, and those who do not have access to the support of information technology will be in a difficult situation of employment (Castells 2001a, b). Of course, the changes in employment patterns that Castells is concerned with are the most direct impacts of information technology on the residents, since employment is the basic means by which the residents earn their incomes, and they provide the economic foundation and basic support for the residents' other social activities. In addition, every resident faces the problem of spatial correlation between employment and residence. Flexible employment allows residents to choose their living space more freely. Leisure facilities and public activities in the vicinity of the residence will further influence the residents' leisure space and activity space. In addition, residents' leisure and activity patterns are also significantly influenced by the Internet. Therefore, it can be said that Internet technology has brought about a major revolution in the lifestyle of urban residents.

In recent years, the development of information and communication technologies, especially smart phones, has brought about significant changes in the way of consume and social interaction of residents. Online shopping, through the choice of massive shopping information and the low-cost characteristics of goods caused by the merchants due to the saving of space rent and the reduction of intermediate links in the operation, not only increases the convenience of residents' shopping and consumption, but also effectively stimulates the residents' desire to consume, so that the geographic space of consumption has been challenged, and the composition of the space for consumption has changed drastically. Various social software, social forms and social spaces relying on information and communication technologies, such as QQ, microblogs, virtual and semi-virtual communities, WeChat and so on, have changed the connotation of the traditional social interaction space, expanded the extension of the social interaction space, and prompted residents to make significant changes in the mode and form of social interaction, which have had an impact on the social interaction of the residents in terms of both quantity and quality. Therefore, it can be said that the technological changes in the new period have led to a great change in the connotation of urban living space, and there is an urgent need for academics to pay attention to urban residents' living space and its evolution in this context and conduct tracking research. Internet technology, information and communication technology and the evolution of cultural concepts have all led to changes in the traditional living space of urban residents and the emergence of so-called "new" forms of living space, so the impact of technological change on the living space of urban residents is an important theoretical topic that deserves attention.

1.1.4 Research Methods and Perspectives on Urban Living Space Need to Be Diversified and Further Explored

Urban living space is aimed at the subject of "urban residents", to explore the spatial characteristics of various social activities related to their lives, and the object of study can focus on both the relatively macroscopic "group" and the microscopic "individual". Traditionally, research on urban space, especially residential space, has favored spatial statistics, such as the most common use of census data or housing survey data to analyze the spatial differentiation of residential spatial structure at the urban scale (Feng et al. 2011, 2016), which is the most typical application of spatial statistics in the study of urban residential space. The focus of such research is on the whole city, and some use the spatial scales of metropolitan areas and urban areas, etc., focusing on the statistical results based on spatial analysis of data, and then obtaining the socio-spatial laws of the whole city scale, which is actually researching the "group". The study of the living space and activity space of urban residents by questionnaire survey is also based on the idea of statistics, and the spatial law obtained is for the "statistical agglomeration", which is for the "group". These researches can be summarized into the research on urban living space at the macro level. Generally speaking, the current research on urban living space in China mainly focuses on the macro level, and there is a lack of micro-level spatial research based on individuals (Feng and Ye 2017). In fact, it is necessary to introduce micro-individual research perspectives and qualitative research methods in the study of urban living space in China, so as to make the study of urban social geography in China more diversified.

Qualitative research is an activity that uses the researcher himself as a research tool to explore the research phenomenon holistically in a natural context using a variety of data collection methods, to analyze the data and form theories using an inductive approach, and to gain an interpretive understanding of the behavior and meaning construction of the research subject through interaction with the subject (Chen 1996, 2002). Qualitative research makes up for the traditional inadequacy of spatial statistics and quantitative analysis in the study of micro-level spatial characteristics, social and spatial interactions, and spatial–temporal evolution mechanisms, and in recent years, it has been introduced by some geographers and used to study socio-spatial issues. Through qualitative research, it is possible to "deconstruct", "map", "tell" and "root" urban social space (Feng and Wu 2011; Feng and Chai 2016). In fact, research methods and research perspectives on urban living space can be realized in the direction of diversification. Spatial statistical analysis relying on data (including big data) often reflects an "outsider's perspective", which focuses on socio-spatial outcomes and has a unique advantage in deriving spatial patterns characteristics. Qualitative research methods relying on individual data, especially in-depth interview methods, reflect an "insider's perspective", which focuses on socio-spatial processes and can play a great advantage in exploring the relationship between space and people and the mechanisms of spatial formation and development.

1.2 Framework of the Book

This book provides an overview of urban living space and its research, starting from identifying concepts related to living space and proposing a research framework. Before discussing urban living space by typology, it focuses on the analysis of urban "lifestyle" and "quality of life", which are the basis for understanding urban living space in China. "Lifestyles" focuses on the traditional lifestyles of Chinese urban residents, the characteristics and influencing factors of the evolution of lifestyles, and the formation of new lifestyles; 'quality of life' focuses on the types and characteristics of quality-of-life differentiation and the evaluation of the quality of the human environment and its spatio-temporal differentiation. On this basis, the authors discuss urban living space in terms of four aspects: living space, interaction space, activity space and leisure space. In addition, the authors also study the living space of different groups and new forms of living space, and look forward to the future study of urban living space in China, in order to have a more comprehensive generalization and grasp of the connotation of living space in the light of the characteristics of the times.

The structure of the research framework of this book is shown in Fig. 1.1.

Overall, the introduction (Chapter 1) and the overview of urban living space (Chapter 2) provide the introduction and basic theoretical padding for the book. Urban lifestyles and their evolution (Chapter 3) and the evaluation of quality of urban life and quality of human habitat (Chapter 4) are important aspects of the study of urban living space, and are the basis for further understanding of various types of urban living space. These two chapters are not purely geographic in the traditional sense, but rather emphasize the intersection of sociology and habitat science with geography. They are an important foundation for understanding urban living space in China, providing a broader disciplinary underpinning for the systematic interpretation of the

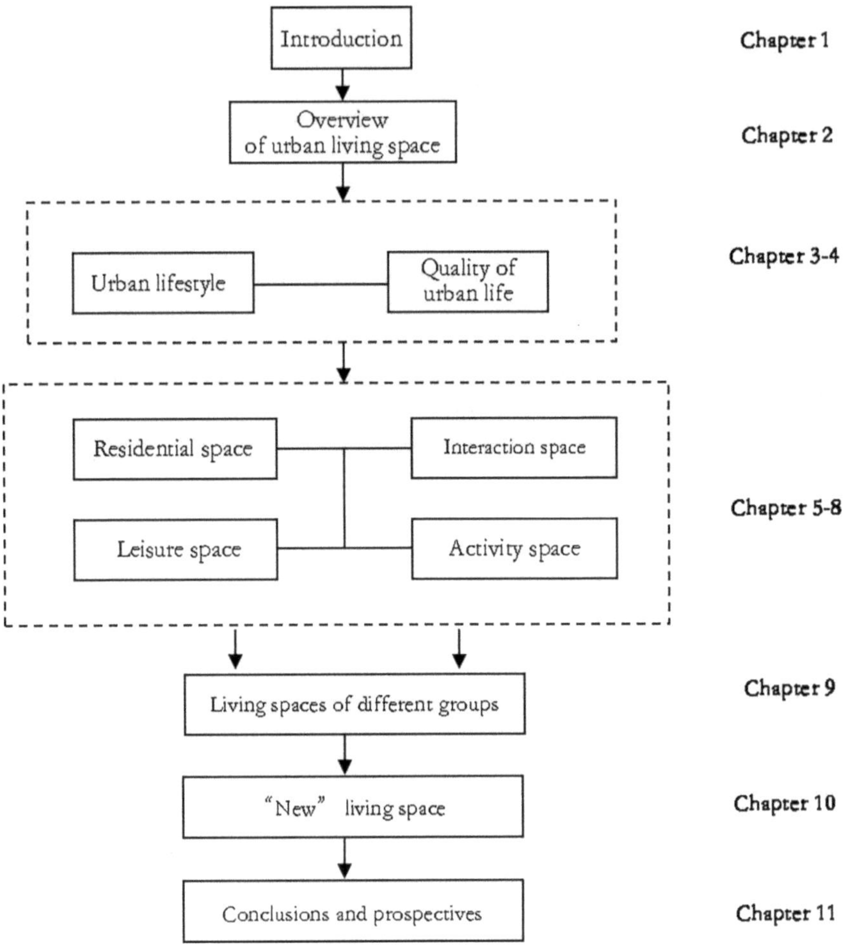

Fig. 1.1 Structure of the research framework of the book

various types of urban living space in this book, and forming an important feature of the book's framework system.

Chapters 5 to 8 are the core components of urban living space, that is, the analysis of Chinese urban living space from four aspects, namely, living space, interaction space, activity space and leisure space, which is also the most important aspect reflecting the perspective of geography. It is worth pointing out that, in the process of the research on urban living space by type, we pay attention to the combination of theoretical generalization and empirical research, the combination of quantitative analysis and qualitative description, the combination of textual analysis and spatial visualization, and the combination of academic discussion and storytelling, and the application of diversified research methodologies and paradigms, so as to strengthen the research norms based on a certain degree of in-depth empirical research, and to summarize the general characteristics of urban living space in China, in order to provide a more detailed understanding of the related fields. We hope that this study will serve as a reference and promotion for related academic research in this field. The living space of different groups: a pluralistic mosaic (Chapter 9) and the "new" living space (Chapter 10) are extensions and supplements to the urban living space, the former fully demonstrating the plurality, complexity, and interest of Chinese urban living space, and the latter embodying the "zeitgeist" of interpreting the latest changes in the living space in close conjunction with technological advances. Chapter 11 summarizes the whole book and gives an outlook on the direction of further research.

1.3 Identification of Relevant Concepts

This book deals with a number of related concepts of urban living space that need to be further analyzed.

The first is the concept of "living space".

Living space refers to the association of residents' social life with space, or its spatial reflection, and is the trajectory left by residents' social activities and behaviors in urban space. The concept of living space is close to the definition of the term "living world" by Johnston et al. in The Dictionary of Human Geography (Johnston et al. 2000). Johnston et al. (2000) argue that "life worlds" refer to the culturally defined spatial and temporal contexts or spheres of everyday life, in other words, the holistic experience of everyday life in which people are directly involved in places and environments. Johnston et al. (2000) also argue that the concept of life worlds in geography has directed attention to the importance of everyday life and the personal geographies and geographies of meaning that are developed and practiced within it. Johnston's argument represents a geographic perspective on the concept of lived space.

It is worth pointing out that there is a difference between the concept of "living space" from the perspectives of geography, sociology and architecture. The concept of living space as defined by geographers includes the meaning of "space", which is a spatial concept based on physical space and public facilities but not limited to physical space, and which may be influenced by the spatial facilities of the region, as well as by the economic strength, consumption level and cultural values of the residents. Sociologists perceive the concepts of social space and living space in an abstract way, as more abstract concepts of space, often discussing the impact of space or spatial production in isolation from physical space, as in the case of the concept of Lefebvre. Architectural scholars go to the other extreme, defining living space as the space enclosed by the building, not only emphasizing the physical space, but also limiting the concept to the physical scope of the building's enclosure and discarding the "living space" extended by the residents in terms of their social life. These conceptual differences actually reflect the different focuses of different disciplines on living space. This book adopts a concept of living space based on a geographic perspective. On the basis of the concept of living space, some secondary concepts, such as urban living space and Chinese urban living space, are formed by adding geographical qualifications, which will be discussed in Chapter 2.

Next, there is the concept of "activity space" and the concept of "daily activity space".

Activity space is the space in which an individual performs most of his or her activities, including the action space that determines a special location (Johnston et al. 2000). While human activities can be carried out at various spatial scales, such as domestic, neighborhood, economic, or urban locations, activity space at larger spatial scales can be discontinuous, often consisting of points that are connected by known paths but separated by areas (Chombart de Lauwe 1952).

The space of daily activities is an important aspect reflecting the interaction between activities and space, which focuses on revealing the spatial characteristics and patterns of various activities and behaviors common to residents in their daily lives, such as work, shopping, leisure, and medical care. Western scholars have used the geography of daily life to describe the daily social life activities and spatial status of urban residents (Lee 1983), which is a better interpretation of activity space.

In addition, there are concepts such as residential space, interaction space and leisure space, all of which belong to the category of living space. Residential space refers to the spatial characteristics of residents' residential behavior, including both the overall urban residential spatial structure and spatial differentiation characteristics in a macro sense, as well as the individual residents' home-buying and site selection behavior and relocation behavior at the micro level. Residential space has social, economic and cultural stability, as well as stability in residents' psychological perception and stability in material composition (Wu 2016). Interaction space refers to the characteristics of residents' social relationship network and neighborhood relationship manifested in a space, interpreting the relationship between residents' social activities and space from the level of social interaction. Leisure space refers to the regularity of residents' leisure activities expressed in space, and also includes the spatial characteristics of residents' utilization of leisure facilities.

1.4 Main Study Areas

Beijing is one of the main empirical research areas of this book. As of the end of 2018, the resident population of Beijing was 21.542 million, of which the floating population was 7.466 million. There are 16 districts in Beijing, but the focus of resident and floating population concentration is in the six urban districts (i.e., Dongcheng, Xicheng, Chaoyang, Haidian, Fengtai, and Shijingshan), and this area is also the one with the highest concentration of urban construction and population problems. Beijing's rapid urbanization process is also more prominent, and urbanization development has caused the city, especially the suburbs, to be undergoing a drastic reconfiguration of population, space and landscape, and the living space of urban residents has become more complex. Traditionally, people's view of Beijing's spatial evolution tends to emphasize the role of governmental forces, as the capital is blessed with the advantage of political resources. In fact, studies have shown that both governmental and market forces are involved in the process of socio-spatial reconfiguration in Beijing, and that there has been a gradual shift from "government-led" to "market-led" since the 1990s (Feng et al. 2008). Either way, there is a consensus on the "complexity" of Beijing's space, which provides useful evidence for understanding China's urban spatial transformation.

Guangzhou is another major empirical research area in this book. Guangzhou is the capital of Guangdong Province, a national central city, a famous international trade center, and a national historical and cultural city. As of the end of 2018, Guangzhou had a resident population of 14.905 million, including a household population of 9.277 million and a floating population of 5.628 million, with an urbanization level of 86.4%. In 2018, Guangzhou's GDP was 228.594 billion yuan, up 6.2% from 2017. The structure of the three industries was 0.98:27.27:71.75, with the contribution rate of the secondary and tertiary industries to economic growth at 26.6% and 73.0% respectively. In 2018, the annual fiscal revenue of Guangzhou was RMB 620.5 billion. Guangzhou has a total urban area of 7434.4 km^2 under its jurisdiction, with municipal co-ordinated districts namely Yuexiu, Haizhu, Liwan, Tianhe, Baiyun, Huangpu and Nansha, which are known as the Old Seven Districts. Dongshan, Fangcun and Luogang were originally one of the old seven districts, which were later abolished due to merger, and Nansha is part of the new old seven districts. The old four districts originally referred to Yuexiu, Dongshan, Haizhu and Liwan, but after the regional restructuring, the old three districts (old urban areas) were adopted, referring to Yuexiu, Liwan and Haizhu; the new four districts are Panyu, Huadu, Conghua and Zengcheng.

Dalian is also one of the main empirical research areas of this book. Located at the southern tip of Liaodong Peninsula, Dalian is situated on the coast of the Yellow Sea and Bohai, backed by the hinterland of northeastern China, and separated from the Shandong Peninsula by the sea, and is an important economic, trade, port, industrial, and tourist city on the eastern coast of China. As at the end of 2018, the household population of Dalian was 5.952 million. In 2018, Dalian's gross regional product was 766.85 billion yuan. Dalian has a total of seven agriculture-related districts/counties, including four city (including Zhuanghe, Pulandian, and Wafangdian), three districts (including Jinzhou, Ganjingzi, and Lushunkou districts), and Changhai County, as well as four national-level pilot zones for opening up to the outside world (including Hi-Tech Park, Bonded Zone, Changxing Island Development Zone, and Huayuankou Economic Zone).

References

Castells M (2001a) The Rise of the Network Society. Translated by Xia, Z. et al. Social Sciences Academic Press, Beijing (In Chinese)

Castells M (2001b) The Informational City. Translated by Cui Baoguo. Jiangsu People's Publishing House, Nanjing (In Chinese)

Chen X (1996) Qualitative research methods in social sciences. Soc Sci China 6:93–10 (In Chinese)
Chen X (2002) Qualitative research methods and social science research. Educational Science Publishing House, Beijing (In Chinese)
Chombart de Lauwe PH (1952) Parisetl' Agglomeration Parisienne. Presses Universitaires de France, Paris
Feng J, Chai H (2016) Application of qualitative GIS in urban social space research. Prog Geogr 35(12):1447–1458 (In Chinese)
Feng J, Wu F (2011) Application of qualitative methods in urban social space research. Geogr Res 30(11):1956–1969 (In Chinese)
Feng J, Ye B (2013) Suburbanization studies from Western socio-spatial perspectives and their implications. Hum Geogr 27(3):20–26 (In Chinese)
Feng J, Ye Z (2017) Transformation of urbanization path and citizenization process in Southern Jiangsu from the perspective of individual life course. Prog Geogr 36(2):137–150 (In Chinese)
Feng J, Zhou Y, Wu F (2008) New trends of suburbanization in Beijing since 1990: from governmentled to market-oriented. Regional Stud 42(1):83–99
Feng J, Wu J, Xie X et al (2011) Interpreting cities through "Population Space": a case study of Wuhan. Urban Dev Stud 18(2):25–36 (In Chinese)
Feng J, Hu X, Su L (2016) Urban population spatial reconstruction and planning response: a longitudinal study of Wuhan. Planners 21(11):24–32 (In Chinese)
Golledge R, Stimson R (1997) Spatial behavior: a geographic perspective. The Guilford Press, London
Gu C, Sun Y (1998) New trends in the development of Chinese metropolises: urban suburbanization. Planners 2:102–104 (In Chinese)
Johnston RJ, Gregory D, Pratt G et al (2000) The dictionary of human geography, 4th edn. Blackwell, Malden
Lee DA (1983) Social geography of city. Harper Row, New York
Wu Q (2016) Theoretical and empirical study on residential spatial differentiation in large cities, 2nd edn. Science Press, Beijing (In Chinese)
Zhou Y (2010) Urban geography explorations: Zhou Yixing's self-selected works. The Commercial Press, Beijing (In Chinese)
Zhou Y (2013) Urban planning pathfinding: Zhou Yixing's critical essays. The Commercial Press, Beijing (In Chinese)
Zhou Y, Cao G (1999) China's urbanization process during 20 years of reform and opening-up. City Plann Rev 23(12):8–14 (In Chinese)
Zhou Y, Meng Y (1998) Suburbanization trend in Chinese large cities. Urban Plann Forum 3:22–27 (In Chinese)
Zhou Y, Meng Y (2000) Beijing's suburbanization and countermeasures. Science Press, Beijing (In Chinese)

Chapter 2
Overview of Urban Living Space

2.1 Urban Living Space and Chinese Urban Living Space

Urban living space is a vital component of urban space. From a functional perspective, urban living space, along with urban production space and urban ecological space, collectively constitutes urban space (Lefebvre 1991). Urban living space, also known as urban social living space, is a composite spatial concept that encompasses both types of urban spaces formed by or involved in various daily life activities and the social spaces created by the observable mobility and activities of individual residents. As a composite spatial concept, it involves not only the material entity space such as buildings, landscapes, environmental elements, infrastructures or public facilities that urban residents access or utilize in all kinds of daily life activities, but also various social behaviors and activities such as living, working, commuting, consuming, and relaxing in order to satisfy the needs of life in daily life, and also social relations and social forms such as order, family, neighborhood, and interpersonal relations in the daily life activities of urban residents. It is worth emphasizing that, although the concept of urban living space under the perspective of geography does not specifically refer to the physical material space, there is no doubt that it cannot be separated from the state of utilization of physical facilities and material space; and although it does not explicitly refer to the social level of the living space, there is no doubt that it focuses more on the social level. On the one hand, the physical living space that leaves the social utilization lacks the significance of geography, and is more the subject of architectural concern; on the other hand, leaving the physical material space and focusing on the abstract social space and living space is not the perspective of geography, but more the perspective of sociology. Therefore, the urban living space under the perspective of geography is the spatial trajectory or imprint left by the social life created and organized by the urban residents who depend on the physical facilities and material space base of the city, and it is a kind of social space based on the physical space, for example, of distance, range, direction, and so on.

J. Feng et al., *Urban Living Space in China*, Urban Sustainability,
https://doi.org/10.1007/978-981-95-0595-1_2

Chinese urban living space is a distinct subset of urban space shaped by China's specific socioeconomic conditions. In comparison to western urban living space, it exhibits specific Chinese characteristics in terms of spatial connotations, historical context, formation processes, constituent elements, spatial structural features and evolutionary mechanisms. The categorization of Chinese urban living space is usually determined by the categories of urban life, and the diversity of daily activities contributes to diverse forms. Urban daily life usually occurs in households, workplaces, places of consumption, non-consumption public places, and the movement between these areas (Zhang 2007), forming distinct elements of urban living. Consequently, Chinese urban living spaces can be categorized into major types based on these elements, including living space, commuting space, working space, shopping space, interaction space, activity space, and leisure space. Among these, working space and living space usually involve unit space in Chinese studies. Unit space is a product of China's traditional planned economy system that affects urban living space. In the planned economy era, the centralized construction of residential areas by work units was common. To enhance the quality of life for their employees, these units also developed various living facilities within the residential areas. A unit functions as a "miniature society," where the diverse living needs of its employees and residents can be satisfied without going outside the unit's living area. Consequently, during the planned economy era, urban living space was inextricably linked to the unit system. As the unit system and unit society in China have weakened, the unit-based workspace, living space, and living service space are undergoing significant transformations. Nevertheless, numerous remnants of the unit society continue to exist within the current urban living space in China.

In addition, Chinese geographers classify urban living spaces in China in various ways, influenced by their distinct research perspectives and focuses. Some scholars categorize urban living space into three subspaces and several sub-subspaces, namely, physical space, economic space, and social space—based on geospatial space (Zhu 1993). Others divide urban living space into family living space, neighborhood interaction space, urban community space, and urban social space (Wang et al. 2000). Additionally, some classifications focus on daily life, dividing urban living space into unit space, consumption space, interaction and leisure space, and residential space (Chai 2000). Furthermore, the conditions of formation, motives for development, and spatial patterns can also be the basis for classifying urban living space types in China. Beyond traditional classifications, advancements in information network technology, virtual reality technology, and new concepts of life have shaped "new" types of urban living spaces tailored for diverse groups, which are also significant components of urban living spaces in China.

The formation of urban living space in China results from the interplay of various elements. Currently, factors such as path dependence, transit-oriented development, sociocultural networks, the transformation of old cities, government regulation and management, industrial restructuring, urban–rural population mobility, as well as scale and filtering effects all play significant roles in shaping urban living space (Wang 2011). Socioeconomic development has continuously promoted the differentiation of urban living space and the evolution of their structures in China, making urban living space a common research hot spot of human geography, urban science, sociology, economics and other disciplines in China. During this transitional period, as urban residents' living standards improve, there is an increasing emphasis on their perceptions and experiences of the surrounding living environment. The rise of humanistic thinking further advocates for attention to human needs (Gu 1994). Consequently, examining and evaluating space from the residents' perspective has become a crucial approach in the study of urban living space in China.

2.2 Urban Life Circle and Physical Urban Living Space

The concept of "life circle" originated in Japan, where the *Rural Living Environment Governance Plan* defines the life circle as the spatial distribution of people's daily production and living activities within the geographic boundaries of a particular geographic and social village, and that the life circle is categorized into hierarchical levels: village—oaza—machi—municipalities—metropolitan area, using the village of a certain population and a defined geographic distance as benchmarks (Zhu et al. 2010). To optimize the spatial structure of urban life and enhance the quality of life for residents, the Japanese government proposed the establishment of a "decentralized network structure" of polycentric cities within Capital region, as outlined in the Third Capital region Plannings. Furthermore, the Ministry of Land, Infrastructure, Transport and Tourism of Japan has also introduced the concept of life circle, which serves as a spatial planning unit tailored to the needs of residents' daily activities. This concept is designed to encompass the essential areas required for one-day living activities, including employment, education, shopping, medical care, and recreation (Chai et al. 2015). Through the implementation of the Japanese government has effectively optimized the spatial structure of urban life and improved residents' quality of life.

Studies on life circle in China are relatively limited, but there has been a noticeable trend of gradual growth in recent years. Chen and Xu (1985) proposed a classification of the living environment into three spheres: a core life circle centered on the home, a basic life circle focused on the neighborhood, and an urban life circle that encompasses the city, based on the surrounding facilities and the distance from the living place. Chai, an early researcher on urban life circle in China, used the case of Lanzhou to propose a three-tiered structure of living space within Chinese cities. This structure includes Basic living circle composed of residential units, Lower-tier living circle formed by homogeneous units, and High-tier living circle based on districts. He argued that this structure emerged under the dual influences of urban administration and residential planning within a planned economy (Chai 1996). Later, he proposed the concept of life circle planning, explored the division method and model of the life circle system, and argued that a comprehensive set of processes and methods related to life circle planning should be established from a scientific perspective to provide a solid foundation for its promotion (Chai et al. 2015, 2019; Chai and Li 2019). Yuan et al. (2005) introduced the concept of the urban "daily life circle" from the perspective of residents' personal lives. This concept refers to the spatial scope encompassing various daily activities of urban residents, such as residence, employment, and education, which collectively represent the urbanized territory of a city. They categorized the "daily life circle" into three types within the urban territorial system: Basic living circle, the fundamental life circle, and the opportunity life circle. Some scholars have summarized existing research and defined the living circle as the behavioral trajectory of individuals moving from their residences to places of work, education, medical care, and other production and living services provision and other settlements in order to satisfy the needs for survival, development, and interaction within the social system of a specific geographical area, which is spatially represented in the form of a circle characterized by directionality and the overlap of adjacent areas (Sun et al. 2012).

American geographer Edward Soja posits that there exists a two-way continuum between physical space and social development in urban space. On one hand, individuals interact with the physical material space, imprinting their characteristics onto the spatial environment and continuously altering and shaping it. On the other hand, this material space serves as a vessel for human life, exerting a continuous influence on and control over social life and human development (Soja 1980). Consequently, urban living space exists not only in the physical material form but also carries specific social connotations and significance. Urban physical living space refers to the tangible material space where urban life occurs. The counterpart is non-physical space, which is the space perceived and utilized by social groups; this includes urban social space formed by communities sharing common values, attitudes, and behaviors. For instance, social space can manifest as symbols and cultures, representing an abstract system composed of values and relationships that describe the characteristics of social structure (Bourdieu 1989). Some scholars have proposed the concept of residential living space morphology, arguing that its composition primarily depends on the distribution patterns, the overall layout characteristics, boundary pattern, the external space pattern of the residence, as well as the architectural pattern on which it relies, of human residential environments and daily life activity spaces within a geographical unit (Xiang, 2010). In comparison, while the concept of residential living space morphology emphasizes the physical material form of living space, it also considers the potential influence of regional culture, particularly regional residential culture.

2.3 Traditional Living Space of Chinese Urban Residents

The traditional living space of urban residents in China predominantly includes residential zones, workplaces, recreational facilities, commercial areas, and public service spaces.

Traditionally, the organization of living spaces, work spaces, and public service areas within Chinese society is centered around the concept of individuals being affiliated with specific units. These units serve as the primary context in which individuals work, reside in housing of Danwei allocation, and access various welfare facilities associated with it (Chai 1996). Units can be classified into two categories: corporate entities, which are profit-oriented (e.g., factories), and institutions, which are non-profit (e.g., schools). The population accommodated by these units can range from a few dozen to tens of thousands of individuals, with spatial dimensions varying from several hundred square meters to several hundred square kilometers (Chai 1996). The typical spatial configuration of a unit consists of a front yard, designated as the production area, and a back yard, which serves as the living area. Privacy levels increase progressively from the front yard to the back yard (Zhang et al. 2009), thereby promoting a cellular residential characteristic in which the living space and workplace of the unit are in close proximity (Gaubatz 2007). In essence, the traditional living and working spaces of urban residents in China are integrated, resulting in minimal separation between work and residence. The housing area of Danwei allocation is relatively small, and there has historically been insufficient investment in urban housing. For instance, in 1994, China's per capita housing floor area was only 6.7 m^2; however, significant improvements in housing conditions have occurred since the marketization of housing, with the per capita housing floor area increasing to 28 m^2 after 2007 (National Bureau of Statistics 2009). Furthermore, the public service spaces available to urban residents are predominantly centered around these units. Many units are equipped with essential facilities such as kindergartens, primary and secondary schools, stores, guest houses, bathrooms, barber shops, hospitals, canteens, and vegetable farms. These comprehensive living facilities enable residents to meet their basic daily needs without having to leave the confines of the unit. Consequently, the unit functions similarly to a small welfare state, and urban residents exist within a relatively enclosed living space (Jie 2000).

The traditional leisure space is characterized by recreational activities conducted within the vicinity of the family residence, with the majority of leisure time spent at home. This includes emotional communication, children's education, and various forms of entertainment such as watching television, listening to the radio, and reading books and newspapers (Chai 2000). Additionally, due to transportation limitations, the geographical distribution of residents' traditional leisure spaces is restricted, centering predominantly around the home, with a radius of activities that typically does not extend beyond the neighborhood (Chai 2000). In addition, there exists a variety of leisure spaces associated with residential units, including chess and card rooms, activity rooms, and playgrounds. Furthermore, a diverse array of sports, cultural, and artistic activities is frequently organized within these units. In urban areas, large-scale leisure spaces are primarily represented by squares and parks. Historically, most urban centers in China were equipped with significant city squares and parks, which primarily served as venues for political gatherings and commemorative events (Yuan 2012). These spaces also evolved into essential leisure spaces for residents. Following the reform and opening-up policies, the rapid advancement of industrialization and urbanization in China has led to a swift expansion of urban leisure activities and spaces. Currently, various demographic groups have access to leisure spaces that range from private residences to external locations (Wu 2012), resulting in a rich diversity of leisure activities. The traditional retail landscape is predominantly characterized by department stores, kiosks, small specialty shops, and vegetable markets. Kiosks, in particular, are typically situated in high-traffic areas of urban space and primarily offer daily necessities. They are characterized by low operating costs, limited scale, a narrow range of products, and widespread distribution. Conversely, large department stores, often state-owned, employ a closed-shelf sales model and feature well-designed store layouts along with exemplary after-sales service, making them a significant shopping destination for urban residents. With the advancement of the market economy and the commodity economy, the variety of shopping spaces in urban areas has expanded considerably. This includes the emergence of large shopping centers and supermarkets of varying sizes, which now offer a more comprehensive range of services, incorporating functions such as exhibitions, dining, and entertainment (Chai 2000).

2.4 The Evolution of Living Spaces for Urban Residents in China

The evolution of living spaces for urban inhabitants in China can be categorized into three distinct phases: the era characterized by a unitary system of living space (1949–1987), the period marked by the decline of unitary living arrangements (1987–2000), and the current phase of diverse residential spaces influenced by a market economy (2000 to present).

In the formative years of the People's Republic of China, the unit system was established as a means to rapidly restore production and consolidate governmental authority. During this phase, the unit emerged as the foundational element of urban life in Chinese cities, serving as a critical link for urban residents to access various social resources. The unit played a multifaceted role in organizing, managing, controlling, cultivating, and safeguarding workers, while also providing them with a sense of identity and fostering community integration. This structure contributed to the development of a sense of place and social belongingness among urban inhabitants (Bo 2014). Despite variations in the scale and layout of unit compounds, the fundamental components remain consistent. Spatially, units often achieve enclosure, closure, and integrity through the use of a production area (work area) and a living area, which includes not only living spaces but also public facilities such as schools, cafeterias, and communal bathrooms. At this stage, the living spaces of urban residents in China exhibited several notable characteristics: first, there was a homogenization of living spaces and a standardization of spatial forms, as land for each unit was allocated directly by the state without differentiation based on the varying benefits of the units. Consequently, spatial planning within the units was relatively uniform, resulting in a balanced quality and density of living spaces for residents across the city (Li and Gu 2011). Second, there was a convergence of work and residential spaces, with the majority of residents' living areas situated in close proximity to their workplaces. Third, the spaces designated for residents' daily activities were characterized by limitations and homogenization, as the units provided essential housing security and basic living service facilities, such as stores and schools, thereby addressing the majority of residents' daily activity needs within the confines of the units.

Following the reform and opening-up policies, the significance of the unit system in China began to diminish. As the reform and opening-up initiatives primarily emphasized economic development, the government-initiated reforms within state-owned enterprises and institutions. The 1984 Decision of the Central Committee of the Communist Party of China regarding the Reform of the Economic System advocated for the establishment of a planning system that consciously integrates the law of value, which subsequently undermined the fundamental role of the unit in the daily lives of residents. Furthermore, numerous enterprises-initiated workforce reductions, and the emergence of a market economy facilitated the growth of private enterprises and individual entrepreneurship, further diminishing the unit's role in employment. By 1998, the provision of welfare housing by units was abolished, resulting in units no longer being responsible for housing and living services. Consequently, urban residents began to adapt to non-unit housing and living services, leading to a decline in the unit's status as the primary unit of life. Nevertheless, as China remains in the early stages of transitioning to a market economy, the unit-based living arrangement continues to persist in certain urban areas.

In the context of increasing diversity in living spaces under the influence of a market economy, urban residents select their residences in locations deemed suitable based on their economic capabilities, in accordance with market principles. This has resulted in a continuous differentiation of living spaces among residents. Concurrently, economic development and enhancements in urban public service facilities have broadened the types and extents of daily activity spaces available to residents. Historically, during the era of the unit system, residential compounds functioned as closed and independent spaces. However, following the transition from this system, the social functions associated with these units have gradually diminished. As a result, residents are now making more informed decisions regarding employment, housing, and consumer behaviors, including shopping and leisure activities. This shift is reflected in the spatial reconfiguration and optimization of urban structures and layouts (Chai et al. 2007, 2008). In terms of living spaces, social stratification has become increasingly pronounced. As the income and living standards of urban residents improve, the socioeconomic divide between affluent individuals and those in middle- and low-income brackets is rapidly widening, thereby facilitating a range of housing options for diverse families. Traditionally, employee housing has been transferred through a process of whereby individuals in these units "filter upwards" renting out their units to urban migrants and subsequently relocating to suburban areas or other locations in search of improved housing conditions, while older individuals or low-income residents remain in the units, passively "filtering downwards" (Ta and Chai 2010). Furthermore, since 2005, China has constructed a significant number of sheltered housing communities to address the housing challenges faced by low- and middle-income families. These sheltered housing communities, along with older unit neighborhoods, have become gathering places for low- and middle-income groups. The movement of different social classes and the maturation of the commercial housing market have contributed to the reorganization and differentiation of inner-city living spaces. During the market economy phase, commercial housing has emerged as the primary component of urban housing supply. Primarily, due to relatively lower prices and a more abundant land supply in suburban areas, these regions have attracted a substantial residential population, both from central urban areas through the centrifugal force of sub-urbanization (Zhou 1996, 2010; Zhou and Meng 1997; Zhou and Feng 2002) and from peripheral distant suburbs and rural areas through the centripetal force of centralized urbanization. The complexity of suburban residential spaces has been heightened by the centripetal force of centralized urbanization, which draws individuals from peripheral suburbs and rural areas, alongside the simultaneous presence of both high-end urban villa districts and affordable housing in suburban locales. Additionally, the separation of work and housing has intensified. Land marketization reform has resulted in the segregation of land designated for various purposes, such as residential, commercial, and industrial. Consequently, many residents have relocated to suburban areas in pursuit of improved living conditions; guided by the dynamics of urban land prices, industrial enterprises have also moved to the outskirts of cities (Feng, J 2002, 2004). These transformations have facilitated the separation of occupational and residential spaces, with the average daily commuting time for residents in many areas exceeding 30 min.

In the context of a market-driven economy, the diversity of living spaces has led to an increasingly varied lifestyle and activity space for urban residents. Central business districts, shopping malls, and a multitude of entertainment and leisure venues have proliferated, resulting in a significant transformation of consumer attitudes. Activities such as shopping, entertainment, dining, and private automobile travel have become essential components of daily life. The diversity, complexity, and even the entertainment value of urban life have begun to align closely with trends observed in Western societies, contributing to the concept of liquid modernity (Bauman 2002). Comprehensive self-service commercial establishments, such as supermarkets and convenience stores, have gradually supplanted traditional markets and kiosks, emerging as the primary venues for residents to purchase everyday items, including food and household necessities. Conversely, large shopping malls located in urban centers have become the predominant sites for acquiring higher-end goods, such as clothing. In recent years, the proportion of daily shopping conducted online has risen significantly, establishing online shopping as a crucial spatial medium for residents' purchasing activities. Moreover, improvements in urban infrastructure and the development of community support facilities have diversified the locations available for residents' leisure activities. These now include community gardens, public squares, expansive shopping malls, shopping centers, bars, and cafes, among other options. The growing number of recreational and leisure venues within urban spaces suggests a notable shift in residents' leisure time from passive activities at home to more active pursuits outside the home.

2.5 Research on Urban Living Spaces in China

With the advancement of the "socio-spatial" turn in urban research in China during the transitional period (Zhong 2013), domestic scholars have established a comprehensive research system and content framework that encompasses various perspectives on the study of urban living spaces. This includes investigations into living spaces through the lens of community living and activity spaces. The former perspective emphasizes the examination of community spaces within living space, and believes that living spaces fundamentally represent the urban habitat and human-centered community environments, with a focus on the structure of community living spaces. Conversely, the latter perspective prioritizes the analysis of individuals' daily life activities, characterizing urban living spaces as those that facilitate these activities. Consequently, these two distinct research perspectives have led to the development of varied research systems concerning urban living spaces. Despite the differences in their approaches, both perspectives fundamentally address the spaces occupied by human daily life.

In the context of research scale, macro-level studies primarily examine urban social spaces, with key areas of focus including patterns of urban residential differentiation, social polarization and spatial differentiation, the interaction between urban form space and social space, and comprehensive evaluations of the quality of the urban living environment. Conversely, micro-level research predominantly centers on urban living spaces, emphasizing the evaluation of community quality of life, community classification methodologies, the structure of urban daily behavioral spaces, and the sustainable development of communities. Currently, the emergence of diverse research perspectives and academic trends—such as humanism, feminism, eco-community concepts, sustainable urban development, virtual space, and new urbanism—signals a transformative era in the study of urban living spaces.

Currently, research on urban living spaces in China primarily concentrates on the following aspects.

2.5.1 Research on the Spatial Structure of Urban Life and Its Changes

Chai (1996) conducted an analysis of the living space structure within Chinese cities, positing that it comprises three distinct levels: Basic living circle, which is formed by individual units; Lower-tier living circle, primarily constituted of homogeneous units; and High-tier living circle, which is organized around districts. Wang (2004) further elaborated on the essence of cities and urban living spaces, emphasizing the significance of interactions among districts (locations or communities) that possess varying interests. He contended that the theoretical foundation of urban socio-living spaces is rooted in the sociological school of Western geography, synthesizing three predominant schools of thought: structuralism, humanism, and behaviorism. In their

2011 study, Wang and Wang (2011) examined the structure of living spaces within urban communities, categorizing it into spatial system elements and humanistic structure constituents. The spatial system elements encompass three categories: urban resources, community resources, and community types. Conversely, the humanistic structure constituents include social fairness, spatial justice, cultural equality, and respect for values. By integrating these two categories, they developed a model for the reconstruction of urban community living spaces. Zhang (2005) analyzed the fundamental characteristics of living space schemas in the information age, identifying six key aspects: leisure, experience, learning, digitalization, green initiatives, and individuality, which reflect the pertinent features of contemporary living spaces. Zhang and Yamamura (2003) investigated the structure of daily life spaces in Lijiang Old Town through the lens of Architectural site theory, concluding that the ritualized daily life spaces in Lijiang Old Town constitute a vital component of local cultural life and should be integrated into the urban protection system. Li et al. (2013) focused on the Huilongguan Cultural Residential Area, a prominent suburban large-scale residential development in Beijing, to explore the characteristics and mechanisms underlying the reconstruction of living spaces in such areas. Their analysis encompassed various types of spaces, including living, working, shopping, and open spaces, while also examining the relationship between the reconstruction of residents' living spaces and the process of suburbanization. Employing qualitative research methods, Wang and Feng (2013) investigated the Huilongguan residential area, specifically analyzing the relocation experiences of urban demolition residents, middle-aged and elderly individuals undergoing institutional transformations, local young people, and young expatriates in Beijing. Their study aimed to uncover the intersections of economic, social, and institutional changes with individual life courses, thereby providing insights into the broader patterns of relocation among suburban residents.

There exists a body of scholarship dedicated to the differentiation of urban living spaces. Following the reform and opening-up policies, the living environments in China have progressively become stratified, leading to the dissolution of the previously balanced distribution of unit communities within urban areas. Scholars such as Gu C.L. and Wu F.L. have engaged in theoretical investigations concerning spatial differentiation within Chinese urban society amidst this transformative context. For instance, Gu et al. developed a conceptual framework addressing the dynamics and processes of socio-spatial transformation in Chinese cities (Gu et al. 2006). Wu introduced the notion of "urban restructuring" in urban studies, which can be examined through three dimensions: political-economic transformation, changes in urban development organizational pattern, and integration into the global economy (Wu 1997). Li and Gu (2011) analyzed the characteristics of urban socio-spatial polarization and the spatial differentiation of living space, identifying categories such as the new urban migrant, the affluent urban class, new urban migrants, and informal urban practitioners and residents. They contend that these emergent "new socio-spatial spaces" represent a distinct form of "Chinese-style social space", of market-oriented reforms and increased openness to the global economy. These "new socio-spatial spaces" are characterized by plurality, heterogeneity, high density, elasticity, and a state of flux and transition. Furthermore, Gu and Sheng (2012) conducted a study on the settlement of low-income university graduates in Beijing, utilizing field research in Tangjialing. Their investigation specifically examined the characteristics of this demographic's settlement in Beijing, the nature of their living environments, and the subsequent spatial implications stemming from the Tangjialing phenomenon.

2.5.2 Research on Urban Community Living Space and Its Planning

Academic research on community living space has predominantly concentrated on the living space of typical inner-city communities and marginalized groups. In the past two decades, urban villages have emerged as a novel type of community situated between urban and rural areas in China. The diverse composition of residents within urban villages warrants a thorough examination of the living space in this unique community type. According to Yan et al. (2004), urban villages represent a distinct living system and social relationship network, essentially functioning as "peasant villages" within a civil society. The transformation of these villages is often more complex than merely the cessation of the peasant identity. Numerous scholars have directed their attention toward the living space in low-income neighborhoods, traditional gated communities, and areas with a high concentration of urban migrants in China. Wei and Li (2006) contend that during the transitional period, low-income groups in Chinese cities can be categorized into "endogenous" and "exogenous" types. The former reside in inner cities and deteriorating unit communities, while the latter typically rent accommodations in urban villages. This segmented housing system has exacerbated residential differentiation among low-income populations. The authors propose enhancing the living space of low-income groups through a model characterized by "reciprocity-diversity-mixed living". Song and Zhu (2009) contend that the emergence of closed communities in China is a negative response to societal spatial differentiation, with the resulting residential segregation reflecting economic disparities. However, this segregation has not yet reached the level of individualized differentiation observed in Western contexts concerning occupations, values, and cultures. Feng and Wang (2008) employed qualitative research methods to investigate the characteristics of social space and its formation mechanisms in residential areas surrounding universities in Zhongguancun, Beijing. Their findings reveal the spatial structure patterns of communities around universities in major cities, shaped by the dual influences of academic institutions and societal factors at the meso-community level. They argue that the flaoting populations and indigenous populations in these communities collaboratively construct a social space characterized by high mobility and loose social networks, providing new insights into the relationship between community mobility, social network characteristics, and the maturity of social space. Based on a questionnaire survey conducted in the Beijing Economic and Technological Development Area, Feng and Xiang (2017) explored the characteristics of residential space and its formation mechanisms within this type of community. They concluded that, amid the transformation of the development area from an "industrial functional territory" to a comprehensive "banized functional territory", the development zone, as both an industrial agglomeration and a commuter town, has contributed to the intricacies of its residential space and the mechanisms underlying its formation.

In addition, numerous scholars have underscored the significance of research pertaining to living space planning and community development. Wang (2000) identified six methodologies for urban living space planning, which include the ecological analysis classification method, geographical analysis classification method, ecological zone boundary determination method, social area analysis method, factor ecological analysis method, and the sensing neighborhood area method. Zhang and Lü (2003) critiqued the prevalent neglect of daily life and its associated challenges within the context of large-scale urban design in China. They proposed an urban design framework centered on daily life spaces, drawing upon philosophical reflections on daily life as a guiding principle for urban planning. Wang (2012) further developed this discourse by establishing a control index system for social living space planning, emphasizing the importance of daily life in urban community planning. This system included an existentialist control index system aimed at enhancing community residents' "sensing-cognition". Yang (2009) examined the planning and design philosophies of Shanghai workers' villages in the 1950s from the "space-institution" perspective, focusing on site selection, spatial structure, household design, and collective living. He argued that the construction of workers' villages represented a dual process of institutionalizing daily life spaces, transitioning from the "neighborhood unit" to the "community unit". The shift from "neighborhood unit" settlement planning to a hierarchical spatial structure in settlement planning marked a trend towards the administrative regulation of daily life. Zhang and Guo (2010) addressed the challenges associated with creation of living space in large-scale residential communities located in urban fringe areas, specifically in the context of residential area planning and design in Dalian. He et al. (2013) analyzed the living space planning issues faced by "transitional communities" inhabited by landless farmers under the government's comprehensive relocation and resettlement policy, proposing strategies for rational planning that consider geographic, living, social, and psychological spaces. Furthermore, scholars have investigated the architectural forms within living spaces during the renewal and transformation of old urban areas from an architectural design perspective (Bao 2010). Feng and Lin (2017) assessed neighborhood residents' satisfaction with living spaces in the old city following its decline, utilizing a questionnaire survey to analyze factors influencing life satisfaction in the old city of Suzhou, and explored the implications for urban planning and transformation. Overall, while research in this domain encompasses a broad spectrum of topics, there remains a need for greater systematic coherence, and a comprehensive theoretical framework has yet to be established.

2.5.3 Research on Daily Activity Space of Urban Residents

The daily lives of urban residents encompass a range of activities, including employment, household responsibilities, leisure pursuits, and shopping (Chai 2000). Consequently, the living space can be conceptualized as the spatial domain within which residents engage in their daily activities. Scholars in China have conducted extensive research in this area.

The shopping behavior, consumption patterns, and leisure activities of urban residents constitute a significant aspect of understanding their living spaces. Chai and his research team previously employed a questionnaire survey methodology to examine the shopping behaviors of urban residents in China, conducting empirical investigations into the spatial and temporal characteristics of shopping activities in major Chinese cities, including Tianjin, Shanghai, and Shenzhen. For instance, Wu et al. (2000) analyzed the spatial, frequency, temporal, and purpose-related characteristics, as well as the travel modes associated with shopping trips undertaken by residents of Tianjin. Similarly, Chai et al. (2004a, b) investigated the frequency, shopping time, spatial characteristics, and structural patterns of shopping behaviors, along with the decision-making factors influencing the consumption of various types of commodities in Shenzhen. Furthermore, Chai et al. (2008) conducted a study on the spatial structure of urban commerce in Shanghai, focusing on residents' shopping and consumption behaviors. Additionally, Chai and Li (2005) examined the spatial characteristics of daily shopping behaviors among elderly individuals in urban China, utilizing survey data from Beijing, Shenzhen, and Shanghai. Moreover, Feng et al. (2007) explored the evolving characteristics of the spatial structure of shopping behaviors among Beijing residents over a decade, identifying the influencing mechanisms based on cognitive distance derived from questionnaire surveys. Their findings indicated that the evolving characteristics of shopping spatial structures in Beijing reflect a centrifugal and polycentricization in the city's commercial development, influenced by macro-environmental changes and shifts in both supply-side (exogenous) and demand-side (endogenous) factors. These three elements interact to facilitate the evolution of shopping behavior spatial structures among Beijing residents. Chen and Feng (2009) investigated the hierarchical structure of urban residents' shopping trips and its evolutionary patterns, discussing the implications of their findings for the future development of urban commerce and spatial organization in Beijing. Chai et al. (2010) examined the relationship between consumer behavior and urban space, highlighting the interconnections among various consumer behaviors. They proposed a research methodology for analyzing the hierarchical structure of commercial centers in Chinese cities, as well as the cognitive hierarchy of these centers based on residents' consumption and cognitive behaviors. Their systematic study addressed the temporal and spatial characteristics of urban residents' service consumption and nighttime consumption behaviors. Chai et al. (2002) investigated the spatial characteristics of leisure activities among urban residents in Shenzhen, Dalian, and Tianjin, focusing on time and space utilization. They aimed to establish a model for the spatial structure of leisure activities for urban residents in China, integrating leisure spaces into the broader research framework of urban life spatial structures in the country. Utilizing community questionnaire survey data, Feng and Xiang (2013) conducted empirical research on the daily activity spaces related to commuting, shopping, education, healthcare, and leisure for residents in the Beijing Economic-Technological Development Area (BDA). Their study illustrated spatial orientation of the separation between work and residential areas, as well as the spatial distribution characteristics of residents' educational, healthcare, and leisure activities.

In the spatial analysis of urban residents' commuting behaviors, research conducted by geographers highlights the spatial dynamics of individuals commuting between their residences and workplaces. This body of work places significant emphasis on understanding the regularities of commuting patterns, which stands in contrast to the predominant focus on the temporal distribution characteristics of commuting within urban transportation planning (Li and Chai 2000). Li and Chai (2000) articulated the distance decay law associated with the commuting behaviors of residents in Dalian, as well as the spatial structural patterns that emerge from these behaviors. Furthermore, Shen and Chai (2013) employed residents' activity logs, GPS positioning data, and GIS 3D visualization technology to simulate the spatio-temporal characteristics of residents' activities across seven theoretical commuting patterns. Additionally, several scholars have investigated the factors contributing to variations in residents' commuting activity spaces. For instance, Zhang et al. (2009a, b) conducted a comparative analysis of commuting behaviors among residents in different residential areas, while Liu et al. (2009) examined the influence of residents' attributes and residential characteristics on commuting distances. Other researchers have explored the evolution of residents' commuting activities in the context of work-life separation. Zhou and Yan (2006) and Meng et al. (2011) identified work-residence separation as a significant factor influencing residents' commuting activities. Conversely, Feng and Zhou (2004) and Xu et al. (2009) reflected on the phenomenon of urban work-residence separation by analyzing changes in commuting characteristics and duration before and after residents' relocations.

2.5.4 Research on the Quality of Urban Living Spaces

Research concerning the quality of urban living spaces primarily focuses on the development of comprehensive urban spatial quality evaluation index systems. This research aims to assess the quality of various living units within urban spaces and to establish a foundation for devising strategies aimed at enhancing the urban living quality.

The primary objective of this research is to evaluate the quality of living spaces in specific types of cities within the region and to conduct comparative analyses among various urban centers. For instance, Wang and Sun, among others, performed a comparative analysis of the quality of living spaces in different provincial capital cities in China, employing Boyer's nine-factor index. Their findings indicated a positive correlation between the high quality of life experienced by urban populations and the size of the city (Sun and Wang 2002; Wang 2004). Additionally, Zeng et al. (2014) utilized spatial analysis to examine the spatial patterns of urban living quality in counties within Jiangsu Province and its economic mechanisms. The study demonstrated that, influenced by the current stage of China's rapid economic development as well as the political and economic system, the quality of urban life is contingent upon the level of economic development in the region. This is evidenced by the observation that cities in economically developed regions generally exhibit a higher quality of life.

Evaluations of the quality of living space across various geographical units within urban environments have been conducted. Ning and Zha (1999) developed a habitat evaluation index system comprising 20 indicators categorized into three domains: living conditions, ecological environment quality, and infrastructure and public service facilities. Their analysis specifically focused on the situation in Shanghai. Similarly, Wang (2004) examined the quality of life in the inner-city community of Xi'an, conceptualizing it in terms of "income space", "safety space" and "health space". In addition, some scholars have investigated the spatial differentiation of quality of life in inner-city neighborhoods. Pan and Wang (1997) assessed the spatial quality of urban living in Xi'an from both objective and perceptual perspectives, concluding that the quality of urban living in Xi'an diminishes progressively from the city center to the suburbs. They identified several factors influencing this spatial quality, including the floating population, social security within the city, transportation conditions in the suburbs, public facilities, children's education, and industrial pollution. Furthermore, Wang (2005) analyzed the spatial distribution of quality of life in Hangzhou, revealing that the quality of life for residents in the city center is significantly higher than that of residents in the suburbs.

The existing body of research on urban spatial quality of life conducted by Chinese scholars indicates a need for enhanced empirical investigations into the spatial disparities in quality of life across various urban types. Furthermore, the methodological challenges associated with data utilization in empirical studies, particularly the integration of big data, represent critical areas for improvement in future research endeavors.

2.5.5 *A Comprehensive Review of Research on Urban Living Spaces in China*

Research on urban social space in China has its origins in the late 1980s, with initial studies concentrating on the macro-statistical significance of community dynamics and residents' daily activities and behaviors at the city level. As socio-spatial research evolved, particularly following the late 1990s, scholars began to focus on various types of urban social spaces catering to diverse populations, with urban living space being a prominent area of interest. To date, significant academic contributions have been made regarding the structure and transformation of urban living spaces in China, encompassing aspects such as urban community living space and its planning, the daily activity space of urban residents, and the quality of urban living space. Notably, geographers have produced substantial research outcomes in this domain. Geography scholars have approached the study of urban living spaces in China from a geographical perspective, addressing the challenges posed by the "transformation of Chinese cities" in the context of rapid urbanization and socio-economic development. Their work has provided valuable insights into achieving and has significantly contributed to the development of urbanization theory with Chinese characteristics, as well as the theoretical framework of Chinese urban space in the contemporary era.

In recent years, urban living spaces in China have emerged as a significant area of inquiry within the discipline of urban geography. Over the past two decades, the predominant research methodologies employed have primarily involved statistical analyses utilizing quantitative approaches. These studies have focused on the characterization of urban spatial structures through relevant regression model parameters, the development of indicator systems for evaluative analyses, and the statistical examination of residents' behaviors and life trajectories based on data collected from questionnaire surveys and life diaries. However, the application of qualitative methods, particularly in-depth interviews, in the investigation of urban living spaces in China is still in its nascent stages. There is a pressing need to explore residents' living space from humanistic and postmodern perspectives. Consequently, it is essential to focus on understanding residents' living spaces through the lenses of humanism and postmodernism. This exploration remains preliminary, particularly in terms of examining living spaces from the standpoint of individual subjects and behaviors, as well as integrating individual life histories and trajectories with Geographic Information Systems (GIS) to achieve a synthesis of visualization and narrative in the study of living space. In conclusion, numerous challenges persist for researchers aiming to apply innovative methodologies to the study of urban living spaces in China. Key considerations include the selection of methodological entry points, the application of new techniques, and the exploration and innovation of the methodologies themselves. Furthermore, there is a notable imbalance in research topics, with a greater emphasis on traditional urban living spaces and insufficient attention to emerging types of living environments, particularly those influenced by advancements in information and communication technologies. While there has been substantial research on the spatial dynamics of various demographic groups, there remains a lack of focus on the individual level and the theoretical implications of living spaces. Additionally, within group living space studies, there is a predominance of research on mass groups, and less attention given to niche populations. From a theoretical perspective, there is a wealth of empirical research conducted on Chinese cities through the application of Western theories; however, there is a scarcity of studies that critically engage with or revise these theories, and even fewer that seek to establish theoretical frameworks for urban living spaces that reflect Chinese characteristics. Therefore, there is an urgent need to develop a comprehensive theoretical framework to guide the study of urban living spaces and their transformations in China.

References

Bao R (2010) Renewal of urban living spaces in old city districts. J Hefei Univ Technol 33(7):1049–1052 (In Chinese)

Bauman (2002) Liquid modernity. Translated by Ouyang J. SDX Joint Publishing Company, Shanghai (In Chinese)

Bo D (2014) The Past and Present of Work Units: Social Space and Governance in Chinese Cities. Nanjing: Southeast University Press

Bourdieu P (1989) Social space and symbolic power. Sociol Theory 7(1):14–25

Chai Y (1996) The structure of urban living space in China based on the Danwei system: a case study of Lanzhou. Geogr Res 15(1):30–38 (In Chinese)

Chai Y (2000) Urban space. Science Press, Beijing (In Chinese)

Chai Y, Li C (2005) Spatial characteristics of daily shopping behaviors of urban elderly in China: cases of Beijing, Shenzhen, and Shanghai. Acta Geogr Sin 60(3):401–408 (In Chinese)

Chai Y, Li C (2019) Urban life circle planning: from research to practice. City Plann Rev 43(5):9–16, 60 (In Chinese)

Chai Y, Liu Z, Li Z et al (2002) The temporal-spatial structure of Chinese cities. Peking University Press, Beijing (In Chinese)

Chai Y, Weng G, Gong H (2004a) Temporal-spatial characteristics of shopping behaviors of Shenzhen residents. Hum Geogr 19(6):79–84 (In Chinese)

Chai Y, Lin T, Gong H (2004b) Analysis of spatial decision factors in shopping behaviors of Shenzhen residents. Hum Geogr 19(6):85–88 (In Chinese)

Chai Y, Chen L, Zhang C (2007) Danwei system transition: a key perspective on China's urban transformation. World Region Stud 16(4):60–69 (In Chinese)

Chai Y, Weng G, Shen J (2008) Study on Shanghai's commercial spatial structure based on residents' shopping behaviors. Geogr Res 27(4):897–906 (In Chinese)

Chai Y et al (2010) Urban space and consumer behavior. Southeast University Press, Nanjing (In Chinese)

Chai Y, Zhang X, Sun D (2015) Urban life circle planning based on spatio-temporal behaviors: a case study of Beijing. Urban Plann Forum 3:61–69 (In Chinese)

Chai Y, Li C, Xia W et al (2019) Delineation model of urban community life circles: a case study of Qinghe subdistrict, Beijing. Urban Dev Stud 26(9):1–8 (In Chinese)

Chen X, Feng J (2009) Hierarchical structure and evolution of urban residents' shopping travel: a case study of Beijing. City Plann Rev 33(1):22–30 (In Chinese)

Chen Q, Xu P (1985) A preliminary study on evaluation methods for urban living environment quality. City Plann Rev 5:52–58, 29 (In Chinese)

Feng J (2002) Spatial diffusion and suburbanization of urban industry in Hangzhou. Urban Plann Forum 2:42–47 (In Chinese)

Feng J (2004) Spatial restructuring of Chinese cities in the transition period. Science Press, Beijing (In Chinese)

Feng J, Chen X, Lan Z (2007) Evolution of shopping behavior spatial structure of Beijing residents. Acta Geogr Sin 62(10):1083–1096 (In Chinese)

Feng J, Lin W (2017) Residential satisfaction and influencing factors in declining old urban neighborhoods: a case study of six communities in Suzhou. Prog Geogr 36(2):159–170 (In Chinese)

Feng J, Wang Y (2008) Social spatial characteristics and formation mechanisms of residential areas around universities in Zhongguancun. Geogr Res 27(5):1003–1019 (In Chinese)

Feng J, Xiang Y (2013) Study on daily activity spaces of residents in development zones. Hum Geogr 27(3):42–50 (In Chinese)

Feng J, Xiang Y (2017) Characteristics and formation mechanisms of residential spaces in development zones: a case study of beijing economic-technological development area. Prog Geogr 36(1):99–111 (In Chinese)

Feng J, Zhou Y (2004) Residential relocation and related spatial behaviors during suburbanization in Beijing. Geogr Res 23(2):227–242 (In Chinese)
Gaubatz P (2007) Globalization and the Development of Urban Central Business Districts in China: Cases of Beijing, Shanghai and Guangzhou. In Wu F, Ma R, and Zhang J. Transition and Reconstruction: Multidimensional Perspectives on China's Urban Development. Nanjing: Southeast University Press (In Chinese)
Gu C (1994) Post-war western schools of urban studies. Acta Geogr Sin 49(4):371–382 (In Chinese)
Gu C, Sheng M (2012) Study on low-income college graduate communities in beijing: the tangjialing phenomenon and its continuation. Hum Geogr 27(5):20–24, 103 (In Chinese)
Gu C, Chan R, Liu G et al (2006) Beijing's socio-spatial restructuring: immigration and social transformation in the epoch of national economic reformation. Prog Plan 66:249–231
He H, Han S, Zhang C (2013) Rational planning of living spaces in "Transitional Communities" in urban-rural fringes: cases from Suzhou Industrial Park. China Ancient City 9:18–23 (In Chinese)
Jie A (2000) Danwei: a unique social living space. J Zhejiang Univ 5:73–80 (In Chinese)
Lefebvre H (1991) The production of space. Wiley Blackwell, Oxford
Li Z, Chai Y (2000) Study on commuting characteristics of Dalian residents. Hum Geogr 15(6):71–76, 63 (In Chinese)
Li Z, Gu C (2011) Transformation of urban social spatial structure in China. Southeast University Press, Nanjing (In Chinese)
Li F, Feng J, Liu J et al (2013) Restructuring of living spaces in suburban large-scale residential areas based on activity types. Hum Geogr 27(3):27–33, 113 (In Chinese)
Liu Z, Zhang Y, Chai Y (2009) Job-housing separation phenomenon and its characteristics in Chinese metropolises: a case study of Beijing. Urban Dev Stud 9:110–117 (In Chinese)
Meng B, Zheng L, Yu H (2011) Changes and influencing factors of commuting time for Beijing residents. Prog Geogr 30(10):1218–1224 (In Chinese)
National Bureau of Statistics (2009) New China in 60 years. China Statistics Press, Beijing (In Chinese)
Ning Y, Zha Z (1999) Evaluation and optimization of metropolitan human settlements: a case study of Shanghai. City Plann Rev 23(6):14–19, 63 (In Chinese)
Pan Q, Wang X (1997) Spatial evaluation of quality of urban life: a case study of Xi'an. Hum Geogr 13(2):29–37 (In Chinese)
Shen Y, Chai Y (2013) Daily activity space of suburban mega-community residents in Beijing based on GPS data. Acta Geogr Sin 68(4):506–516 (In Chinese)
Soja EW (1980) Socio-spatial dialectic. Ann Assoc Am Geogr 70(6):207–225
Song W, Zhu X (2009) Gated communities in China: negative spatial response to social differentiation. Planners 25(11):84–88 (In Chinese)
Sun F, Wang X (2002) Current status and trends of urban living space and community sustainable development research in China. Prog Geogr 21(5):491–499 (In Chinese)
Sun D, Shen S, Wu T (2012) Allocation of public service facilities at county level from the perspective of life circle theory: case study of Pizhou, Jiangsu Province. Planners 28(8):68–72 (In Chinese)
Ta N, Chai Y (2010) Study on the transformation of danwei communities in Chinese cities from filtering perspective. Hum Geogr 25(5):6–10 (In Chinese)
Wang X et al. (2000) Research on urban social spatial structure in China. Beijing: Science Press (In Chinese)
Wang X (2004) Research on urban living space structure in China. Science Press, Beijing (In Chinese)
Wang W (2005) Quantitative evaluation of quality of urban life in Hangzhou. Acta Geogr Sin 60(1):151–157 (In Chinese)
Wang W (2011) Comprehensive evaluation of urban residents' quality of life in Hami City. Contemp Econ 4:78–81 (In Chinese)
Wang L (2012) Controlled indicator system for urban community living space planning. Mod Urban Res 2:45–54 (In Chinese)

Wang Y, Feng J (2013) Reconstruction of suburban residents' relocation behavior from life course perspective. Hum Geogr 27(3):34–41, 50 (In Chinese)
Wang L, Wang X (2011) Deconstruction and quality reconstruction of urban community living space structure. Scientia Geogr Sin 31(1):22–28 (In Chinese)
Wei L, Li Z (2006) Housing dilemmas and improvement models for China's urban low-income class. Urban Plann Forum 2:53–58 (In Chinese)
Wu F (1997) Urban restructuring in China's emerging market economy: towards a framework for analysis. Int J Urban Reg Res 21:640–663
Wu Z (2012) Evolution of leisure activities in China from human-time-space perspective. Econ Geogr 32(2):149–153, 176 (In Chinese)
Wu Q, Ren D, Yang Y (2000) Theoretical basis and research levels of urban residential spatial differentiation. Hum Geogr 15(3):1–5 (In Chinese)
Xiang B (2010) Study on urban residential space form in Northern Shaanxi from regional cultural perspective: case study of Ganquan County, Yan'an. Xi'an University of Architecture and Technology, Xi'an (In Chinese)
Xu D, Wang Z, Zhu X et al (2009) Social spatial structure of Nanjing in transitional period: factor ecological analysis based on the fifth census data. Geogr Res 28(2):484–498 (In Chinese)
Yan X, Wei L, Zhou R et al (2004) Coordinating urban-rural relations in rapidly urbanizing areas: case study of 'urban village' redevelopment in Guangzhou. City Plann Rev 28(3):30–38 (In Chinese)
Yang C (2009) Institutionalization of everyday living space: analytical framework for Shanghai worker new villages in the 1950s. J Tongji Univ 6:38–45 (In Chinese)
Yuan J (2012) Evolution of urban squares in modern China. Chinese National Academy of Arts, Beijing (In Chinese)
Yuan J, Sun Z, Zhang N et al (2005) Reconstruction of China's urban regional system based on 'daily life circle.' Scientia Geogr Sin 25(1):17–22 (In Chinese)
Zeng W, Zhang X, Xiang L et al (2014) Spatial pattern and economic analysis of quality of urban life at county level in Jiangsu Province. Econ Geogr 34(7):28–35 (In Chinese)
Zhang G (2005) Spatial schemata of human living spaces in the information age. City Plann Rev 29(10):29–36 (In Chinese)
Zhang X (2007) Study on daily life space: formation of Shanghai's urban daily life space. Tongji University, Shanghai (In Chinese)
Zhang M, Guo Z (2010) Research on creation of living space and lifestyle guidance in large-scale residential areas at urban fringe: case study of Dalian Longpan Jinquan Phase II & III. Chin Overseas Architect 1:62–66 (In Chinese)
Zhang J, Lü J (2003) From large-scale urban design to 'everyday living space.' City Plann Rev 27(9):39–44 (In Chinese)
Zhang T, Yamamura T (2003) Analysis of daily life space structure in Lijiang ancient town. J Peking Univ 39(4):33–39 (In Chinese)
Zhang Y, Chai Y, Zhou Q (2009) Spatiality and transformation of Chinese danwei compounds: case study of Beijing Jingmian No. 2 Factory. Urban Plann Int 24(5):20–27 (In Chinese)
Zhong X (2013) Social space and social transformation: the socio-spatial turn in urban studies during transitional China. Soc Sci Abroad 2:14–21 (In Chinese)
Zhou Y (1996) Suburbanization of Beijing and its implications. Scientia Geogr Sin 16(3):198–206 (In Chinese)
Zhou Y (2010) Urban geography explorations: Zhou Yixing's self-selected works. The Commercial Press, Beijing (In Chinese)
Zhou Y, Feng J (2002) Issues in applying "main city" concept. City Plann Rev 26(8):46–50 (In Chinese)
Zhou Y, Meng Y (1997) Suburbanization of Shenyang: a comparison with Western suburbanization. Acta Geogr Sin 52(4):289–299 (In Chinese)
Zhou S, Yan X (2006) Residential-employment spaces and their impact on travel behavior in Guangzhou. City Plann Rev (5):13–18, 26 (In Chinese)

Zhu W (1993) Space, symbol, city: a theory of urban design. China Architecture & Building Press, Beijing (In Chinese)
Zhu C, Wang D, Ma L (2010) Allocation of urban-rural public service facilities based on life circle: a case study of Xiantao. In: Proceedings of 2010 China urban planning annual conference (Planning Innovation) (In Chinese)

Chapter 3
Urban Lifestyles and Their Evolution

3.1 Urban Lifestyles and Their Characteristics

With the advancement of socio-economic development and the improvement of people's living standards, the concept of people-oriented urban planning and urban development has become increasingly ingrained in public consciousness, so understanding the lifestyles of urban residents and their relationship with living spaces has become critically important. The concept of lifestyle originated in sociology and has been explored by numerous Western sociologists, including Karl Marx, Max Weber, and Thorstein Veblen. With the rapid increase of economic development and social transformation speed in China, the lifestyles of Chinese urban residents have also undergone dramatic changes, and understanding this change is the prerequisite for comprehending urban living spacc in China.

3.1.1 Lifestyles in Sociology

3.1.1.1 Studies on Lifestyles in China and the West

Marx and Engels frequently used the concept of "lifestyle" in their works such as *The German Ideology* and *The Eighteenth Brumaire of Louis Bonaparte*. Their discussions on lifestyle laid the foundation for sociological research on this subject. To summarize, the concept of "lifestyle" in the works of Marx and Engels generally present the following two meanings: either as a crucial indicator for distinguishing social classes, or a comprehensive term encompassing all aspects of life, including production activities (Gao 1998). Marx and Engels established the theoretical foundation for lifestyle studies. Since the nineteenth century, sociological research on lifestyles has primarily developed along three theoretical approaches.

J. Feng et al., *Urban Living Space in China*, Urban Sustainability,
https://doi.org/10.1007/978-981-95-0595-1_3

Sociologists, represented by Max Weber, were the first to pioneer research on lifestyles. They conducted research along the first sense of lifestyle proposed by Marx and Engels, where lifestyle was seen as an important symbol for recognizing the social status of a group (Ma and Xia 2004). Thorstein Veblen also followed this line of thought, arguing that the conspicuous consumption of the upper classes—intended to showcase their quality of life and wealth—constituted the entirety of their lifestyles (Tian 2011). Lifestyle continued to be mentioned as an indicator of class status.

After World War II, a new trend in lifestyle research emerged, shifting the focus toward consumption patterns, with scholars advocating for the study of commodity consumption as a means of understanding and defining lifestyles. By the late 1970s, rapid economic development had led to a dramatic improvement in living standards, making it difficult for single indicators commonly used in the past, such as private car ownership and housing conditions, to accurately reflect differences in people's lifestyles any longer (Tao 2006). Stimulated by the new socio-economic development, scholars began to explore new research perspectives, which led to the emergence of the analysis of lifestyle types. For instance, American scholar Mitchell (1983), based on the survey data, categorized the lifestyles of Americans into four main types: demand-driven, outward-oriented, inward-oriented, and a mixed outward-inward type.

Reviewing the evolution of lifestyle research abroad, lifestyle has gradually developed from a subordinate and marginal concept to an independent concept, and eventually became an independent field of study. In China, lifestyle research developed in parallel with the restoration and reestablishment of sociology, beginning in the 1980s. During this period, scholars, corresponding to the needs of China's modernization, believed that the construction of lifestyles should also be an important part of modernization, and proposed guiding principles and specific frameworks for constructing the socialist mode of production with this goal in mind. However, since the early 1990s, lifestyle research in China has fallen into a period of stagnation. Although there has been some progress in foundational theories of lifestyle as well as researches on leisure, consumption, and family lifestyles, a large number of scholars have moved on to other research fields (Wang 2013).

In the studies of lifestyles within both Chinese and Western sociology, different scholars at different stages varied significantly in their understanding and conceptualization of lifestyle, but in general, definitions of lifestyle have consistently been grounded in the variability of objects (Gao 1998). Initially, differences in lifestyles were believed to be dominated by differences in social class or stratum, where people of different classes or strata had clear and fixed patterns of life, and common attributes of a group overrode individual characteristics. However, as social stratification became more complex, the boundaries of groups defined by class or stratum became ambiguous. Individuals' lifestyles no longer strictly aligned with group identities, making it impossible to form an in-depth understanding of the differences between classes or strata solely based on lifestyles. Today, with highly developed socio-economic levels, abundant material supplies and diverse cultural values, lifestyles are increasingly revealing differences between individuals. Personal values and socio-economic status together dominate lifestyle choices of the individuals, leading to the formation of diverse lifestyle patterns, and people have begun to focus on the interaction between specific types of lifestyles and their daily lives.

3.1.1.2 The Significance of Returning to Everyday Life

Today, the conditions of the era have changed significantly. In the context of globalized and post-industrialized society, lifestyle research, which emphasizes a return to everyday life, has become even more relevant and valuable.

From a macro point of view, lifestyle serves as an effective entry point for revealing the way society operates. Traditional grand generalized theories are no longer able to fully and accurately portray entirety of social operation. Only by precisely grasping human activities within their broader environmental and spatial contexts can we avoid reducing society to a mere interplay of variables (Li 2007). Therefore, the rediscovery of daily life provides a new entry point for a more comprehensive and concrete understanding of the nature of social operation.

From a micro point of view, lifestyle research itself provides an answer to the question of *how people live*. Interpreting and reconstructing the contemporary world of life can help clarify the pathways of how to construct a *good life*. For China, which is facing changes in the times and transformation in society, the intervention of lifestyle research has pointed out a new direction for solving the problem of imbalance between material and spiritual life.

3.1.2 Urban Lifestyles

3.1.2.1 Understanding Urban Lifestyles from Different Perspectives

After the Industrial Revolution, industrialization accelerated urban development, and urban life began to take on many new characteristics that were different from those of the rural areas. The rapid pace of urbanization has forced people to pay attention to the characteristics, developmental trends and potential impacts of urban lifestyles.

The sociologist Georg Simmel was the first to focus on modern metropolitan life in the West. In *The Metropolis and Mental Life*, he argued that metropolitan life, unlike rural life, significantly causes tension in individuals' mental life. Influenced by Simmel's theories, Louis Wirth, a sociologist of the Chicago School, published *Urbanism as a Way of Life* in 1938, in which the research framework for urban lifestyles was introduced for the first time. He pointed out that urban lifestyles derive from population size, residential density, and the heterogeneity of residents and group life, and spatial characteristics are the fundamental reason why urban lifestyles can be recognized as a distinct type of lifestyle. In his work, Wirth suggests that urban life can be examined from three perspectives, based on three spatial characteristics: (1) as a physical structure comprising demographic, technological and socio-ecological order; (2) as a system of social organization that includes a distinctive social structure, a set of social institutions and a typical pattern of social relationships; and (3) as a particular set of attitudes, values, and numerous personalities emerged in the forms of typical collective behaviors and regulated by special mechanisms of social control (Wirth 2007).

Although Wirth considered spatial characteristics as the basic variables defining urban lifestyles, he also emphasized that urban lifestyles are not confined to the city; they exist to varying degrees in any area influenced by urbanization. After Wirth, Herbert Gans found that the characteristics of urban life are difficult to be fully generalized by spatial characteristics and that lifestyle is not perfectly aligned with type of residence. Instead, he suggested understanding urban lifestyles from the perspective of cultural specificity. Claude Fischer, on the other hand, affirmed the importance of space in the formation of urban lifestyles, while also introducing the concept of subcultures, making both space and culture key factors influencing urban lifestyles (Zhang 2009).

Overall, urban lifestyle, as a type of lifestyle, is a mode of living formed in the context of industrialization and urbanization which is different from rural areas. Cities are not only the geographical background of the occurrence of this kind of lifestyle, but their spatial characteristics and several unique features qualify and influence people's daily lives, so that urban residents, actively or passively, share similar living conditions. Urban areas are distinguished by high population density, significant industrial concentration, advanced social division of labor, well-established public infrastructures, heterogeneous populations, constant interaction between diverse cultures and values, etc. They have not only changed people's mode of production, but also reshaped people's ideology and spiritual life. Urban residents' daily behaviors, such as consumption, leisure, and social interaction, are fundamentally different from those in rural areas: labor lifestyles have developed from centralized and planned structures to autonomous and creative ones; consumption lifestyles have evolved from merely meeting basic needs to being more fashion-oriented; leisure lifestyle have transitioned from passive relaxation to more fulfilling activities; and social interaction have moved from closed network to more open and dynamic forms, and so on (Xiao and De 2003).

3.1.2.2 Characteristics of Urban Lifestyles

One of the distinctive features of urban areas is the significant heterogeneity of their populations, which inevitably leads to distinct lifestyle characteristics among different social groups. Nevertheless, when considering urban residents as a whole and comparing urban lifestyles with rural ones, urban lifestyles continue to present a number of universal characteristics. Based on the discussions of scholars such as Lu (1987), Xiao and De (2003), these characteristics can be summarized as complexity, openness, mobility, and diversity, as detailed below.

(1) Complexity. The city is a complex system, and the living demands of a large-scale population have led to the development of multiple functions such as production, commerce, culture, education, and recreation. The diversity and high concentration of functions provide urban residents with a variety of choices, leading to the development of urban lifestyles that vary across different dimensions and levels.
(2) Openness. The city is an open system. Frequent movement of the population introduces diverse cultures and lifestyles to urban areas. Urban residents are generally more open-minded and tolerant, which leads to a greater acceptance of diverse lifestyles. As a result, urban lifestyles are not closed and static; their openness contributes to a constant state of flux.
(3) Mobility. The mobility of urban lifestyles is an inevitable consequence of their openness. Changes in economic, cultural and political factors lead to shifts in lifestyles that are influenced by these factors. Movement of the population also keeps the urban lifestyles in a constant state of conflict and integration of the old and the new. Furthermore, vertical movement of the population and structural changes influence residents' living standards; consumption, leisure, and social interactions are also subject to adjustment.
(4) Diversity. The city is a concentration of diverse lifestyles. The richness of the demographic structures provides a foundation for a wide range of lifestyle variations. People are free to choose different lifestyles based on different occupations, ethnicities, social classes and geographic regions. The general pursuit of urban residents to express their personalities also drives them to create and develop their own unique lifestyles.

3.2 Traditional Lifestyles of Chinese Urban Residents

Traditional lifestyles are lifestyles that have been standardized, customized and solidified in society (Wang 1988). Some aspects of the traditional lifestyles of Chinese urban residents have persisted throughout history, while others, though no longer existing in their original forms, have been integrated into modern lifestyles and continue to endure. Thus, traditional lifestyles are not in direct opposition to modern lifestyles; rather, they serve as the prerequisite and foundation for understanding modern lifestyles. Given the broad scope of urban residents' lifestyles in China, it is difficult to provide a comprehensive discussion of all aspects. Therefore, this section focuses on summarizing the characteristics of traditional lifestyles among Chinese urban residents in terms of social interactions, consumption behavior, recreation, and traveling activities.

3.2.1 Daily Interactions Based on Neighborhood and Sense of Place Community

Neighborhoods serve as basic residential units, and neighborhood relations are close interpersonal relationships formed by the proximity of residential space. In rural China, where life is relatively stable and spatially constrained, a wide range of "acquaintance societies" are established based on kinship and geographical ties. These connections, based on trust and emotional bonds, contribute to the formation of stable and long-lasting neighborhood relationships, resulting in highly overlapping social networks among rural residents. However, compared to rural areas, urban areas are large in scale and densely populated, with urban life characterized by complexity, openness, mobility and differentiation, so that urban residents are more likely to experience individual strangeness and alienation (Tao 1987; Xiao and De 2003). Within this context, neighborhood relations in urban areas tend to be more limited in scope than in rural areas. They typically occur within smaller spatial scales and are more commonly established through geographical and occupational ties in mixed-surname residential settings with strong connections between work and daily life. Nevertheless, it is undeniable that neighborhood relations still remain a crucial part of the social network of urban residents (Ta and Chai 2013). Good neighborhood communication and mutual assistance can deepen emotional bonds, strengthen residents' sense of identity and belonging to the community in which they live, and thus contribute to greater social cohesion and harmony.

Neighborhood interactions are closely related to the form of housing, and traditional housing types such as Siheyuan, Dazayuan, Lilong housing, and Danwei fostered stable, familiar, and intimate neighborhood relationships among urban residents. Among them, Siheyuan is a typical representative of traditional Chinese houses, which was gradually developed and matured during the Ming and Qing dynasties. Its architectural layout consists of four independent houses enclosing a spacious central courtyard. This structure fulfills the residential space needs of a large family, and the internal quadrangle space not only carries the intimacy and harmony of blood relations, but also demonstrates the order and strictness of etiquette, norms, and patriarchal laws. Meanwhile, the isolation of the external courtyard wall strengthens the closure and privacy of the residence (Cui 2013). After the founding of the People's Republic of China, rapid influx of urban population forced numerous Siheyuan to transform into Dazayuan, where multiple households cohabited in a shared courtyard. Although the environment inside the courtyard is dilapidated and far worse than before, the courtyards together formed an important social space for interactions and activities through the tandem of neighborhoods, where adults can chat while doing housework and children can play together. Different families have cultivated deep neighborhood feelings and solid cohesion through decades of mutual support, making harmonious courtyard life a key component of local identity and memory. In coastal cities such as Shanghai and Tianjin, there has been a transition from Siheyuan to Lilong housings. While the traditional courtyard shrunk into smaller patios, which remained vital spaces for family activities, the interconnected alleys served as crucial semi-public spaces. These alleys not only facilitated daily movement but also enhanced neighborhood interactions and strengthened residents' local attachment (He 2008; Cui 2013). From the traditional Siheyuan to the later Dazayuan and Lilong housing, the bloodline basis of urban neighborhoods has gradually been replaced by geographic and mixed-surname patterns, and with the changes in China's political system and socio-economic development, the changes in the form of housing have led to new changes in the composition of urban neighborhoods.

During the planned economy period, numerous cities in China adopted the Danweilization model of community construction and welfare housing allocation system. Residential areas were constructed adjacent to workplaces, and employees were allocated housing by their Danwei. A typical feature of this system is that the diversified functions of Danwei production, daily life, recreational activities all take place within the organizational form of the Danwei (Zhang and Yi 2019). The physical scope of the Danwei is defined by courtyard walls and roads, with one or two gates, so that people could meet most of their daily needs within the compound (Zhang 2009). Neighborhood relationships in Danwei are primarily based on occupational ties, as residents are mostly colleagues at work who have close relationships and frequent interactions with each other, thus forming a strong sense of community emotion. However, the nature of Danwei enclosure is the forced agglomeration of population, so the restriction of personnel mobility and similarity of identity made the social relations of the residents relatively homogeneous. The spatial closure of "work-life integration" further limited the establishment of a broader and more diversified network of social relations among the residents.

After the Reform and Opening-Up, the implementation of the market economic system and housing system reform led to the acceleration of urban housing investment and construction. Housing has been gradually commercialized and socialized, and the multi-tiered housing market has met the diversified needs of different families, so that the urban residents can choose the appropriate housing based on their own incomes and preferences. This broke down the monolithic, closed form of housing under the Danweilization and promoted the differentiation of urban living space. Residential communities replaced Danwei, and common community management systems and cultural construction formed the basis for a new type of neighborhood relationships and sense of local community.

However, in today's individual, unit-type residential communities, urban residents' neighborhood relationships are facing a crisis. High levels of anonymity, distant interactions, and widespread social indifference have become common phenomena among urban residents (Hu 2015; Feng and Lin 2017). This is because accelerated urbanization and internal migration within cities have significantly weakened the kinship and work-based relationships in traditional communities, leading to an increase in the heterogeneity and anonymity of the population. In addition, virtual social interaction in the digital age has radically transformed the way people interact with each other, and the pursuit of utilitarianism under the anxiety of the postmodern society has had a lasting impact on trust and emotions. The weakening and disintegration of neighborhood relationships can negatively impact residents' sense of community security, identity, and belonging (Shi 2014). Therefore, revitalizing neighborly interactions and enhancing the quality of urban life have become important issues in contemporary urban development.

3.2.2 *Consumer Behavior Based on the Development and Utilization of Physical Commercial Spaces*

The Chinese term for "city(城市)" is a combination of Chinese character "*cheng*(城)" and "*shi*(市)". In ancient China, *cheng* referred to a walled defensive area for external protection, while *shi* referred to the marketplace where commercial activities took place. This shows that the behavior of urban residents in a specific commercial space for larger-scale sales and consumption behavior has existed since ancient times. In addition, the spatial organization of ancient Chinese *shi* underwent a fundamental change during the Song Dynasty. Before the Song Dynasty, *fang*(坊) and *shi* were strictly separated, with clear boundaries between residential and commercial areas. Until Song Dynasty gradually abolished the confined *fang-shi*(坊市) system, allowing shops to open along streets without restrictions. Morning markets and night markets merged seamlessly, with mixed expansion of residential and commercial areas, leading to the rapid growth of the commodity economy.

Entering the twentieth century, under the strong influence of trade activities and Western culture, consumption spaces in Chinese cities has increased in number, size and variety. From banks, cinemas, cabarets, cafés, and department stores introduced from the modern West, to state-owned department stores implemented during the planned economy period after the founding of the People's Republic of China, and later to large chain supermarkets, shopping malls, and urban complexes that emerged one after another in the context of the rapid development of retail and service industries following the Reform and Opening-Up, the spatial forms and types of urban consumption have continuously evolved, and their functions have also become increasingly compounded and efficient (Yang and Guo 2018), meeting the diversified consumption needs of urban residents with greatly increased income levels, in both materials and services (Lu 1987). Within this context, people's needs are closely tied to places: shopping for clothes requires visiting retail streets, watching movies necessitates going to the cinema, dining with family, friends, or colleagues takes place in restaurants, and purchasing groceries and daily necessities happens in supermarkets, etc. Such consumption concepts have become deeply internalized in the perceptions of urban residents and have externalized into the general consumption patterns of urban residents.

Geographers have found that the spatial patterns of urban residents' daily shopping activities follow the rule of distance attenuation centered around their homes. Generally speaking, the higher the level of commodity, the farther the average distance of residents' shopping travel, and the lower the frequency of travel. Urban residents of contemporary China have formed a three-tiered shopping travel structure centered on their residences, which are the neighborhood consumption zone constituted by convenience stores, the community consumption zone constituted by community supermarkets and farmers' markets, and the urban consumption zone constituted by large supermarkets, department stores, and shopping malls (Wu et al. 2013). This behavior of selecting suitable modes of travel based on one's own shopping purpose, preference and purchasing power, and visiting physical spaces of different types, locations, scales and centrality for consumption is the traditional consumption pattern of urban residents in China.

Although the rapid rise of online shopping in the Internet era has significantly impacted physical commercial spaces due to its convenience and freedom from spatial and temporal constraints, the consumption mode relying on physical spaces still holds irreplaceable advantages. In physical spaces, consumers can truly perceive, experience and evaluate the information of products and the display environment, especially for one-stop commercial centers that integrate shopping, dining and entertainment which more conveniently meet the diverse needs of people's consumption, socialization and leisure. Additionally, the experiential consumption venues that create immersive environments and themed atmospheres have become a dominant trend (Yang and Guo 2018). For instance, retail stores bring surprises to consumers through high-quality, creative and personalized spatial designs, immersing them in a multi-sensory spatial experience involving vision, smell, etc. This kind of experience-focused and consumer-centered shopping experience cannot be replicated in the cyberspace. Furthermore, as the growth rate of Internet users in China is slowing down and the size of the customer base of the online retail industry tends to stabilize, the "new retail" model, which deeply integrates online, offline, and logistics, is expected to shape the future of the retail industry and play a role in restructuring consumption patterns of urban residents (Wang 2019). In short, the rich physical commercial spaces constitute an important dimension of urban living space. The bustling and lively commercial activities are an indispensable part of urban life, and even contributes to cultural branding and urban appeal. It is difficult to imagine Paris without Champs-Élysées or New York without Broadway. Likewise, the development of urban culture in China cannot be separated from the careful cultivation and construction of physical commercial and consumption spaces.

3.2.3 Collective Leisure and Entertainment Activities Based on Public Spaces

The socio-economic development and the improvement of productivity have led people gain more leisure time, making recreation and entertainment an important aspect of lifestyle. In ancient times, urban residents' leisure activities were mostly characterized by ethnicity, including cultural activities such as music, chess, calligraphy and painting, sports activities such as horse racing and boating, folk activities such as temple fairs and incense, and marketplace activities such as teahouses and theaters (Wu 2009). In modern era, with the introduction of new Western lifestyles and the transformation of traditional urban society in China, people's psychology, behavior and value systems have changed. The establishment of Western-style venues such as cinemas, parks, theaters, museums, and libraries, along with the rise of mass media such as newspapers, radio, and television, has further diversified urban residents' recreational activities, creating a fusion of Chinese and Western, as well as traditional and modern, elements (Fu 2007). Similar to consumption behavior, the leisure and recreation behavior of urban residents in China also relies on physical venues, and is often characterized by public and collective nature.

During the planned economy period, especially in the special era of the Cultural Revolution, people's leisure and entertainment options are limited and monotonous in content, and public venues related to those activities were insufficient. However, under the influence of the prevailing ideology of collectivism, a large number of mass activities relying on the public spaces and of a collective and large-scale nature emerged, shaping the traditional forms of urban leisure and entertainment since the second half of the twentieth century. For example, the nationwide promotion of radio gymnastics in the 1950s and 1960s turned it into a widely influential and enduring mass sports and entertainment activity in China. making it a habitual part of daily life, with streets and alleys transforming into exercise fields (Xu 2019). In the 1960s and 1970s, dancing *zhongziwu* (loyalty dance) and watching *yangbanxi* (revolutionary opera) together in large venues became key activities during people's leisure time. Sports competitions, collective dances, cultural performances, and tea parties organized by factories enriched the collective entertainment life of workers. Urban squares, grand auditoriums, stadiums, and other large-scale landmark public places have also become symbols of unity and collective honor because of their functions of gathering and rejoicing.

Today, urban public service facilities are more comprehensive than ever. The leisure and entertainment life of urban residents has become richer and extended into various aspects including culture, sports, technology, and the arts. While cyberspace has reshaped people's lifestyles and personalized leisure experiences have gained greater popularity, some traditional forms of collective recreation still remain. For instance, in recent years, square dancing has become the most favored recreational activity among middle-aged and elderly people. Whether in urban or rural areas, it is common to see groups of people performing synchronized dance movements to loud music in streets and plazas during the evening (Wang 2015). Similarly, in the early mornings, it is common to see middle-aged and elderly residents in nearby communities spontaneously forming organizations in urban parks, gathering together to practice Tai Chi and perform gymnastics as a means of physical exercise. These collective leisure and recreational activities taking place in public spaces reflect the revival of mass culture (Xu and Chen 2019). Some may interpret this phenomenon as a form of collective nostalgia in contemporary society, while others criticize it as being at odds in today's postmodern society that emphasizes diversity and individuality. However, its positive socio-cultural effects still shouldn't be ignored. In the urban life where interpersonal relationships are becoming increasingly distant, collective leisure and recreational activities are undoubtedly a good remedy to enhance the sense of belonging and identity.

3.2.4 Convenience and Frequent Travel Activities Based on Public Transportation Systems

The evolution of transportation is a microcosm of the changing times. In ancient society, apart from short-distance travel on foot, long-distance travel and freight transportation had to rely on animal or human power. The underdeveloped transportation and communication meant that most of the human activities were largely confined to the limited spaces of family, neighborhoods, and villages, reflecting a narrow regional scope. The lack of communication and connections between cities and between urban and rural areas made them more like isolated islands. As society entered the modern era, under the influence of Western technology, the pace of transportation innovation accelerated significantly. Traditional modes of transportation such as horse-drawn carriages, sedan chairs and rickshaws are gradually replaced by modern vehicles such as trams, buses, bicycles, automobiles, trains and ships. At the same time, the planning and construction of urban infrastructure also provided more convenience for the residents' travels (Bai 2017).

Although new forms of rail transportation, such as subways and high-speed rail, as well as airplanes have become widely available in large cities, and there have even been bold visions of future transportation, buses—efficient, punctual, affordable, and with extensive routes—remain the primary non-human-powered mode of transportation in cities, and constitute the traditional means of travel for urban residents in China. The public transportation system allows residents to conveniently carry out daily commuting, shopping, recreational and social activities within the city, as well as short trips between the city and the suburbs. This has greatly increased travel frequency of the residents, largely expanded the range of activity spaces, and significantly promoted both population mobility and urban vitality.

With the expansion and extension of bus routes and road systems within the city, the spatial structure of urban areas has also changed and even been reshaped (Li 2014). Transportation connections between different regions have been significantly strengthened, instead of being isolated and alone, facilitating the exchange of information and the flow of resources, transforming the city into a networked, organic whole. Some marginal underdeveloped areas have also improved their location conditions thanks to the passage of bus routes or the establishment of bus stops, leading to new development opportunities.

3.3 Impact of Economic and Social Changes on Residents' Lifestyles

Economic and social changes have had a profound impact on the lifestyles of urban residents. This section primarily discusses these effects from five aspects: institutional reforms, capital forces, technological innovations, cultural development, and social transformations.

3.3.1 Institutional Reforms

Reforms at the institutional level have fundamentally changed the way in which urban spatial resources are organized, allocated and utilized, thereby profoundly changing the lifestyles of urban residents. Under the planned economy system, economic and social resources were allocated in advance by central planning departments, while in contrast, the market economy system allocates resources based on the economic rationality of the actors, with accompanying regulatory measures from the government (Wu 1991). In terms of urban residents' lifestyles, during the planned economy era, people's daily life resources were often exchanged based on allocated resource coupons, such as "food stamps" and "meat stamps", and the circulation of economic and social resources required centralized state coordination. In the market economy, however, money plays a more important role, and these past lifestyles have been replaced by the freely flowing resources in the market economy.

Closely related to the economic system reforms is housing system reform. During the planned economy era, housing for urban residents was mainly in the form of public housing, with allocation largely controlled by Danwei, which also played a key role in the organization of urban spaces. After the socialization and commercialization of housing, the previous system of Danwei-allocated housing was gradually replaced by the purchase of housing through money, and the social space organization of Danwei in the planned economy era has also been gradually replaced by new types of residential communities and gated communities (Xu et al. 2018). "Housing prices" have also become a hotly debated social topic in the new century, and renting has become an important way for young people to meet their housing needs.

Another aspect of the planned economy era that continues to deeply affect the lives of urban residents is the hukou system. After the founding of the People's Republic of China, the hukou system was established, marking the formalization of the dual system of urban and rural areas. Before the Reform and Opening-Up, migration from rural areas to urban areas was strictly limited (Wen 2002). After the Reform and Opening-Up, starting with allowing rural residents to bring their own food rations when moving to cities, the urban and rural hukou system gradually relaxed. This led to a gradual acceleration of urbanization, urban expansion, and the rise of migration to cities for work, with *nongmingong* (rural migrant workers) becoming a significant urban lifestyle in the new era. First-tier cities have become important battlegrounds for young people seeking career success, but hukou system—along with the welfare benefits tied to it, such as housing and education—still have a profound impact on the lives of urban residents.

3.3.2 Capital Forces

After the implementation of the market economy, capital forces have gradually emerged in shaping, transforming, and renewing urban space, significantly impacting urban residents' lives. Differential land rents shape urban spaces, leading to the formation of regional housing prices (O'Sullivan 2015). Real estate development dominated by large capital has expanded rapidly, and with the reforms in the hukou and housing systems, demand for commercial housing has surged, causing urban housing prices—particularly in first-tier cities—to rise sharply. this influence, purchasing homes through loans and renting has become an important lifestyle for the new generation of urban residents. The enormous pressure of living costs, such as housing prices, has also given rise to slogans like "escape from Beijing, Shanghai and Guangzhou", with "new first-tier cities" quickly emerging in response to these circumstances.

The rise of capital power has had a major impact on urban residents' shopping behavior. During the planned economy era, food and vegetables were often distributed uniformly in the form of coupons, whereas, since the Reform and Opening-Up, local vegetable markets around residential communities were the main places for urban residents' daily shopping. As urban capital increasingly penetrated spatial production, differential rents led to the development of central business districts (CBDs), and large shopping centers and supermarkets—driven by economies of scale— began to significantly impact traditional markets. Over time, vegetable markets gradually faded from urban residents' lives, replaced by large supermarkets, shopping centers, and even online shopping (O'Sullivan 2015).

At the consumption level, capital shapes the consumption culture in all aspects of urban life and changes people's consumption habits and ideas. Harvey's "spatial fix" theory, proposed in the 1980s, suggests that capital alleviates crises through time delay and geographical expansion (Wang 2015). From this perspective, credit card consumption is a typical example of capital's "spatial fix" —from the perspective of urban residents, they satisfy their current consumption needs by advancing future income. In the new era, with the maturation of the Internet, online loans and credit services like *Huabei*(花呗) have become important ways for young people to consume.

3.3.3 Technological Innovations

Since the twentieth century, the development of communication technology has greatly improved the efficiency of communication and reduced the time and special costs of information transmission. The emergence of telephones and computers has dramatically improved the efficiency of information transmission, leading to new restructuring of urban spatial location. Then, the emergence of instant messaging software and smartphones has given rise to many new lifestyles. For example, social networking applications have shaped the virtual social spaces that react to the physical urban spaces—activities like "meeting online friends in real life" have become one of the lifestyles among young people. Online shopping has transformed traditional in-store retail experiences, and activities such as receiving package and leaving positive reviews have become a new lifestyle in urban life. Additionally, technologies such as mobile office have expanded possibilities to company locations, enabling more people to work remotely or pursue careers as freelancers.

The emergence of early Internet technology has greatly expanded people's living spaces. With access to vast online information resources and search engines, individuals significantly enhanced their ability to obtain information. Online shopping provided cost-effective goods, while social networking applications facilitated quick connections with new friends and the discovery of groups of interest. As internet technology continued to evolve, it became increasingly integrated into people's daily life, and the advent of the "Internet+" model has further strengthened the integration of online and offline activities. For instance, services such as *Meituan delivery*, *Meituan grocery shopping*, supported by advanced logistics and distribution networks, have gained widespread popularity. The proliferation of mobile payment has accelerated China's transition—particularly in urban areas—toward a cashless society. From large shopping centers to small street-side retail stores, and even in public transportation and taxis, mobile payment has rapidly become popular, significantly reducing residents' frequency of cash usage. Internet technology, represented by mobile payment, has also expanded into various aspects of urban life, including hospital registration and utility bill payments, profoundly changing the lifestyles of urban residents.

The maturity of big data technology is closely related to the advancement of Internet technology. Intelligent algorithms based on massive datasets enabled personalized recommendations and precise marketing strategies tailored to individual consumers (Ruan 2016). Correspondingly, big data technology has also given rise to issues such as "big data price discrimination", where high-spending users are identified and charged higher prices (Xu et al. 2019). Big data technology can be applied in urban management to develop intelligent transportation systems and establish big data urban planning platform. Congestion avoidance and smart mobility solutions are changing the activity spaces of urban residents, forming the mechanism of interaction between individuals and urban spaces.

3.3.4 Cultural Development

The Internet has greatly accelerated cultural exchanges between China and the rest of the world. Under the dual influence of foreign cultures and the Internet itself, online culture has emerged and gradually become an important form of culture in the modern era. Early rise of Internet literature (e.g., *First Intimate Encounters)* formed a large scale of Internet literature fandom. Over time, Internet literature has expanded beyond youth demographics, gaining widespread popularity across various age groups. Based on this, internet-driven "new cultural" forms continue to interact with physical urban spaces, with anime conventions, gaming expos, and cosplay gradually becoming an important part of entertainment spaces. Meanwhile, Internet- and TV-based variety shows, and movies have become essential elements of residents' leisure activities.

Under the influence of foreign cultures, traditional Chinese culture once has experienced a tortuous development, even facing periods of decline. However, with people's critical reflection on foreign cultural influences and the establishment of national and cultural confidence, traditional Chinese culture has regained public attention. Cultural scholars represented by Yi Zhongtian and Yu Qiuyu have played a key role in reigniting interest in *guoxue* (classical Chinese studies) through the media of television and books. This revival has extended beyond online engagement to offline activities, such as a new type of book clubs held in bookstores. As the cultural life of urban residents becomes richer, their interest in traditional Chinese culture continues to grow.

With rapid economic development, Chinese cities, particularly first-tier cities which are relatively more developed, are gradually showing the characteristics of the post-industrial era. The rise of consumer culture, has transformed consumption and leisure from mere entertainment or lifestyle into crucial means of enhancing life enjoyment and deriving personal meaning. Consumption is no longer just an act of purchasing, it has become a reflection of personal characteristics and quality of life. The popularity of luxury goods, along with emerging trends such as "entry lux", "niche" and "personalized customization", indicates that the consumer culture is shaping new cultural values and lifestyles—embedding itself deeply into urban life.

As previously mentioned, foreign cultures have entered China with the help of media such as the Internet. In the new era, foreign cultural elements have played significant roles in residents' lives, especially in urban settings. For example, bar has become an important place for young people to spend their weekends. Similarly, coffee and Western cuisine, once regarded as novel and exotic by the public, have gradually been localized.

3.3.5 Social Transformations

Since the Reform and Opening-Up, Chinese society—especially urban society—has undergone a significant transition from collectivization to individualization. Traditional collectivism has gradually weakened, giving way to a more individualized society (Feng 2014). The once-unified rhythms and lifestyles of the past have broken down, leading to greater diversity in people's life paces and consumption choices. Personalized choices are presented in clothing, diet, and daily consumption. Compared with the past, people pay more attention to self-development, and the lifestyles of urban residents have become increasingly diverse and individualized.

During the era of planned economy systems, the state centrally allocated resources, but after the Reform and Opening-Up, the market economy developed rapidly, and the power of the public gradually came to the fore. Public participation and bottom-up social organizations became increasingly prominent (Guo 2011). The rise of civil organizations, non-governmental organizations (NGOs) and non-profit organizations (NPOs), etc., have played an important role in the establishment of social organization and relationship (Yu 2013). Private institutions such as private schools have taken on increasingly crucial responsibilities. Organizations such as clubs and community activities, which are based on grassroot social relations, have played important roles in urban society. On one hand, people have highly individualized needs; on the one hand; numerous civic organizations contribute to constructing collective meanings.

With the development of the Internet, people are paying more attention to their online lives. Diverse social groups have gradually formed in the cyberspace. On one hand, online social groups break down limitations of identity and gender, bringing people together in virtual spaces. On the other hand, these social groups are closely related to their social classes, with people often gathering based on interests, tastes, hobbies, and other common bonds. These groups engage in various online and offline activities (Zhang 2013; Zhang 2014). At the same time, under the influence of major events, online groups tend to organize to form a huge force (e.g., phenomena such as "*Tieba* fights"). Communication tools represented by WeChat have greatly reduced the cost of social organization through features like group chats, contacts, and moments. Traditional social means such as exchanging phone numbers or business cards have gradually been replaced by adding contacts on WeChat and forming group chats.

3.4 Characteristics of the Evolution of Urban Residents' Lifestyles in China

3.4.1 Stages of Evolution in Urban Residents' Lifestyles in China

From a longitudinal perspective, the evolution of urban residents' lifestyles in China can be divided into four stages.

3.4.1.1 1949 to the 1980s

After 1949, influenced by the political system and economic and social development policies of China, the overall lifestyle of urban residents retained the characteristics of the planned economy era. During this period, residents depended mainly on the unified allocation of the state to obtain living resources, such as food stamps, meat stamps, etc. The pace of life was highly uniform, a feature most evident among the youth involved in the *Shangshan Xiashang* ("Down to the Countryside Movement"): from work to meals to rest, their daily rhythm of life was strikingly consistent. From the perspective of urban residents' activities, due to the restriction of hukou system, the population movement between urban and rural areas was minimal. Additionally, the system of welfare housing meant that people's living spaces were closely related to their working spaces, leading to a high proportion of socializing within familiar circles, and a high degree of overlap in people's activity spaces. During special periods such as the Cultural Revolution, residents' lives were influenced by ideology, with political activities occupying an important place in daily life. Collectivism prevailed, and once political groups were defined, there was a marked unity in terms of ideology, political activities, and even daily life (Wang and Feng 2014). At this stage, due to the relatively lower material standards, people's consumption choices were limited, and due to the influence of the housing and hukou system, there were few noticeable disparities in material living standards. Consumption choices are primarily focused on meeting basic needs of life, meaning that the personalized consumption choices were rare. Influenced by the international situation of the Cold War and China's own national conditions, foreign cultural exchange was limited, traditional and local cultures forming the main sources of entertainment and cultural life for urban residents during this period.

3.4.1.2 1980s to 2000

During this period, with the implementation of the market economy system, the power of market capital became a crucial force influencing urban space and residents' lives. The reform policies, represented by housing system reform and state-owned enterprise reform, fundamentally changed the mechanism governing urban residents' production and lives. Housing system reform has made housing resources into scarce commodities, leading to rapid urban land expansion and soaring housing prices. At the same time, along with the reform of the hukou system, urban–rural population mobility intensified, with migrant workers and *nongmingong* becoming an essential part of the urban workforce (Wu 2011). Shopping in the city and traveling to the rural countryside have also gradually become significant lifestyles. Moreover, with the advancement of Reform and Opening-Up, the influx of foreign commodities and cultures has had a great impact on China's old urban lifestyle. From oversized sunglasses and bell-bottoms to Adidas and Nike, and later to domestic brands like Li-Ning and Warrior, the consumption choices of urban residents have gone through multiple changes, evolving towards personalization and diversification, showing a combination of local and foreign influences in their consumption choices. During this period, due to the influence of market economy, capital gradually became the dominant force shaping urban spaces. From the construction of large-scale shopping malls and the rise of urban CBDs, to the rapid expansion of urban land, the proliferation of new types of communities and gated communities, and the dismantling of the Danwei community model (Liu and Chai 2013), people's everyday shopping and residential spaces have undergone great changes. Simultaneously, the new trend of Reform and Opening-Up has gradually eliminated the mechanical collectivism of the past, leading to significant transformations in both daily life and ideological consciousness. A heightened sense of individual independence emerged, along with an expansion of public participation and self-organization, which became key forces in structuring and organizing urban residents' lives (Xu 2004).

3.4.1.3 2000–2010

Since 2000, Internet technology has gradually emerged. The Internet at this period takes PC-based Internet as the main carrier. Online shopping, initially perceived as a low-cost alternative, gradually became the dominant mode of consumption. In contrast, the traditional physical shopping spaces gradually shifted toward experiential shopping and integrated entertainment after experiencing the impact of network shopping, realizing the new transformational development. Meanwhile, with the reform of hukou system and housing system, housing resources became increasingly scarce, and prices have continued to rise, reaching a relatively high level; the population of first-tier cities at this stage was huge and continued to expand. Along with the rapid expansion of urban population scale, the process of suburbanization accelerated, making private car gradually becoming an indispensable mode of transportation. Urban activity space and reachable range further expanded, with residential, work and leisure spaces further extending. Additionally, with China's accession to the World Trade Organization (WTO) and the hosting of the Beijing Olympics, China's interactions with foreign countries became increasingly close, accelerating the process of globalization. Products from around the world, once considered rare and exclusive, became commonplace consumer choices, particularly for urban residents in their daily lives.

3.4.1.4 2010 to Present

A notable feature of this stage is the further deepening of the Internet and the rise of mobile Internet. The rise of smartphones and mobile Internet strengthened the integration of Internet into daily life, making Internet applications more lightweight and routine. Mobile payment systems, represented by WeChat Pay and Alipay, accelerated the transition toward a cashless society, permeating nearly every aspect of urban residents' lives—from large supermarkets and retail stores to street vendors, from public transportation to hospital appointments. Mobile payments replaced traditional cash transactions in numerous aspects, offering greater flexibility and efficiency (Wang 2015). At the same time, the Internet placed greater emphasis on integrating online and offline experiences. Platforms such as Meituan delivery and Meituan grocery shopping deeply connected offline, physical spaces and online, virtual spaces. Didi, Ofo, etc. expanded people's mobility options, significantly improving efficiency of urban residents' daily travel. The technology represented by mobile Internet and mobile payment has profoundly transformed residents' daily life, continuously shaping new urban lifestyles.

In addition, the population scale and housing prices of first-tier cities have reached a high level, rising the pressure and cost of living in those cities. The emergence of "escape from Beijing, Shanghai and Guangzhou" and "new first-tier cities" has further changed people's tendency, prompting individuals to reassess urban life quality and personal aspirations. With the in-depth development of Internet technology and the gradual reduction of social mobility, Chinese society during this period is gradually exhibited a mixture of modern and post-modern characteristics. On the one hand, the high pressure of life and the "996" work schedule are squeezing the living spaces of urban youth; on the other hand, the rise of consumer culture and various subcultures provided young people with new possibilities for entertainment and recreational spaces.

3.4.2 *Characteristics of the Evolution of Urban Residents' Lifestyles in China*

3.4.2.1 Collision, Restructuring and Development of Foreign, Consumer and Traditional Cultures

Since the Reform and Opening-Up, the market economy has facilitated a gradual resurgence of cultural exchange between China and the rest of the world. With China's accession to the WTO and the hosting of the Beijing Olympics at the beginning of the twenty-first century, based on the rapid development of the Internet, foreign cultures began to collide, interact, reconstruct and integrate local traditional cultures in China. In the early years of the Reform and Opening-Up, foreign commodities and cultures were seen as novel existences to Chinese residents. Many foreign commodities would lead the trend as soon as they were imported, and become the new fashion of urban life. From early foreign clothing styles to Western fast-food chains like KFC, and later to Western cuisine and coffee, each wave of foreign influence initially sparked widespread enthusiasm and consumer fervor.

However, with the gradual deepening of global interactions, as well as the rapid economic and social development of China, a growing sense of cultural self-confidence has gradually emerged. People began to adopt a more rational perspective on foreign cultures. Simultaneously, traditional Chinese culture has regained significance after experiencing a period of decline, prompting people to reflect on its development and reconsider its value. From the *Guoxue* Fever that emerged at the beginning of the twenty-first century to the rise of *Guochao* (Li-Ning, Warrior, Huawei, Xiaomi), local Chinese culture has gained renewed significance, drawing increased public attention and appreciation. At the same time, foreign commodities and cultures have undergone processes of localization and adaptation while taking root in China. Alongside the development of the market economy, consumer culture is gradually becoming an important ideology in society. Under the influence of consumerism, consumption is no longer simply a means to satisfy people's basic needs, but a means to seek meaning in life and construct personal identities (e.g., the concepts of financial freedom, private customization, etc.). In the ongoing interaction between local and foreign cultures, consumer cultures continue to evolve.

3.4.2.2 The Joint Shaping of Urban Life by Institutional and Market Forces

After the planned economy era, the power of market capital has gradually become the dominant force shaping urban life. The urban rent gradient has shaped the spatial structure of urban areas, thus shaping people's activity patterns. Economies of scale have made large shopping malls and supermarkets the mainstream choices for shopping, while Internet giants leveraging their capital and technological power to reshape people's daily lives. It is worth noting that institutional factors continue to profoundly affect urban life. The housing system reform fundamentally altered the living spaces of urban residents and dismantled the traditional Danwei model, thus changing people's social spaces (Bian et al. 1996). Under the influence of this institutional changes, housing prices surged, and purchasing houses through loans or renting houses have become the mainstream lifestyle of young people. The pressure of survival in first-tier cities increased, and slogans such as "escape from Beijing, Shanghai and Guangzhou" and "The North doesn't believe in tears" have reflected the pressure of life in these cities. In terms of the hukou system, the loosening of hukou restrictions has led to an increase in rural–urban migration, but some resources tied to the hukou system, such as education and healthcare, have not truly become mobile. Meanwhile, in terms of entertainment and social life, foreign cultural and social products have been widely used in China, but on the other hand, due to censorship and regulatory systems in China, they have developed into distinct forms and usage patterns (e.g., censorship of movies, etc.) from those of other countries.

3.4.2.3 The Blending of Modern and Postmodern Lifestyles

After experiencing the planned economy era, the Reform and Opening-Up and the new century, the lifestyle of urban residents in China is presenting a mix and fusion of traditional, modern and post-modern lifestyles. With the gradual development of consumer culture and the improvement of material and culture standards, tourism and leisure have become important components of people's daily life, and cultural and spiritual consumption have become indispensable parts of people's lives. In this context, industrialized consumption choices and post-modernized personalized consumption choices together constitute people's consumption culture. On the one hand, mass-produced commodities can relatively affordably meet people's daily needs; on the other hand, personalized consumption choices that signify meaning and express tastes are increasingly welcomed. Correspondingly, the attitudes of urban residents towards life reflect this obvious blending. The "996" work schedule and staying up late have become the norms for young people, while at the same time, practices like health preservation and the humorous mixing of "beer with goji berries" reflect a return to traditional ways of life in postmodern society. Creating and seeking meaning during the daily consumption and life has become defining characteristics of contemporary urban youth.

3.5 New Lifestyles of Chinese Urban Residents

Reform and Opening-Up has brought about dramatic changes in China, leading to rapid economic development and a significant improvement in people's living standards. The changes in people's lifestyle, such as clothing, food, housing and transportation, serves as a microcosm of this era of change. Under the reform of the socialist market economy system, commercial housing gradually replaced

Danwei from the planned economy era, becoming the primary form of residential and commercial property for urban residents. Abundant material conditions have greatly satisfied people's daily needs, and beyond the evolution of the "three big items", the widespread adoption of private automobiles profoundly changed urban residents' transportation patterns.

In the 1990s, China seized the opportunities presented by the Third Technological Revolution, with the development and widespread adoption of the internet revolutionizing the lifestyles of urban residents. The application of information technology reshaped various aspects of urban residents' daily lives, from work and consumption to personal entertainment and social interactions. The wave of economic globalization not only introduced global commodities and advanced production technologies to China but also influenced people's values, consumption patterns, and cultural activities. Transnational capital brought Western cultural and leisure practices to China, which were embraced by urban residents for their perceived "fashion" and "novelty." As a result, symbol-based consumption in cultural and leisure activities became a new urban trend. While material abundance increased, global climate change and environmental issues also gained attention within Chinese society. As China pursued modernization and urbanization, it continuously adjusted its industrial structure in adherence to the principles of sustainable development. Today, many urban residents advocate for a "low-carbon lifestyle," integrating sustainable development concepts into various aspects of daily life, including transportation, shopping, and work.

Overall, the lifestyles of urban residents in China have been continuously evolving, leading to the emergence of various new ways of living. These lifestyles exhibit distinct characteristics, which this section will focus on through the following five key aspects: flexible mobility enabled by private vehicles public transportation system; convenient shopping facilitated by the Internet and highly efficient logistics services; diverse cultural and leisure activities shaped by symbolic consumption; new social interaction models driven by information technology; low-carbon lifestyles rooted in the principles of sustainable development.

3.5.1 Flexible Mobility Enabled by Private Vehicles and Public Transportation System

Since the 1970s, with the rapid economic development and the strengthening of national comprehensive strength, the living standards of urban residents in China have been greatly improved. In the 1980s, the state began to deregulate automobile industry, allowing car imports and initiating joint ventures with foreign enterprises. By the 1990s, purchasing private cars had become an emerging trend among urban residents. During the period from the Reform and Opening-Up era to the late twentieth century, China's automobile industry represented advancements in technology and played a crucial role in the country's industrialization and economic development.

After China joined the WTO, the industry transitioned into a market-oriented development phase, leading to a substantial increase in the number of private cars, which became a primary mode of transportation for urban residents. Relevant data indicates that the number of household-owned automobiles per 100 households nationwide rose from 0.25 in 1998 to 29.7 in 2017. Notably, after 2000, private car ownership in China experienced rapid growth.

For a long period of time, automobile has become a symbol of both personal identity and national status under the marketing: private cars signified individual success as well as the rising stature of the nation (Bao 2006). As household sedans transitioned from luxury items to daily necessities, private vehicles became a crucial mode of transportation for urban residents, significantly affecting people's travel patterns. Car ownership provided urban residents with greater flexibility and convenience in commuting, and the widespread adoption of private vehicles facilitated the spatial separation of residential and work locations, reinforcing the interplay between automobilization and suburbanization, which gradually altered the traditional pattern of urban residents concentrating in central districts (Wang 2009). In addition, private car significantly influenced shopping behaviors, closely linking car travel to the rise of large shopping malls, which usually equipped with a certain capacity of parking lots. This consumption pattern, catering to car-owning households, meant that large shopping centers did not necessarily have to be located in the city center. Many Chinese cities, for instance, built outlet malls in suburban areas. The proliferation of private cars also introduced a new mode of leisure travel—self-driving tourism, which are free from the constraints of fixed schedules and routes, and enhance the enjoyment and experiential quality of the journey (Wang 2009). Self-driving tourism has become a fashionable and increasingly popular choice for urban families.

The significance of private cars in urban areas is reflected both in the convenience they provide and in how taxis and ride-hailing services supplement public transportation. Taxis offer residents flexible, point-to-point travel in cities, allowing even those without private cars to enjoy the "hail a cab" service. Meanwhile, ride-hailing services, which emerged in 2010, use mobile apps to provide car travel services by reservation to urban residents. Today, urban residents can simply use their smartphones to book a ride at designated locations, addressing gaps in public transportation.

Private cars provide point-to-point travel, but their excessive use often results in traffic congestion, while urban public transportation offers the advantages of high passenger capacity and low fares. The combination of both provides urban residents with flexible travel option. The development of urban public transportation systems has always been a key component of urban infrastructure construction, and buses are an essential traditional public transportation mode in cities. With the improvement of urban public transportation systems and the development of rail transit, by 2018, 36 Chinese cities, including Beijing and Shanghai, had opened subway lines. In order to effectively address urban traffic congestion, cities like Beijing, Guangzhou, and Urumqi have established Bus Rapid Transit (BRT) systems, which, along with subway transfers and walking, provide urban residents with efficient and high-quality public transportation services. The development of urban public transportation has now extended beyond city limits, moving toward integrated public transport systems in line with regional integration. For example, Guangzhou and Foshan jointly built the Guangfo Metro to create a "Guangzhou-Foshan Living Circle", allowing urban residents of both cities to use the metro for cross-city commuting and consumption.

3.5.2 Convenient Shopping Facilitated by the Internet and Highly Efficient Logistics Services

The rapid development of the Internet in China since the 1990s has profoundly changed people's daily lifestyles, the most prominent example of which is online shopping. Internet shopping in China emerged in the 1990s, and the Internet played a key role in the development of e-commerce. However, the logistics industry became one of the bottlenecks for its early development. Over the first decade of the twenty-first century, with improvements in policies, comprehensive development of logistics infrastructure, and advancements in information and communication technologies, online shopping in China entered a period of rapid growth and prosperity. According to the "China E-commerce Report 2018" published by the Department of E-commerce and Informatization of the Ministry of Commerce (2019), the total e-commerce transaction scale in China reached 31.63 trillion yuan in 2018, with online retail sales exceeding 9 trillion yuan, and physical goods online retail sales surpassing 7 trillion yuan. This accounted for 18.4% of the total retail sales of consumer goods, illustrating the significant impact of online shopping on the purchasing behavior of residents. Due to differences in infrastructure, purchasing power, and other factors between urban and rural areas, the number of urban residents participating in online shopping is 11 to 13 times higher than that of rural residents (Mao and Huang 2014). Therefore, urban residents are the main drivers of internet shopping, and online shopping has become a prominent feature of the new lifestyle of contemporary urban residents in China.

Relying on the Internet and efficient logistics distribution, coupled with the convenience of mobile payments, urban residents today are able to easily break through the constraints of time and space. They can purchase goods and services online anytime and anywhere to meet various needs in daily lives. In many cities, it is no longer necessary to go to stores in person to buy fresh food, groceries, or medicines. With just a mobile order, residents can have these items delivered within a designated time frame. Similarly, products such as textiles and home appliances can be purchased through e-commerce platforms, allowing consumers to compare prices and receive deliveries at home. Meanwhile, urban residents are also able to purchase a wide range of services, from housekeeping services to online courses, Internet shopping has almost covered the physical consumer goods and services needed by urban residents. Unlike traditional shopping behavior, Internet shopping is convenient, free and flexible. Urban residents can buy a full range of products at a lower price than physical stores, saving time on shopping trips and receiving products from other regions or even from abroad in a short time. For example, Northern city residents can order *Qingtuan* (a traditional Chinese rice dumpling) made in the Jiangsu and Zhejiang regions for the Qingming Festival simply through an e-commerce platform.

Internet shopping has led to changes in the consumption habits of urban residents in China, as consumption can take place anywhere with Internet connection, and payment for purchases can even be made during working hours. As face-to-face interactions in traditional shopping are reduced, people increasingly rely on online credit mechanisms and rating systems to select products and services. E-commerce platforms not only utilized traditional Chinese festivals for promotional activities, but also created "shopping festivals" and "e-commerce festivals" such as "Double Eleven". These "shopping frenzies" are promoted through widespread advertisements in cities, while residents carefully use coupons and group buying to make online purchases, forming new social integration through consumption (Li 2010). Residents' consumption values are also quietly changing, pursuing new trends, prone to impulsive consumption and moving towards luxury. Of course, the carbon emissions brought about by express delivery are also issues that need to be addressed.

3.5.3 Diverse Cultural and Leisure Activities Shaped by Symbolic Consumption

The cultural and recreational activities of contemporary urban residents in China are becoming increasingly diverse. As a result of the rise in economic standards, urban residents have increased their expenditures on those activities. Along with the economic globalization and the flow of multinational capital, the Western culture and leisure forms of entertainment in Chinese cities are initially embraced by young people who favor the new trend. Activities in the Western culture and entertainment places such as dance halls, bars, cafés, Western restaurants, and golf courses became a new cultural trend in urban life. Bars, as a kind of personalized leisure place, emerged in major cities in China after the Reform and Opening-Up, initially related to the distribution of embassy districts and foreigner neighborhoods. By the 1990s, bars had become significant recreational spots in major cities (Weng and Chai 2003). These "imported" cultural leisure activities are more of a symbolic consumption for Chinese people. Initially, such consumption was not cheap for most residents, and those who were able to accept Western-style entertainment were typically young people with higher educational levels and a greater acceptance of Western culture. Thus, choosing Western leisure activities became a symbol of being trendy, open-minded, and financially capable.

The Western style of cultural recreation has evolved over the past two to three decades and is no longer the unfamiliar thing it was at first. Along with the expansion of consumer groups, Western-style leisure activities in China have developed into a localized form, blending both Western and Chinese cultural elements. The first bar to move into Nanluoguxiang in Beijing was based on the architectural form of the

Siheyuan and the "Beijing flavour" (Li 2009), and Starbucks has also ventured into the business of mooncakes and zongzi. Urban residents in these Western-style consumption places can not only immerse themselves in the atmosphere of Western culture, but also reconnect with local traditions through cultural elements (e.g. Siheyuan, ceramics, opera masks, festive foods, etc.). This fusion of Chinese and Western elements offers new imagery and symbols for urban spaces (Lin and Wang, 2011).

Contemporary urban residents in China exhibit a trend of symbolic consumption in their cultural leisure activities. Personalized cultural leisure activities based on commercial spaces have partially replaced the collective cultural leisure activities traditionally centered around public spaces. Meanwhile, the social interaction roles once fulfilled by families and communities are now provided by the market: urban residents engage in cultural leisure activities through symbolic consumption, using it to express their individuality, identity, status, and taste, in order to gain recognition from their social groups and participate in social activities. For example, cafés label consumers as "elites" or "white-collar workers", live houses affirm musical taste, bars serve as gathering places for "trendy" individuals, teahouses represent an appreciation for and preservation of traditional culture, and golf courses symbolize "aristocratic sports." Urban residents place great emphasis on the spatial symbols of cultural leisure venues, seeking to express their personalities through these activities and find suitable spaces for consumption where they can meet friends and engage in social gatherings.

3.5.4 New Social Interaction Models Driven by Information Technology

The development of the Internet has opened up new possibilities for social interactions among urban residents in China. While traditional social interactions in home, community and workplace spaces have shifted to commercial spaces, they have also increasingly moved online, forming a new mode of social interaction driven by information technology. The rise of social network services such as QQ, Weibo, and WeChat has broken the spatial and temporal constraints of traditional face-to-face communication. The simultaneous development of social media application and smartphones has revolutionized the mode of social interaction, and the scope, frequency, quality and modes of social interaction of urban residents have undergone significant changes. New languages and symbols have been created and adopted by residents in digital interactions, and the online communities have challenged the superficial, anonymous and temporary nature of traditional urban social relationships.[1] Take WeChat for example, as of 2018, WeChat data indicated that more than one billion users remain active on a monthly basis, and by 2015 WeChat had already achieved near-total penetration in China's first-tier cities, with adoption rates

[1] In 1938, Louis Wirth published his influential paper Urbanism as a Way of Life, in which he argued that traditional urban social relationships are superficial, anonymous, and temporary.

reaching 93% in Beijing, Shanghai, Guangzhou and Shenzhen.[2] For most urban residents, social media applications such as WeChat or QQ are simple to use yet highly functional—adding new friends is a prerequisite for establishing connections, maintaining family ties and emotional bonds can be facilitated through video or voice calls, and work-related matters often require group chats for discussions and notifications. Even those unable to visit relatives during the New Year can send blessings and digital red envelopes via WeChat. This new type of social interaction based on information technology strengthens the existing "circle" in real life through online interactions, and on the other hand, meeting strangers online expands the scope of social interaction and forms new communities, which is not only different from the social relations formed in the collective life under the planned economy, but also different from the loosely knit social relations traditionally found in urban settings.

There are various platforms for online social networking, including open forums, social networking apps for acquaintances, and stranger-based social platforms, all of which cater to the personalized social needs of internet users. This new mode of social interaction has facilitated the formation of online communities based on hobbies, consumption preferences, etc., while also enabling more precise communication between individuals. WeChat, for instance, is based on real-life social relationships, allowing traditional community-based social interactions to shift into WeChat groups. In contrast, platforms such as Weibo and Tieba function as semi-anonymous forums where users can express opinions, engage in discussions, and connect with like-minded individuals. In recent years, stranger-based social platforms have emerged, offering users more private communication environments. Overall, this new form of social interaction is more complex and personalized than traditional modes of socializing. Urban residents can now tailor their online and offline social engagements to align with their individual needs and preferences.

The new mode of social interaction driven by information technology places greater emphasis on self-expression and subjectivity in online social interaction. Urban residents can participate in the discussions of urban public affairs, share their personal life, and exchange resources and information through digital interactions, all of which constitute essential components of the public life of contemporary urban residents. In addition, these new types of social interactions influence urban residents' perceptions and understanding of the city. Some location-based social network services reinforce the significance of physical urban spaces, allowing urban residents to access relevant geographic information through online social interactions. At the same time, their perception of places is shaped by others' evaluations and specific labels within digital social networks.

[2] Data source: http://www.techweb.com.cn/data/2015-12-14/2240879.shtml.

3.5.5 Low-Carbon Lifestyles Rooted in the Principles of Sustainable Development

In an era of material abundance, people have also begun to realize that environmental protection and economic development are challenges shared by all of humanity. In 1987, the World Commission on Environment and Development elaborated the concept of sustainable development for the first time in *Our Common Future*, establishing it as a widely accepted principle within the international community. Sustainable development refers to development that meets the needs of the present without jeopardizing the ability of future generations to meet their own needs. China, as a developing country, has also recognized its responsibility and obligation to achieve sustainable development. Efforts have been made to accelerate industrial restructuring, control environmental pollution, and promote a transition toward scientific, green, and low-carbon lifestyles and consumption patterns under the framework of ecological civilization (Wu and Wu 2012). "Low-carbon life", as a response to the concept of sustainable development attitude and lifestyle, has gain widespread public support. Today, "environmental protection" and "low-carbon" have become synonymous with healthy and trendy lifestyles, encouraging more individuals to embrace sustainable urban living.

Low-carbon transportation is an important aspect of a low-carbon lifestyle. In response to low-carbon mobility initiatives, bike-sharing programs have been introduced in many cities with relatively developed economies, and urban residents have reduced the use of private cars in favor of public transportation or cycling for commuting. More environmentally friendly electric vehicles have also received policy support, while cities have continued to improve pedestrian infrastructure, promoting a return to healthier, more sustainable, and safer modes of transportation.

How to achieve low-carbon and environmentally friendly practices in shopping and consumption has also become a topic of concern. As early as 2007, the Chinese government issued the *Notice of the General Office of the State Council on Restricting the Production, Sale, and Use of Plastic Shopping Bags* to limit plastic bag usage and curb "white pollution." Reusable cotton shopping bags have since become a popular alternative, and urban residents have increasingly rejected single-use tableware. Frugality and resource conservation are being emphasized, leading to initiatives such as the "Clean Plate Campaign" in restaurants to reduce food waste. Meanwhile, second-hand trading platforms and flea markets have gained popularity as trendy shopping venues.

In urban construction, the aesthetic principles of the "city beautiful movement" have been criticized, while low-carbon architecture and landscape design based on ecological safety patterns have gained emphasis (Yu and Li 2003). An increasing number of urban residents are participating in the development of low-carbon communities.

3.6 Perspectives on the Living Space of Urban Residents from the Evolution of Lifestyles

Since Georg Simmel, traditional urban sociology has generally regarded urban lifestyles as determined by factors such as space and culture. However, more recent urban sociological research and practice emphasizes human agency, arguing that individuals in cities can actively choose their life. The aspiration to transcend existing lifestyles can drive people to reshape their living spaces. The New Urbanism movement, which emerged in the United States in the late twentieth century, is a practical movement for the transformation of urban space from the pursuit of lifestyle change. The rise of the New Urbanism movement indicates that new lifestyles are becoming a significant force for changes in cities and can lead to the transformation of urban space.

Living space is the tangible and experiential space of everyday life, encompassing the totality of places where various daily activities occur. It represents the spatial projection of different types of activities and social relationships that constitute people's daily lives (Wang 2011). Since the founding of the People's Republic of China, economic and social transformations driven by institutional, capital, technological, cultural, and social forces have subverted the traditional lifestyles. The collision between foreign culture, consumer culture, mass culture and traditional culture has been intense; the dissolution of the Danweilization and the development of the market economy have expanded individuals' choices regarding their lifestyles. The emergence of convenient transportation methods has enabled people to move more freely within physical space, and the development of the Internet and mobile technology has further extended living space from the physical realm to the virtual. In this context, the evolution of urban lifestyles has unsettled and reconstructed original urban living spaces. This section examines representative new lifestyles that are significant and characteristic of the times as a lens to explore the transformation of urban living spaces driven by societal changes. It also aims to reflect the reciprocal influence and interaction between lifestyle evolution and spatial transformation.

3.6.1 The New Urban Living Space After the Dissolution of the Danweilization

Danweilization is a system of urban organization that emerged within the political and economic context of socialism and planned economy system. The basic needs of urban residents were guaranteed by Danwei in which they live, and the state exercised out social management, resource allocation, and social integration (Li 2013). After the founding of the People's Republic of China, the country adopted a strategy of prioritizing heavy industry, and the urban industrial system was established based on the Danwei structure (Zeng 2015). In the 1950s and 1960s, nearly all urban residents lived in the spatial framework of Danweilization. It was not only a place of employment for urban residents, but also a fully functional "miniature society" that provided housing, education, and a range of public services. During this period, people's daily activities were confined to the Danwei space, resulting in relatively homogeneous and monotonous activity spaces and types, with no significant differentiation in living spaces. Moreover, the enclosed nature of Danweilization space also made people's social communication relatively simple, as colleagues were also neighbors, leading to close and deep interpersonal relationships.

After the Reform and Opening-Up, especially with the establishment of the market economy system, Danweilization began to dismantle under the reforms in the housing and land systems. Danweilization was no longer the sole provider of housing and living services, marking the end of its role as a complete living unit. People began to freely choose their places of residence in cities based on their economic capacity, moving from Danweilization neighborhoods to commercial housing communities. The original community of familiar individuals began to disband, with social interaction moving towards the Western urban style characterized by greater social distance. At the same time, people with different economic levels and interests developed diverse lifestyles, actively expanding the range of their living spaces. Urban living spaces exhibited distinct differentiation and heterogeneity.

3.6.2 New Leisure Lifestyles in the Context of Mass Culture

During the process of urbanization in various countries, a new cultural form adapted to metropolitan life has quietly emerged, gradually replacing traditional folk culture. This culture, which arises in modern urban industrial and consumer societies, targets the modern urban masses as its consumers and is disseminated through contemporary mass media. It is characterized by being popular, ephemeral, consumable, superficial, and standardized, and is referred to as mass culture (Gao 1996; Zhang 2007). Today, mass culture has become the mainstream form of contemporary culture, profoundly influencing people's lifestyles, particularly altering their consumption preferences and leisure activities.

Mass culture has driven the emergence of various new public leisure and entertainment spaces in urban areas. The popularity of mass culture has led to the widespread embrace of modern, internationalized forms of leisure and entertainment, giving rise to a series of new public entertainment venues imbued with symbols of fashion and high status, such as bars, cinemas, art galleries, and cafés. These leisure spaces are not only sites of material consumption but also contribute to the diversification of the urban space structure through their clustered or dispersed spatial distribution across the city. People engage in leisure, entertainment, and social interactions within these spaces, creating new types of living environments.

3.6.3 New Lifestyles in the Age of the Internet

Before the widespread adoption of the Internet, in order to meet the needs of various activities, people had to shuttle between various functional spaces within the city. Some scholars argue that during this period, there was a close and simple linear relationship between the activities of urban residents and specific spaces (Yuan and Pang 2016). People had to spend a certain amount of time and resources traveling to reach living spaces that provide specific functions. However, the virtual spaces created by the Internet allow people to fulfill their leisure, consumption, and social needs without physical movement, breaking the traditional connection between urban living spaces and their specific functions.

The Internet has weakened the constraints that space once imposed on people's daily lives and has disrupted the strict correlation between living spaces and their specific functions, shaping urban living spaces with overlapping functions. People can now engage in leisure, shopping, and social interactions within their living spaces, allowing them to meet their lifestyle needs more freely. However, in the Internet era, the blurred boundaries of spatial functions also make it difficult to completely separate people's private affairs from their work. Leisure spaces can be converted into workspaces at any time, making it difficult for people to maintain a purely leisure-oriented life.

In addition, the Internet era has given rise to a new lifestyle characterized by "*zhai* (staying at home)", where human-to-human communication is replaced by human–machine (computers/smartphones) communication. "*Zhai*" can satisfy their daily needs without leaving their homes. While the Internet creates and extends people's virtual living spaces, it also makes the physical living spaces shrink. New lifestyles based on the Internet have changed the definition of daily life and has driven continuous adjustments in real urban spaces to meet the demands of the online world.

3.6.4 New Lifestyles in the Era of Flexible Mobility

The level of development of transportation tools determines the upper limit of people's mobility. Before the advent of automobiles, people's living spaces were extremely limited in range, and people had to carry out all the activities necessary for daily life within enclosed areas. The invention and popularization of the automobile transformed people's travel habits, allowing them to break free from the constraints and limitations of time and space, and realize the free flow of the body in space (Lin 2012). It can be said that the development of modern transportation tools has opened up an era of flexible and convenient mobility, making many lifestyles possible.

There is no doubt that the invention of the automobile marked an important point in the transformation of travel methods. The greatest change that automobiles brought to human life was the significant enhancement of people's mobility. With the help of urban buses, people could reach farther places with less time and energy expenditure, expanding their activity radius and incorporating previously unfamiliar areas into their living spaces. The emergence of automobiles also facilitated the complete separation of employment and residential spaces, giving rise to long-distance commuting and altering people's daily temporal and spatial arrangements. In addition, the decline in the cost of automobiles and the improvement of people's living standards have led to the increasing popularity of automobiles. Compared to the time constraints imposed by public transportation, private cars offer greater freedom of movement, allowing individuals to choose their travel times and destinations according to their needs, making urban living spaces increasingly diversified and decentralized.

In recent years, people have begun to recognize the negative environmental impact of automobiles and other modes of transport, leading to another round of changes in modes of travel, mainly the promotion of low-carbon modes of travel. Low-carbon travel encourages people to reduce car usage and instead rely more on walking, cycling, and other travel methods that place less burden on the environment. With the promotion of this new lifestyle concept, urban planning—such as road traffic organization patterns, public service facility distribution, urban green space design, and the scale of neighborhoods and residential areas—has increasingly focused on creating cities that are more conducive to low-carbon travel. As low-carbon lifestyles are more widely recognized and accepted, the urban living spaces are also undergoing corresponding adjustments to meet people's aspirations and pursuits of this new way of life.

3.7 Chapter Summary

Lifestyle is an effective entry point for revealing the mechanisms of social operation, and it has been a traditional research topic in sociology. The academic research on it has evolved from being a subordinate, marginal concept to an independent field of research, accompanied by a shift from emphasizing grand, overarching theories to a renewed focus on everyday life—an evolution of significant theoretical and practical importance. With rapid advancement of China's economic development and the acceleration of social transformation, the lifestyles of urban residents in China have undergone profound changes. Understanding these changes is a prerequisite for comprehending urban living spaces in China.

Traditional lifestyles are not entirely opposed to of modern lifestyles, but rather a foundation and context for understanding them. When examining the distinctions between traditional and modern lifestyles from the perspectives of social interactions, consumption behavior, leisure activities, and mobility patterns, the most prominent characteristics of traditional urban lifestyles in China can be identified as follows: daily interactions based on neighborhood relationships and a sense of local community, consumption behaviors based on the development and use of physical commercial spaces, collective leisure and entertainment based on public places, and convenient and frequent travel activities based on public transportation systems. Economic, social and cultural transformations have profoundly influenced the lifestyles of urban residents, with institutional reforms, capital forces, technological innovation, cultural development and social change having particularly significant impacts. From a historical perspective, the evolution of urban residents' lifestyles in China can be divided into four phases: the planned economy phase from 1949 to the 1980s; the phase from the 1980s to 2000, during which market forces became a crucial factor shaping urban spaces and residents' lives; the phase from 2000 to 2010, marked by the growing influence of internet technologies; and the phase from 2010 to the present, characterized by the deepening development of the internet and the rise of mobile internet technologies. Overall, the transformation of urban lifestyles in China exhibits the following key trends: the collision, reconstruction, and evolution of traditional culture, consumer culture, and foreign influences; the joint shaping of urban life by institutional frameworks and market forces; and the coexistence of modern and postmodern lifestyle characteristics.

In the 1990s, China seized the opportunities brought by the Third Technological Revolution, and the development and widespread adoption of the Internet revolutionized the lifestyles of urban residents. The application of information technology has introduced a new landscape for urban residents, transforming aspects of life ranging from work and consumption to personal entertainment and social interactions. Meanwhile, the wave of economic globalization not only brought global commodities and advanced production technologies to China but also influenced people's values, consumption patterns, and cultural activities. Against this backdrop, the new lifestyles of Chinese urban residents have exhibited several distinct characteristics, including flexible mobility supported by private cars and public transportation systems; convenient shopping enabled by the Internet and efficient logistics; diverse cultural and leisure activities driven by symbolic consumption; new modes of social interaction facilitated by information technology; and a low-carbon lifestyle guided by the principles of sustainable development.

Since the founding of the People's Republic of China, economic and social transformations brought about by institutional, capital, technological, cultural and social forces have subverted the traditional lifestyles. Foreign culture, consumer culture, mass culture and traditional culture have collided fiercely, with the dissolution of Danweilization and the development of market economy broadening people's choices in lifestyle. The emergence of convenient mobility options has enabled greater freedom of movement in physical space, while the development of the Internet and mobile technologies has further shifted people's living spaces from the physical to the virtual. In this context, the evolution of urban lifestyles has disrupted and restructured the original urban living space. The new urban living space after the dissolution of Danweilization, the new leisure life in the context of mass culture, the new lifestyle in the age of the Internet and the age of flexible mobility all offer critical insights into how lifestyle evolution has profoundly impacted the spatial experiences of urban residents in China.

References

Bai Y (2017) A study on the evolution of social lifestyles of urban residents in modern Shenyang (1905–1931). Northeast Normal University, Changchun (In Chinese)

Bao Z (2006) Rational reflections on contemporary metropolitan life in China. Soc Sci Nanjing 9:1–7 (In Chinese)

Bian Y, Logan J, Lu H et al (1996) Danweilization" and housing commercialization. Sociol Res 11(1):83–95 (In Chinese)

Cui S (2013) Research on neighborhood interaction space planning in old city renewal. Tianjin University, Tianjin (In Chinese)

Department of E-commerce and Informatization, Ministry of Commerce (2019) China E-commerce Report 2018. China Commerce and Trade Press, Beijing (In Chinese)

Feng L (2014) Individualization trend in contemporary Chinese society and its political implications. Soc Sci 12:20–27 (In Chinese)

Feng J, Lin W (2017) Residential satisfaction and influencing factors in declining old urban neighborhoods: a case study of six communities in Suzhou. Prog Geogr 36(2):159–170 (In Chinese)

Fu X (2007) On the Changes in Urban Cultural and Entertainment Lifestyles in Modern China. J Southwest Jiaotong Univ (Social Sciences Edition), (5):111–117 (In Chinese)

Gao B (1996) Elite culture, popular culture, and folk culture: group differences and transformations in Chinese culture. Soc Sci Front 2:108–113 (In Chinese)

Gao B (1998) A brief review of theoretical developments in Western lifestyle studies. Sociol Res 3:61–72 (In Chinese)

Guo F (2011) Research on the evolution and development of China's urban social management system since 1949. Central China Normal University, Wuhan (In Chinese)

He C (2008) Impact of modern communication methods on urban housing evolution. Hunan University, Changsha (In Chinese)

Hu Y (2015) Research on community interaction spaces in the information age. Beijing University of Civil Engineering and Architecture, Beijing (In Chinese)

Li C (2007) Research on urban leisure plaza environmental design centered on humanization. Hefei University of Technology, Hefei (In Chinese)

Li Y (2009) Nanluoguxiang: cultivating "Beijing-Style" characteristics. China Today 4:62–64 (In Chinese)

Li P (2010) New trends in urban lifestyles: online consumption and leisure—an analysis based on cultural lag theory. Admin Law 6:57–60 (In Chinese)

Li L (2013) Transition and research of the "Danwei System." Jilin Univ J Soc Sci Edn 53(1):11–14 (In Chinese)

Li P (2014) Modern public transportation and urban lifestyles: pre-war "Capital" Nanjing. Lanzhou Acad J 9:67–74 (In Chinese)

Liu T, Chai Y (2013) Research progress on China's urban danwei system. Areal Res Dev 32(5):13–21 (In Chinese)

Lin X (2012) Urban, automobile and spatial reconstruction of life world. Acad Rev 3:87–94 (In Chinese)

Lin G, Wang L (2011) Locality and spatiality of bars under globalization: a case study of Guangzhou. Sci Geogr Sin 31(7):794–801 (In Chinese)

Lu X (1987) Main characteristics and functions of urban lifestyle. Sociol Res 4:116–122 (In Chinese)

Ma S, Xia J (2004) Theoretical review of Western lifestyle studies. Jiangxi Soc Sci 1:242–247 (In Chinese)

Mao Y, Huang Z (2014) Investigation on the development of China's online shopping market. Econ Rev 8:82–86 (In Chinese)

Mitchell A (1983) The nine American lifestyles: who we are and where we're going. Warner, New York

O'Sullivan (2015) Urban economics. Peking University Press, Beijing (In Chinese)

Ruan L (2016) Research on precision marketing in JD.com under Big Data Era. University of Electronic Science and Technology, Chengdu (In Chinese)
Shi Y (2014) The dissolution and reconstruction of urban grassroots communities. China Dev Observer 10:66–69 (In Chinese)
Ta N, Chai Y (2013) Geographical interpretation of urban lifestyles. Chin Urban Stud 00:232–244 (In Chinese)
Tao Y (1987) Preliminary exploration of urban lifestyle in transition. J Anhui Provinc Party School (3):63–66, 69 (In Chinese)
Tao Y (2006) Typological study of lifestyles and its implications. Sociology 4:44–50 (In Chinese)
Tian F (2011) Consumption, lifestyle and social stratification. Heilongjiang Soc Sci 1:88–97 (In Chinese)
Wang Y (1988) A comprehensive view of traditional Chinese lifestyle. Sociol Res 6:104–110 (In Chinese)
Wang G (2009) Automobilized lifestyles of urban residents. Lanzhou Acad J 9:111–114 (In Chinese)
Wang K (2011) Review on urban living space research. Prog Geogr 30(6):691–698 (In Chinese)
Wang Y (2013) Contemporary significance of lifestyle research: Experiences and implications from 30 years of studies in China. Sociol Rev China 1(1):22–35 (In Chinese)
Wang J (2015) Cohort and collectivism construction in square dance: Field study based on three sites in S City. East China University of Science and Technology, Shanghai (In Chinese)
Wang S (2015) Mobile payment market and its development in China. Northwest University, Xi'an (In Chinese)
Wang X (2015) Critical theory research on David Harvey's urban "spatial fix." Nanjing Normal University, Nanjing (In Chinese)
Wang Z (2019) Transformation of retail industry from new retail perspective. China Bus Trade 21:3–5 (In Chinese)
Wang X, Feng T (2014) Collective action in collectivist era and individualized era. Shandong Soc Sci 5:45–51 (In Chinese)
Wen T (2002) How we lost the freedom of migration. China Reform 4:24–25 (In Chinese)
Weng G, Chai Y (2003) Shenzhen residents' bar consumption behavior and spatial characteristics. In: Proceedings of the 9th symposium of young Chinese geographers on building a moderately prosperous society in all aspects. (In Chinese)
Wirth L (2007) Urbanism as a way of life.// Sun Xun, Yang Jianlong, Urban Culture Studies, vol 3: Reading Cities. Shanghai Joint Publishing Press, Shanghai (In Chinese)
Wu J (1991) On planning and market as resource allocation mechanisms. Soc Sci China 6:125–144 (In Chinese)
Wu C (2009) Leisure and entertainment in ancient China. J Handan College 19(2):58–61 (In Chinese)
Wu C (2011) Review on peasant behavior studies during collectivization era. Vicissitudes (1):96–97, 105 (In Chinese)
Wu Y, Wu Y (2012) Urban lifestyle transformation and sustainable urban development. J Southwest Minzu Univ 33(6):117–122 (In Chinese)
Wu Q, Huang S, Cui W (2013) Development trends of urban consumption space from retail format evolution perspective. Mod Urban Res 5:114–120 (In Chinese)
Xiao X, De P (2003) Conflict and integration: transition of urban lifestyle. Acad Forum 3:123–126 (In Chinese)
Xu G (2004) Transformation of social value paradigms in China since reform and opening-up. Explor Free Views 5:24–26 (In Chinese)
Xu Q (2019) Political representation of square dancing: a diachronic analysis of its development in Chinese cities. Central China Normal University, Wuhan (In Chinese)
Xu Y, Chen H (2019) Cultural functions and social benefits of square dancing. People's Tribune 3:140–141 (In Chinese)

Xu M, Peng K, Yang Z et al (2018) Local public resources and contractual communities: study on gated communities under polycentric governance theory. City Plann Rev 42(12):67–75 (In Chinese)

Xu Y, Wang Q, Dong R et al (2019) Current situation, problems and countermeasures of "big data price discrimination." Reform Openning 1:15–20 (In Chinese)

Yang S, Guo Y (2018) From consumption to experience: development and evolution of Chinese commercial space. Literary Life 12:272–278 (In Chinese)

Yu M (2013) Analysis of the development mechanism of grassroots non-profit organizations in China. Central China Normal University, Wuhan (In Chinese)

Yu K, Li D (2003) The road to urban landscape: conversations with mayors. China Architecture & Building Press, Beijing (In Chinese)

Yuan Y, Pang H (2016) Urban living space under the 'Internet+' background. Urban Architect 9:329 (In Chinese)

Zhang X (2007) Study on daily life space: formation of Shanghai's urban daily life space. Tongji University, Shanghai (In Chinese)

Zhang J (2009) Study on the relationship between metropolitan life and city space. J Tongji Univ 20(4):51–58 (In Chinese)

Zhang W (2009) A trend in neighborhood relationship development: enlightenment from traditional danwei compounds and SOHO communities. J Henan Normal Univ (Philosophy and Social Sciences Edition) 36(3):93–95 (In Chinese)

Zhang P (2013) Research on online group buying based on community identity. Central South University, Changsha (In Chinese)

Zhang L (2014) Resonance of fragments: language communication patterns in online communities. Fudan University, Shanghai (In Chinese)

Zhang X, Yi Y (2019) Structural changes of neighborhood relations in China from historical perspective. J Yibin Univ 19(10):75–84 (In Chinese)

Zeng W (2015) Urban residents' living spaces during transitional period: a case study of Nanjing. Nanjing Normal University, Nanjing (In Chinese)

Chapter 4
Evaluation of Quality of Urban Life and Human Settlement Quality

4.1 Research on the Quality of Urban Life and Its Development in China

With the rapid development of social economy, the acceleration of urbanization and the accumulation of social wealth, the improvement of the quality of life has received more and more attention. China has proposed new requirements for urbanization development, emphasizing adherence to the people-centered principle. Focusing on human development is both a requirement of the times and a core tenet of the Scientific Outlook on Development (Liu and Dong 2012); people-centered development has become the goal of the overall development of society (Zhang et al. 2014). The essence of social development extends beyond economic growth, prioritizing the degree of satisfaction people get in their lives and how to improve the quality of urban life. The government's ultimate goal is to continuously improve the people's quality of life, ensuring workers live with dignity and decency, and enabling all citizens to live a good life (Liang 2014). Obviously, the ultimate goal of social development is not only to improve people's material living standards, but also to focus on improving the quality of spiritual and cultural life. At the same time, with the rapid development of the economy, the connotation of quality of urban life is constantly enriched, and under the premise of developing material and spiritual life, attention is also paid to the development of urban ecological environment, living conditions and social services. Therefore, the concept and evaluation system of quality of urban life have continued to develop, making it an important aspect of socio-economic development.

J. Feng et al., *Urban Living Space in China*, Urban Sustainability,
https://doi.org/10.1007/978-981-95-0595-1_4

4.1.1 The Concept of Quality of Urban Life and Its Development

Quality of life (QOL) is a relatively abstract concept, which is understood differently by different disciplines and has not yet formed a unified concept (Zeng et al. 2014). The quality of urban life not only covers the material life and spiritual and cultural life of individuals or groups, but also includes the requirements for the ecological environment and social services, etc. The quality of urban life can be studied from different contents, levels and perspectives.

Since the 1970s, Chinese scholars have discussed the theory of quality of life. It was not until the mid-1980s that extensive research began, with relevant departments and scholars exploring the connotation of quality of urban life. In 1986, the Beijing Academy of Social Sciences, in its project report "Capital Social Development Strategy", proposed that "the quality of life is a metric for assessing living standards, which not only includes the material level, but also the spiritual and moral content, while the material conditions are the basis of the quality of life". Li (1986), in *Socialist Political Economy*, defined quality of life from the perspective of resource allocation. He argued that the foundation of quality of urban life is people's actual income, while factors such as culture, environment, services, public security, and social norms serve as additional elements that enhance it. As social conditions improve, welfare increases, leading to a higher quality of life. In 1990, Ye Nanke, through a study of domestic and international quality of life, proposed that quality of life is a dynamic system characterized by multiple factors and levels. It is closely related to residents' consumption capacity and lifestyle (Ye 1990). Lu and Wei (1991, 1992) viewed quality of life as a measure of living standards. They argued that the level of material production is the fundamental and determining factor of quality of life. As material production develops, it continuously shapes people's spiritual needs, thereby improving quality of life. In 1999, Chen Yiping argued that quality of life consists of two aspects: the supply of life and the demand for life. This refers to the extent to which society provides for citizens' lives and the degree to which citizens' needs are satisfied (Chen 1999). Later, Zhou (2001, 2003) defined the quality-of-life system from the perspective of residents' satisfaction and sense of identity. He argued that improving national life satisfaction and adequacy is based on a certain material foundation and is closely related to residents' identity of themselves and their social environment.

In the 1990s, the concept of quality of urban life became more comprehensive, emphasizing holistic development. Based on studies from over 20 countries, the World Health Organization defined quality of urban life as an individual's perception of their position in relation to their concerns, goals, expectations, and standards they value within different cultural and value systems (World Health Organization 1993). This is a broad definition that encompasses an individual's physical and mental health, social relationships, degree of independence, and characteristics within their surrounding environment.

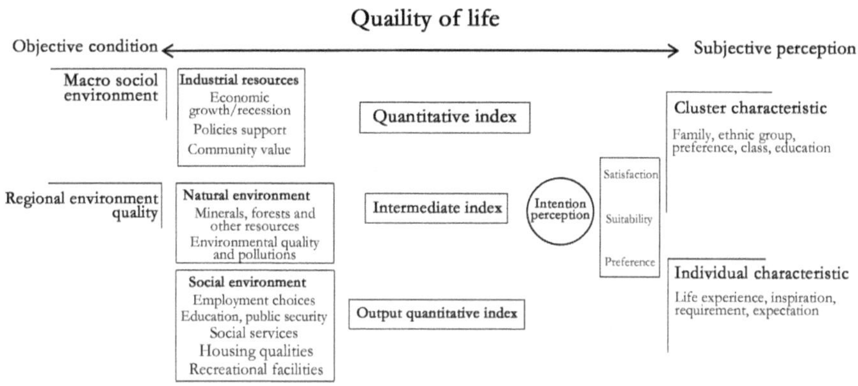

Fig. 4.1 Conceptual diagram of quality of urban life

Overall, research on quality of urban life is characterized by an interdisciplinary approach, involving fields such as economics, sociology, demography, psychology, geography, and medicine. The definition of quality of life can be summarized into three categories (Fig. 4.1):

1. **Subjective perception of quality of life.** This focuses on understanding quality of life through subjective perceptions such as happiness, satisfaction, suitability, presence, and preferences.
2. **Objective conditions of quality of life**. Initially, relevant departments and scholars in China understood quality of life from the perspective of objective conditions. They believed that quality of life is based on material conditions, and the development of material production leads to improvements in living conditions, thereby enhancing quality of life.
3. **Quality of life combining subjective perception and objective conditions**. In later research, Chinese scholars gradually viewed the combination of people's satisfaction with social life and the adequacy of resources provided by society as the standard for evaluating quality of life. In this view, quality of life consists of both material and spiritual aspects.

4.1.2 The Stages of Research on Quality of Urban Life and the Rising of Research on Quality of Urban Life in China

In 1927, American scholar William Ogburn first studied urban life as a sociological indicator and proposed various factors related to urban living. Later, quality of urban life became a specialized indicator for measuring people's living conditions, and with the development of socio-economic conditions, its connotation, indicator system, and research methods continued to expand. The development of quality of urban life research can be divided into three main stages.

First stage: the emergence of quality-of-life research (1920s–1960s). After quality of life was first introduced as a sociological indicator, the 1933 book *Recent American Trends*, written by scholars from the Hoover Institution in the United States, further explored the development dynamics and trends of various aspects of urban life. In 1958, American scholar J.K. Galbraith formally introduced the concept of quality of life in his book *The Affluent Society* (Galbraith 2009). In 1960, the U.S. "President's Commission on National Goals" report and the "Second National Planning" document by scholars such as Harvard Business School professor Bauer, officially presented quality of life as a specialized term, which helped establish it as an independent research field. Subsequently, in 1966, Bauer officially introduced the concept of "quality of life" in his book Social Indicators, becoming a pioneer in the field of quality-of-life research. From this point, research on "quality of life" was separated from the study of social indicators and became an independent field of study.

Second stage: rapid development of quality-of-life research (1970s–1980s). Starting in the 1970s, quality of life research gradually expanded from Europe and the United States to other countries, becoming a global research field. China entered the popular phase of quality-of-life research in the late 1980s. In 1989, Lin Nan and Lu Hanlong conducted the first study on the quality of life of Chinese residents using a questionnaire survey. Following this, Chinese scholars began a series of quantitative empirical studies on quality of urban life, though most were still focused on the field of sociology (Feng 2007). During this period, research on quality of life primarily focused on objective indicators based on socio-economic conditions, with Chinese scholars paying more attention to material living conditions and social services.

Third stage: stable development of quality-of-life research (1990s–present). With the rapid development of the socio-economic landscape and the acceleration of urbanization, people's material living standards have significantly improved, leading to a greater pursuit of spiritual fulfillment. As a result, scholars in this stage realized that using only objective indicators to measure quality of urban life could not reflect its diversity. After a certain level of socio-economic development, improvements in objective conditions have less of an impact on enhancing people's quality of life (Xia 2002). Starting in the 1990s, Western developed countries entered a stage of higher economic development, where people not only sought the continuous improvement of material living standards but also placed greater emphasis on the sense of satisfaction and happiness derived from social life. Therefore, it became necessary to measure quality of life from subjective perception. At this time, the North American school, represented by the United States and Canada, advocated using subjective indicators to study quality of urban life, while many European countries emphasized using living conditions as measurement indicators. In the 1990s, with the initial successes of China's Reform and Opening-up, people increasingly pursued material needs. As a result, quality of urban life research in China primarily focused on objective indicators. For example, Pan and Wang (1997) conducted a study on the spatial evaluation of quality of urban life in Xi'an. Fan (2006) developed an evaluation system consisting of three levels—evaluation goals, evaluation factors, and evaluation indicators—with a total of 30 indicators. This system was used to measure the quality of life of residents in 31 cities across China. With the rapid socio-economic development and acceleration of urbanization, residents' living standards have continuously improved. At this point, people were no longer satisfied with material life alone but began to pursue spiritual needs as well. Chinese scholars started to conduct extensive research on quality of urban life by combining both subjective and objective indicator systems. A typical example of this is the systematic study on the quality of life in 35 cities across China, conducted annually since 2011 by the Quality of urban life Research Center at the China Economic Experiment Research Institute. The research combines both subjective satisfaction indices and objective economic indices. The objective economic index primarily includes socio-economic indices, while the subjective satisfaction index covers five areas: living standards, cost of living, human capital, social security, and health levels. Notably, the satisfaction indices for social security and health levels have undergone changes since 2017. The former shifted from focusing on medical care, pension security, and urban safety satisfaction to a questionnaire designed to address residents' satisfaction with personal burdens related to medical insurance. The latter replaced the living experience satisfaction index with the residents' health level satisfaction index. These changes have made the subjective satisfaction index more relevant to current realities (Lang et al. 2017). Since 2018, the subjective satisfaction index of quality of urban life has included four indices: consumer confidence, quality of education, health levels, and medical services (Zhang et al. 2019).

In general, quality of urban life research has formed three comprehensive indicator systems:

1. Objective indicators based on various living conditions;
2. An indicator system based on people's subjective perception in life and work (satisfaction and happiness);
3. A comprehensive indicator system that considers both objective conditions and subjective perception simultaneously.

Starting in the 1970s, China began preliminary research on quality of urban life, and by the mid-to-late 1980s, it entered the fast phase of quality of urban life research. During this period, research primarily focused on objective indicators based on socio-economic conditions, with Chinese scholars paying more attention to material living conditions and social services. In the early twenty-first century, with the rapid socio-economic development and acceleration of urbanization, people's material living standards significantly improved. As a result, people increasingly sought spiritual fulfillment, making it necessary to measure quality of life from subjective perception. Thus, during this stage, quality of urban life research in China entered a phase where both objective and subjective indicators were combined in comprehensive indicator systems. After long-term development, quality of urban life has become a widely studied research area in China, attracting attention from various disciplines, including sociology, economics, psychology, medicine, philosophy, geography, and urban and rural planning.

4.1.3 Methods for Evaluating Quality of Urban Life and Its Assessment in Chinese Cities

4.1.3.1 Methods for Evaluating Quality of Urban Life

The evaluation of quality of urban life primarily employs two methods: hedonic pricing research and indicator system evaluation.

(1) **Hedonic pricing research on quality of urban life**

International hedonic pricing research on quality of urban life mainly focuses on housing prices, wage relationships, and related aspects. Scholars have developed theoretical models with significant reference value, making this field relatively mature. Representative studies include those by Roback (1982), Berger et al. (2008), and Glaeser (2011).

Roback (1982) examined quality of urban life through labor market research. In workers' process of choosing a city, factors such as housing prices and wage levels reflect labor mobility trends and the selection of "advantageous cities" that attract workers. As society develops, quality of urban life, wage levels, and housing prices tend to reach a state of equilibrium across all cities. Roback's (1982) theory on labor and quality of urban life has become a key research topic in urban economics. Berger et al. (2008) applied this labor market and housing price theory to Russia's transitioning economy. Their findings confirmed that, during the economic transition, the relationship between wage levels, housing prices, and quality of urban life aligned with Roback's theoretical framework. Glaeser (2011) further demonstrated that cities with higher economic development levels and better quality of life tend to have higher housing prices while simultaneously attracting high-skilled labor.

Overall, research on the hedonic pricing of quality of urban life primarily focuses on housing prices, wage levels, and labor market dynamics. As socio-economic development progresses, a city's economic level and quality of life influence housing price trends. Labor mobility between cities depends on the relative balance between wages and housing costs. Additionally, factors such as the surrounding environment, transportation accessibility, public services, and education quality play a crucial role in workers' housing choices.

(2) **Indicator system evaluation of quality of urban life**

Research on indicator systems for quality of urban life is conducted by establishing comprehensive evaluation frameworks. These systems primarily consist of three types of indicators: subjective indicators, objective indicators, and comprehensive indicators that combine both subjective and objective factors.

International subjective indicator systems mainly focus on people's perceptions of quality of urban life, including life happiness, comfort, and job satisfaction. These indicators reflect how different populations perceive the advantages and disadvantages of urban living. The evaluation primarily examines aspects such as life, work, environment, personal growth, self-identity, and happiness as social indicators to study quality of urban life (Hancock et al. 1999). Research on objective indicator systems primarily includes factors such as economic development, health status, job opportunities, income levels, infrastructure services, education, public safety, medical conditions, and environmental quality. Some scholars also develop comprehensive quality of urban life indicator systems based on various aspects of daily life and welfare conditions, including living expenses, food, housing, transportation, social services, and the surrounding environment. The comprehensive quality of urban life indicator system that combines both subjective and objective aspects is composed of social quality indicators based on economic metrics and indicators related to life satisfaction and happiness. Subjective indicators are typically gathered through questionnaire surveys and interviews, including individuals' perceptions of life satisfaction, happiness, welfare, and other personal perceptions. These indicators are used to improve life quality rather than being part of the evaluation process itself.

4.1.3.2 Evaluation of Quality of Urban Life in China

In terms of implicit price research on quality of urban life, there has been relatively little research in China. Most studies focus on the relationship between housing prices, the labor market, and wage levels to assess quality of urban life. For example, Zhou (2009) conducted a study on 233 Chinese cities, examining the relationship between housing prices, wage levels, and urban comfort. The study found that the impact of urban comfort on housing prices and wage levels varies regionally. Zheng et al. (2011) conducted a study on the labor and housing markets in 84 Chinese cities. They found that, under the premise of controlled wage level differences between cities, housing costs were positively correlated with factors such as good climate conditions, transportation, greenery coverage, healthcare services, and years of schooling. In contrast, housing costs were negatively correlated with air pollution levels (such as SO2 emissions and industrial dust emissions) and traffic congestion. Clearly, quality of urban life is directly related to wage levels and housing costs. Labor forces with higher wage levels are more likely to choose areas with better environments, convenient transportation, and complete surrounding facilities, opting for higher-quality housing conditions.

Chinese scholars have conducted extensive research on quality of urban life evaluation. Early quality of urban life indicator systems primarily focused on objective indicators, but later evolved to establish comprehensive indicator systems that combine both subjective and objective factors. China has been studying quality of urban life through the indicator-based method since the late 1980s. Sociologist Lin Nan, through questionnaire surveys, identified key indicators that significantly influence quality of urban life. His research revealed that not all socio-economic indicators have a significant impact on quality of urban life (Lin and Lu 1989). When the Chinese Academy of Sciences research team began studying quality of urban life, they selected 12 indicators, including residential consumption expenditure, wage income, food, housing, utilities, convenience of life, energy consumption, price index, and spiritual life (Wu et al. 1991). By 1992, the quality of urban life indicator system was expanded to include 23 indicators. After entering the twenty-first century, Chinese scholars conducted a large number of studies evaluating quality of urban life in cities such as Beijing, Shanghai, Hangzhou, Xi'an, and Dalian using indicator systems. Objective indicators primarily included aspects like economic base, infrastructure services, healthcare, cultural education, employment, leisure and entertainment, living conditions, transportation and communication, and living environment (Yang and Zhou 2000; Wang 2005; Wang and Wang 2010). Subjective indicators mainly involved satisfaction and happiness indicators related to living standards, living costs, interpersonal relationships, human capital, social security, and health levels (Lin 2010; Wang 2011; Feng and Zhang 2012). Additionally, some studies incorporated comprehensive indicator systems combining both subjective and objective factors (Yi and Wang 2009; Zhou 2013; Zhang et al. 2014). Building on this foundation, some scholars have quantitatively evaluated quality of urban life using socio-economic environmental indicators such as population density, residential land prices, and the proportion of highly educated individuals, as well as biophysical environmental indicators like Built-up area percentage, land surface temperature, and NDVI (Normalized Difference Vegetation Index) (Yang and Wu 2012).

In China, the research on quality of urban life indicator system evaluation has evolved from assessing the quality of life in large cities (Wang 2005; Zhang et al. 2012) to studying regional differences in the quality of life in prefecture-level cities (Zhang et al. 2014; Wang et al. 2015). The research methods have included statistical methods and mathematical models such as fuzzy comprehensive evaluation (Liu and Dong 2012), principal component analysis (Zhang et al. 2014), analytic hierarchy process (Zhou 2013), element analysis (Liao and He 2002), factor analysis (Li et al. 2012), distance comprehensive evaluation method (Liu and Chen 2009), coefficient of variation method, as well as questionnaire surveys, and a combination of GIS and RS (Geographic Information Systems and Remote Sensing) techniques. The research focus has gradually shifted from overall urban residents' evaluation to the quality-of-life assessment of special groups, such as the elderly (Shi et al. 2018) and the poor (Xue et al. 2017). The evaluation indicators have also shifted from objective indicators to a focus on subjective indicators, and a comprehensive indicator system combining both subjective and objective indicators has been established (Wang et al. 2019). The subjective indicators mainly include residents' life satisfaction, comfort, job satisfaction, etc., which are collected through questionnaire surveys. The objective indicators include income and consumption (such as food, housing, utilities, transportation, etc.), social services, education, medical conditions, living environment, and public policies. Additionally, the comprehensive indicators also take subjective factors into account such as employment and unemployment rates, as well as objective factors like transportation accessibility, surface temperature, and building environment.

Overall, research on quality of urban life in China primarily focuses on indicator system evaluation. It has gradually shifted from initially focusing on objective indicators to establishing subjective indicator systems, and now reflects the overall level of quality of urban life through a comprehensive set of indicators that combine both subjective and objective aspects.

4.2 Types and Characteristics of Differentiation of Quality of Urban Life

The meaning of differentiation is often understood in different ways. For example, some scholars emphasize that differentiation is a developmental and transformative process of social organization, culture, and forms, evolving from simplicity to complexity and from homogeneity to heterogeneity (Wu 2001). Other scholars believe that differentiation refers to the continuous changes in social and economic attributes that create social distance among individuals, leading to their separation from groups. It is a dynamic process of separation and alienation (Yang 2006). Social spatial differentiation refers to the relative concentration of wealthy and impoverished groups as a result of social stratification, where housing, work, and consumption tend to converge within these concentrated areas (Boal 1977). Quality of urban life differentiation falls under the category of social element differentiation. The rapid economic growth and continuous development of cities have brought about a series of urban social spatial issues, such as regional differences in the ecological environment, transportation, social welfare, living conditions, and regional conflicts (Pan and Wang 1997). These issues reveal the differentiation of quality of urban life. The research on the social space of quality of urban life in China has gradually shifted from focusing on economic spatial factors to emphasizing social spatial factors in the evaluation, beginning to pay more attention to the improvement of human-centered quality of urban life. A great deal of research has been achieved in this area. In a certain sense, the differentiation of quality of urban life itself is a manifestation of the social spatial differentiation of cities.

4.2.1 Types of Differentiation of the Quality of Urban Life in China

Since the Reform and Opening up, especially with the transition to a market economy system, there has been a clear stratification of social classes in China. The living needs and value orientations of residents have become increasingly diversified and complex, and urban residential spaces have gradually become more differentiated. These factors all contribute to the differentiation of urban residents' quality of life. Therefore, the types of differentiation in quality of urban life in China can be further analyzed from aspects such as space, social hierarchy, value orientation, and demand characteristics.

4.2.1.1 Spatial Differentiation of the Quality of Urban Life

The essence of quality of urban life is the process of obtaining urban living resources, which include employment, income and consumption, residential space (housing and environment), power and status (social status and social relations), and public safety. Selectivity and constraints (opportunities for obtaining resources and ability constraints) exist in the process of acquiring urban living resources. These selectivity and constraints determine the inequality and spatial differentiation in quality of urban life. Feng and Zhong (2020), using Changzhou as an example, comprehensively assessed the spatial characteristics of urban residents' quality of life from two aspects: internal living environment (housing conditions) and external living environment (facility accessibility). They found that the spatial structure of urban residents' quality of life exhibits a clear overlay of concentric and fan-shaped patterns.

There is a certain coupling relationship between the spatial structure of quality of urban life and residents' socio-economic attributes. That is, there are significant differences in the quality of life across different resident attribute areas, and each resident attribute area has a corresponding quality of life area with a high degree of overlap. This spatial relationship provides valuable insights for urban planning and management.

Duan et al. (2019), using Xi'an as an example, studied the living space quality of urban public rental housing communities. They found that there is a certain degree of spatial deprivation in the accessibility of community resources, and the overall quality of living space in public rental housing communities is low. The main reasons for this are twofold: First, the spatial layout of public rental housing communities is relatively remote, with resources such as hospitals, commercial areas, and recreational facilities being in a state of low accessibility. Second, community residents have low satisfaction with the housing area, housing quality, property management, and community greening. As a result, there is weak community belonging and low willingness for long-term residence.

4.2.1.2 Social Hierarchical Differentiation of Quality of Urban Life

The quality of life of urban residents is closely related to life opportunities, which are often constrained by factors such as social class, status, power, and lifestyle. The inequality of life opportunities causes social hierarchical differences in the quality of life across different groups within the city. In empirical studies on the quality of life of urban poor groups in China, it was found that the general perception of their living conditions was predominantly negative and pessimistic. Dissatisfaction with life was mainly closely related to work conditions and salary levels, while life satisfaction was mainly associated with factors such as neighborhood relationships and transportation conditions (Xue et al. 2017). For the affluent groups in Chinese cities, surveys indicate that most of them own more than one property, with the size of their properties continuously expanding. They show an optimistic attitude toward their future lives and tend to favor foreign brands. They view luxury brands as symbols of power, success, and quality of life. The majority of them are willing to support public welfare activities, and a significant portion supports environmental protection efforts (Li and Gu 2011).

4.2.1.3 Value Oriented Differentiation of Quality of Urban Life

In terms of the perception of quality of life, urban residents have different value orientations, which are mainly influenced by factors such as personality, expectations, sense of place, religion, and community emotions. These factors affect an individual's or group's choices regarding the quality of life in a city. As social and economic development progresses, people's understanding of themselves and society continuously evolves, and their value orientation toward quality of life changes accordingly. Therefore, the "position" of residents in the social hierarchy is not fixed; it is in flux according to an individual's or group's perception of life. The measurement standards and spatial evaluation of quality of urban life constitute a dynamical system. The value orientation of urban life is primarily reflected in residents' satisfaction with their living conditions, happiness, and other aspects in daily lives. Zhan et al. (2014) conducted a survey of Beijing residents and found that housing conditions, residential environment, supporting facilities, and transportation were the main factors influencing residential satisfaction. Similarly, Feng and Lin (2017), through a survey of six typical communities in Suzhou's old city, concluded that urban residents' satisfaction is primarily affected by housing conditions, community environment, supporting facilities, and social networks. Dang et al. (2018) studied the intercity differences in residents' happiness and found that residents in the most economically developed cities had the lowest levels of happiness. City size and environmental pollution negatively affected residents' happiness, while income levels, good public security, and a favorable cultural environment had a positive impact. Similarly, there is a connection between urban residents' leisure activities and subjective well-being. Leisure time, participation in leisure activities, and leisure satisfaction significantly and positively influence residents' subjective happiness (Wang and Sun 2019). The subjective well-being of the elderly is directly related to the built environment of their daily activity locations. Moreover, due to differences in the duration, frequency, and purpose of their stays at various activity locations, the impact of the built environment on the subjective well-being of the elderly also varies (Zhou et al. 2019).

4.2.1.4 Differentiation of Demand in Quality of Urban Life

From the perspective of living needs, different groups have varying expectations for quality of life. Generally, residents' economic status, educational background, and lifestyle influence the spatial structure of quality of urban life. For example, residents with higher economic status typically choose residential areas that are safe, have a pleasant environment, minimal noise disturbance, and well-equipped service facilities. In contrast, average residents tend to prioritize areas with convenient transportation, comprehensive surrounding services, and relatively reasonable living conditions. Clearly, different types of urban residents have different requirements for the quality of urban living spaces. The quality of urban living spaces is a multi-level social spatial system that meets the diverse living needs of different residents. Wang et al. (2019), using data from the 2018 Tianjin residents' transportation and housing survey along with urban economic geography big data, measured residential quality of life and concluded that: Highly educated professionals and the creative class have a higher willingness to pay for quality of life and prefer access to subways, educational and medical resources, libraries, sports facilities, and a good ecological environment; The floating population has a lower willingness to pay for quality of life; Different age groups have distinct preferences for urban service facilities and the environment. Younger groups prioritize cultural, leisure, and entertainment facilities, whereas older individuals place greater emphasis on the accessibility of universities, medical facilities, and parks/green spaces in relation to quality of urban life.

4.2.2 *The Socio-Spatial Characteristics of Differentiation of Urban Life Quality in China*

China is currently undergoing a period of political, economic, and social transition. With the rapid development of cities, the spatial optimization of key production factors such as land, resources, and labor has become a priority. This process has led to the differentiation of urban social space, which in turn is the primary factor driving the socio-spatial differentiation of urban life quality. This phenomenon is not only evident in economically advanced large cities but has also become a widespread trend in urban development across the country. To some extent, the socio-spatial differentiation of urban life quality is a spatial manifestation of social stratification or differentiation in life quality. Therefore, social differentiation and spatial differentiation are interconnected and mutually restrictive. In China, the socio-spatial differentiation of urban life quality is mainly reflected in three aspects: residential differentiation, differentiation in public service facilities, and differentiation in public and cultural spaces.

1. Residential differentiation in urban life quality. Residential spatial differentiation is the most typical manifestation of the socio-spatial differentiation of urban life quality. It refers to the clustering of similar groups in specific residential spaces based on differences in occupation, income level, and cultural background. This results in relatively independent, concentrated, and stratified residential spaces (Ye 2001). It can also be defined as "the residential spaces where different social classes relatively cluster due to variations in occupation, income, and cultural background" (Wan and Wang 2003). Obviously, the phenomenon of urban residential spatial differentiation reflects the housing choices, willingness, and quality requirements of different social classes. This ultimately leads to the clustering of similar groups in specific residential areas, forming a

spatial distribution pattern of residential environment quality based on residents' choices.

Foreign scholars have conducted early research on urban residential spatial differentiation. Theories such as Howard's "Garden City" (Zhang 2016), Wright's "Broadacre City," Le Corbusier's "Radiant City," and Perry's "Neighborhood Unit" have explored residential spatial differentiation from different perspectives. In addition, the rapidly developing concepts of New Urbanism and urban human settlements theory in recent years have further examined the phenomenon of residential spatial differentiation in Western cities (Register 1987).

The phenomenon of residential spatial differentiation in Chinese cities also emerged relatively early. During the feudal society, residential segregation occurred due to the influence of class opposition and a strict hierarchical structure. However, after the founding of the People's Republic of China, the government's policies of unified construction and allocation led to a redistribution of housing. This resulted in the formation of Danwei communities, where people from the same industry or Danwei lived together, effectively eliminating the general concept of residential spatial differentiation.

Since the Reform and Opening-up policy, especially after the housing system reform in 1998, the commercial housing system has been established, leading to the gradual formation of urban residential spatial differentiation, which has accelerated and become more polarized. With the rapid development of the social economy, income disparities among different social groups have widened, and social stratification based on wealth and income has become increasingly evident. Meanwhile, the central and local governments have begun to emphasize the economic benefits of land, using differential land rent and land-use revenue to promote urban development. Under the combined influence of these factors, residential spatial differentiation based on group economic strength has intensified, leading to the emergence of different types of residential areas with distinct spatial characteristics, such as mid-to-high-end residential areas and mid-to-low-end residential areas. These two types of residential areas exhibit significant differences in terms of living environment, resource allocation, infrastructure, and regional culture, comprehensively reflecting disparities in quality of urban life. For example, the residential spatial differentiation characteristics in the central urban area of Shanghai show that the city center is primarily composed of high-end commercial housing, with mid-to-high-end residences acting as a transitional zone. The city's outskirts feature mid-to-low-end residential areas, while the edge areas are characterized by modern residential communities, such as detached houses and high-end villa areas (Huang 2001).

For example, in Dalian, the spatial differentiation of commercial housing prices shows a single-core concentric circle distribution (Fig. 4.2). Sporadic high-price centers appear near the coast, further dividing Dalian's residential space into five types: low-end, mid-to-low-end, mid-to-high-end, high-end, and ultra-high-end residential areas (Li et al. 2004b; Li and Tang 2007). In summary, with the rapid development of cities in China, the phenomenon of residential differentiation is evident in

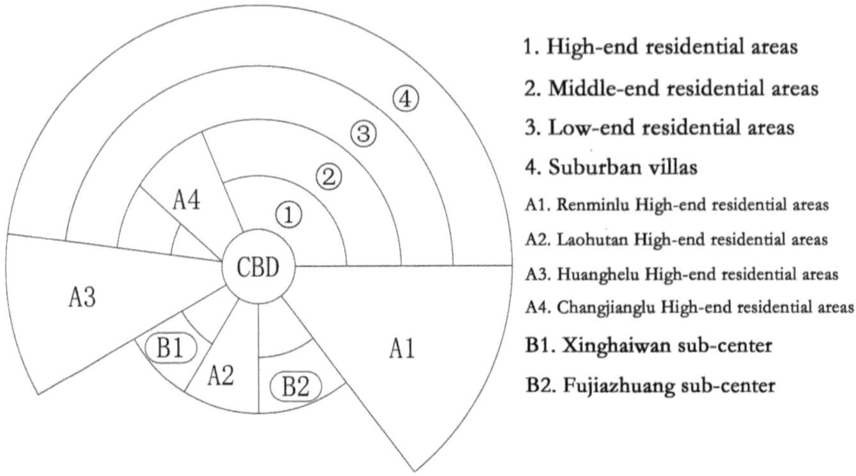

Fig. 4.2 Urban residential spatial differentiation model of Dalian

large and medium-sized cities, and the degree of differentiation is gradually intensifying. The phenomenon of homogeneous individual units and heterogeneous overall social spaces is a typical feature of urban residential space in China. Against the backdrop of diversified community development, differences in residential environments, infrastructure services, and community services all contribute to the differentiation of residential spaces and the unequal distribution of residential resources.

2. Differentiation of public service places in quality of urban life. Urban public service places mainly include commercial spaces, medical institutions, modern cultural and educational institutions, entertainment venues, etc., which are also important indicators reflecting quality of urban life. Commercial facilities, as urban consumption spaces, are usually concentrated in rapidly developing central areas of the city. Currently, high-end consumption spaces are increasingly separating from general consumption spaces, resulting in a stratification of various urban consumption places. This phenomenon is becoming more pronounced in an increasing number of cities.

High-end consumer places, such as large shopping malls, star-rated hotels, upscale clubs, and top office buildings, tend to occupy certain districts or streets in cities, becoming hubs for the urban wealth class. Due to the premium services and high-end consumption characteristics, the general public is often deterred and avoids these areas, thus creating a natural barrier within the city. The land prices around these areas are higher, and the costs for buying or renting properties are often several times greater than in other areas, leading to a differentiation in the social space of urban life quality.

Currently, in China's commercial real estate development, property developers intentionally cater to the consumption needs of upper class for their own benefit. They segment commercial services into various levels, offering different goods and services to different social and wealth classes. This strategy promotes the spatial differentiation of commercial facilities. In addition to the spatial differentiation of commercial facilities, other public service facilities in cities are also showing signs of differentiation. For example, high-quality educational resources, medical institutions, and modern cultural and entertainment venues (such as theaters and libraries) are often concentrated around the city's economic development centers, resulting in uneven distribution of public service facilities. This leads to issues in some relatively underdeveloped areas where services are inadequate. City residents in different regions enjoy significantly different levels of basic cultural education and public services. The spatial differentiation of these city service places and public service facilities ultimately contributes to the broader phenomenon of urban social spatial differentiation and the spatial differentiation of living quality.

3. Differentiation of public space and cultural space in quality of urban life. Public spaces in cities are mainly places for residents to relax, entertain, communicate, and exercise, including parks, squares, and the like. Usually, parks or large squares have better infrastructure. They not only serve the purpose of beautifying the environment but also provide convenience for nearby residents to relax and enjoy their leisure time. Therefore, when residents choose their living areas, they often consider the distribution of surrounding parks or squares. It is worth noting that there are differences in the activity places and public services required by social groups with different income levels. High-income groups typically choose high-end venues such as clubs and gyms for activities, while lower-income and economically disadvantaged residents are more likely to choose public spaces such as urban parks and squares for their activities (Zhao et al. 2013). Clearly, there is a spatial differentiation phenomenon in urban public interaction spaces and public services. The main factors driving this social spatial differentiation are related to the differentiated needs of different income groups and the imbalance between the supply and demand of service facilities (Zhou et al. 2013). In addition, cultural spaces are also worth attention. In the government-led rapid urbanization process, urban cultural spaces exhibit significant spatial differentiation, particularly in areas where rural residents are transitioning to urban areas. The cultural differences between these residents and the local population create obstacles to integrated development. The adaptation difficulties of new citizens to urban culture and the lagging cultural development in suburban areas are also manifestations of the disparity in living quality reflected in cultural space differences (Jiang Bin and Li, 2007). In conclusion, during the process of urban development, the differentiation of public and cultural spaces has increasingly affected the differences in the quality of life among urban residents, which is an issue that urban planning and management departments should pay attention to.

4.2.3 Factors Influencing the Socio-Spatial Differentiation of Quality of Urban Life in China

Regarding the causes of social spatial differentiation in Chinese cities, scholars have explored the issue from different perspectives. For example, Jing (2013) believes that factors such as history, government, markets, family, and individual behavior all contribute to the social spatial differentiation in urban areas. Wei and Li (2006) analyzed the phenomenon of social spatial differentiation from the perspectives of economic conditions, the housing market, social mobility, and population mobility.

Overall, in the process of urban development in China, the social spatial differentiation of quality of urban life is influenced by various factors such as policy, market economy, population migration, and family and individual factors. Among them, policy factors play a greater role, especially the current situation of resource redistribution in China's urban development process has profoundly affected the layout of urban residential space. Therefore, this section summarizes the influencing factors of social spatial differentiation in quality of urban life in China from the perspectives of organizational level, individual level, and other factors.

1. Organizational level: At the organizational level, the mechanism of social spatial differentiation in quality of urban life is mainly divided into two categories: government departments and private sectors.

Government departments primarily control urban space through their authority. During the process of urbanization, China's top-down government policies are the main driving force behind the formation of urban development models and spatial structures. The government's macro policies, including land policies, economic reform policies, social security policies, and urban development concepts, are the dominant factors that control urban social spatial differentiation (Wu et al. 2002).

The government controls the overall spatial layout of cities by formulating Urban Master Plans, which dictate the city's development model and direction. In China, the state is the legal owner of land, and its management and administration are carried out through national land management departments and subordinate agencies. Land managers allocate land to urban development based on different goals, leading to varying land use patterns in different urban areas. This, in turn, results in the differentiation of residential spaces across different regions of the city. Land managers or owners can also influence the construction and development of urban residential spaces through various taxation systems. Urban development concepts such as old city reconstruction, development zones, and high-tech zones create favorable external spatial conditions that contribute to the differentiation of urban residential spaces (Wang 2006). With strong government funding, various new urban districts in China have been developed to alleviate the pressure on old cities. In the process of urban construction, differences emerge between the old and new districts in terms of residential environments, public service facilities, transportation conditions, and the urban landscape, resulting in spatial disparities in residents' quality of urban life.

The government's control over local finances also affects the ability of local governments to modify urban social spaces, which is an indirect influence. Government departments safeguard public interests through urban planning, municipal construction, public space development, and coordination of spatial interests.

Therefore, various stakeholders at the organizational level, driven by both public interest and their own interests, influence the differentiation of urban social spaces, which further impacts the spatial differentiation of urban life quality.

The private sector plays a key role in creating urban wealth and promoting urban development. From the perspective of the social spatial differentiation of life quality, attention is mainly focused on the real estate development and financial sectors. With the acceleration of urbanization, the real estate development industry continues to grow and take on diverse forms (state-owned, collective, private, Sino-foreign joint venture, sole proprietorship, etc.). By controlling capital elements, it has become a significant force influencing urban spatial differentiation and layout (Zhou et al. 2005).

After real estate developers obtain the development rights to land, they influence the urban structure through the material environment and social characteristics of housing. Housing construction, based on market demand, typically focuses on quickly providing the most cost-effective and affordable residences for different groups. As a result, different-sized high-density residential clusters emerge within the city and on its outskirts. In addition, the housing commercialization reform has promoted the emergence of urban residential space differentiation. The financial sector (such as banks, exchanges, etc.) participates in urban spatial differentiation through trust loans, which is one of the indirect factors influencing urban spatial differentiation. The implementation of trust loans has made it possible for an individual to afford a house. Factors such as housing prices, location, social attributes, family composition, and personal economic strength have become decisive factors in residents' housing choices.

Therefore, the private sector in cities mainly influences urban spatial differentiation through capital factors, with the goal of their own profitability. This plays a significant role in the social spatial differentiation of urban life quality. Monetized housing allocation system has made housing choices for urban residents relatively free, leading to differences in housing selection between different social strata and promoting the phenomenon of residential space differentiation.

2. Individual level. After the Reform and Opening-up, China's social economy entered a rapid development phase. People gradually shifted from focusing on material life to seeking spiritual needs. Urban residents, due to differences in professions, experience disparities in social status and income, which in turn affect their choices of living environment, creating differences in living and social circles, as well as differences in their subjective perceptions of life. Ultimately, these factors lead to the social spatial differentiation of urban life quality. The main factors here include profession, reputation, living environment, transportation, work location, and subjective feelings.

Occupation is the primary source of economic life and an important factor in social relationships, also a means of accessing rights. Throughout history, occupation has been a main basis for societal class differentiation. Residents with different occupations often have distinct daily lives, social relations, and values, which can divide them into different social groups. In Chinese society, occupation is one of the key factors determining an individual's position and overall social-economic status (Lu 2003).

Prestige is also crucial in the spatial differentiation of urban life quality. Prestige refers to an inner admiration for others, reflecting a distinction from ordinary social status. In a society, residents with equal prestige and different incomes may choose the same residential environment; however, those with equal incomes and differences in prestige may form residential separation.

Living environment, work location, and transportation are critical in the spatial differentiation of urban life quality. Comfortable living and working environments, convenient transportation, and nearby infrastructure are primary considerations when choosing a residential area. Different living habits, work environments, and the ability to grasp good living opportunities contribute to spatial differences.

Subjective perception, particularly individuals' judgments on their profession, living space, and social circles, as well as their hobbies, happiness, and satisfaction with life, all influence the social space differentiation of urban life quality.

3. Other factors. Other factors influencing the social spatial differentiation of urban life quality include the following:

First, income disparity among urban residents. With the rapid development of China's economy, the income gap among urban residents has widened, leading to social class differentiation. Residents with similar social classes and attributes tend to choose residential areas with similar quality, resulting in spatial differentiation in living quality (Li and Tang 2007).

Secondly, the impact of social aging. The family is the basic unit of society, and its structure influences urban residents' choices of residential space. In particular, the housing preferences of elderly family members cannot be overlooked. There is a distinct difference in housing preferences between young and elderly people. Elderly individuals, considering their physical condition, prioritize the convenience of medical facilities and accessibility to public transportation, which is different from the preferences of younger people. As a result, elderly housing communities are often found in older urban areas, while younger individuals tend to choose newly developed residential areas in suburban locations. With the rapid aging of China's population, age has become an important factor influencing the social spatial differentiation of life quality.

Finally, the impact of population mobility. As China's urbanization process continues to accelerate, a large number of surplus rural laborers have moved into cities. Due to lower wages and poorer economic conditions, they typically choose to live in urban fringe areas or at the urban–rural interface, where rent is relatively cheap but the quality of the living environment is lower. It can be said that population mobility has played a role in the formation of spatial differentiation in the quality of residential environments between urban and suburban areas (Gu and Chris, 1997; Shan 2011).

4.3 Typologies of Urban Human Settlements and Influencing Factors

4.3.1 Concept of Urban Human Settlements

In the 1960s, Greek urban planner C. A. Doxiadis, in his book *Ekistics: An Introduction to the Science of Human settlements*, first introduced the term "Human settlements." He believed that the geographical space where humans survive and live, including both urban and rural settlements, constitutes human settlements (Li and Jin 2012).

In the 1990s, Professor Wu Liangyong from Tsinghua University further developed this concept based on Doxiadis' work. He stated that the Human settlements is the place where people live together, the terrestrial space where humans survive, and the base on which humans rely for survival in the natural world (Wu 2001b). The core of Human settlements is "people," and a good living environment is built to meet the various needs of people. The five main components of Human settlements are the natural system, human system, social system, residential system, and supporting system (Wu 2001a).

With the development of Human settlements science, scholars have continuously enriched the theory of Human settlements from various fields such as urban planning, geography, ecology, morphology, resource studies, environmental economics, and sustainable development (Howard 1946; Doxiadis 1977; Yanitsky 1987; McHarg 1992; Qi et al. 2007). Some scholars have also interpreted Human settlements from both broad and narrow perspectives. In the broad sense, Human settlements refer to the integration of various material and non-material elements related to the development of individuals, society, or humanity as a whole. In the narrow sense, Human settlements refer to the geographical space related to human habitation and survival activities, which is an artificial environment built upon the natural environment (Li et al. 2005). In conclusion, Human settlement is a complex, evolving, and open system that will expand and deepen with the continuous development of society, economy, and advancements in science and technology (Zhou and Zhu 2002).

As an essential component of Human settlements, the concept of urban Human settlements is understood differently by various scholars. Li et al. (1999) categorized urban Human settlements into traditional and comprehensive perspectives. The traditional perspective interprets urban Human settlements at different scales—some studies analyze urban spatial structures from a large-scale geographical viewpoint, while others explore residential planning theories from a smaller-scale architectural perspective. In contrast, the comprehensive concept of urban Human settlements refers to the environment created to support residents' work, living, education, culture, entertainment, and healthcare within the multi-dimensional development of a city.

Ning and Zha (1999) categorized urban Human settlements into hard and soft environments based on its components. The hard environment refers to the physical foundation that supports residents' activities, including various service facilities such as housing conditions, infrastructure quality, ecological environment, and public services. The soft environment encompasses non-material aspects formed through residents' interaction with the hard environment, such as living comfort and aesthetics, sense of security and belonging, social order, and information exchange.

Li (2001) proposed that urban Human settlements are human-centered environments composed of natural, social, and architectural elements, serving as both living spaces for urban residents and areas for secondary and tertiary industries. Li et al. defined urban Human settlements as places centered on people, integrating cultural, natural, and spatial elements. They not only reflect material and cultural living standards but also serve as key indicators of a country or region's economic and social development (Li 2001; Li and Liu 2003; Li and Li 2005).

Clearly, urban Human settlements are human-centered spaces for urban residents' living and activities, consisting of natural, cultural, and spatial elements. Their development not only affects the quality of life of urban residents but also plays a crucial role in the sustainable development of cities.

4.3.2 Types of Urban Human Settlements in China

Human settlements, based on regional systems, can be divided into urban and rural Human settlements.

Regarding the types of urban Human settlements in China, different scholars have proposed various classification methods. The main classifications include the following two types.

1. Based on regional hierarchy and spatial scale, urban Human settlements can be classified into proximate residential environments, community environments, and urban environments. Some scholars have even expanded the research scale to national and regional environments. Nevertheless, all Human settlements studies are fundamentally based on urban environments. The proximate residential environment, also known as the microscale urban Human settlement, mainly includes housing and neighborhood environments. It encompasses not only indoor living spaces but also surrounding external activity spaces and the visual environment within the residents' line of sight. This residential environment is based on the integration of human-nature and human-social interactions, serving as a space that influences human emotions and activities. Its core function is to accommodate various activities closely related to residents' daily lives, such as living, sleeping, dining, entertainment, social gatherings, and neighborhood interactions, both within and outside the home. Based on factors influencing urban

Human settlements—such as housing quality, neighborhood relations, natural environment, living convenience, light rail transportation, education and healthcare, the proximate residential environment can be categorized into five types: low-quality housing type, balanced residential system type, high-quality housing with well-developed public facilities type, superior natural environment type, and close-knit neighborhood type (Li et al. 2014).

Community environment, also known as the meso-scale urban Human settlement, serves as the primary setting for residents' social activities, with its geographical scope corresponding to that of a residential district. Its development and utilization foster the formation of social groups and localized communities. This environment encompasses residents' lifestyles, social interactions, and cultural characteristics of the community, as well as activities such as commuting, schooling, daily shopping, and seeking medical treatment for common illnesses.

Urban environment, also known as the macro-scale urban Human settlement, encompasses the entire urban system, including ecological, architectural, and transportation environments, as well as the social environment shaped by urban planning and management. Its primary function is to meet residents' various social needs, facilitate activities and interactions, and ensure social security and the efficient operation of urban activities. As a macro-scale Human settlement, the urban environment integrates all aspects of proximate residential and community environments. Therefore, urban planning and management departments must take a broader and long-term perspective, comprehensively considering urban development and diverse needs to improve the urban Human settlement in multiple aspects.

2. According to the form of existence, the urban Human settlement can be divided into hard and soft environments, a classification that aligns with the understanding of urban Human settlement by Ning and Zha (1999). According to Ning et al. and Wang et al., the urban Human settlement's hard environment refers to the totality of all material aspects related to human settlement, consisting of spatial and physical elements.

The three components of the urban Human settlement's hard environment are: Housing conditions, which include the size, quality, and facilities of the residence, measured by the value of the housing itself. Quality of urban ecological environment, which refers to indicators that reflect the urban ecological environment, such as water, air, surrounding noise, and the level of urban greenery. Infrastructure and public service facilities, including commercial, educational, and service facilities, as well as squares, roads, transportation, and various activity spaces.

The urban Human settlement's soft environment is an intangible, non-material form of environment, also referred to as the social environment of human settlement. It encompasses residents' subjective physical and psychological experiences of comfort, interest, social security, sense of belonging, and information exchange (Wang et al. 2005). The level of the soft environment can be assessed through field surveys and questionnaires. Urban Human settlement is a combination of both hard and soft environments. The hard environment forms the material foundation for residents' daily life and work, while the soft environment ensures the proper functioning of the hard environment. Both are interdependent, interconnected, and mutually reinforcing (Wang et al. 2005).

4.3.3 Influencing Factors of Urban Human Settlements in China

Urban Human settlement is a multi-layered and multi-factor complex system centered around humans, and as such, its influencing factors are diverse and multidimensional (Li and Liu 2003). Overall, the factors influencing urban human settlement are mainly related to its five subsystems, each of which has varying degrees of influence on the human settlements. This results in regional disparities in the quality of human settlements (Li et al. 2014).

The five subsystems of the urban human settlements are the natural system, social system, human system, residential system, and supporting system (Wu 2001a). The natural system includes factors such as topography, landforms, geology, hydrology, climate, resource environment, and land use; the social system includes population, economy, culture, social relationships, health, welfare, public management, and public services; the human system encompasses the material needs of the inhabitants, as well as their physiological, psychological, and behavioral characteristics; the residential system includes the living environment, housing, community services, culture, education, healthcare, and community atmosphere; the supporting system includes infrastructure and public service systems such as water, electricity, gas, heating, transportation, and communication (Wu 2001b).

The natural system is a fundamental factor affecting the urban human settlements and largely determines the quality of the environment. Among these, hydrological and climatic factors are particularly active. If a city suffers from water shortages, severe water pollution, or has an unsuitable climate for human habitation, the overall quality of the urban human settlements will be impacted, regardless of how well other indicators are developed. Therefore, the natural system plays a crucial role in shaping the urban human settlements and is one of the key factors influencing its overall quality.

The social system plays a decisive role in improving the urban human settlements, directly affecting the standard of living in the city. Among these, the level of urban economic development is the key factor determining the quality of the urban human settlements, while public management and services, as well as health and welfare, play supportive roles.

The human system primarily reflects people's various needs, psychological characteristics, and behavioral traits in urban social life, such as satisfaction with life, sense of presence, and comfort in the living environment. Subjectively, these factors influence the level of the urban human settlements and indirectly reflect the quality of other influencing factors. Generally, in cities with higher economic levels, such as first-tier cities where competition is intense, housing prices and living costs are high, and survival pressures are greater, the comfort of life is affected. In contrast, residents in provincial capitals, sub-provincial cities, and second- and third-tier coastal cities may enjoy relatively more relaxed lives, leading to a higher livability level and thus a higher human settlements quality (Wang and Feng 2011).

The residential system and supporting system provide the foundational conditions for urban human settlements from the perspectives of living conditions and public service facilities. The community environment and facilities offer convenient living conditions for residents, influencing comfort and quality of life. Factors such as culture, education, and healthcare have become increasingly important to residents, serving as major determinants in the choice of cities and residential areas. The infrastructure and public service systems are the basic guarantees of residents' lives, while the transportation system is a decisive factor in the accessibility and convenience of travel within the city. Therefore, both the residential system and the supporting system are crucial factors in determining the quality of urban human settlements. Cities in developed regions have an absolute advantage in public facilities and social services such as employment, healthcare, education, and social opportunities, making them the main factors that attract labor and determine the overall quality of urban human settlements.

Clearly, the role of different influencing factors on urban human settlement quality varies, and the same factors have different effects in human settlements at different city levels. However, overall, the five subsystems all play crucial and irreplaceable roles in urban human settlement.

4.4 Evaluation of Quality of Urban Human Settlement and Its Spatio-Temporal Differentiation

4.4.1 Evaluation of Quality of Urban Human Settlement

The environment is the external world on which humans rely for survival. The quality of the environment directly affects people's living standards and quality of life. Environmental quality evaluation serves as the foundation for environmental research (Li et al. 2014). It can accurately reveal the current state of the environment, help understand environmental evolution patterns, identify environmental issues, explore influencing factors, and investigate feasible pathways for environmental management. Traditional environmental quality evaluation primarily focused on natural ecological aspects such as air quality, water quality, and atmospheric conditions. However, with the expansion of subfields like urban geography, social geography, and economic geography, environmental quality evaluation has gradually been applied to socio-economic development and cultural studies, making it more comprehensive and holistic.

The commonality between urban human settlement quality evaluation and general environmental quality evaluation lies in their use of scientific methods to accurately reflect the value characteristics of environmental quality and provide targeted recommendations for improving and addressing environmental issues. Therefore, environmental quality evaluation serves as the foundation for environmental research, forming the basis for regional planning and management, resource development and restoration, and environmental pollution control.

The distinction between urban human settlement quality evaluation and general environmental quality evaluation lies in the following aspects: 1. General environmental quality evaluation primarily studies objective outcomes, often focusing on single or multiple independent objects while overlooking the holistic and integrated nature of the environment. It also tends to ignore the internal interconnections and coupling relationships within the research subjects, leading to variations in evaluation indicators and results. 2. General environmental quality evaluation often neglects the central role of "human" as the primary subject. However, environmental changes are largely the result of human agency, making it essential to incorporate human factors into the assessment process.

According to the theories of Marx and Engels, there is a bidirectional interaction between humans and the environment. Neglecting or overemphasizing either side leads to an incomplete and inaccurate understanding (Yuwen and Yang 2016). The evaluation of environmental quality in urban human settlements places greater emphasis on the objective relationship between human needs and the state of the environmental system. By using quantitative results, it reveals the current state of human settlements within a region and their comprehensive characteristics. This approach effectively identifies existing issues, predicts future development trends, summarizes spatio-temporal differences in human settlements, analyzes underlying mechanisms, and uncovers Spatio-temporal patterns. Such evaluations contribute to the improvement and development of urban human settlements, informing relevant policies and promoting their scientific and coordinated advancement.

The acquisition of indicator data for human settlement quality evaluation has evolved over time. Initially, it relied on survey questionnaires reflecting residents' subjective perceptions. Later, objective data from government statistical yearbooks and reports became the primary source. More recently, a comprehensive evaluation system has emerged, integrating both subjective and objective data to form a complete ecological structure for assessing human settlement quality.

In recent years, the rise of big data driven by advancements in internet and computing technology has led to a more diversified and comprehensive indicator system for evaluating urban human settlement quality. For example, the emergence of satellite remote sensing data has made urban land-use classification more precise. The widespread adoption of Geographic Information Systems (GIS) has enabled the analysis of various points of interest using digital maps. Additionally, the vectorization of road networks has facilitated the inclusion of accessibility metrics for healthcare, shopping, and education within the evaluation system. Researchers can integrate residents' subjective preferences with objective statistical data to conduct more precise and multidimensional evaluations of human settlement quality. Meanwhile, advancements in spatial science and technology have made research methods more intelligent and data-driven.

The methods for evaluating the quality of human settlements have continuously evolved with the development of mathematical models, geographic information systems, and emerging tools such as big data. Currently, the main evaluation methods include subjective approaches such as the Analytic Hierarchy Process (AHP), Principal Component Analysis (PCA), and Fuzzy Comprehensive Evaluation, as well as more objective methods like the Entropy Weight Method, Grey Relational Model, and Geodetector Model. When evaluating the quality of human settlements in practice, it is important to select the appropriate method based on the evaluation object, in order to avoid negative impacts and ensure that the conclusions align with objective reality. After establishing an indicator system and calculating the environmental quality score, it is necessary to classify the urban human settlement quality standards for different locations and historical periods, in order to explore their spatio-temporal differences.

4.4.2 Indicator System for Urban Human Settlements in China

Wu Liangyong academician introduced the theory of "Ekistics" to China based on the country's actual situation and developed it into the "Human settlements Science" theoretical system. He divided human settlements into five major systems: natural system, human system, residential system, social system, and supporting system. The five systems are not isolated or independent; instead, they are interconnected and permeate one another. The coordination and balance between these systems are crucial for maintaining the healthy and stable development of human settlements.

The selection of evaluation indicators for urban human settlements directly affects the results of the entire evaluation system. Therefore, during the process of selecting indicators, it is important to follow principles such as comprehensiveness, scientific accuracy, representativeness, and feasibility to reflect the overall characteristics of human settlements. Since the introduction of human settlement science, over the past 30 years of development, a variety of human settlement indicators have been developed to suit China's specific characteristics. In research, the geographic spatial scale has gradually expanded, and in the construction of indicator systems, local conditions have been taken into account, combining the functions, characteristics, and structure of the research objects. This approach aims to fully reflect both the subjective and objective characteristics of the city, striving for comprehensive and scientific research results. Meanwhile, different from the evaluation of environmental quality in the general sense, the process of selecting indicators for urban human settlement quality evaluation should focus on reflecting the elements associated with human beings, so as to reflect the subjective and objective feelings and needs of human beings towards the environment. Ning and Zha (1999) made pioneering contributions to the indicator system and evaluation methods of human settlement theory.

Other scholars have supplemented and improved the indicator system and evaluation methods for human settlement quality from different perspectives (Chen 2000; Zhang 2007; Li and Li 2008).

In the human settlements, the evaluation of natural system elements focuses on factors related to urban green space ecology, land use expansion, and resource and energy consumption. Indicators such as per capita park green space area, per capita wastewater discharge, per capita sewage treatment rate, per capita industrial air emissions, per capita industrial solid waste production, and air quality are commonly used for assessment. Urban ecological natural construction and protection have been highly emphasized in China, with relevant policies introduced to strengthen the protection of natural ecosystems, laying the foundation for building a beautiful China and achieving healthy and sustainable development. In the evaluation of natural system quality, not only should the current status of urban natural resources and green space ecology be assessed, but future development trends should also be predicted to meet the requirements of sustainable development.

Human beings are the core of human settlement science. Without the element of "people," a spatial entity cannot be called a "human settlements." The human-centered perspective is a key characteristic of human settlement research. The study of human systems focuses on changes in human behavior, needs, and desires within a regional space, and aims to analyze the related mechanisms and principles. The human system indicator system generally includes factors such as sex ratio, urban population density, age composition of the population, occupational structure, and the incidence of criminal cases. Human needs are the fundamental driving force behind the coordinated development of the human settlement system. In the process of selecting indicators for the human system, it is important to fully reflect the human-centered nature of human settlement theory. The indicators should not only consider the "biological human's" demand for natural resources but also pay attention to the "social human's" need for cultural and social aspects.

Chinese scholars have already explored the perspective from image of the city to study human settlement, with questionnaire as the carrier, picture identification and cognitive map as the main body, and human imagery connotation as the theme (Li and Li 2006).

The social system elements include indicators such as GDP per capita, urban registered unemployment rate, urban labor force participation rate, average wage of employed workers, consumer price index, per capita disposable income of urban residents, Engel's coefficient of urban residents, per capita retail sales of social consumer goods, and per capita car ownership rate. In the process of human settlement, to meet people's living needs, different individuals collaborate and form a social spatial system. The composition of human groups within this system creates internal social environmental differences due to factors such as age, social class, and family. On the other hand, regional differences arise externally due to varying levels of production efficiency and development. Chinese government departments and society are committed to eliminating poverty, investing significant effort in addressing issues such as imbalanced wealth development. The construction and development of the human settlements focus on social equity, and achieving balanced intra- and inter-city development is an effective means of eliminating urban development imbalances.

Housing is the most basic fundamental function of urban units. Residential system elements include indicators such as the price of commercial housing, the age of buildings, per capita housing floor area, and building density. The residential system is not only the result of the coordinated relationship between the overall urban layout, housing supply, and residential choices, but the improvement in economic development also drives the increase in housing costs. Therefore, the residential system is, to some extent, a reflection of the urban socio-economic development.

In the context of spatial heterogeneity in urban development, housing prices and living conditions have become important factors for job seekers in choosing a city. Therefore, macroeconomic regulation of housing prices and policy planning for residential facilities have become effective measures for city government to retain and attract talent.

The supporting system indicators include the urban water supply coverage rate, urban gas supply coverage rate, per capita electricity consumption, household internet penetration rate, per capita urban road area, and the number of hospital beds per 1,000 people, among others. The supporting system connects the operations of various systems, so it has a significant impact on other systems and levels. The development of the supporting system directly affects the quality of life for urban residents. In the context of economic globalization and rapid urban expansion, people are no longer confined to small-scale regional activities. High-speed trains, as an important transportation tool in China, have made significant contributions to efficient transportation between cities and within urban areas, promoting coordinated development. Therefore, in the study of urban human settlements between regions, accessibility-focused indicators should not be overlooked. The intensification of population mobility and the improvement of living standards have raised the demands for urban infrastructure construction. Therefore, high-quality urban human settlements are often accompanied by the development of high-quality and balanced public infrastructure.

It is important to note that the overall changes in urban human settlement quality stem from the dynamics of the five major systems. The final evaluation results are constrained and influenced by these five systems. At the same time, each subsystem is not only specific but also dynamically changing. The changes in subsystems not only alter the overall evaluation of urban human settlement but also impact the coupling and coordination between the systems. For example, the most noticeable effect of socioeconomic improvement is the increase in per capita income. Higher income means that people have higher demands and expectations for the city's natural and cultural environment. This, in turn, changes the residential system and supporting systems. Such a chain effect can enhance the overall quality of the urban human settlements. However, when the city's development speed does not match the high expectations of high-income groups, or when high costs of housing and consumption behaviors that are inconsistent with income arise, failing to meet or support human activity needs, people develop a willingness to relocate. Once this willingness accumulates to a certain extent, migration behavior occurs. Migration and population flow, for cities in a healthy development cycle, promote the introduction of talent, increase urban vitality, and thus lead to an "even better" urban environment. For cities with imbalanced systems and poor quality of urban human settlements, it is inevitable that population migration and outflow lead to a slowdown in GDP growth, slow urban construction, and weak industries. This polarization effect between cities results in extremely uneven development, which in turn affects the living standards of urban residents. Therefore, maintaining the balance and healthy development of the five major systems is the direction of human settlement quality research.

The five major systems are built on the foundation of multiple related disciplines, not limited to environmental science and geography, but also the result of interdisciplinary integration and intersection. In the process of selecting human settlement quality evaluation indicators, indicators that are appropriate to the situation and characteristics of the research area should be chosen based on its positioning, functions, and features. A comprehensive approach should be adopted, incorporating theoretical knowledge from multiple disciplines to address the complex phenomena of urban human settlements. For example, in the study of resource-based cities' human settlements, in addition to selecting indicators that reflect the general living and residential conditions of urban residents, it is also necessary to choose indicators that reflect the characteristics of the city's secondary and tertiary industries to represent the effects of urban transformation and industrial functional characteristics, thereby providing decision-making and practical services for the city's healthy development. In the study of tourism cities, the impact of indicators related to tourism industries, such as urban infrastructure, ecological environment and third-sector employment should not be overlooked.

In the nearly 30 years of research in human settlement science, many evaluation indicator systems have been published. During the establishment of these systems, to adapt to the dynamic development of urban human settlement, the evaluation indicator system must align with the characteristics of urban development in different periods. It should both fully reflect the features of urban human settlement quality and specifically measure the spatial distribution patterns, as well as the interrelationships and coupling mechanisms during the temporal evolution process, in order to reflect the comprehensive nature of human settlement. The "Evaluation Indicator System for the China Human settlements Award" issued by the Ministry of Housing and Urban–Rural Development of the People's Republic of China in 2016 (Table 4.1) can serve as a reference for selecting and grading urban human settlement quality indicators (for brevity, the indicator standards are omitted here and can be found in the original table from the Ministry of Housing and Urban–Rural Development of the People's Republic of China).

The evaluation of quality of urban human settlement should not only focus on the results but also explore the causes such as driving forces, influencing factors, determinants, and negative externalities. In addition to traditional weight-determination methods, tools such as geographical detectors, spatial statistical models, and geographic information technology have significantly advanced researchers' ability to analyze and study these underlying causes.

Table 4.1 Evaluation indicator system for the China human settlements award

Tier 1 indicators	Tier 2 indicators	Tier 3 indicators
A. Residential environment	A1. Housing and community	Housing Security for permanent residents (qualitative indicator)
		Completion rate of affordable housing projects
		Renovation of slums, urban villages, and old communities (qualitative indicator)
		Residential district development (qualitative indicator)
		Construction of convenient community service circles
	A2. Municipal infrastructure	Urban public water supply coverage rate
		Urban water supply quality (qualitative indicator)
		Urban water supply network leakage rate
		Urban gas supply coverage rate
		Urban sewage treatment
		Urban drainage (qualitative indicator)
		Sponge city construction
		Urban appearance environment
		Urban broadband construction
	A3. Transportation and mobility	Average commute time
		Public transportation mode share
		Pedestrian and bicycle traffic system construction
		Urban road network density
		Urban parking (qualitative indicator)
	A4. Public services	Rational distribution of nine-year compulsory education schools (qualitative indicator)
		Campus safety (qualitative indicator)
		Public cultural and sports facility land area per capita
		Number of health service centers (stations) per 10,000 people
		Number of hospital beds per 10,000 people

(continued)

Table 4.1 (continued)

Tier 1 indicators	Tier 2 indicators	Tier 3 indicators
		Number of books in public libraries per 10,000 people
B. Ecological environment	B1. Urban ecology	Ecological environment protection and restoration (qualitative indicator)
		Urban biodiversity (qualitative indicator)
	B2. Urban greening	Green coverage rate in built-up areas
		Green area rate in built-up areas
		Per capita park green space area
		Park green space service radius coverage rate
		Rate of tree-lined road promotion
	B3. Environmental quality	Urban air quality
		Urban surface water quality
		Average noise level in urban areas
C. Social harmony	C1. Social security	Social insurance fund collection rate
		Urban minimum livelihood guarantee (qualitative indicator)
	C2. Elderly care	Elderly care service system construction
	C3. Disability services	Disability services and protection system (qualitative indicator)
		Barrier-free facility construction (qualitative indicator)
	C4. Urbanization of migrants	Urbanization policies for migrant populations (qualitative indicator)
	C5. Public participation	Public participation in planning, construction, and management (qualitative indicator)
	C6. Historical culture and urban character	Preservation of historical and cultural heritage (qualitative indicator)
		Urban landscape and characteristics (qualitative indicator)

(continued)

Table 4.1 (continued)

Tier 1 indicators	Tier 2 indicators	Tier 3 indicators
D. Public safety	D1. Urban management and infrastructure safety	Digital urban management and safe operation of municipal infrastructure
		Comprehensive management of underground pipelines and underground integrated utility tunnels
	D2. Social safety	Road traffic accident mortality rate
		Criminal case incidence rate
	D3. Disaster prevention	Per capita fixed shelter area in urban areas
		Urban public firefighting infrastructure integrity rate
		Urban flood control (qualitative indicator)
	D4. Urban emergency	Urban emergency system construction (qualitative indicator)
E. Economic development	E1. Income and consumption	Per capita disposable income of urban residents
		Engel's coefficient
	E2. Employment level	Urban registered unemployment rate
F. Resource conservation	F1. Energy conservation	Energy consumption per unit of GDP
		Proportion of energy-efficient buildings in existing buildings
		Residential heating metering in northern heating areas
		Proportion of renewable energy in total energy consumption
	F2. Water conservation	Water usage per 10,000 Yuan of GDP
		Urban sewage reuse rate
		Industrial water reuse rate
	F3. Land conservation	Urban population density
	F4. Green buildings	Proportion of newly built green buildings

(continued)

Table 4.1 (continued)

Tier 1 indicators	Tier 2 indicators	Tier 3 indicators
		Use of green building materials and construction of modular buildings
Comprehensive negative item	Cities that have experienced major accidents related to safety, pollution, ecological damage, or illegal construction within the past two years, causing significant negative impacts, will face a one-vote veto	

Source Ministry of Housing and Urban–Rural Development of the People's Republic of China. Notice on Issuing the Evaluation Indicator System for the China Human settlements Award and the Themes for the China Human settlements Model Award (Document No. 92 [2016], Urban Construction Department), May 20, 2016

4.4.3 Spatio-Temporal Differentiation of Urban Human Settlements in China

Spatio-temporal differentiation characteristics are a key research focus in human settlement science. The evolution and interconnections of human settlements across different locations and scales form the foundation of this research. Over time, various systems and elements exhibit spatial heterogeneity, with different research subjects displaying distinct types of spatial heterogeneity. The study of Spatio-temporal differentiation in human settlements not only improves the quality of urban life, but also explores the coordination between population and socio-economic factors, urban economic and industrial transformation, and the relationship between jobs and housing separation.

The spatial changes in a city's outline are reflected through structural transformations, while temporal changes are shown through the evolution of processes. The combination of the two collectively results in Spatio-temporal differentiation. In the study of Spatio-temporal differentiation of human settlements across multiple cities, the first step is to determine the city's hierarchy, function, and positioning. Then, based on the city's characteristics, the impact of spatial structure on process evolution is analyzed, along with how process evolution alters spatial structure. By examining spatial structure and process evolution, the study aims to reveal urban location characteristics, uncover development patterns, functional differences, and transportation network structures, and clarify intercity interactions, ultimately serving regional coordination and sustainable development.

The study of China's urban human settlements in terms of temporal evolution and spatial dimensions has significant advantages. As one of the world's oldest civilizations, China has a diverse history of human settlement development, accumulating extensive experience over time. This provides valuable insights for summarizing the evolution of human settlements, understanding the formation and characteristics of regional settlement cultures, identifying development patterns, and offering decision-making references for future urban planning.

The development of human settlements has evolved over time from simple regional compartmentalization with a single mode of production and a low level of productivity to a functional industrial structure characterized by rapid population agglomeration and expanding land use. It has further transformed into a multifaceted spatial model that integrates economy, society, culture, technology, education, and entertainment.

Today, relying on advanced and complex infrastructure networks, highly integrated urban complexes—urban agglomerations—have taken shape. These urban agglomerations have become new regional growth poles in China. Additionally, China's vast territory, rich ethnic cultures, and extensive natural zonation have led to distinct spatial industrial structures and cultural practices. Examples include the agricultural planting industry chain and heavy industry base in the northeast, tea culture in the south, agroforestry and animal husbandry in the northwest, and livestock farming in the north. Through combining the unique advantages of regional locations, people have created diverse human settlement patterns and cultures in their production and daily life. Moreover, the spread and integration of these distinctive cultures have given rise to new human settlement development models.

Since the Reform and Opening up, China's urbanization has entered a period of rapid development. Along with the deepening of research on urban human settlements theory and the emphasis placed by government departments on urban infrastructure services, there has been a significant change in the spatial and temporal development of urban human settlement quality. The overall level has gradually improved, and the gap between cities is gradually narrowing, presenting a positive development trend.

1. From the perspective of spatial distribution, cities with higher quality of human settlement are concentrated in the eastern region of China, and the improvement of human settlement construction has been relatively fast in terms of time sequence. The central region mainly consists of mid-level cities, with the quality of human settlements in this area in a phase of improvement. The western region has a lower level of urban human settlement quality, with overall development being slow. So the western region still shows a clear gap compared to the eastern and central regions.

In terms of the north–south spatial distribution (with the Qinling-Huaihe line as the natural boundary between northern and southern China), the quality of urban human settlements in the southern regions is generally better than in the northern regions, and the differences in urban human settlements within the southern regions are smaller than those in the northern regions.

2. Among different types of urban functions, the quality of human settlements in provincial capitals and tourism cities is higher than that of other types of cities (Zhang 2019). Provincial capitals, as political and cultural centers, have the ability to integrate resources and development factors, incorporating advantages such as party and government agencies, higher education institutions, and well-established medical facilities. At the same time, provincial capitals inject new vitality into the urban system through population concentration, investment

attraction, and purchasing power. Additionally, provincial capitals usually serve as transportation hubs connecting different regions, playing a supporting role in freight transportation and population flow.

In different types of cities, tourism cities, with their abundant tourism resources, have inherent advantages in terms of ecological environment and urban infrastructure compared to resource-based cities. At the same time, to accommodate large volumes of tourists, tourism cities typically have well-developed services such as hotels, shopping, dining, and bustling transportation networks. The huge flow of people and beautiful scenic views provide strong momentum for urban image building and the development of the tertiary economic sector.

3. Cities with higher quality human settlements in China tend to be distributed in a "cluster" pattern, primarily concentrated in the eastern coastal regions, especially in the Bohai Rim, Yangtze River Delta, and Pearl River Delta regions. Representative cities in the Bohai Rim include Beijing, Tianjin, Dalian, Qingdao, Qinhuangdao, Weihai, and Panjin; representative cities in the Yangtze River Delta include Shanghai, Hangzhou, Nanjing, Suzhou, Ningbo, Changzhou, Wuxi, Wenzhou, and Shaoxing; representative cities in the Pearl River Delta include Guangzhou, Shenzhen, Zhuhai, Foshan, Huizhou, and Zhongshan.

Some inland areas along rivers and railroads in China also have cities with good human settlement quality, although they are mostly distributed sporadically. As the demands for cities increase with socio-economic development, the pace of human settlement development accelerates, and urban growth has shifted from isolated, singular cities toward multi-city agglomerations. City clusters, consisting of megacities as the core and large cities as constituent units, represent a more advanced stage of urban development.

Among the three world-class city clusters of Beijing-Tianjin-Hebei, the Yangtze River Delta, and the Hong Kong-Zhuhai-Macao Greater Bay Area, the Hong Kong-Zhuhai-Macao city cluster has the highest quality of human settlement development, followed by the Yangtze River Delta. The cities within these two clusters exhibit relatively balanced development levels, whereas the Beijing-Tianjin-Hebei city cluster shows significant quality disparities between its cities. (Zhang 2019).

The emergence of urban agglomerations plays a significant role in urbanization and socio-economic development (Li et al. 2014). The dense and developed supporting systems between cities create compact spatial organization and strong economic connections, ultimately achieving high urban integration and unification. This, in turn, drives the collaborative and high-quality development of surrounding areas through regional linkage, establishing a new mechanism characterized by green and coordinated development, openness and shared benefits, and mutually beneficial cooperation.

Overall, there is a significant spatio-temporal disparity in the quality of urban human settlements in China, with considerable quality differences between cities. The distribution pattern shows a zonal trend, with quality decreasing from the eastern to the central and western regions, similar to the economic development levels.

Additionally, there is a north–south distribution pattern, with higher quality in the south and lower quality in the north (Li and Jin 2012).

The quality of urban human settlements is strongly geographically linked to the economy of the city. In the eastern regions, the higher economic development level, along with greater investments in urban construction and ecological environment, results in better quality urban human settlements. In contrast, the central and western regions, with lower economic development levels and relatively lower investments, face lower quality urban human settlements. Therefore, enhancing the city's economic strength and optimizing its economic structure are key guarantees for improving the quality of urban human settlements. A high level of economic development can effectively promote urban infrastructure construction, ecological environment optimization, and improvements in the living environment, achieving comprehensive, high-quality, and high-level coordinated development of urban human settlements.

At the same time, the overall spatial distribution of urban human settlements in China also mirrors the distribution of urbanization levels. There is a strong consistency between the level of urbanization and the quality of urban human settlements: the higher the level of urbanization, the better the quality of urban human settlements; conversely, the lower the level of urbanization, the worse the quality of urban human settlements (Li et al. 2004a; Li and Tian 2015).

For the improvement and enhancement of urban human settlements, the development model of other cities should not be blindly adopted. It is essential to fully consider regional characteristics and implement differentiated approaches for different cities. On one hand, urban human settlements theory should serve as an important support for decision-making. On the other hand, in practical application, the theory of urban human settlements should be further developed and enriched, thus creating a positive interactive cycle.

4.5 Outlook on the Quality of Urban Life Reconstruction in China Based on Improvement of Human Settlement

4.5.1 Background and Connotation of Restructuring of Quality of Urban Life

China is currently in the rapid stages of industrialization and urbanization. In the past, academic research on quality of urban life mainly focused on sociological perspectives. However, with the shift in social realities and research concepts, the focus of quality of urban life research has transitioned from "emphasizing material aspects, neglecting culture" to "constructing a complete quality of urban life ecosystem." From the perspective of its emergence, the rapid development of urbanization has led the entire society to shift toward post-industrialization. This process is primarily reflected in the social and cultural aspects as "how to connect quality of urban life with the quality structure of urban living spaces, and how to incorporate human-centered principles into the restructuring of quality of urban life" (Yang et al. 2010). From its application background, quality of urban life was first applied in social development research. In terms of China's current stage of development, the goal of "Comprehensively Building a Moderately Prosperous Society in All Respects" is one of the applications of quality of urban life research. Broadly speaking, the application of quality of urban life not only includes basic economic indicators but also encompasses aspects such as spiritual, ecological, and cultural dimensions. Furthermore, various aspects of social development indicate that quality of urban life research holds social significance (Fan 2006). From the perspective of individual needs, economic and social development has largely fulfilled physiological, safety, and emotional belonging needs, while the realization of self-actualization demands higher expectations from society (Sun 2013). Quality of urban life not only reflects the life conditions of residents but also indicates their psychological satisfaction with their economic and social realities. Whether at the societal or individual level, the construction of high quality of urban life is essential. Therefore, the restructuring of quality of urban life holds significant social importance.

Material and spiritual life together constitute the quality of life in a city. From the relationship between the two, the improvement of spiritual quality of life is contingent upon the fulfillment of material quality of life. However, they do not necessarily improve simultaneously. The key to enhancing quality of urban life lies in coordinating the development of both aspects. Quality of urban life is influenced by various factors, including the economic foundation, social fairness and efficiency, and social security. The economic foundation is the prerequisite, serving as the basis for meeting the material and cultural needs of the people, and it is also the means for human development (Hu et al. 2013). The coordination between social fairness and efficiency is an important driving factor: the former facilitates capital accumulation, while the latter ensures the realization of individual rights. The coordinated development of both helps unify individual rights and opportunities. Social welfare is a supporting factor, as the quality of residents' living conditions is an important reflection of quality of urban life.

Currently, there is a consensus in academia regarding the connotation of quality of urban life. It is primarily reflected in aspects such as population density, income level, housing conditions, healthcare facilities, education level, and environmental conditions (Wang 2005).

By reconstructing quality of urban life from the perspective of human settlements, it is possible to organically combine and analyze both subjective and objective aspects, as well as the soft and hard aspects of the living environment. This approach also explores the influencing factors and mechanisms, providing effective suggestions for improving quality of urban life.

4.5.2 The Significance, Concept, and Key Focus of Restructuring of Quality of Urban Life in China

Looking at the trajectory of urban development in China since the founding of the People's Republic, cities serve not only as crucial carriers of individual quality of life but also as key indicators of urban development and residents' happiness. The unique characteristics of China's social stages and development path have shaped a distinctive urbanization process. This transformation has led to a shift from a predominantly rural society to an urban society where the majority of the population resides in cities. As a result, cities have gradually evolved into comprehensive entities that integrate residents, society, nature, economy, and culture.

These interconnected components inevitably influence and interact with each other during development, sometimes even leading to conflicts. The effects of these interactions and conflicts, in turn, impact urban development and quality of urban life. This cyclical process drives the transformation of China's urban development model and the restructuring of quality of urban life. Enhancing quality of urban life through improvements in human settlements plays a crucial role in optimizing urban functions, maximizing development potential, and ultimately improving residents' quality of life.

In 2005, the China Economic Experiment Research Institute released the *China Quality of urban life Report*, which identified three main factors influencing quality of urban life in China: basic necessities (clothing, food, housing, and transportation), life cycle needs (birth, aging, illness, and death), and socio-economic stability (secure living and fulfilling employment). Consequently, research on quality of urban life covers a wide range of topics, and its restructuring must follow certain fundamental theoretical principles.

The first is the theory of humanism. The *Urban and Rural Planning Law of the People's Republic of China*, promulgated in 2008, legally guarantees the application of humanism in urban spatial planning. Therefore, the restructuring of quality of urban life should also adhere to the core principle of " putting people first," highlighting the subjectivity of the human being in this process. The ultimate goal of restructuring of quality of urban life is to enhance people's satisfaction with various living conditions and their perception of that satisfaction.

The second is the theory of social development stages, which serves as the practical foundation for restructuring of quality of urban life. Based on the fundamental national conditions of the primary stage of socialism, it is essential to grasp the phased characteristics of China's social development and formulate a reasonable framework of restructuring and improvement steps. The goals of restructuring should neither be set too low nor exaggerated beyond feasibility. Instead, a thorough exploration of the characteristics of the current socio-economic development stage is necessary to ensure a more rational and effective restructuring of quality of urban life.

Additionally, the restructuring of quality of urban life is a systematic project that involves various aspects of China's socio-economic development and thus needs to follow the principles of systems engineering theory. Based on the current development conditions of different cities across regions, it is essential to adopt a human settlements perspective to coordinate relationships among material and spiritual wealth production, logistics services, government organization and management, and the legal environment. Adhering to these intrinsic connections to achieve optimal operation among various systems is an inherent requirement of restructuring of quality of urban life.

Finally, the concept of sustainable development plays a crucial role. Since the Reform and Opening-up in 1978, China has been a firm advocate of sustainable development, elevating it to a key guiding principle for economic growth and further formulating the Scientific Outlook on Development based on this idea. Therefore, it is necessary to incorporate the concept of sustainable development into the restructuring of quality of urban life, balancing present and future development to achieve the goal of sustainable urban growth.

The key aspects of reconstructing quality of urban life based on improvements in human settlements lie in the coordinated restructuring of human settlement factors, the balanced development of both the physical and social environment, and the rationalization of urban living spaces. The construction of quality of urban life varies in focus across national, societal, and individual levels. In terms of the coordinated development of human settlement factors, the rapid and continuous advancement of China's economy and society has led to issues such as insufficient supply and imbalanced distribution of urban living resources, which have become limiting factors in improving quality of urban life. Achieving a rational allocation of urban living resources through infrastructure, educational resources, and social resources is one of the fundamental bases for reconstructing and enhancing quality of urban life by establishing both macro and micro quality of urban life systems. In terms of the restructuring of both soft and hard human settlement environments, the focus lies in enhancing residents' perception of the livability of their living spaces and improving their satisfaction with urban living environments or convenience.

In the restructuring of urban living space, it is important not to discuss the restructuring without considering the spatial dimension (Lu and Liu 2018). Urban living space is essentially the urban human settlement, an organic combination of material and non-material elements (Sun and Wang 2002). Due to significant differences in urban development levels across various regions (such as between cities in Eastern and Western China, or coastal and inland cities), the restructuring of quality of urban life should be conducted with focus and specificity, tailored to different places, locations, and social classes.

4.5.3 Outlook on the Restructuring of Quality of Urban Life in China

Building a good human settlement is an important goal of China's social development. In the twenty-first century, China's urban development plans introduced a "new human-centered" development approach, which emphasizes a greater focus on both the individual and social value of urban and rural residents. This approach aims not only to meet the growing material needs but also to enrich the spiritual well-being of individuals. It also strives to reduce the disparities in quality of urban life caused by unequal distribution of resources, with the ultimate goal of improving quality of urban life across the board.

As the geographical entity and basic unit where humans reside, the human settlement will become the fundamental unit for improving quality of urban life in China. Therefore, during the restructuring of quality of urban life, by clarifying city types, human settlement types, urban resource allocation, and residents' emotional attachment, and considering factors such as regional hierarchy, class composition, equitable distribution of urban resources, residents' value levels, and differences in lifestyle, we can better understand the social and cultural characteristics of cities from both macro and micro perspectives. This will effectively improve both material and spiritual quality of life, ultimately enhancing overall quality of life.

From an economic perspective, the restructuring of quality of urban life will, to some extent, influence the spatial location choices of individuals and businesses, as well as impact the overall quality of urban development in China. As an important indicator, quality of urban life can be used to measure the effectiveness of local governments in fulfilling their roles and policy implementation. It can also assess the effectiveness and impact of government public spending, while providing valuable insights for the development of urban planning policies. From the perspective of urban development, the restructuring of quality of urban life has a positive effect on the growth of cities.

In terms of China's current urban development model, a high quality of urban life will attract both human resources and capital, which are key factors in inter-city competition at this stage. Considering the social context of China's development, social progress will drive an increase in individual income levels, while economic structural transformation and environmental issues will raise residents' awareness of environmental protection. As a result, quality of urban life will become the primary factor in the location choices of individual residents and businesses in the future.

As the willingness of residents and businesses in China to pay for quality of urban life increases, more attention will be given to improving the quality of life in regions and cities, leading to a simultaneous enhancement of both quality of urban life and the competitiveness of regional and urban development.

The restructuring of quality of urban life from the perspective of improving human settlements emphasizes refined urban management, ensuring that Chinese residents live more comfortably and securely. Looking ahead to the restructuring of China's quality of urban life based on the improvement of human settlements, the following characteristics should define this process: 1. "People-centered" and "nature-centered" should coexist harmoniously, which should be a permanent issue in the restructuring of China's quality of urban life. 2. The shift from optimizing a single system to coordinating and optimizing multiple systems is the fundamental approach to reconstructing quality of urban life in China. 3. The transition from a tendency to emphasize development over planning to one that places equal emphasis on both development and planning, with more attention to humanism in urban development planning and a greater focus on sustainable development.

From a disciplinary perspective, the restructuring of quality of urban life in China will promote the integration of fundamental principles from various disciplines, such as the merging of humanities and urban planning concepts. The restructuring of quality of urban life moves away from an excessive focus on material conditions and begins to incorporate human-centered needs into the research scope. The resulting "urban society-life space quality evaluation system" and "urban life space planning concept system" will play a crucial guiding role in the future development of quality of urban life in China.

From the perspective of elements, the restructuring of quality of urban life will usher Chinese urban planning into the post-industrial era. By understanding urban material elements and constructing the smallest living units of the city, it will promote connections between units and between elements, forming a postmodern approach to urban planning.

The restructuring of quality of urban life in China's future will, under the concept of post-industrial urban planning, achieve a sustainable state of urban development. Under the concept of new humanism, it will achieve a coupled and coordinated state of elements and systems. Under the concepts of social equity and spatial justice, it will achieve optimal allocation of urban resources. Under the principles of social value protection and spatial dignity, it will reach the best state of individual quality of life.

In conclusion, achieving the restructuring of quality of urban life in China through improvements in human settlements is feasible and holds significant practical and contemporary value. Research in this area has both academic and practical value. Coordinating the various elements of quality of urban life and promoting its enhancement will be a top priority in future research on urban living spaces.

4.6 Chapter Summary

As China's urban construction continues to progress and the overall level of urbanization gradually improves, the focus of urban construction work has shifted from quantitative accumulation to qualitative enhancement. Against this backdrop, the improvement of quality of urban life has gained increasing attention from scholars and experts across China. As the geographical entity and basic unit of human habitation, the human settlement serves as the practical platform for enhancing quality of urban life.

Research on quality of urban life in China is highly correlated with the study of urban human settlement science, as both serve as essential foundations for improving people's livelihoods and enhancing social and economic development. By applying the theories of human settlement science to improve quality of urban life, the material living conditions and happiness index of urban residents can be elevated. At the same time, human settlement science advances through case studies and decision-making processes aimed at addressing practical urban problems and responding to environmental changes.

This chapter focuses on five key aspects: the theoretical framework development of quality of urban life in China, the characteristics of social-spatial differentiation, the types and influencing factors of China's urban human settlements, the evaluation and Spatio-temporal differentiation of China's urban human settlement quality, and the restructuring of quality of urban life based on human settlement improvement. Additionally, it summarizes the Spatio-temporal distribution characteristics of quality of life and quality of human settlements in Chinese cities.

Quality of urban life is a comprehensive assessment of material conditions, spiritual well-being, and environmental quality, all centered on human needs. As Chinese researchers adopt more diverse perspectives and deepen their studies on quality of urban life, the evaluation index system has been established. Scholars have gradually incorporated residents' subjective perceptions alongside objective material conditions, resulting in a comprehensive evaluation system that integrates both subjective and objective factors.

The quality of urban life in China exhibits variations in spatial distribution, social stratification, value orientation, and demand characteristics. Currently, the spatial pattern of quality of urban life results from the optimized allocation of key production factors such as land, resources, and labor. This optimization leads to urban social spatial differentiation, which in turn becomes the primary factor shaping the social-spatial disparities in quality of urban life.

The social-spatial differentiation of quality of urban life reflects how social stratification or division manifests in spatial distribution. Therefore, social differentiation and spatial differentiation are Interconnected and mutually restrictive.

In the spatial differentiation of quality of urban life in China, cities like Beijing, Shanghai, Nanjing, and Dalian have become focal points for researchers. Empirical studies have identified a hierarchical model of high-end, mid-range, and low-end living spaces. This differentiation is typically reflected in variations in residential area, price, and type, shaping distinct spatial patterns within urban environments.

The urban human settlement directly impacts residents' quality of life and is also closely linked to sustainable urban development. It serves as a human-centered space for living and activities, formed by the integration of natural, cultural, and spatial elements. Based on regional hierarchy and spatial scale, China's urban human settlements can be categorized into proximate residential environments (micro-level urban human settlements), community environments (meso-level urban human settlements), and city environments (macro-level urban human settlements).

Based on the existing forms of urban human settlements in China, they can be categorized into hard and soft environments. The five major subsystems—natural system, social system, human system, residential system, and support system—along with their interactions and differences in coordination, are the primary factors influencing changes in urban human settlements.

The evaluation of quality of urban human settlement differs from general environmental quality assessments in that it is human-centered and integrates the five major systems to analyze the current state of urban human settlements and explore the characteristics of internal urban systems. Assessing urban human settlement quality helps identify spatial and temporal patterns of change, providing empirical support for improving settlement quality, advancing high-quality urban construction, and promoting sustainable development. The construction of China's human settlement quality evaluation index system is primarily based on the five major systems as the criteria layer, with a focus on the coupling and coordination within and between systems to quantitatively reveal the overall development level of urban human settlements. The spatial and temporal differentiation characteristics are a key research focus in urban human settlement quality studies. Currently, there is a significant spatial and temporal differentiation in the quality of urban human settlements in China, with an overall trend of improvement. Tourism cities and provincial capitals tend to have higher human settlement quality, and overall, high-quality human settlements exhibit a cluster distribution pattern. This is a characteristic of the gradual maturity of urban development as China's urbanization process advances. A more advanced spatial organization form—urban agglomerations—has emerged between cities. The integration and development of urban agglomerations have led to an upgrade in the research on urban human settlements. Enhancing economic strength and improving urban basic public services are effective ways to improve the quality of urban human settlements in China and mitigate the differentiated development between cities.

Quality of urban life not only reflects the living conditions of city residents but also serves as a macro-level indicator of a country or region's development level and social progress. Through the improvement of human settlements and by following theories such as humanism, social development stages, and systems engineering, urban elements, the hard and soft aspects of human settlements, and the rationalization of urban living spaces can be coordinated, thereby reconstructing quality of urban life. Clarifying the significance of reconstructing China's quality of urban life and promoting a balance between material life and spiritual culture is very crucial. This approach plays a key role in achieving both material and spiritual well-being for individuals, reducing regional disparities in quality of urban life, and enhancing the overall quality of urban life in China.

High-quality urban development is a multidimensional endeavor. As an integral component of enhancing the quality of China's cities, quality of urban life must be advanced through innovation-driven technological advancement, coordinated progress in both material and cultural-ethical spheres, and a consistent focus on residents' subjective well-being. Amidst the dynamic processes of urban expansion and evolving functional strategies, sustained economic growth should be leveraged to continuously elevate material living standards and foster a sense of urban happiness. This approach serves to strengthen social governance and improve people's livelihoods, aligning with the overarching goals of national development.

The theory of quality of urban life and human settlements in China has effectively guided urban planning, construction, and management practices, providing scientific theoretical support for urban development. At the same time, this theory and its methods have been further developed through practical applications, creating a positive feedback loop.

References

Berger MC, Blomquist G, Sabirianova K (2008) Compensating differentials in emerging labor and housing markets: estimates of quality of life in Russian cities. J Urban Econ 63:25–55

Boal FW (1977) Ethnic residential segregation. In: Herbert DT, Johnston RJ (eds) Social Alea in cities, volume 1, spatial processes and form, volume 2, spatial perspectives on problems and politics; Ann Assoc Econ Geogr 41–79

Chen F (2000) Evaluation of urban residential environments and satisfaction. City Plan Rev 24(7):25–27, 53. (In Chinese)

Chen Y (1999) Rethinking the evaluation of quality of life. Soc Sci Res 1:83–87 (In Chinese)

Dang Y, Zhang W, Chen L et al (2018) Analysis of intercity differences in residents' well-being and influencing factors: a multi-scale modeling approach. Geogr Res 37(3):539–550 (In Chinese)

Doxiadis CA (1977) Ecology and ekistics. Elek Boods Ltd., Athens

Duan Z, Wang X, Zhang T et al (2019) Study on the quality of living space in urban public rental housing communities: a case study of Xi'an. Hum Geogr 34(4):81–88 (In Chinese)

Fan B (2006) Construction and measurement of urban residents' quality of life evaluation system in China. J Zhejiang Univ (Humlties Soc Sci Ed) 36(4):122–131 (In Chinese)

Feng D, Zhang X (2012) Subjective quality of life indicator system and evaluation for residents in Xi'an-Xianyang region. J Xi'an Polytech Univ 26(1):117–120 (In Chinese)

Feng J, Lin W (2017) Residential satisfaction and influencing factors in declining old urban neighborhoods: a case study of six communities in Suzhou. Prog Geogr 36(2):159–170 (In Chinese)

Feng J, Zhong Y (2020) Spatial structure of urban residents' quality of life based on residential environment: a case study of Changzhou. Acta Geogr Sin 75(6):1237–1255 (In Chinese)

Feng X (2007) Thirty years of quality of life research: retrospect and discussion. Soc Sci Res 6:1–8 (In Chinese)

Galbraith JK (2009) The affluent society. Jiangsu People's Publishing House, Nanjing (In Chinese)

Glaeser E (2011) Cities, productivity, and quality of life. Science 333:592–594

Gu C, Chris C (1997) Factors Influencing Beijing's Social Spatial Structure and Its Evolution. City Planning Review 21(4):12–15 (In Chinese)

Hancock T, Labonte R, Edwards R (1999) Indicators that count! Measuring population health at the community level. Can J Public Health 20:22–26

Howard E (1946) Garden cities of tomorrow. Faber and Faber, London

Huang J (2001) Evolution of residential spatial structure in Shanghai's Central City. Urban Problems 4:30–34 (In Chinese)

Hu T, Du S, Li Z (2013) Quality of life-oriented urban planning: significance and characteristics. Urban Plan Int 28(1):7–10. (In Chinese)

Jiang B, Li X (2007) Spatial differentiation of urban cultural spaces under rapid urbanization. Prog Geogr 26(5):111–117. (In Chinese)

Jing X (2013) A review of domestic and international urban spatial research from a sociological perspective. Urban Dev Stud 20(3):44–49 (In Chinese)

Lang L, Zhang L, Zhao J et al (2017) Stable quality of life index and better-than-expected health index: report on quality of life in 35 Chinese cities. Econ Perspect 9:4–19 (In Chinese)

Li Y (1986) Socialist political economy. The Commercial Press, Beijing (In Chinese)

Li X, Liu J (2003) Comprehensive evaluation of climatic factors for livability in major Chinese cities. Econ Geogr 23(5):656–660 (In Chinese)

Li L (2001) Urban human settlements. China Light Industry Press, Beijing (In Chinese)

Li X (2001) Research on Dalian's urban human settlements. Jilin People's Publishing House, Changchun (In Chinese)

Li X, Li W (2005) Quantitative analysis of coordinated development between urban human settlements and economy in Dalian since the 1990s. Econ Geogr 25(3):383–390 (In Chinese)

Li X, Li J (2006) Analysis of Dalian's urban spatial image. Acta Geogr Sin 61(8):809–817 (In Chinese)

Li X, Tang X (2007) Quantitative analysis and mechanisms of residential spatial differentiation in Dalian. J Liaoning Norm Univ (Nat Sci Ed) 30(2):223–225 (In Chinese)

Li X, Li M (2008) Evaluation of urban human settlements in major Chinese cities based on self-actualization needs. Sci Geogr Sin 28(6):742–747 (In Chinese)

Li Z, Gu C (2011) Transformation of urban social spatial structure in China. Southeast University Press, Nanjing (In Chinese)

Li X, Jin P (2012) Characteristics and spatio-temporal differences of urban human settlement quality in China. Sci Geogr Sin 32(5):521–529 (In Chinese)

Li X, Tian S (2015) Geographical scale research on human settlements in China. Sci Geogr Sin 35(12):1495–1501 (In Chinese)

Li W, Ye X, Sun Y (1999) Evaluation of urban human settlements: a case study of Hangzhou. Econ Geogr 19(2):39–44 (In Chinese)

Li X, Zhang C, Zhang X et al (2004a) Quantitative study on urbanization and urban human settlements: a case study of Dalian. China Popul Resour Environ 1:93–98 (In Chinese)

Li X, Zhang X, Zhang C (2004b) Spatial differentiation of commercial housing prices in Dalian. AR Res Dev 23(6):35–39 (In Chinese)

Li H, Xu R, Gao Z et al (2005) Human settlement quality evaluation at the city scale: a case study of Nanjing. Hum Geogr 20(1):1–5 (In Chinese)

Li Z, Chen M, Pan L (2012) Measurement and evaluation of residents' quality of life based on factor analysis. Northwest Popul J 33(2):22–26 (In Chinese)

Li J, Zhang W, Sun T et al (2014) Agglomeration characteristics and economic performance of urban clusters in China. Acta Geogr Sin 69(4):474–484 (In Chinese)

Li X, Zhang Y, Gao J (2014) Types and spatial patterns of urban human settlements: a case study of Shahekou District, Dalian. Sci Geogr Sin 34(9):1033–1040 (In Chinese)

Li Z, Yuan X, Yang W (2014) Current status, challenges, and prospects of environmental quality evaluation research. Resour Sci 36(1):175–181 (In Chinese)

Liang Z (2014) Evaluation of quality of urban life in Guangdong Province based on Hedonic model. South China University of Technology, Guangzhou (In Chinese)

Lu X (2003) Stratification and mobility in contemporary Chinese society. Jiangsu Soc Sci 4:2 (In Chinese)

Lin X (2010) Regional differences in urban residents' quality of life in Zhejiang: a cluster analysis based on objective indicators. Northwest Popul J 31(3):95–100 (In Chinese)

Liao X, He C (2002) Application of factor analysis in quality of life evaluation: A STUDY on American quality of life status. J Xiangtan Inst Technol (Soc Sci Ed) 4(4):27–32 (In Chinese)

Lin N, Lu H (1989) Social indicators and quality of life structure model: a study on Shanghai urban residents' life. Soc Sci China 4:75–97 (In Chinese)

Liu L, Chen Q (2009) Urban residents' quality of life evaluation based on distance comprehensive evaluation method. Shanghai Manag Sci 31(1):87–89 (In Chinese)

Liu R, Dong W (2012) Fuzzy comprehensive evaluation of residents' subjective quality of life in Shaanxi Province. Futur Dev 1:95–98 (In Chinese)
Lu J, Liu H (2018) Review and prospect of quality of life research: from the perspective of urban economics. Jiangsu Soc Sci 2:89–95. (In Chinese)
Lu S, Wei L (1991) Comparative study on relationship between quality of life and demographic characteristics. J Peking Univ (Philos Soc Sci) 3:58–69 (In Chinese)
Lu S, Wei L (1992) Mechanism of subjective and objective indicators in quality of life measurement. Soc Sci China 1:122–137. (In Chinese)
Mcharg IL (1992) Design with nature. Natural History Press, New York
Ning Y, Zha Z (1999) Evaluation and optimization of metropolitan human settlements: a case study of Shanghai. City Plan Rev 23(6):14–19, 63. (In Chinese)
Pan Q, Wang X (1997) Spatial evaluation of quality of urban life: a case study of Xi'an. Hum Geogr 13(2):29–37 (In Chinese)
Qi X (2007) Towards limited community: a social network analysis of urban residential neighborhoods. Capital Normal University Press, Beijing (In Chinese)
Register R (1987) Ecocity Berkeley: building cities for a healthier future. North Atlantic Books, Berkeley
Roback J (1982) Wages, rents, and the quality of life. The Journal of Political Economy 90(6):1257–1278
Shan J (2011) Impacts and countermeasures of residential spatial differentiation and poverty concentration. Urban Dev Stud 10:19–23 (In Chinese)
Shi Mi, Chen C, Chen Y et al (2018) Quality of life and influencing factors among elderly home-based caregivers in urban Beijing. Pract Geriatr 32(7):639–642. (In Chinese)
Sun L (2013) Several issues in urban residents' quality of life research. Party Hist Gard 8:65–66. (In Chinese)
Sun F, Wang X (2002) Current status and trends of urban living space and community sustainable development research in China. Prog Geogr 21(5):491–499. (In Chinese)
Wan Y, Wang L (2003) Urban residential spatial differentiation and planning strategies for residential areas. Urban Probl 6:76–79 (In Chinese)
Wang W (2005) Quantitative evaluation of quality of urban life in Hangzhou. Acta Geogr Sin 60(1):151–157 (In Chinese)
Wang H (2006) Development zones and socio-economic spatial polarization in Xi'an. Acta Geogr Sin 61(10):1011–1024 (In Chinese)
Wang K (2011) Review on urban living space research. Prog Geogr 30(6):691–698 (In Chinese)
Wang W, Feng Y (2011) Comprehensive evaluation and influencing factors of urban human settlements in China based on factor analysis. Ecol Econ 5:174–177 (In Chinese)
Wang Y, Wang Y (2010) Preliminary discussion on constructing evaluation index system for residents' quality of life. Inn Mongla Sci Technol Econ 8:28–30 (In Chinese)
Wang X, Sun J (2019) Leisure and subjective well-being of urban residents: case study of Guangzhou. Geogr Res 38(7):1566–1580 (In Chinese)
Wang C, Wang H, Chen S (2005) Evaluation and optimization measures of urban human settlement environment: case study of Suzhou. Yunnan Geogr Environ Res 17(1):41–45 (In Chinese)
Wang Z, Cheng Y, Ma J et al (2015) Measurement and spatial analysis of urban livelihood quality in Northeast China. Sci Geogr Sin 35(2):190–196 (In Chinese)
Wang L, Ji Y, Wang Y (2019) Measurement and influencing factors of quality of urban life: empirical analysis based on micro-survey and big data in Tianjin. Urban Dev Stud 26(4):79–94 (In Chinese)
Wei L, Li Z (2006) Housing dilemmas and improvement models for China's urban low-income class. Urban Plan Forum 2:53–58 (In Chinese)
World Health Organization (1993) Measuring quality of life. Geneva: World Health Organ 2(2):77–81
Wu L (2001a) Introduction to human settlement environment science. China Architecture & Building Press, Beijing (In Chinese)

Wu L (2001b) Exploration of human settlement environment science. Planners 17(6):5–8 (In Chinese)
Wu Q (2001) Theory and practice of residential spatial differentiation in large cities. Science Press, Beijing (In Chinese)
Wu H, Zhu Q, Wu J (1991) Social development and social indicators. China Society Press, Beijing (In Chinese)
Wu Q, Zhang J, Zhu X (2002) Theoretical study on differentiation mechanisms of urban residential space in modern China. Hum Geogr 17(3):26–30 (In Chinese)
Xia H (2002) Quality of life research: review and evaluation. Mark Demogr Anal 8(1):67–75 (In Chinese)
Xue D, Lü Y, Huang J et al (2017) Subjective quality of life of urban poor groups: case study of typical communities in Xi'an. Sci Geogr Sin 37(4):554–562 (In Chinese)
Yang S (2006) Evolution of social space in Chinese metropolises. East China University of Science and Technology Press, Shanghai (In Chinese)
Yang M, Zhou C (2000) Significance of constructing quality of life indicator system. Econ Inf 2:9–12 (In Chinese)
Yang W, Wang X, Zhang D et al (2010) Review and prospect of quality of urban life and living space quality research. Hum Geogr 4(3):20–23. (In Chinese)
Yang Z, Wu S (2012) Evaluation of urban residential quality of life. J Beijing Jiaotong Univ (Soc Sci Ed) 11(4):71–74 (In Chinese)
Yanitsky O (1987) Social problems of man's environment. Nauka, Moscow
Ye N (1990) New dimensions of social development: a brief review of quality of life studies. Soc Sci Rev 4:35–38 (In Chinese)
Ye Y (2001) Preliminary exploration of residential spatial differentiation. Planners 3:94–97 (In Chinese)
Yi H, Wang P (2009) Analysis of subjective leisure quality of life among urban residents in China. Stat Inf Forum 1:81–86 (In Chinese)
Yuwen L, Yang X (2016) Marx and Engels' theory on 'human-environment' relationship and its application in ideological education. Ideol Educ Res 5:26–30 (In Chinese)
Zeng W, Zhang X, Xiang L et al (2014) Spatial pattern and economic analysis of quality of urban life at county level in Jiangsu Province. Econ Geogr 34(7):28–35 (In Chinese)
Zhan D, Meng B, Zhang W et al (2014) Residential satisfaction perception and behavioral intention of Beijing residents. Geogr Res 33(2):336–348. (In Chinese)
Zhang W (2007) Indicator system and methodology for evaluating urban residential environment. Sci Geogr Sin 27(1):17–23 (In Chinese)
Zhang W (2016) Core framework for livable city construction. Geogr Res 35(2):205–213 (In Chinese)
Zhang W (2019) Basic patterns and optimization strategies of human settlements in China. Think Tank: Theory Pract 4(4):68–70, 74. (In Chinese)
Zhang L, Lang L, Zhao J et al (2019) Stable quality of life index and better-than-expected health index: 2019 quality of life report of 35 Chinese cities. Econ Perspect 9:3–17 (In Chinese)
Zhang Y, Wu Q, Liu X (2012) Evaluation and analysis of Chinese residents' quality of life. Stat Decis 24:106–108 (In Chinese)
Zhang L, Zhao X, Zhang S (2014) Evaluation and spatial pattern analysis of urban residents' quality of life in Anhui. Econ Geogr 34(4):84–90 (In Chinese)
Zhao L, Hong Z, Yan H (2013) Reflections on constructing urban public communication spaces. Urban Dev Stud 20(1):80–85 (In Chinese)
Zheng S, Fu Y, Ren R (2011) Residents' preferences for quality of urban life: from housing cost dynamics and convergence perspectives. World Econ Pap 2:35–51 (In Chinese)
Zhou C (2001) Social development and quality of life. Social Sciences Academic Press, Beijing (In Chinese)
Zhou C (2003) China's quality of life: current status and evaluation. Social Sciences Academic Press, Beijing (In Chinese)

Zhou J (2009) Regional disparities in urban amenities, housing prices and wage fluctuations. J Financ Econ 35(9):80–91 (In Chinese)
Zhou D (2013) Research on objective quality of life of urban residents in Jiangsu Province. J Southeast Univ 6:42–46 (In Chinese)
Zhou Z, Zhu W (2002) Review of human settlement research. Nanjing Soc Sci 12:84–88 (In Chinese)
Zhou C, Chen S, Luo Y (2005) Housing spatial structure and its formation mechanisms in Guangzhou built-up areas. Geogr Res 24(1):77–88 (In Chinese)
Zhou C, Jiang H, Gao J (2013) Formation mechanisms of socio-spatial differentiation in urban public services: a case study of parks in Guangzhou. City Planning Review 37(10):84–89 (In Chinese)
Zhou S, Peng Y, Liu L et al (2019) Impacts of built environment of daily activity places on subjective well-being of the elderly. Geogr Res 38(7):1625–1639 (In Chinese)

Chapter 5
Residential Space

5.1 Fundamental Units of Urban Residential Space in China

5.1.1 Fundamental Units of Urban Residential Space from the Perspective of Residential Function

Urban residential space constitutes a critical category of urban space, intrinsically linked to the daily lives of urban inhabitants. Historically, in China, the adage "to live in peace and work in contentment" underscores the notion that a stable and secure living environment is essential for fostering future professional development. This highlights the paramount importance of residence in the lives of individuals within urban settings. The residents of each city reside within the fundamental unit of urban living environments. The fundamental unit of urban residential space is defined as the smallest spatial entity within an urban context that exhibits homogeneity and completeness in its spatial structure, social function, and administrative planning and management. This unit is considered indivisible at the urban scale. It encompasses a specific scale of construction and population, primarily serving residential purposes while simultaneously offering social and economic services. In general, the fundamental unit of urban residential space is comparable in scale to the administrative jurisdiction and number of households overseen by community neighborhood committees. Typically, this unit encompasses 2–3 residential complexes, comprising between 500 and 1,000 households.

The fundamental unit of urban residential space is intricately linked to the evolutionary dynamics of urban residential restructuring in China. During the era of a planned economy, the "*Danwei*" served as the essential unit for urban operations, the daily lives of residents, and economic advancement. Within the framework of the *Danwei* system, the core unit of urban space in China is organized around the

J. Feng et al., *Urban Living Space in China*, Urban Sustainability,
https://doi.org/10.1007/978-981-95-0595-1_5

Danwei community, characterized by significant functional and demographic diversity among units, yet marked by a lack of interconnectivity. Residents within these units exhibit stronger social ties, while the architectural design and living conditions tend to be highly uniform, often described as "average construction, average orientation, average set." (Dou 2011) By the conclusion of the twentieth century, the reform of the housing system coincided with a significant advancement in the marketization of housing. The fundamental unit of urban residential space, previously shaped by the *Danwei* system, began to experience increasing influence from market dynamics. As a result of this marketization, the emergence of residential segregation and the spatial differentiation of social classes became increasingly prominent (Fang 2014). The fundamental unit of urban residential space has shifted from a focus on the *Danwei* to a configuration shaped by the differentiation of urban social spaces influenced by market dynamics. From the perspective of the interactions among these fundamental units, the previously distinct separation of units, a consequence of the unit system, is gradually diminishing. However, this transformation is accompanied by the emergence of new class distinctions among spatial units, driven by capital influences. Regarding the internal structure of these fundamental units, the urban residential space is no longer characterized as a singular functional core providing comprehensive services for both work and daily life. In the context of urban development in China, the internal functions of these fundamental units in the contemporary period exhibit a greater degree of homogeneity, while also incorporating various social services and life-sustaining functions. Furthermore, the fundamental unit of urban residential space is increasingly interconnected with the urban commercial center.

The fundamental unit of urban residential space serves as a significant spatial entity within the urban social framework. In the context of the contemporary era, cities are experiencing an increase in demographic, cultural, and nationality diversity. Additionally, characteristics such as spatial heterogeneity, fragmentation, multiculturalism, and the rapid emergence of subcultures have come to define the landscape of modern urban social spaces (Feng 2005b; Feng and Liu 2007; Feng and Zhou 2008). The significant escalation of housing prices in China since the onset of the twenty-first century has rendered housing costs, along with the associated economic income levels, a critical determinant of the characteristics of the fundamental unit of urban residential space. Additionally, the availability of educational resources serves as another pivotal factor influencing this fundamental unit. The limited accessibility to educational resources enhances the integration of the urban residential space with other social resources. Furthermore, in numerous megacities across China, the fundamental unit of urban residential space also serves as a key indicator of the spatial differentiation between migrant and local populations. Geographically, this fundamental unit is delineated by specific spatial boundaries, which encapsulate the spatial concentration of educational and economic resources, as well as social stratification, within a defined scale. The fundamental unit of urban residential space serves as a fundamental component that encapsulates the structure of residential space within the broader context of urban social space (Feng and Zhou 2008). Urban residential space constitutes a critical element in the examination of urban social spatial structures. Consequently, the fundamental unit of urban residential space plays a pivotal role in shaping and influencing the development and transformation of the broader urban spatial framework.

It is essential to differentiate between the residential unit and the fundamental unit of urban residential space. The term "residential unit" can possess varying interpretations across different academic disciplines and scales. Some researchers equate residential units with residential neighborhoods, which are defined as self-contained urban residential space that include corresponding public amenities and green spaces, delineated from external environments by specific barriers such as roads and buildings, and characterized by morphological similarities among the structures and landscapes within the vicinity (Cheng et al. 2006). This notion aligns with the "urban residential unit" concept introduced by Le (2005), which posits that each individual possesses an independent garden and living room, thereby constituting a self-sufficient residential unit interconnected via public corridors. The concept of the residential unit varies in significance across the domains of urban and rural planning, architecture, and remote sensing.

5.1.2 Fundamental Units of Residential Space Based on a Spatial Statistical Perspective

From a research standpoint, various spatial scales and types are utilized to represent urban residential space. The street township scale is the most prevalent, with streets serving as the jurisdiction of the street office, an administrative unit equivalent to a township. Additionally, there exist smaller spatial units, such as neighborhood committees and village committees. Some researchers also employ postal districts, which are defined by the postal service, or traffic survey districts, which are delineated by transportation or planning departments for the purpose of conducting specialized urban traffic studies. Among these four categories of spatial units, street townships are the most frequently utilized due to the accessibility of census data and the relative ease of obtaining associated electronic maps. In contrast, neighborhood and village committee spatial units are more finely delineated, yet data acquisition is more challenging, and analyzing the maps of these units in relation to administrative changes poses significant difficulties. Furthermore, postal districts and traffic survey districts lack administrative significance, making it challenging to correlate them with statistics derived from fundamental administrative units, thereby limiting their broader applicability.

It is important to note that, alongside pertinent data such as census information, the investigation of residential spaces necessitates access to maps that depict grassroots administrative divisions. Specifically, these maps should clearly outline the boundaries of streets (townships) and the delineations of neighborhood committees (village committees). In general, maps delineating street (township) boundaries can be sourced from either the civil affairs department (regional toponymic office) or the database associated with the Second National Land Use Survey. Conversely, obtaining maps that illustrate the boundaries of neighborhood committees (village committees) presents greater challenges. In certain municipalities, only the boundaries of peripheral villages are provided, while the boundaries of neighborhood committees within urban areas may not be disclosed. However, in some instances, it is possible to access distinct layers representing the boundaries of neighborhood committees (village committees) within the national land use survey database, urban planning maps (status quo maps), or relevant geographic information systems (GIS). However, it is worth emphasizing that the maps in question represent static depictions for a specific year. To effectively monitor alterations in urban and rural boundaries, it is essential to examine the changes in the aforementioned maps of grassroots administrative districts over an extended timeframe. This examination is necessary to ensure comparability of data before and after the specified period, which entails conducting a survey of the adjustments made to grassroots administrative districts. The primary agency responsible for overseeing the modification of administrative divisions at the grassroots level is the District Zoning and Geographical Names Office within the Civil Affairs Bureau. Typically, Civil Affairs Bureaus across various regions maintain a District Zoning and Geographical Names Office tasked with managing adjustments to administrative divisions and changes in geographical nomenclature within the city. Consequently, this office serves as a valuable resource for investigating the evolution of grassroots administrative divisions over a defined period. For instance, when examining the transformations in urban residential spaces between 2000 and 2010, it is essential to analyze the adjustment of administrative divisions of the city, as well as the alterations in place names throughout this decade. Subsequently, utilizing the most recent map of these basic-level administrative divisions, one can reconstruct the map corresponding to the year 2000 or any other specified year. The underlying principle of this methodology involves a thorough investigation of the grassroots administrative units affected by the zoning changes, identifying the nature of these changes—such as mergers, separations, abolitions, and renamings—to ensure that the relevant areas are comparable across the time periods in question. Following this analysis, the latest base maps can be processed to generate maps from other years that maintain comparability. Such maps are deemed "spatio-temporal continuity" (Feng 2012). By digitizing these base maps, characterized by spatial and temporal continuity, researchers can produce maps that are applicable for studies concerning urban residential spaces.

5.2 Urban Residential Spatial Structure and Residential Regional Patterns

5.2.1 Research on Urban Residential Spatial Structure and Identification of Issues

Since the 1990s, China has enacted a series of policies, including the paid land use system, tax-sharing system reforms, housing system reforms, and the establishment of housing monetization and guarantee systems. These measures have prompted local governments to pursue revenue generation from the housing and land markets amid the ongoing process of housing privatization. Consequently, the real estate market has experienced rapid growth. Notably, locational disparities within urban areas remain a significant factor influencing variations in urban housing prices, which in turn contribute to the differentiation of residential spaces. Thus, the development of urban residential spatial patterns is intricately linked to the evolution of the real estate market. Furthermore, the emergence of a market economy since the 1990s has dismantled the previous social structure characterized by cadres, workers, and peasants, leading to a gradual establishment of social differentiation primarily driven by occupational distinctions. This occupational differentiation subsequently results in disparities in socioeconomic status, which also influences residents' housing choices. It can be posited that the interplay between the spatial differentiation of housing resulting from the marketization of housing post-1990s and the variations in urban residents' housing preferences stemming from socioeconomic status differentiation will culminate in a distinct pattern of urban residential space differentiation. This phenomenon is expected to be prominently illustrated in the urban socio-spatial structure as evidenced by data from the sixth national census conducted in 2010.

National census data is a survey of individual residents and household attributes based on place of residence, and provides important data support for identifying the social spatial structure of cities based on residence. Research that uses census data at the neighborhood level (streets and townships) to reflect the statistical meaning of residential space is also known as "social area" research. Some scholars also use the term "urban social space research". Regardless of the concept used, this type of research actually reflects the urban residential spatial structure.

National census data constitutes a comprehensive survey of individual residents and household characteristics, grounded in their respective places of residence. This data serves as a crucial resource for elucidating the social spatial structure of urban environments. Investigations that employ census data at the neighborhood level—encompassing streets and townships—to interpret the statistical significance of residential space are commonly referred to as "social area" research. Alternatively, some scholars designate this field of study as "urban social space research." Regardless of the terminology employed, such research fundamentally aims to illuminate the urban residential spatial structure.

Significant advancements have been observed within the academic realm regarding urban social space studies. Researchers have employed census data to perform empirical analyses in major metropolitan areas, including first-tier cities such as Beijing, Shanghai, and Guangzhou (Feng and Zhou 2003a; Gu et al. 2005; Wu and Li 2005; Feng et al. 2007; Tang et al. 2016), as well as other prominent urban centers like Nanjing, Hefei, and Changchun (Xu et al. 2009; Huang and Huang 2013; Li and Zhang 2015). Additionally, studies have focused on cities with significant minority populations, such as Urumqi (Zhang et al. 2012). These investigations have identified key determinants influencing urban social space in China, including occupational categories, income levels, educational attainment, and housing conditions. Research conducted in Hong Kong and Taiwan has similarly indicated that occupational differentiation and income disparity within the labor market are principal factors shaping social space (Lo 1975, 1986, 2005; Hsu and Pannell 1982). Beyond traditional analyses of social districts, a considerable body of literature has utilized housing as a focal point to examine residential spatial differentiation, as evidenced by factors such as housing tenure, costs, choices, and migration patterns (Lim and Lee 1993; Li and Siu 2001; Wu 2002; Feng and Zhou 2004; Li and Wu 2008; Yang et al. 2015). Recent investigations have also highlighted the impact of school district gentrification on residential space formation in China (Wu et al. 2014), with increasing attention being paid to the spatial differentiation of housing and middle-class residential space (Li et al. 2010; He et al. 2010; Zhou et al. 2016a, 2016b). The study of urban social space in international contexts commenced in the 1940s and 1950s, notably through the work of Shevky and Bell on American cities (Shevky and Williams 1949; Shevky and Bell 1955). Research on urban social space in foreign contexts originated in the 1940s and 1950s, notably through the work of Shevky and Bell, who conducted studies on American cities (Shevky and Williams 1949; Shevky and Bell 1955). They identified socioeconomic status, family structure, and race as the three fundamental factors influencing urban social areas, which also serve as primary indicators for assessing the spatial differentiation of urban residential environments. However, their descriptive methodology faced criticism for its inadequacy in elucidating the specific spatial distribution patterns associated with these sociological variables (Mark and Ray 2011). The advent of computer technology subsequently facilitated the incorporation of factor ecological analysis into the study of social space. Research employing this analytical method has revealed that the spatial configurations of social districts, shaped by the aforementioned variables, exhibit distinct patterns. Specifically, socioeconomic status tends to create fan-shaped social districts, family status is often organized in concentric circles, and ethnic influences typically manifest as dispersed clusters, resulting in a polycentric spatial arrangement. The interplay of these three factors contributes to the overarching spatial structure model of urban society (Xu et al. 1989). Since the 1960s, the evolution of globalization and advancements in information technology have diversified research on urban social space structures in Western contexts. For example, research has been conducted from a global perspective on the social spatial structure (Schnell and Benjamini 2005), on residential spatial differentiation from the perspective of residential segregation of immigrants (Dupont 2004; Icel and Sharp

2013), and on the characteristics and evolution of urban social spatial structure from the perspective of behavioral geography (Pais 2017).

The decade spanning from 2000 to 2010 witnessed significant urbanization in China. By 2011, the urban population constituted 51% of the total population, marking the completion of China's transformation from a "rural society" to an "urban society" (Xu and Shi 2016; He and Qian 2017). This transformation was accompanied by substantial alterations in urban infrastructure, land utilization, and other built environments, alongside a reconfiguration of the social landscape, including demographic and economic shifts within urban populations. Concurrently, China's Gini coefficient reached 0.415 in 2011, a figure that, according to international benchmarks, indicates a considerable income disparity and suggests an escalation in social differentiation within the country (Zhou et al. 2015). During this period, Beijing also experienced rapid urbanization. By the end of 2010, the city's resident population had reached 19.619 million, with urban residents numbering 16.864 million, thereby comprising 86% of the total resident population. This represented an increase of 8.5 percentage points from 77.5% in 2000 (Beijing Municipal Bureau of Statistics 2011). Notably, the urbanization of land outpaced that of the population, as the proportion of built-up area in Beijing rose from 6.4% in 1991 to 21.4% in 2011, reflecting an increase of approximately 15 percentage points (Cong and Liu 2015). This section aims to investigate the changes that occurred in the structure of urban residential space during this period and to identify the factors that have driven these transformations. Additionally, the ecological-economic gradient theory of population migration posits that the interplay between an individual's ecological price and the ecological conditions of a region influences population movement, thereby affecting spatial distribution (Cao et al. 2003). An individual's ecological price is influenced by factors such as socioeconomic status and age, while the ecological conditions of a region are shaped by various elements, including economic development. Furthermore, the family unit, which exists between the individual and the region, plays a crucial role in determining population distribution, as factors such as family size and life stage significantly influence residential choices. These theoretical considerations warrant further empirical investigation.

This section focuses on Beijing, the capital of China, utilizing data from the sixth national census conducted in 2010. It selects indicators that reflect three levels of analysis: individual, family, and regional. Employing factor ecological analysis, the study aims to identify the types and characteristics of residential spaces in Beijing as of 2010. A comparative analysis with data from 2000 (Feng and Zhou 2003a; Feng 2004) is conducted to examine the spatial restructuring of residential space in Beijing over the preceding decade. Additionally, a differentiation index is calculated to elucidate the evolution of social spatial differentiation based on residential patterns.

Social districts are employed to represent the urban spatial structure at a macro level in a statistical context, while the differentiation index serves to illustrate the spatial trends of social factors among various residential populations at a micro level. The integration of these two analytical approaches provides insights into the developmental trends of urban residential spaces and facilitates an examination of the mechanisms underlying the structure and differentiation of residential spaces.

This analysis aims to uncover the patterns of urban residential space differentiation during a period of rapid urbanization. It is noteworthy that Beijing exemplifies one of the most complex socio-economic developments and possesses a substantial urban population in China, making it a representative case among the country's mega-cities. The empirical investigation of residential spaces in Beijing seeks to illuminate the overarching characteristics of urban residential space development across China.

5.2.2 Study Area, Data and Methodology

5.2.2.1 Study Area and Data

The empirical research area in this section is based on the area surrounded by the 6th ring road in Beijing. On the premise of ensuring the integrity of the administrative boundary, 12 districts (Dongcheng District, Xicheng District, Chaoyang District, Fengtai District, Shijingshan District, Haidian District, Mentougou District, Fangshan District, Tongzhou District, Shunyi District, Changping District, and Daxing District) were selected from the 16 districts of Beijing. In other words, the four districts of Huairou, Yanqing, Miyun and Pinggu were not included in our study. The area covered by the above 12 districts has been regarded as the scope of the Beijing metropolitan area (Sun 1992; Feng and Zhou 2004). Although Beijing's rapid development has caused the scope of the metropolitan area to exceed the city's administrative boundaries in some directions, there is no doubt that the scope of the above 12 districts is still the main scope of the Beijing metropolitan area. It should be emphasized that the empirical analysis and corresponding maps of Beijing's internal urban structure in this chapter are all based on this scope. For the convenience of data processing, the data of a small number of townships and street units were combined and processed, and the combined units are referred to as "neighborhood units". Finally, 246 neighborhood units were obtained (Fig. 5.1). The main source of research data is the data from the 6th National Population Census of Beijing in 2010. Previous related research mainly considered personal attribute variables that reflect socioeconomic status, as well as family attribute variables that reflect family structure and housing conditions. However, regional demographic changes are an important factor affecting the evolution of residential space. For example, the regional population density can reflect the comfort of living, which in turn affects the housing choice behavior of individuals or families; the regional population growth rate can reflect the strength of population attraction, which indirectly reflects important factors affecting residential location choice, such as employment opportunities and housing conditions in the region. This section calculates the population density of each neighborhood and the average annual population growth rate over a 10-year period, in order to more comprehensively reflect the evolution of the structure and differentiation of residential space.

In the process of variable selection, various combinations of variables were tested to ascertain their effectiveness. Initially, it was observed that when percentage data

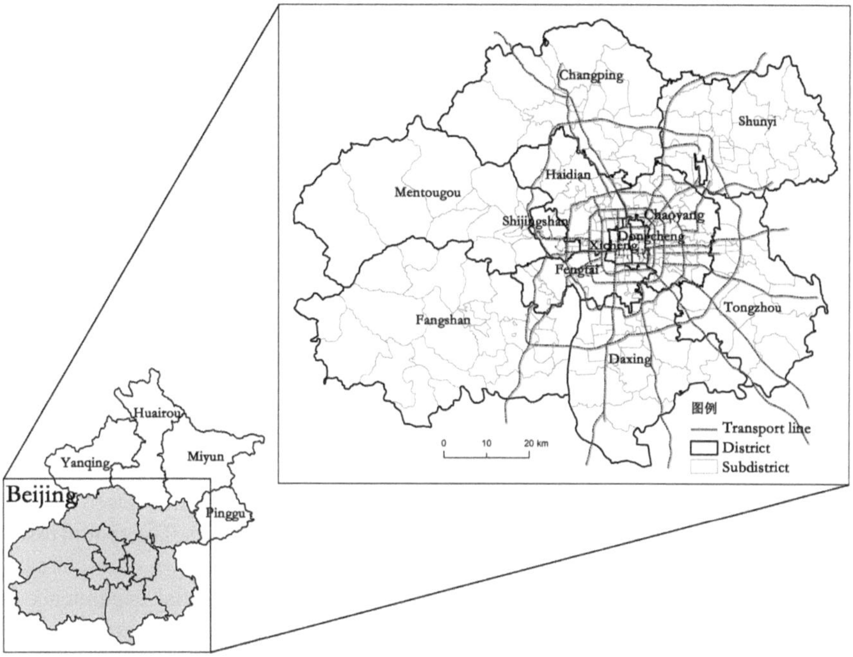

Fig. 5.1 Schematic diagram of the study area

is utilized for analysis, the cumulative variance contribution rate derived from factor analysis is notably low, approximately 50%, indicating that the primary factors extracted convey limited variability. While the use of percentage data may be advantageous when all variables are homogenous, this study encompasses a diverse array of variable types, including demographic factors, housing characteristics, population density, and population growth rates. Consequently, the application of absolute value data is deemed more appropriate. Furthermore, the objective of this section is to emphasize the extent of spatial differentiation, and absolute value data facilitates a more direct examination of the disparities in various indicators across different neighborhoods (Li and Wu 2006). Therefore, the decision was made to utilize absolute value data for the analysis.

The classification results of residential space utilizing various combinations of variables exhibit notable similarities. The primary distinction lies in the fact that certain variable combinations categorize the neighborhoods of the old city and Mentougou District into a single group. This phenomenon can be attributed to the relatively small total population of these two regions. Despite the old city exhibiting a high population density, its neighborhoods are compact, resulting in a limited overall population. Mentougou District, designated as an ecological conservation and development zone within Beijing, imposes restrictions on population growth, leading to a similarly modest total population. Consequently, many indicators in these areas yield relatively low values, prompting their amalgamation into one category during the clustering analysis. Additionally, the last principal component extracted, which pertains to population density, contributes minimally to variance, thereby exerting a negligible influence on the final delineation of residential spaces. To elucidate the distinctions between the old city and Mentougou District, it is advisable to incorporate housing-related variables, such as per capita housing floor area and the average number of rooms per household, alongside family-related variables, including family size and generational types. Ultimately, a total of 88 variables reflecting individual, familial, and regional characteristics were selected, which, in conjunction with the 246 neighborhood units, culminated in the creation of an original data matrix of dimensions 88×246.

5.2.2.2 Methodology

In addition to the traditional factor analysis and cluster analysis methods used in social space analysis, spatial autocorrelation analysis can also be introduced to assist in analysing the spatial distribution characteristics of factor scores. The spatial distribution map of factor scores can only reflect the spatial distribution characteristics of the attribute values of each factor, but cannot reflect the spatial dependence or spatial correlation characteristics of spatial things. Therefore, further spatial autocorrelation analysis of factor scores can reflect the spatial correlation characteristics and agglomeration status of each main factor, and identify the "cold spots" and "hot spots" of factor scores in space.

Spatial autocorrelation analysis can be divided into global spatial autocorrelation and local spatial autocorrelation, which are commonly measured using the *Moran's I* value. The calculation formula is shown in Table 5.1 (Wang 2006). Given a significance level, *Moran's I* > 0 indicates positive spatial autocorrelation, with observations in spatial units tending to cluster together; *Moran's I* < 0 indicates negative spatial autocorrelation, with observations in spatial units distributed randomly; *Moran's I* = 0 indicates no spatial autocorrelation, with observations in spatial units distributed randomly.

Firstly, factor analysis was performed on the original data matrix. By observing the scree plot and eigenvalues, five main factors were extracted, with a cumulative variance contribution of 73.97%. Due to the vague factor structure, a direct oblique rotation method was used to obtain the rotated factor loading matrix, based on which the main variables reflected by each factor were determined. The scores of each factor for the research units were then calculated and analysed for spatial autocorrelation to assess the spatial clustering of each main factor, thereby identifying the spatial cold and hot spots of each main factor.

Secondly, cluster analysis was carried out based on the factor scores of each research unit. A hierarchical clustering method was used, employing the sum of squared deviations to calculate the squared Euclidean distance between clusters, dividing the residential space into six categories, and then calculating the mean and squared mean values of each category's scores on the main factors to determine the characteristics of each type of residential space. By comparing the characteristics of

Table 5.1 Improving factor analysis with spatial autocorrelation analysis

Traditional method	Improvement	Model	Additional notes	Implications for improvement
Factor score	Global spatial autocorrelation of factor scores	$I = \frac{n}{\sum_{i=1}^{n}\sum_{j=1}^{n} W_{ij}} \times \frac{\sum_{i=1}^{n}\sum_{j=1}^{n} W_{ij}(x_i-\overline{x})(x_j-\overline{x})}{\sum_{i=1}^{n}(x_i-\overline{x})^2}$ (1)	I indicates the global *Moran's I* index, n is the number of research objects, x_i and x_j are the attribute values of the research areas i and j, respectively, and $\bar{X}$ is the mean value of all attribute values in the sample; W_{ij} is the weight matrix that measures the relationship between spatial things, which is determined by spatial topology and distance	The spatial autocorrelation auxiliary analysis can reveal the spatial correlation characteristics of each principal factor, that is, whether there is a trend of convergent aggregation. Among them, global spatial autocorrelation can test whether factor scores are spatially clustered, describing their spatial characteristics throughout the study area
	Local spatial autocorrelation of factor scores	$I_i = Z_i \sum_j W_{ij} Z_{ij}$(2)	I_i represents the partial *Moran's I* index, Z_i is the normalized form of x_i, with a mean of 0 and a variance of 1, and W_{ij} is the distance weighting matrix	Local autocorrelation Specific distribution of factor scores in space Hot spots (clusters of high values) and cold spots (clusters of low values)

urban residential spaces in 2000 and 2010, the evolution pattern of urban residential space structure over the ten years was analysed.

5.2.3 *Residential Spatial Structure and Regional Pattern of Residential Space*

5.2.3.1 Principal Factor Characteristics of Urban Residential Spatial Structure

In the context of factor structure, a principal component analysis was conducted on the original data matrix, resulting in the extraction of five principal factors. The loadings associated with each factor, along with the corresponding variable information, are presented in Tables 5.2 and 5.3. According to established geographical methodologies, indicators with loadings exceeding 0.4 are deemed significant (Li and Wu 2008). This threshold has been applied in the current analysis, with factors exhibiting loadings above 0.4 highlighted in bold in Table 5.2. The five principal factors are categorized as follows, based on the variable information represented by each factor: local white-collar workers (factor 1), migrant blue-collar workers (factor 2), agricultural population (factor 3), intellectuals (factor 4), and population density (factor 5).

The main variable information reflected in each principal factor is as follows.

Factor 1: Local white-collar workers. This factor, which contributes 39.02% of the variance, mainly reflects the variables of higher education, working in the tertiary industry such as finance, real estate, public administration and social organizations, being in charge of state organs, party and mass organizations, enterprises and public institutions, Professionals and technicians, Office personnel and related personnel, better housing conditions, medium household structure, and permanent registered population.

Factor 2: Migrant blue-collar workers. This factor explains 21.64% of the variance and reflects information on variables such as lower educational attainment, secondary industries such as Manufacturing and Construction, the circulation department, people working in commerce and Services, operators and related personnel in production and transport equipment, smaller family structures, poorer housing conditions, faster population growth, lower dependency ratios, young and middle-aged people, and migrants.

Factor 3: Agricultural population. This factor has a variance contribution of 6.69%, and mainly reflects information on variables such as the illiterate population, those engaged in agriculture, larger family structures, self-built housing, and larger housing areas.

Factor 4: Intellectuals. This factor has a variance contribution of 14.76%, and mainly reflects information on variables such as a high level of education, those engaged in education, scientific research and technical services, and Geological

Table 5.2 Principal factor load matrix of residential spatial structure in Beijing in 2010

Characteristics			Variable	Factor loadings				
				1	2	3	4	5
Personal property	General	Gender and age	Sex ratio (female = 100)	**–0.561**	**0.655**	0.015	0.185	–0.190
			0–14 years old	**0.757**	0.351	0.144	0.067	–0.003
			15–64 years old	**0.635**	**0.407**	0.075	0.258	0.060
			65 years old and above	**0.752**	–0.154	0.090	0.257	0.382
		Household and nationality	Permanent registered population	**0.796**	–0.017	0.171	0.282	0.196
			People who were originally from the neighborhood but are now abroad and do not have household registration	–0.017	–0.239	–0.010	**0.757**	0.313
			Migrants	**0.426**	**0.678**	–0.015	0.157	–0.050
			Mongol population	**0.568**	0.178	0.030	**0.514**	–0.094
			Hui population	**0.490**	–0.084	0.031	0.142	0.150
			Uyghur population	0.088	0.009	0.087	**0.555**	–0.104
			Miao population	0.271	0.237	0.081	**0.602**	–0.115
			Zhuang population	0.326	–0.033	0.066	**0.768**	–0.060
			The Korean nationality	0.397	–0.046	–0.056	0.136	–0.266
			Manchu population	**0.675**	0.192	0.003	0.314	0.081
			Tujia population	0.248	0.306	0.074	**0.539**	–0.076
	Socioeconomic characteristics	Education level	Illiterate population	**0.489**	0.248	**0.577**	–0.144	0.296
			Primary school	**0.573**	**0.515**	0.314	–0.020	0.154
			Junior secondary school	0.291	**0.778**	0.174	0.017	0.104

(continued)

Table 5.2 (continued)

Characteristics			Variable	Factor loadings				
				1	2	3	4	5
			Senior secondary school	**0.747**	0.311	0.006	0.060	0.202
			College	**0.864**	0.100	–0.053	0.153	–0.002
			Undergraduate coursework	**0.684**	–0.122	–0.031	**0.546**	–0.036
			Graduate coursework	0.320	–0.171	0.003	**0.820**	0.008
				1	2	3	1	5
Personal property	Socioeconomic characteristics	Employed by sector	Farming, forestry, animal husbandry, and fisheries	–0.190	–0.022	**0.852**	0.063	0.003
			Mining and quarrying	0.022	–0.044	–0.037	0.124	–0.126
			Manufacturing	0.189	**0.599**	0.240	–0.008	–0.101
			Electric power, gas, and water	**0.813**	0.007	–0.082	–0.123	0.076
			construction	0.320	**0.721**	–0.016	–0.065	–0.097
			Transportation, storage and postal	0.395	**0.668**	0.112	–0.184	0.164
			Information transmission, computer services and software	**0.663**	0.028	–0.036	0.345	–0.223
			Wholesale and retail trade	**0.451**	**0.637**	–0.043	0.056	0.113
			Accommodation and food services	**0.513**	0.274	–0.094	0.305	0.252
			Finance	**0.846**	–0.089	–0.117	0.141	0.147
			Real estate	**0.787**	0.228	–0.102	0.094	–0.018
			Leasing and business services	**0.757**	0.148	–0.125	0.200	0.063

(continued)

Table 5.2 (continued)

Characteristics			Variable	Factor loadings				
				1	2	3	4	5
			Scientific research, technical services and geological prospecting	**0.566**	–0.174	–0.002	**0.464**	0.076
			Water conservancy, environment and public facilities management	**0.556**	**0.431**	–0.011	–0.042	0.003
			Resident services and other services	**0.486**	**0.586**	–0.095	0.064	0.074
			Education	**0.555**	–0.017	0.048	**0.601**	–0.111
			Health care, sports, and social welfare	**0.842**	0.001	–0.014	0.046	0.166
			Culture, sports and entertainment	**0.843**	0.020	–0.157	0.006	0.078
			Public administration and social organizations	**0.837**	0.060	–0.022	–0.139	0.105
			International organizations	**0.531**	–0.148	–0.270	–0.050	0.024
		Occupational types	Managers (government agencies, political parties, businesses)	**0.564**	0.179	–0.063	0.145	–0.127
			Professionals and technicians	**0.840**	–0.028	–0.042	0.309	–0.080
			Office personnel and related personnel	**0.855**	0.044	–0.064	0.184	0.051
			Business and Services personnel	**0.473**	**0.626**	–0.045	0.092	0.123

(continued)

Table 5.2 (continued)

Characteristics			Variable	Factor loadings				
				1	2	3	4	5
			Agricultural, forestry, animal husbandry, fishery, and water conservancy production personnel	–0.190	–0.015	**0.856**	0.054	–0.003
			Production and transportation equipment operators and related personnel	0.071	**0.819**	0.227	–0.071	0.003
		Reason for being unemployed	Not working after graduation	**0.758**	0.201	0.173	0.016	0.116
			Lost job due to company reasons	**0.749**	0.125	–0.018	0.010	0.131
			Lost job due to personal reasons	**0.634**	0.275	0.016	0.139	–0.180
			Contracted land was requisitioned	0.095	0.343	0.166	–0.082	–0.136
			Retired	**0.466**	–0.092	–0.090	–0.031	**0.419**
			Caring for family	**0.487**	**0.432**	0.124	–0.136	–0.125
				1	2	3	1	5
	Family structure		Number of one-person households	**0.525**	**0.573**	–0.109	0.093	0.110
			Number of two-person households	**0.723**	**0.421**	–0.015	0.042	0.085
			Number of three-person households	**0.863**	0.173	0.040	0.065	0.106

(continued)

Table 5.2 (continued)

			1	2	3	1	5
		Number of four-person households	**0.762**	0.272	0.273	0.062	0.161
		Number of five-person households	**0.719**	0.272	0.388	0.086	0.078
		Number of six-person and above households	0.399	0.254	**0.507**	0.112	0.391
		One-generation households	**0.612**	**0.541**	–0.063	0.058	0.066
		Two-generation households	**0.844**	0.190	0.075	0.068	0.147
		Three-generation households	**0.762**	0.121	0.318	0.141	0.243
		Four-generation and above households	0.174	0.300	**0.797**	0.033	0.089
		Overall dependency ratio	0.192	**–0.689**	0.066	–0.325	0.217
		Child dependency ratio	0.245	–0.221	0.261	**–0.499**	–0.334
		Elderly dependency ratio	0.101	**–0.658**	–0.043	–0.136	0.386
		Average household size of family households	–0.046	**–0.432**	**0.719**	–0.037	–0.025
		Percentage of family households in total households	0.296	–0.248	0.050	**–0.810**	0.063
Family property	Housing property	Renting low-rent housing	0.109	**0.596**	–0.033	0.028	0.237
		Renting other housing	0.339	**0.730**	–0.121	0.055	0.230
		Building own house	–0.255	0.313	**0.784**	–0.094	0.076
		Purchasing marketable housing	**0.849**	0.126	–0.116	–0.081	–0.410
		Purchasing second-hand housing	**0.848**	–0.046	–0.091	0.011	–0.271
		Purchasing affordable housing	**0.659**	0.054	0.008	0.004	–0.180
		Purchasing former public housing	**0.484**	–0.367	–0.116	0.345	**0.458**
		Household monthly rental cost less than 100 yuan	0.043	–0.084	–0.119	–0.008	**0.604**

(continued)

Table 5.2 (continued)

			1	2	3	1	5
		Household monthly rental cost 100–200 yuan	0.125	**0.622**	0.134	–0.037	0.237
		Household monthly rental cost 200–500 yuan	0.157	**0.846**	–0.133	–0.010	0.023
		Household monthly rental cost 500–1000 yuan	0.200	**0.546**	–0.095	0.186	0.269
		Household monthly rental cost 1000–1500 yuan	**0.603**	0.334	–0.038	–0.018	0.044
		Household monthly rental cost 1500–2000 yuan	**0.728**	0.268	–0.109	–0.017	–0.002
		Household monthly rental cost 2000–3000 yuan	**0.720**	0.029	–0.105	0.238	0.109
		Household monthly rental cost 3,000 yuan or more	**0.458**	–0.082	–0.132	0.363	0.104
		Average housing floor area per capita/m^2	0.188	–0.284	**0.632**	0.022	**–0.451**
		Average number of rooms per household / room/household	–0.217	–0.181	**0.726**	0.023	–0.150
Regional property	Population density	Population densities (persons per sq km^2)	0.359	–0.283	–0.311	0.251	**0.476**
	Population growth	Average annual population growth over the past 10 years	0.396	0.537	–0.243	0.049	–0.371

Table 5.3 Information on the main variables reflected in each principal factor

Variable			Factor 1	Factor 2	Factor 3	Factor 1	Factor 5
Personal property	General	Gender and age	Sex ratio (negative), Permanent registered population of all ages,	Gender ratio, 15–64 years	–	–	–
		Household registration and nationality	Permanent registered population, migrant population, Mongolian nationality, Hui nationality, Manchu nationality	Migrants	–	Originally from the neighborhood, now abroad, no household registration, Mongol population, Uyghur population, Hmong population, Zhuang population, Tujia population	–
	Socioeconomic status	Education level	Illiterate population, Primary school, Senior secondary school, College, Undergraduate coursework	Primary school, Junior secondary school	illiterate population	Undergraduate coursework, Graduate coursework	–

(continued)

Table 5.3 (continued)

Variable			Factor 1	Factor 2	Factor 3	Factor 1	Factor 5
		Industry type	Departments that serve production and life, industries that serve the improvement of scientific and cultural standards and the quality of the population, and industries that serve social needs	Manufacturing, Construction, Transportation, Storage and Postal Services, Wholesale and Retail Trade, Water Conservancy, Environment and Public Facilities Management, Resident Services and Other Services	agriculture	Industry serving the improvement of scientific and cultural levels and the quality of the population	–
		Occupation type	Managers of state organs, party and mass organizations, enterprises and institutions; professionals and technicians; office personnel and related personnel; and personnel in commerce and Services	Commercial and services personnel, production, transportation equipment operators and related personnel	Agricultural, forestry, animal husbandry, fishery, and water conservancy production personnel	–	–

(continued)

Table 5.3 (continued)

Variable			Factor 1	Factor 2	Factor 3	Factor 1	Factor 5
		Reason for being unemployed	Not working after graduation, lost job due to employer, lost job due to personal reasons, retired, taking care of family matters	Caring for the household	–	–	Retired
Family property	Family structure		Number of households with one to five people, one to three generations	Number of one-person households, number of two-person households, one-generation households, elderly dependency ratio (negative), total dependency ratio (negative), average household size (negative)	Households of 6 or more, households of 4 or more generations, average household size	Child dependency ratio (negative), percentage of one-person households in total households (negative)	–

(continued)

Table 5.3 (continued)

Variable		Factor 1	Factor 2	Factor 3	Factor 1	Factor 5
	Housing property	Purchase various types of housing, monthly rents of 1,000 to 3,000 yuan	Rental housing, monthly rent 100–1,000 yuan	Self-built housing, per capita housing construction area, average number of rooms per household	–	Purchasing commercial housing (negative), purchasing original public housing, monthly rent below 100 yuan, per capita housing construction area (negative)
Regional property	Population density	–	–	–	–	population density
	Population growth	–	Annual population growth rate over the past 10 years	–	–	–

prospecting and census work, as well as those of minority nationality and a low child support ratio.

Factor 5: population density. This factor, which explains 5.45% of the variance, reflects a high population density, the purchase of former publicly owned housing, very low housing rents, and a small living area.

Based on the spatial distribution of the factor scores, the global spatial autocorrelation *Moran's I* value was calculated for each factor (Table 5.4). At the significance level of 0.001, the *Moran's I* index of each factor is significantly positive, indicating that there is a tendency for factor scores to cluster. Among them, the clustering trend of factor 3 (agricultural population) is the most prominent, and the clustering trends of factor 4 (intellectuals) and factor 5 (population density) are also more obvious.

The score space distribution map and spatial autocorrelation distribution map of each factor are shown in Fig. 5.2.

Factor 1: Local white-collar workers. The neighborhoods with higher scores are mainly large residential areas, including Huilongguan and Dongxiaokou in Changping District (which was divided into four in 2012 and now includes the four neighborhoods of Tiantongyuan North, Tiantongyuan South, Huoying and Dongxiaokou), as well as streets in the suburbs with a large population, including Lugouqiao, Xincun, Huaxiang, Dahongmen and Nanyuanxiang in Fengtai District. In addition, the streets where the government seats of the surrounding districts and counties are located also have high scores.

As can be seen from the spatial autocorrelation map, the hotspots with high factor 1 scores are mainly concentrated in the large residential areas in the north immediately adjacent to the Third Ring Road, as well as in the southwestern area with a large population immediately adjacent to the Second Ring Road. The coldspots with low scores are mainly distributed in the peripheral districts and counties, which are farther away from the city centre.

Factor 2: Migrant Blue-Collar Workers. The spatial distribution of factor scores reveals a distinct ring structure. Areas exhibiting higher scores are predominantly located between the Fourth Ring Road to the north and the Sixth Ring Road to the south, as well as between the Third Ring Road to the south and the Sixth Ring Road to the north. Notable neighborhoods within this concentration include Xibeiwang in Haidian District, Beiqijia (Building Materials City) in Changping District, Cuigezhuang and Shibalidian (which host home improvement building materials and logistics warehouses) in Chaoyang District, Dahongmen (a clothing market), Nanyuan Township, Xincun, Huaxiang (a flower market), and Lugouqiao in Fengtai District, along with Jiugong (Lin Yizhuang), Xihongmen, and Huangcun in Daxing District. These neighborhoods serve as focal points, with a trend of decreasing factor

Table 5.4 *Moran's I* value for each principal factor

	1	2	3	4	5
Moran's I	0.378	0.296	0.608	0.575	0.484

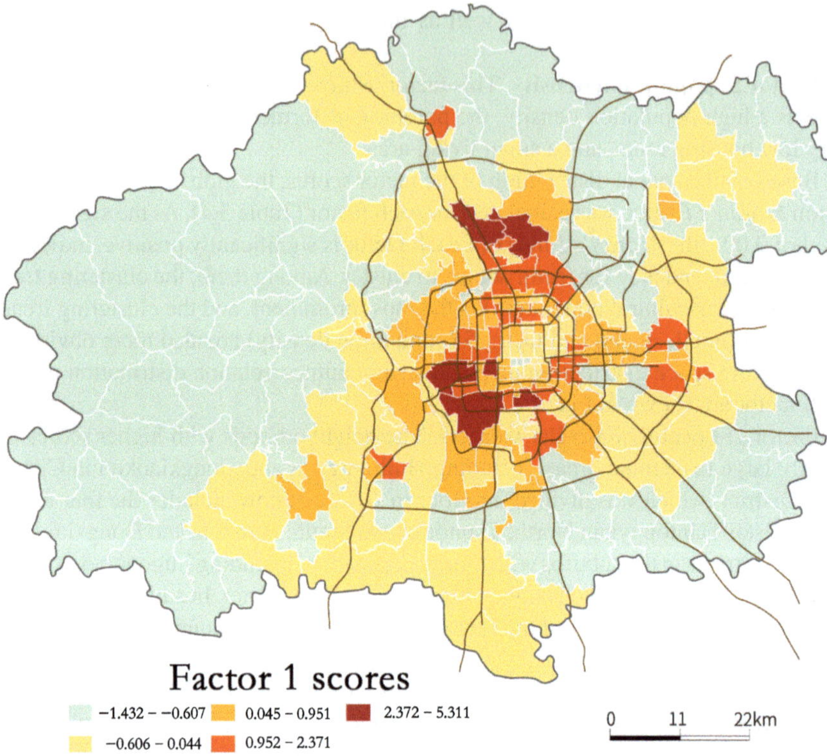

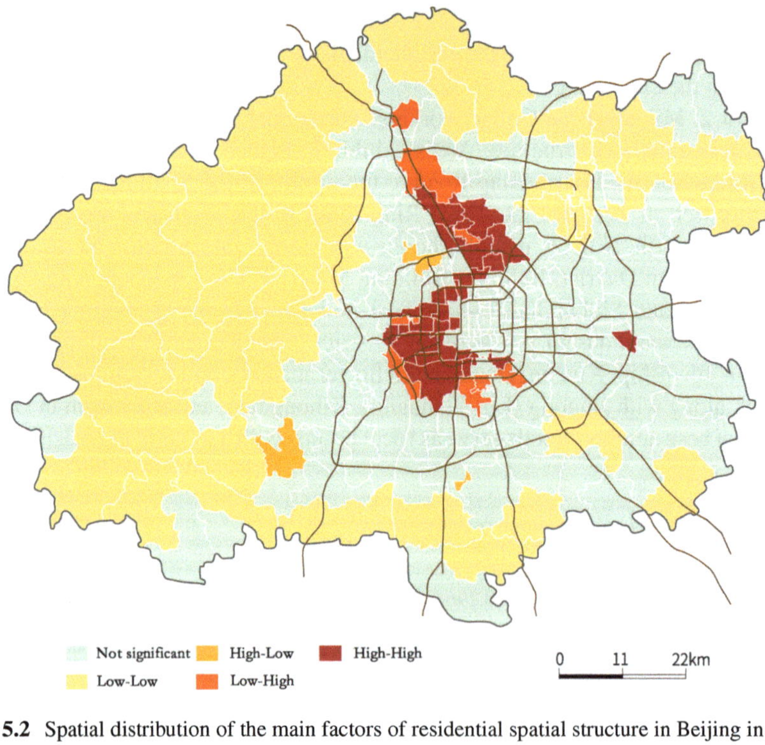

Fig. 5.2 Spatial distribution of the main factors of residential spatial structure in Beijing in 2010. *Note* Both the distribution pattern of the scores of each factor and the spatial autocorrelation of each factor are given

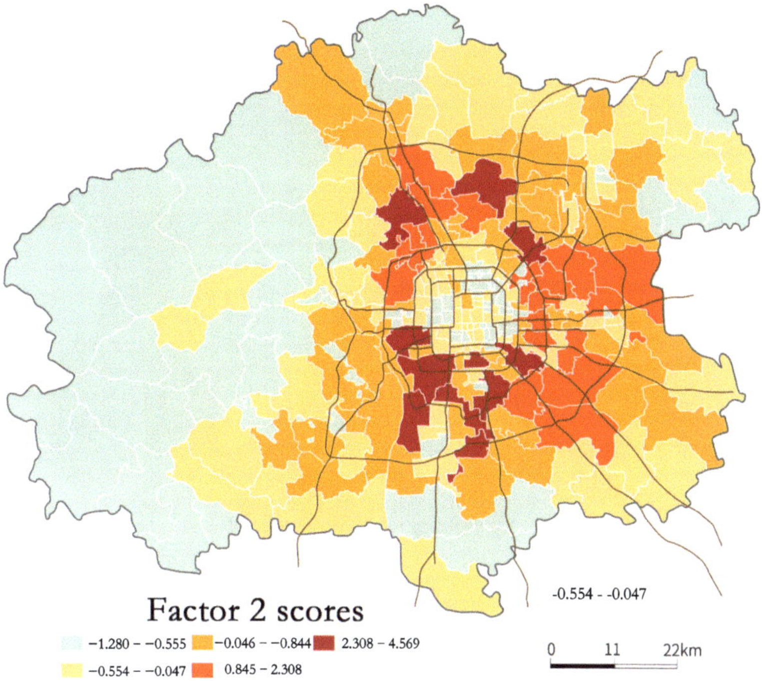

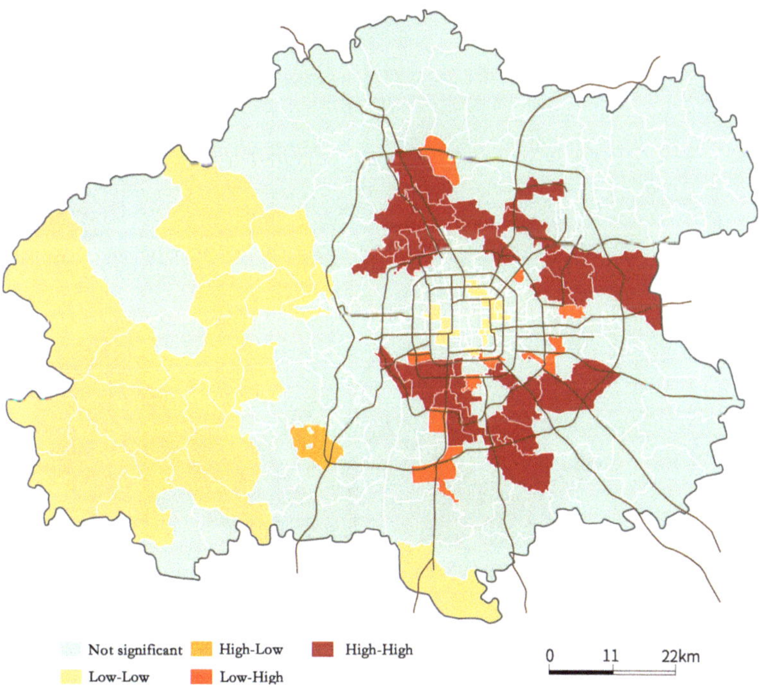

Fig. 5.2 (continued)

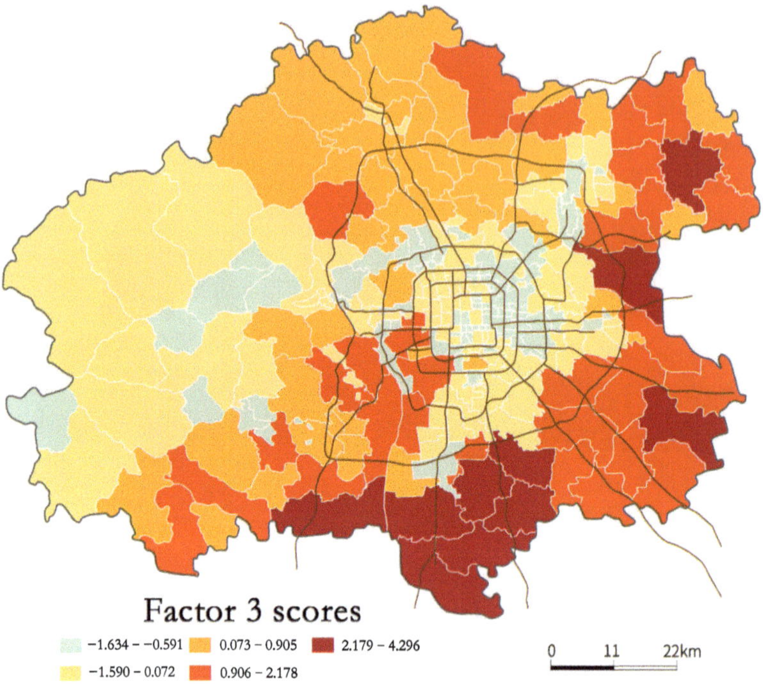

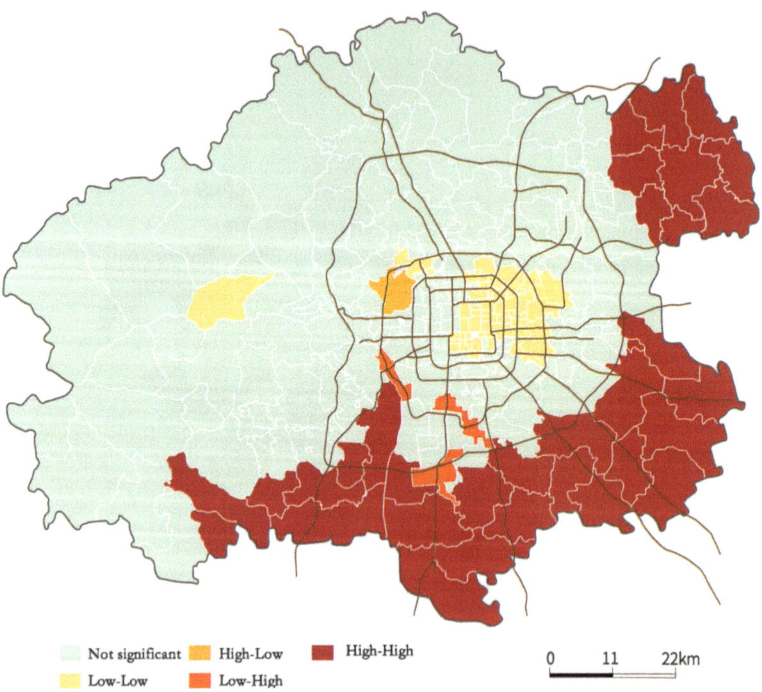

Fig. 5.2 (continued)

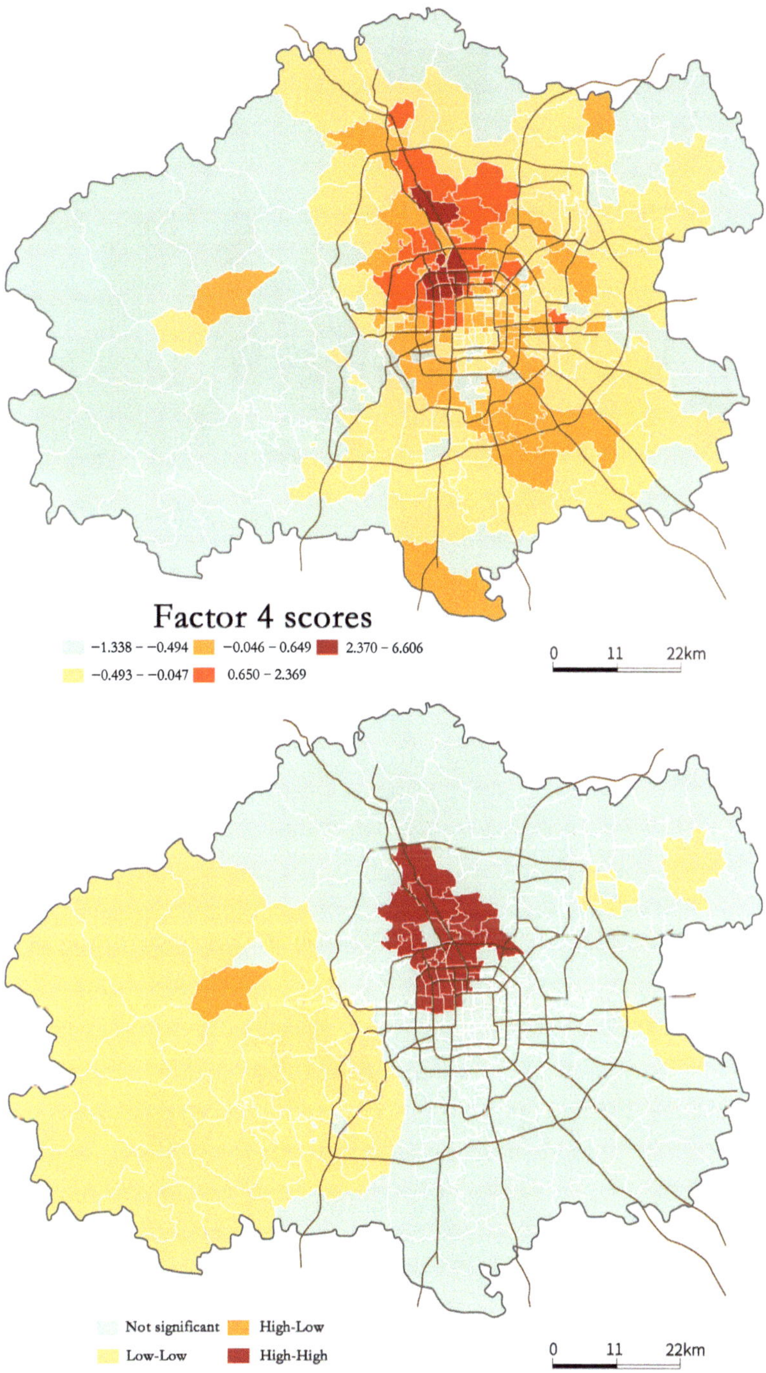

Fig. 5.2 (continued)

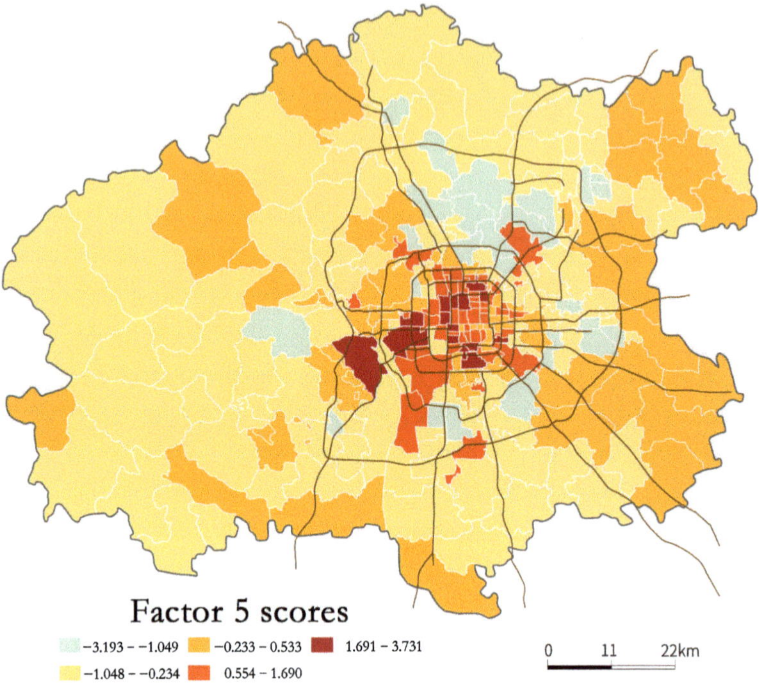

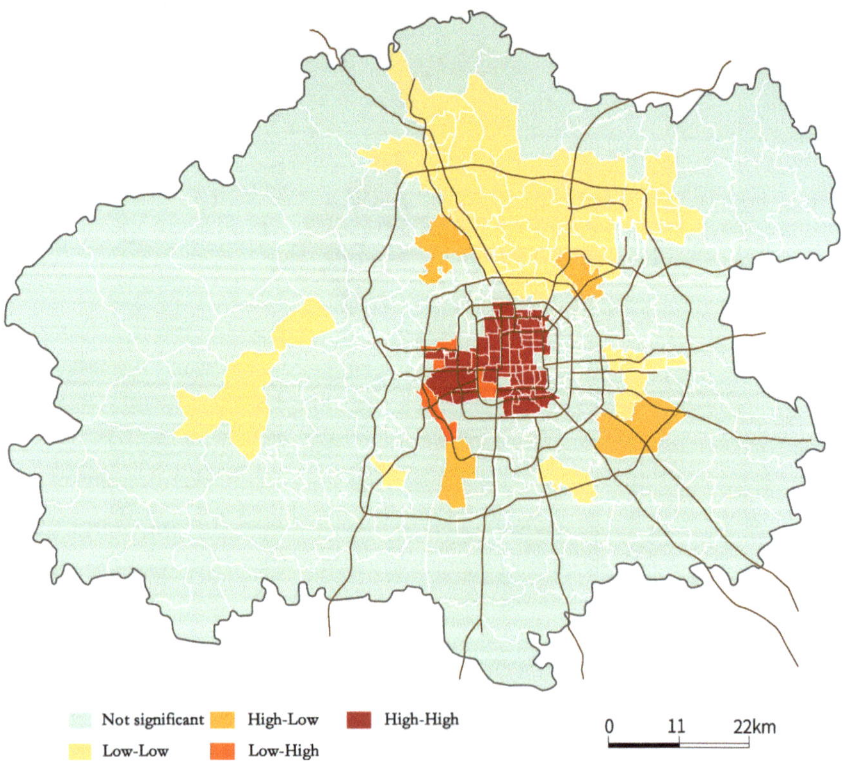

Fig. 5.2 (continued)

scores as one moves inward toward the historical city center and outward toward the suburban areas.

The spatial autocorrelation map shows that the hotspot neighborhoods with high factor 2 scores are mainly distributed in the northern and eastern areas between the 5th and 6th rings, as well as in the southern area between the 3rd and 6th rings. The coldspot neighborhoods with low scores are mainly distributed in Mentougou and Fangshan districts, as well as some neighborhoods near the 2nd and 3rd rings.

Factor 3 pertains to the agricultural population. The spatial distribution of factor scores exhibits a declining trend from the periphery towards the center. Neighborhoods characterized by elevated scores are predominantly located in the outer districts, particularly in the eastern and southern regions. Notable areas include Yangzhen in Shunyi District, Songzhuang Town and Huo County Town in Tongzhou District, as well as Qingyundian Town, Weishan Zhuang Town, Anding Town, Panggezhuang Town, Lixian Town, and Yufa Town in Daxing District.

The spatial autocorrelation map further illustrates that neighborhoods identified as hotspots, which exhibit high scores for factor 3, are primarily situated in the peripheral counties and districts. This includes the eastern section of Shunyi District and the southern regions of Tongzhou District, Daxing District, and Fangshan District. Conversely, neighborhoods classified as coldspots, characterized by low scores, are mainly found in the historical city center and the northeastern area adjacent to it.

Factor 4: Intellectuals. Neighborhoods exhibiting elevated scores predominantly reside within the Haidian District, encompassing areas such as Xueyuan Road, Tsinghua University, Haidian, Zhongguancun, Zizhuyuan, Beixiaoguan, Huayuanlu, and Beitaipingzhuang. Additionally, Huilongguan in the Changping District also demonstrates a high score.

The spatial autocorrelation analysis indicates that the regions identified as hotspots with significant scores for Factor 4 are primarily situated in the Haidian District, while coldspots characterized by lower scores are largely found in the Mentougou and Fangshan Districts.

Factor 5: Population Density. Neighborhoods with high scores for this factor are chiefly located within the old city and in certain adjacent areas. Notable examples include Hepingli, Beixinqiao, and Yongdingmenwai in the Dongcheng District, as well as Shichahai, Xinjiekou, Zhanlanlu, Yuetan, Guang'anmennei, and Baizhifang in the Xicheng District. Furthermore, neighborhoods such as Xiluyuan, Lugouqiao, Dahongmen, and Nanyuanxiang in the Fengtai District, Panjiayuan in the Chaoyang District, and Wanshoulu in the Haidian District also register high scores.

The spatial autocorrelation map reveals that neighborhoods identified as hotspots for Factor 5 are predominantly located within and surrounding the old city. Conversely, coldspot neighborhoods with lower scores are primarily situated to the north of the old city, beyond the Fourth Ring Road.

5.2.3.2 Residential Spatial Structure and Residential Regional Patterns (2010)

Cluster analysis based on factor scores divides Beijing's residential space into six types, and the spatial distribution of each type is shown in Fig. 5.3. The mean value and mean square of each type of residential space for each factor score are calculated (Table 5.5) to determine the characteristics of each type of residential space.

(1) High-Density, Crowded Old City (Type 1): The mean value and mean square sum of the factor score for this type of residential space are higher than those of the other factors. It is typically characterized by high-density, crowded central city areas. It is concentrated within the Third Ring Road, and all the neighborhoods in the Dongcheng District and Xicheng District within the Second Ring Road belong to this category. Most of the neighborhoods between the Second and Third Ring Roads, and some of the neighborhoods between the Third and Fourth Ring Roads also fall into this category.

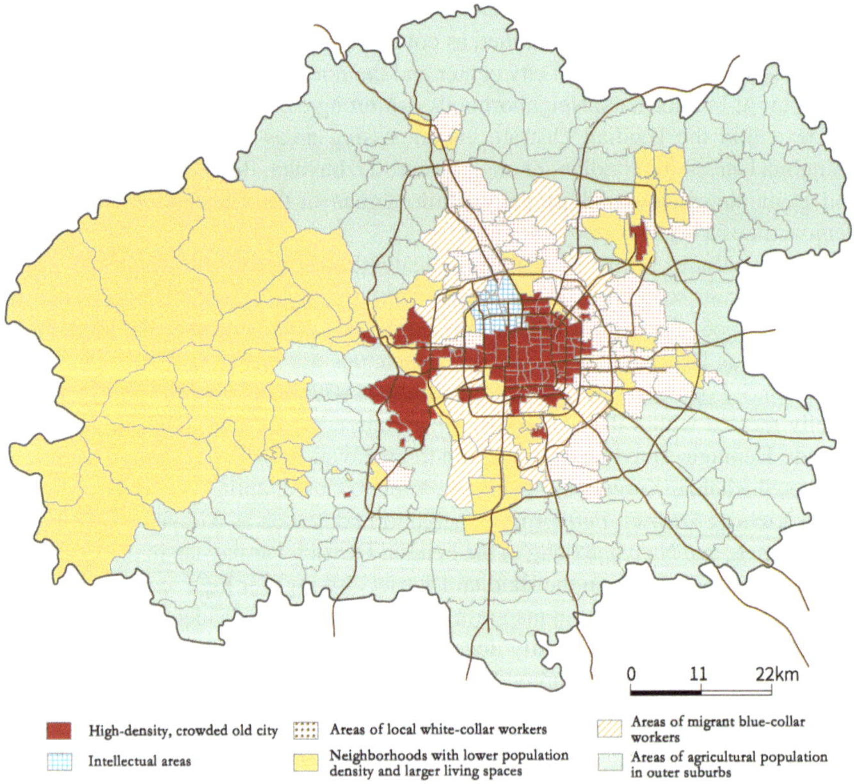

Fig. 5.3 Distribution of residential space in Beijing by type (2010)

Table 5.5 Discriminant table for characteristics of residential space in Beijing in 2010

Type	Number of neighborhoods included	Statistical value	1	2	3	4	5
Type 1	71	Mean value	0.286	−0.521	−0.557	−0.007	**1.005**
		Sum of squares Mean value	0.701	0.385	0.507	0.234	**1.580**
Type 2	59	Mean value	−0.409	−0.336	−0.528	−0.452	**−0.622**
		Sum of squares Mean value	0.493	0.317	0.375	0.381	**0.626**
Type 3	31	Mean value	**0.840**	0.891	−0.345	0.482	−0.880
		Sum of squares Mean value	**2.722**	1.200	0.451	0.992	1.491
Type 4	12	Mean value	1.181	**3.287**	0.300	0.333	0.900
		Sum of squares Mean value	4.195	**12.638**	0.938	0.460	2.952
Type 5	64	Mean value	−0.620	−0.124	**1.267**	−0.424	−0.336
		Sum of squares Mean value	0.468	0.375	**2.569**	0.262	0.252
Type 6	9	Mean value	0.367	−0.254	−0.367	**3.925**	0.366
		Sum of squares Mean value	1.253	0.129	0.264	**19.402**	0.623

(2) Neighborhoods with lower population density and larger living spaces (Type 2). The mean value of the scores for each factor for this type of residential space is negative, with factor 5 being particularly prominent. This type of social space is found in neighborhoods with lower population density and better living conditions. It is mainly distributed in the northern Mentougou District and Fangshan

District, as well as in many urban–rural fringe areas on the outskirts of Inner suburb.

(3) Areas of local white-collar workers (Type 3). The mean value and mean square sum of factor 1 scores for this type of residential space are the highest, indicating a relatively high concentration of white-collar workers in this social area. It is mainly distributed in the northern and eastern areas between the 4th and 6th rings, with Chaoyang District as the main body, and also involving some neighborhoods in Haidian District, as well as neighborhoods near the district government sites in Changping District, Shunyi District, Tongzhou District, Daxing District and Fangshan District.

(4) Residential areas with a high concentration of migrant blue-collar workers (Type 4). The mean value and mean sum of squares of factor 2 scores for this type of residential space are higher than those of the other factors, indicating that this type of social area has a high concentration of migrants. It is mainly distributed between the 3rd and 6th rings, with Fengtai District as the main area, and also involving some neighborhoods in Haidian District, Chaoyang District, and Daxing District.

(5) Areas of agricultural population in outer suburbs (Category 5). The absolute value of the mean value of factor 3 is higher than that of the other factors, and it is the only positive value. The mean value of the squared sum is also higher than that of the other factors, indicating that this social space is mainly residential areas for the agricultural population. It is mainly distributed in the peripheral areas of Changping District, Shunyi District, Tongzhou District, Daxing District and the southeastern part of Fangshan District.

(6) Intellectual areas (Category 6): The mean value and mean square of factor 6 have the highest scores for residential spaces, indicating that this social area is mainly inhabited by intellectuals. It is mainly distributed in the neighborhoods of Tsinghua Campus, Yan Yuan, and Xueyuan Road in Haidian District, where many universities in Beijing are concentrated.

5.2.4 *Restructuring of Urban Residential Space (2000–2010)*

5.2.4.1 Evolution of the Main Factors of Residential Space (2000–2010)

In 2000, the main factors of residential space in Beijing and their corresponding variance contribution rates were (Feng and Zhou 2003a): 27.41% for the general working class, 21.89% for migrants, 8.71% for the agricultural population, 20.92% for the intellectual class and minorities, and 10.69% for living conditions. By 2010, the main factors and their variance contributions were 39.02% for local white-collar workers, 21.64% for migrant blue-collar workers, 6.69% for the agricultural population, 14.76% for intellectuals, and 5.45% for residential conditions. A comparison of the main factors and their spatial distributions in 2000 and 2010 shows the following:

First, there has been a shift in employment structure, with a clear occupational differentiation. The factor that has changed the most has shifted from the general salaryman factor to the local white-collar workers factor, mainly due to the upgrading of Beijing's industrial structure and the increase in the proportion of the tertiary industry, which has led to a shift in the employment structure of the population. The variance contribution rate of this factor has also increased significantly, making it the most important factor affecting residential space, indicating that occupational differentiation has led to increased social segregation.

Second, the development of the real estate market has reduced the power of former public housing. The variance contribution rate of the population density factor has decreased to half of the original residential conditions factor. The population density factor in this book mainly reflects the former public housing with low population density, low rent, small housing area, and concentrated in the central area. With the advancement of urban renewal and old city restructuring projects, the number of former public housing has gradually decreased. With the development of the real estate market, the proportion of former public housing in all types of housing has decreased. At the same time, commercial housing has spread from the city center to the suburbs, resulting in a gradual decrease in the dominant influence of the former public housing with poor living conditions on the urban spatial structure. The above reasons have led to a significant decrease in the variance contribution rate of this factor.

Third: The spatial distribution of migrants shows a clear trend of suburbanization. Its role in the formation of residential space has not changed much, and its variance contribution rate has basically remained the same. However, compared with 2000, the spatial distribution of the main factor of migrants has a further trend of spreading to the suburbs.

Fourth, there is path dependence in the spatial distribution of intellectuals, and there is a clear trend towards spatial agglomeration. Due to the general improvement in educational qualifications, the status of intellectuals in social differentiation has declined, and their variance contribution rate has decreased. However, their spatial distribution has further concentrated in the northwest corner of the city center.

Fifth, the proportion of agriculture in the urban industrial structure has decreased, and the influence of the agricultural population on urban society has declined, and their variance contribution rate has also decreased. The spatial distribution of the agricultural population main factor has not changed much, and it is still concentrated in the outer suburbs of the city.

5.2.4.2 Evolution of Urban Residential Space and Residential Regional Patterns (2000–2010)

A comparative analysis of the urban residential spatial structure and regional patterns in Beijing between 2000 and 2010 (refer to Figs. 5.4 and 5.5, and Table 5.6) reveals that the predominant type of residential space within Beijing's urban social space has remained relatively stable. Notably, a significant transformation has occurred in the categorization of residential space, wherein the residential zones previously associated with the population in remote suburbs have been reclassified. These areas now amalgamate with white-collar workers engaged in various tertiary industries, primarily comprising permanent registered residents, thus forming a new category termed "Areas of Local White-Collar Workers." The principal alterations in the distribution of residential spaces are outlined as follows (see Table 5.6): (1) High-density, congested urban areas are predominantly located in the central urban zone and its adjacent neighborhoods, exhibiting only a marginal expansion; (2) The extent of suburban areas characterized by low population density and spacious living conditions has diminished, primarily due to a decrease in the number of neighborhoods on the outskirts of the near suburbs classified within this social category; (3) Residential space in remote suburban towns have been partially supplanted by both suburban areas with low population density and large living spaces, as well as by areas designated for local white-collar workers. The spatial distribution of "Areas of Local White-Collar Workers" is notably more concentrated, primarily situated in the eastern and northern regions of the inner suburbs; (4) The distribution of migrants has become more widespread, with a tendency towards concentration and expansion into peripheral areas; (5) The spatial distribution of the agricultural population remains largely unchanged, continuing to be predominantly located in the outer suburbs; (6) The distribution of intellectual areas has also become more widespread, with a notable increase in spatial concentration.

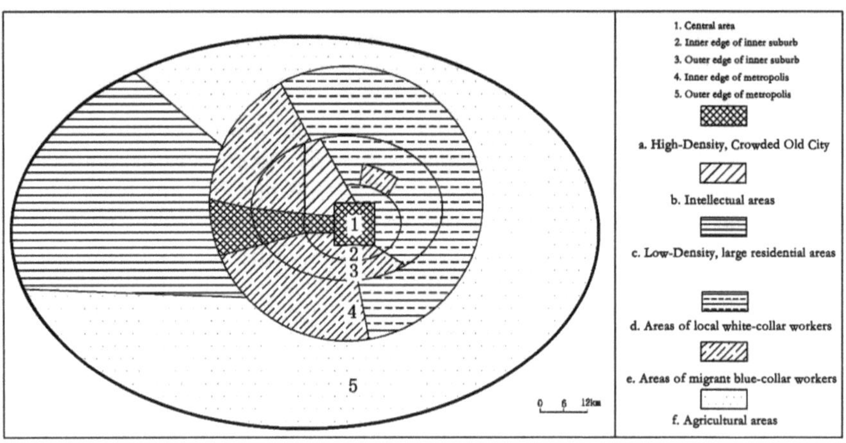

Fig. 5.4 Regional pattern of residential spatial structure in Beijing (2010)

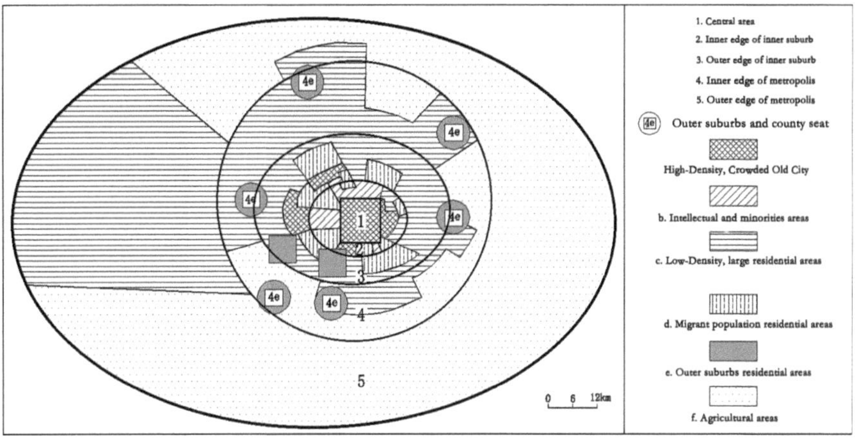

Fig. 5.5 Regional pattern of residential spatial structure in Beijing (2000) (*Source* Feng and Zhou 2003a)

5.2.5 *The Process of Forming Urban Residential Spatial Structure and Its Mechanism of Restructuring*

5.2.5.1 The Formation Process of Urban Residential Spatial Structure and Its Mechanism

The urban residential spatial structure serves as a spatial representation of the social elements associated with the urban residential population, reflecting the materialized choices of residential spaces made by various social classes. The interplay of individual differences, family characteristics, and regional disparities contributes to the establishment of a distinct urban residential spatial structure (see Fig. 5.6). Individual differences manifest primarily through factors such as gender, age, household registration, nationality, educational attainment, and employment status, all of which collectively shape an individual's socio-economic standing. Family differences are predominantly evident in aspects such as family size and generational composition, with the socio-economic status of family members further influencing the overall socio-economic status of the household. Regional differences are primarily characterized by variations in industrial structure and housing markets, which provide residents with diverse employment opportunities and living conditions, thereby resulting in significant disparities in population density and growth. Concurrently, advancements in regional industrial structures have transformed the existing employment landscape, thereby reshaping the socio-economic status of individuals. The housing market plays a crucial role in determining the type, quantity, and location of housing supply, thereby constraining the residential choices available to residents. Ultimately, the differences in residential preferences among individuals of varying socio-economic

Table 5.6 Comparison of residential space in Beijing in 2000 and 2010

Type	Residential space in 2000		Residential space in 2010		Major changes from 2000 to 2010
	Name	Characteristics	Name	Characteristics	
Class 1	High-density, crowded old city	Mainly distributed in the central urban area and its immediate eastern periphery; in addition, there is also scattered distribution in some Inner suburb areas to the south and west of the central urban area	High-density, Crowded old city	Concentrated in areas within the Third Ring Road. All neighborhoods within the Second Ring Road in the Dongcheng and Xicheng districts fall into this category, as do most neighborhoods between the Second and Third Ring Roads and some neighborhoods between the Third and Fourth Ring Roads	They are basically concentrated in the central urban area and its surrounding neighborhoods, with a slightly expanded scope
Class 2	Low-density, spacious city Suburbs	Mainly distributed in the outer periphery of the Inner suburbs, the inner periphery of the metropolitan area, some areas in the western part of the outer periphery of the metropolitan area, and local areas close to the seat of government in Outer suburbs	Areas with lower population density and larger residential areas	They are mainly distributed in Mentougou District and the northern part of Fangshan District, as well as in many urban–rural fringe areas on the outskirts of Inner suburb	The number of neighborhoods on the outskirts of inner suburbs that fall into this category of social areas has decreased

(continued)

Table 5.6 (continued)

Type	Residential space in 2000		Residential space in 2010		Major changes from 2000 to 2010
	Name	Characteristics	Name	Characteristics	
Class 3	Residential areas in outer suburbs	With the exception of Changxindian Township (including Changxindian Street) and Huaxiang Township (including Xincun) in Fengtai District, the remaining six neighborhoods are the government seats of the six counties in the outer suburbs. Although these two neighborhood units in Fengtai District are not part of the outer suburbs, they are already on the outer edge of the inner suburbs and are adjacent to the Outer suburbs in terms of location. Therefore, this type of social area basically reflects the concentrated residential areas of the population in the towns and cities in the outer suburbs	Areas of local white-collar workers	Mainly distributed in the northern and eastern areas between the 4th and 6th rings. Chaoyang District is the main area, and also involves some neighborhoods in Haidian District, as well as neighborhoods near the county seats of Changping District, Shunyi District, Tongzhou District, Daxing District and Fangshan District	In fact, this category of area has been partially replaced by "urban suburbs with low population density and large living areas" and partially by "Areas of local white-collar workers." Among them, the spatial distribution of "Areas of local white-collar workers" is more concentrated, and they are mainly distributed in the eastern and northern parts of the inner suburbs
Class 4	Areas with a high concentration of migrants	Mainly located in the Inner suburbs and not far from the Central District	Areas of migrant blue-collar workers	Mainly distributed between the 3rd and 6th rings. Fengtai District is the main area, and also involves some neighborhoods in Haidian District, Chaoyang District and Daxing District	The migrants are more widely distributed, but their spatial distribution is more concentrated, and there is a tendency for them to spread to the outskirts

(continued)

Table 5.6 (continued)

Type	Residential space in 2000		Residential space in 2010		Major changes from 2000 to 2010
	Name	Characteristics	Name	Characteristics	
Class 5	Areas with an Agricultural Population	Mainly distributed in: ① the north, east and south along the urban area, except for the county seat of the 6 districts in the far suburbs and the surrounding neighborhoods; ② the vast majority of the outer edge of the urban area, except for some neighborhoods in Mentougou District and its neighboring Fangshan District	Areas of agricultural population in outer suburbs	Mainly distributed in the outer areas of Changping District, Shunyi District, Tongzhou District, Daxing District and the southeastern part of Fangshan District	The spatial distribution of the agricultural population has not changed much, and it is still basically concentrated in the outer suburbs
Class 6	Intellectual Areas and Areas with a High Concentration of Minorities	Basically all close to the city center, concentrated in two areas immediately north and west of the city center. The area north of the city center is a concentration of universities in Beijing, while the west of the city center is not only home to many universities, but also the famous "Xinjiang Village"	Intellectual areas	Mainly distributed in the streets of Haidian District such as Tsinghua University, Yanyuan, and Xueyuan Road, where many universities in Beijing are concentrated	Intellectual areas are more widely distributed, and their spatial distribution is also more concentrated

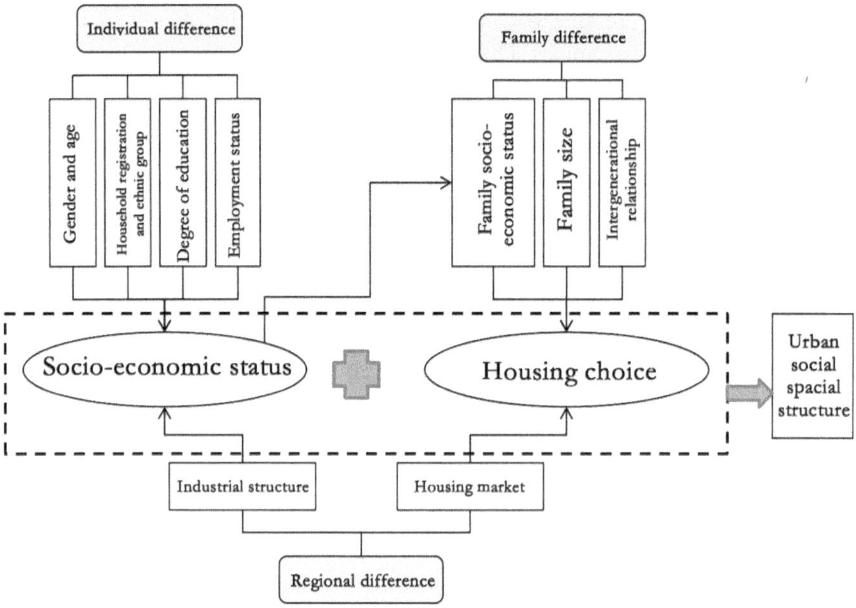

Fig. 5.6 Evolution of social space structure and differentiation in Beijing

statuses lead to spatial differentiation in residential areas across social classes, culminating in the phenomenon that "birds of a feather flock together", which in turn establishes a specific urban residential spatial structure.

Firstly, differences in individuals' socioeconomic status determine their ability to pay and residential preferences. These include gender and age, household registration, ethnicity, education level, employment status, and other aspects. In terms of gender and age, different types of jobs have specific gender preferences. For example, the textile industry tends to recruit women, while Manufacturing, Construction, and other industries prefer to recruit men. Young workers have an absolute advantage in the job market over other age groups. Household registration and nationality affect people's living conditions through a series of welfare policies attached to household registration, such as medical treatment and schooling. Nationality also plays a role in determining where people live, as people of the same nationality tend to live close to each other. Education level largely determines an individual's skill level, which in turn affects their position in the employment market. Employment status includes the industry, occupation and employment status, which directly affect people's income level. The above aspects comprehensively determine the socio-economic status of individual residents.

In addition, family size, generational type and family socioeconomic status determine the housing choices of families. The family differences that can be reflected in census data are mainly reflected in family size and generational type. Family size is an important factor affecting the size of a family's housing, while generational type determines the pressure of supporting children and the elderly in a family, and affects its housing choices. Different housing choices are mainly reflected in housing sources and housing costs. Housing sources include self-built, purchased or rented housing, the purchase of commercial housing, second-hand housing, Affordable Housing, former public housing, rental of low-rent housing or rental of other housing. Housing costs include the cost of purchasing a home or the cost of renting a home. In addition, the differences in the individual socioeconomic status of the members of a family also affect the family's socioeconomic status, their ability to pay and their choice preferences, which in turn affect the family's housing choices.

Finally, the industrial structure and housing market of a region further lead to the formation of residential space differentiation, which can be reflected in the data on industry, occupation, and housing in the national census data. Changes in the jobs provided by the upgraded regional industrial structure affect residents' migration behavior, leading to changes in their demographic composition and corresponding changes in their socioeconomic status. The development of the regional housing market directly affects the spatial distribution of housing, including its type, price, and quality. The agglomeration of housing of the same type attracts residents with the same type of affordability, that is, the same type of socioeconomic attributes.

5.2.5.2 Mechanism of Restructuring of Urban Residential Space

The interaction of administrative, market, and social forces exerts a multifaceted influence on individuals, families, and regions, giving rise to distinct variations across these levels. Driven primarily by socioeconomic status and shaped by the dynamics of housing space distribution, these forces collectively contribute to the formation of a specific urban residential spatial structure (Fig. 5.7).

(1) **Administrative power**

The government plays a pivotal role in shaping the evolution of urban social space by guiding and regulating urban development through strategic planning and the establishment of relevant policies and institutional frameworks. In the case of Beijing, urban planning has been closely tied to urban construction initiatives. The Beijing Urban Master Plan (2004–2020), for example, emphasizes the focused development of three key new towns: Tongzhou, Shunyi, and Yizhuang. Specifically, Tongzhou is designated as a supplementary area supporting the central city's administrative, financial, and trade functions; Shunyi is positioned to foster modern manufacturing and airport logistics; while Yizhuang is transitioning from an economic and technological development zone into a comprehensive industrial new city. These new towns are tasked with relieving population pressures and functional burdens from the central city, emerging as primary nodes for industrial and population agglomeration in Beijing's suburban areas. Notably, the significant divergence in the segregation index values between the population engaged in secondary industries and those in production and transportation industries from 2000 to 2010 can be partly attributed to this strategy.

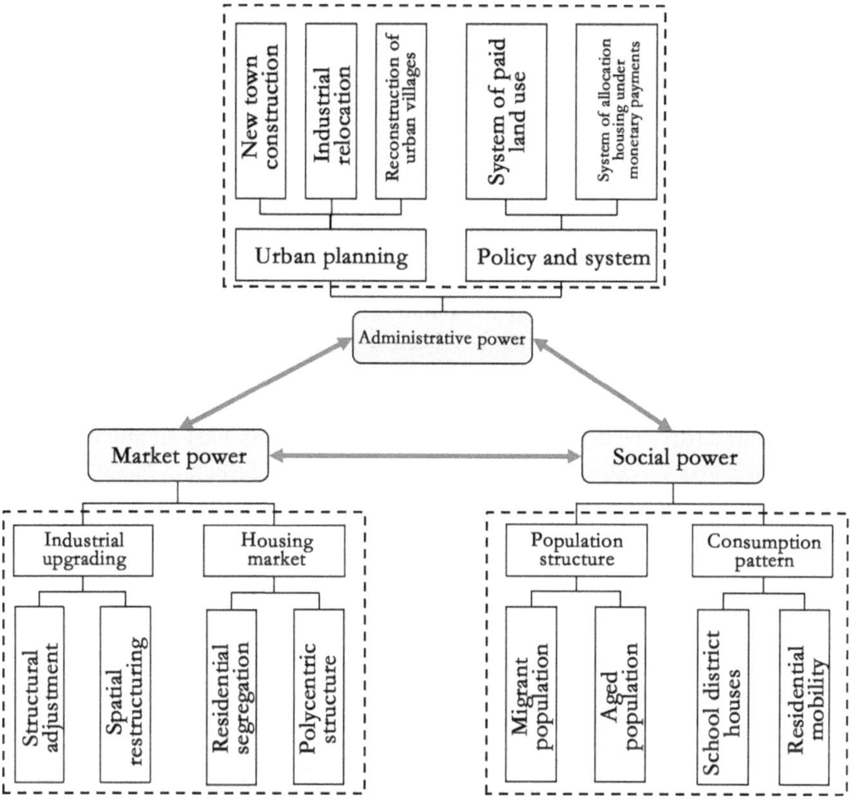

Fig. 5.7 The mechanism of restructuring of Beijing's residential space

Additionally, industrial land relocation and transformation initiatives were undertaken to improve air quality and prepare for the 2008 Olympics. The relocation of traditional industries—such as those in Shougang, the Tonghui River South Chemical Industry District, and Fatang—not only reduced pollution but also prompted the displacement of a large number of associated workers. Furthermore, concerted efforts to renovate "urban villages" aimed to enhance the city's overall quality and environment. However, as these areas often provided low-cost housing for migrant populations, their demolition and redevelopment inevitably resulted in the displacement of a substantial number of migrants, contributing to the restructuring of Beijing's sociospatial landscape.

At the institutional level, the introduction of the paid land use system in 1992 and the implementation of the housing monetization policy in 1998 facilitated the commodification of housing, laying the groundwork for the rapid development of the real estate market. Between 2000 and 2010, Beijing's real estate sector expanded markedly, impacting not only traditional work-unit residential compounds established under the welfare housing system but also intensifying the social spatial differentiation among residents of varying socioeconomic status.

(2) **Market forces**

Market forces, particularly industrial restructuring and housing market development, have played a crucial role in reshaping Beijing's urban residential spatial structure. On the one hand, industrial restructuring and the spatial reorganization of industries have influenced both the composition of the employed population and their spatial mobility, thereby reshaping the city's spatial configuration.

Firstly, industrial restructuring has been a key driver. In 2000, the primary, secondary, and tertiary industries accounted for 2.5, 32.7, and 64.8% of Beijing's GDP, respectively. By 2010, these proportions shifted to 0.9, 24.0, and 75.1%, indicating a marked decline in the primary and secondary sectors and a significant expansion of the tertiary sector. This adjustment in industrial structure has triggered a corresponding transformation in the labor market and employment patterns. Specifically, the proportion of employees in the primary, secondary, and tertiary industries changed from 11.8, 33.6, and 54.6% in 2000 to 6.0, 19.6, and 74.4% in 2010 (Beijing Municipal Bureau of Statistics 2011). The increasing dominance of the tertiary sector has elevated the local white-collar workforce as a distinguishing feature of Beijing's labor force, differentiating it from other employment groups.

Secondly, spatial reorganization has accompanied this industrial shift. Beyond the directives of urban planning, enterprises themselves—motivated by cost minimization—have increasingly relocated production facilities to suburban areas, accelerating the suburbanization of industries. This relocation has, in turn, altered the spatial distribution of the residential population, as employment opportunities and housing demands shifted outward. For instance, the peak distribution of both migrant populations and workers in secondary industries exhibited a clear trend toward suburban areas during this period.

On the other hand, the development of the housing market constitutes a direct force driving residential spatial differentiation and the formation of the current urban spatial structure. One prominent feature is the evident residential segregation shaped by housing prices. As housing of similar price brackets becomes increasingly concentrated, individuals from different socioeconomic classes tend to cluster in distinct locations, reinforcing patterns of spatial stratification and facilitating the emergence of so-called "rich people's areas" and "poor people's areas" (Wang et al. 2014).

Furthermore, the spatial distribution of housing prices has evolved into a polycentric structure. Between 2001 and 2005, Beijing's housing price distribution exhibited early signs of polycentricity, with new subcenters such as Zhongguancun, the CBD, and the Olympic Center emerging alongside the traditional administrative core around Tiananmen (Qin and Jiao 2010). From 2005 to 2012, this trend became more pronounced, as subcenters including the Asian Games Village, Wanliu-Xiangshan, Zhongguancun, Fuxingmen, and the CBD solidified their positions, further reinforcing the polycentric nature of the city's housing and residential landscape (Wang et al. 2014).

(3) **Social forces**

Social forces have significantly shaped Beijing's urban residential spatial structure, particularly through demographic shifts and evolving lifestyle patterns.

On the one hand, changes in the distribution of different residential populations have altered the spatial composition of the city. First, there has been a substantial increase in the migrant population, accompanied by shifts in the social stratification of residents. The number of migrants in Beijing rose from 2.568 million in 2000 to 7.045 million in 2010, with their share of the total population increasing from 18.9 to 35.9%. Remarkably, approximately three-quarters of the city's population growth during this period was attributable to the influx of migrants. Spatially, the migrant population is typically categorized into three regions: the central urban area (Dongcheng and Xicheng districts), the inner suburbs (Chaoyang, Haidian, Fengtai, and Shijingshan districts), and the outer suburbs (the remaining districts and counties). In 2000, migrants accounted for 11.6, 61.6, and 26.8% of the populations in these respective areas. By 2010, these proportions had shifted to 7.8, 53.8, and 38.4%, reflecting a marked increase in the migrant population in the outer suburbs and a decline in the central city and inner suburbs.

Secondly, population ageing is particularly pronounced in the inner city and rural periphery. Across both 2000 and 2010, the elderly population (aged over 60) comprised approximately 12.5% of the study area's total population. Notably, in the inner city, the proportion of elderly residents remained consistently higher at around 16.9%, primarily due to the concentration of retired individuals residing in former public housing units. Additionally, the rural periphery also exhibits a high proportion of elderly residents, largely a consequence of younger populations migrating outward for employment, thereby increasing the relative share of elderly among the remaining population.

On the other hand, shifts in lifestyles and consumption patterns have further contributed to the restructuring of urban social space. A key factor is the heightened emphasis placed by young families on children's education, coupled with the scarcity of high-quality educational resources. This has fueled the phenomenon of "school district housing" in the central city, where property prices are significantly elevated. For instance, in Dongcheng District, the price of housing with access to key primary schools is on average 14.716% higher than that of properties zoned for ordinary schools (Dong and Dong 2017). These elevated prices function as a filtering mechanism based on socioeconomic status, influencing the spatial distribution of young families.

Moreover, the increasing purchasing power of residents has led to a rise in residential migration aimed at improving living conditions. This has resulted in the widespread settlement of local white-collar populations in Beijing's inner suburbs. A case in point is the Huilongguan residential area, where spontaneous migration is prevalent; approximately 50% of economically advantaged residents relocate specifically to seek a better living environment (Chai and Feng 2014).

5.2.6 Conclusion and Discussion: The Latest Characteristics of Restructuring of Urban Residential Space in China's

The empirical case of Beijing offers valuable insights into the general characteristics of urban residential space evolution during China's rapid urbanization. First, the transformation of industrial structure and industrial spatial patterns has driven corresponding changes in residential space and its regional configuration. Occupational differentiation and spatial occupational segregation have become increasingly evident. Whereas wage earners once played the primary role in shaping residential patterns, they have gradually been supplanted by local white-collar workers, who now exert a dominant influence on residential space types. The growing presence of white-collar workers has amplified their impact on the city's sociospatial structure. Notably, the spatial distribution of workers in the primary and secondary industries—particularly those employed in production and transportation sectors—is more geographically concentrated, reflecting both the decline of the agricultural population in metropolitan areas and the trend of urban manufacturing and logistics industries toward spatial consolidation.

Secondly, the distribution of the migrant population has expanded and shifted outward toward the suburbs. This trend is driven by several factors. On the one hand, the substantial increase in the migrant population over the past decade, coupled with saturation in the central city's housing market and labor market, as well as rapidly rising housing prices, has compelled migrants to seek alternative living and working spaces in the suburbs. On the other hand, improved transportation infrastructure and the development of suburban amenities have provided favorable conditions for suburban living. Additionally, the burgeoning suburban population has generated growing demand for services. The suburbanization of industries and the flourishing suburban housing market have, in turn, created diverse employment opportunities for migrants in sectors such as commerce, services, manufacturing, and construction—further encouraging their suburban settlement.

Furthermore, a noticeable trend toward the spatial concentration of intellectuals has emerged. The knowledge spillover effect from universities and research institutes has attracted the establishment of numerous science and technology parks. The agglomeration and scale advantages enjoyed by enterprises have reinforced the clustering of high-tech industries, resulting in a more centralized distribution of intellectuals. Although the overall educational level of the city has improved, distinct spatial differentiation persists. The inner suburbs exhibit the highest educational attainment, with a growing proportion of residents holding university degrees or higher qualifications across various distance ranges. This trend peaks in several sub-districts of Haidian District's inner suburbs, where colleges and research institutes are densely concentrated.

In analyzing the driving mechanisms behind the formation and evolution of urban residential spatial structures, three key levels emerge: individual differences, family differences, and regional differences. These levels collectively shape the process of urban sociospatial differentiation. Underpinning this process is the dynamic interplay of administrative forces, market forces, and social forces. Administrative forces refer primarily to the government's role in influencing urban spatial development through urban planning, policy-making, and regulatory frameworks that guide the growth of the urban economy and society. Market forces involve the impact of industrial restructuring and spatial reorganization, which shape the composition and spatial mobility of the employed population. Social forces pertain to changes in the structure and distribution of both the migrant and elderly populations, which in turn transform the city's social fabric. These three forces—government, market, and society—interact continuously and exert influence across all levels of urban social space, ultimately shaping the specific configuration of residential space and driving its ongoing evolution.

In addition to the empirical research conducted in Beijing, recent developments in China's urban residential space can also be understood by examining academic studies on the social space of cities such as Guangzhou, Suzhou, and Changzhou, based on data from the sixth national census (2010) (Zhou et al. 2016a, 2016b; Feng and Zhong 2018, 2020; Yang and Feng 2019). These studies reveal that whether examining the evolution of urban residential space from 2000 to 2010 or focusing on the static landscape in 2010, a common trend emerges: the transformation of the urban employment structure has driven increased occupational differentiation, which in turn has led to class stratification and the formation of distinct residential areas. For instance, Changzhou exhibits a middle-class residential pattern, while Beijing shows clear distinctions between areas dominated by local white-collar workers and those inhabited by migrant blue-collar workers. Similarly, Guangzhou's residential areas reflect the proportion of middle-income versus low-income classes—all underscoring this trend of social and occupational differentiation. Zhou et al. (2016b) specifically analyzed the spatial differentiation and formation mechanisms of middle-class residential areas in Guangzhou. They identified four sub-categories—education, occupation, income, and mixed—and argued that factors such as social stratification, the real estate market, globalization, the persistence of traditional social spaces, individual agency, and urban construction have all played pivotal roles in shaping and evolving these middle-class areas. Moreover, the role of household registration (hukou) began to visibly influence residential patterns by 2010. In Beijing, the differentiation between local white-collar workers and migrant blue-collar workers reflects the intersection of occupation and household registration. Similar patterns are observed in Suzhou (settled employment population agglomeration), Changzhou (local household registration population clusters), and Guangzhou (mixed residential areas of migrants and locals), all illustrating the significance of hukou status in shaping residential space. The influence of age as a factor in urban social structure has also become increasingly apparent. Examples include agglomeration areas of the non-working-age population in Suzhou and suburban middle-aged clusters in Changzhou. It is anticipated that as aging intensifies over time, more pronounced elderly residential clusters will emerge. In fact, early signs of this trend are already visible in Suzhou's non-working-age agglomerations. The spatial concentration of elderly populations is primarily driven by two factors: first, in many large Chinese cities, the aging of the registered population in old urban centers is significantly higher, leading to the "aging" characteristics of these areas; second, the rise of the elderly care industry—particularly the proliferation of nursing homes and elderly-oriented real estate—has intensified the clustering of elderly populations, especially among the oldest age groups, in specific urban locales.

Overall, the recent evolution of residential space in China's major cities highlights the increasingly influential roles of both market forces and social forces, as well as their interaction with urban space. Market forces primarily include the real estate market and labor market dynamics influenced by industrial upgrading. Social forces encompass shifts in population structure—particularly aging—as well as evolving lifestyles and consumption patterns, all of which have significantly shaped the new characteristics of urban residential spatial structures across China's major cities.

5.3 Differentiation of Urban Residential Space and Differentiation of Social Groups of Residential Population

5.3.1 Research on Differentiation of Urban Residential Space and Proposal of Research Questions

Urban socio-spatial differentiation pertains to the pronounced uneven distribution of social elements within urban environments (Feng and Zhou 2003b; Feng 2005a; 2005b), with residential spatial differentiation serving as a fundamental aspect of this phenomenon. Given that census data are predicated on residential locations, the spatial differentiation assessed through such data at the neighborhood level (including streets and towns) fundamentally reflects the statistical significance of urban residential patterns.

The examination of intra-urban spatial differentiation, particularly residential spatial differentiation, constitutes a foundational element of geographic analysis. This analysis, akin to national and continental geographic differentiation, offers a comprehensive descriptive framework that elucidates basic utility and delineates intra-urban areas based on the criteria of distinctiveness and relative homogeneity in performance (Knox and Pinch 2000). Scholars in Western urban social geography have notably focused on the exploration of differences and inequalities, as well as the resultant patterns of urban spatial structure. They contend that these topics are critical within urban social geography due to their capacity to reveal a complex, multifaceted urban landscape characterized by segregation, overlap, and polarization (Knox and Pinch 2000; Kitchin and Tate 2000; Pacione 2001a, 2001b).

In the aftermath of World War II, the morpho-functional transformation of urban landscapes has led to a fluid interplay between diversity and polarization within metropolitan spaces and communities, thereby reshaping group identities. The contemporary urban environment is increasingly marked by demographic diversity, cultural plurality, and nationality variation. The heterogeneity and fragmentation of post-suburban settings, alongside the emergence of multiculturalism and rapidly evolving subcultures, have become defining features of modern urban landscapes. Additionally, the actualization and privatization of public spaces, as well as the proliferation of walled and gated communities, have gained significant traction (The Ghent Urban Studies Team 1999).

In light of globalization, the international global city is progressively evolving into a segmented, fragmented, polycentric, and postmodern social space, particularly within the context of post-Fordism, flexible labor markets, and the expansion of the producer services sector (Friedmann and Wolff 1982; Garreau 1991; Sassen 1991; Coffey and Shearmur 2002; Wu and Li 2005; Feng et al. 2008; Gu 2011; Feng et al. 2012; Feng and Zhao 2016). Consequently, the investigation of increasingly diversified, fragmented, and subdivided urban residential spatial differentiation, along with its restructuring characteristics, holds substantial significance.

In recent years, the accessibility of census data, particularly at the neighborhood level, has led to a notable increase in empirical research examining the spatial structure and differentiation of urban society in China during its transitional period. This body of work includes studies focused on major cities such as Beijing, Shanghai, Guangzhou, Xi'an, and Nanchang (Wang 2000; Feng and Zhou 2003a, 2004; Feng 2004; Gu et al. 2005; Li and Wu 2006; Xuan et al. 2006; Zhou et al. 2006; Zhong and Feng 2017). The prevailing conclusion from these studies indicates the presence of a distinct socio-spatial structure within large Chinese cities during the transition period, characterized by heterogeneity and the phenomenon of socio-spatial differentiation.

Building upon previous research, it is imperative to further investigate several key areas in light of emerging phenomena and challenges: (1) The urban socio-spatial structure, as derived from the study of social districts, represents a relatively macroscopic and generalized differentiation. Therefore, it is essential to analyze and illustrate the spatial differentiation patterns and characteristics of various social elements, including different types of residential populations, employed populations, and housing conditions in greater detail; (2) An examination of the spatial differentiation patterns and characteristics of residential spaces in Chinese cities over time, particularly during the transition period; (3) The spatial differentiation of urban structures based on social districts should not only be viewed as a macroeconomic and generalized phenomenon but also as a significant one. Additionally, there is a notable gap in research regarding the restructuring process of spatial differentiation in urban housing in China over time, especially during the transition period, necessitating further exploration of these restructuring characteristics through quantitative methods; and (4) The potential establishment of a theoretical framework for understanding the restructuring of socio-spatial differentiation or spatial differentiation in urban housing in China during the transition period, particularly in relation to institutional, market, and cultural changes. In light of these considerations, this section employs data from Beijing to enhance the understanding of socio-spatial differentiation in urban China during the transition period.

5.3.2 *Study Area, Data and Methodology*

5.3.2.1 Study Area and Data

The empirical study area delineated in this section encompasses the primary administrative region within the 6th ring road of Beijing, comprising 12 of the city's 16 districts: Dongcheng District, Xicheng District, Chaoyang District, Fengtai District, Shijingshan District, Haidian District, Mentougou District, Fangshan District, Tongzhou District, Shunyi District, Changping District, and Daxing District. The boundaries of the study area are illustrated in Fig. 5.1. Historically, Dongcheng District and Xicheng District have been identified as the central urban area, which prior to 2010 included the four districts of Dongcheng, Chongwen, Xicheng, and Xuanwu. In the concentric structure of Beijing, Dongcheng and Xicheng Districts are typically recognized as the core urban center, with the former Dongcheng and Chongwen Districts amalgamated into the current Dongcheng District, and the former Xicheng and Xuanwu Districts merged into the present Xicheng District. This central area is often referred to as the "urban area," while the districts of Chaoyang, Fengtai, Shijingshan, and Haidian are classified as inner suburb districts. The remaining ten districts and counties within the urban region are classified as outer suburb.

This section of the research utilizes sub-neighborhood data derived from the Third (1982), Fifth (2000), and Sixth (2010) National Population Censuses conducted in Beijing. It is important to distinguish the concept of neighborhoods in this context from the architectural notion of "neighborhoods," which pertains to street offices and administrative units at the township and commune levels. The study investigates the alterations in the administrative divisions of neighborhoods within the Beijing metropolitan area from 1982 to 2000. In order to facilitate comparability, the populations of the relevant neighborhoods were aggregated based on the findings of this examination. A spatial database was subsequently established by integrating the collected neighborhood data with corresponding neighborhood maps. Population redistribution not only signifies changes in spatial location but also reflects the differentiation of various social groups within urban spaces, thereby revealing the characteristics of multi-layered spatial overlap within the city (Wu 2005). Consequently, census data at the neighborhood level serves as a robust instrument for elucidating the characteristics of the city's residential space

The term "Tri-industry 2" encompasses sectors that facilitate production, which includes extractive industries, the generation and supply of electricity, gas, and water, as well as geological surveying and water management. "Tri-industry 3" pertains to sectors that support life, such as finance, insurance, real estate, and social services. "Tri-industry 4" is associated with sectors that promote the advancement of science and culture. Similarly, "Tri-industry 5" also refers to sectors dedicated to the enhancement of science and culture. The designation "third estate 4" identifies sectors aimed at improving scientific and cultural standards and enhancing population quality, which includes health, sports, social welfare, education, culture and the arts, as well as radio, film, television, scientific research, and comprehensive technical services. Conversely, "third estate 5" pertains to sectors that address the public needs of society, encompassing state institutions, party and government bodies, social organizations, and various other industries. The term "third sector" specifically refers to sectors that fulfill societal public needs, including state organs, party and government entities, social organizations, and additional industries.

5.3.2.2 Methodology and Models

The GIS software ArcGIS is used to extract the area and coordinates of the center of mass (geometric center) of each neighborhood in the Beijing metropolitan area, and the distance between the geometric center of each neighborhood and Tiananmen Square is measured as the distance of each neighborhood from the city center. On the basis of the spatial database, the spatial display function of the GIS is used to generate a spatial pattern map and summarize the spatial differentiation characteristics of each type of social elements.

In order to accurately characterize the restructuring of spatial differentiation in urban society, standard deviation, extreme deviation, information entropy, absolute differentiation index, relative differentiation index and segregation index are calculated respectively. The model of information entropy is expressed as follows:

$$H = -\sum_{i=1}^{n}\left(\frac{X_i}{\sum_{i=1}^{n} X_i}\right)\ln\left(\frac{X_i}{\sum_{i=1}^{n} X_i}\right) \tag{5.1}$$

In the model, H is the information entropy; X is a social element and i is a neighborhood sample.

The differential index, commonly referred to as the localization index, quantifies the segregation degree of residents among various subpopulation groups within a broader geographic area (Johnston et al. 2000). This index serves as a metric for assessing the differential of one type of factor in relation to another. The model is articulated as follows:

$$I_d = \frac{1}{2}\sum_{i=1}^{n}\left|\frac{X_i}{\sum_{i=1}^{n} X_i} - \frac{Y_i}{\sum_{i=1}^{n} Y_i}\right| \tag{5.2}$$

In this equation, I_d represents the index of differentiation, while X and Y denote two distinct categories of social elements, and i signifies the sample of neighborhoods. The value of I_d ranges from 0 to 1, where a value of 0 indicates no differentiation and a value of 1 signifies complete differentiation. A higher value of I_d corresponds to a greater degree of differentiation between the elements X and Y.

The derivation of the index of variation model presented in this text is twofold. Firstly, when Y is assigned a value of 1, the total sample size is denoted as N, and Eq. (5.2) is interpreted as the "absolute index of variation," which quantifies the extent of variation in relation to the average distribution pattern. Secondly, Y when Y signifies the total neighborhood population, the total sample size again represents the metropolitan area's population, and Eq. (5.2) is understood as the "relative index of variation," which measures the degree of variation in relation to the distribution pattern of the residential population.

The segregation index, as articulated by Johnston et al. (2000), incorporates percentage weight in addition to the relative differential index. The model is defined as follows:

$$I_s = \frac{I_d}{1 - X/Z} \tag{5.3}$$

IS denotes the segregation index; *la* represents the relative dissimilarity index in relation to the total population of the metropolitan area; and the denominator serves as a weight that reflects the proportion of the remaining population after a specific demographic segment has been excluded from the total population.

$$1 - \frac{X}{Z} = 1 - \sum x_{ai} / \sum x_{ni} \tag{5.4}$$

In Eqs. (5.3) and (5.4), X and $\sum x_{ai}$ signify the total number of individuals within a particular category in a metropolitan area, while Z and $\sum x_{ni}$ denote the total population of that metropolitan area. The segregation index is frequently employed by Western researchers to assess racial segregation within urban environments. In the context of China, where racial segregation is not prevalent, the segregation index can be utilized to evaluate the clustering of specific demographic groups and the integration of the urban populace. A higher segregation index for a particular demographic indicates a stronger tendency for that group to cluster, resulting in diminished interaction with the broader population.

In addition to the aforementioned models, this section seeks to establish a conceptual framework aimed at elucidating the spatial restructuring of urban socio-spatial differentiation in a comprehensive manner. The conceptual model approach has been extensively employed in Western urban social geography as a means of generalization (Knox and Pinch 2000). Furthermore, to facilitate the visualization of the evolution of social indicators from 2000 to 2010, the values of these indicators for the year 2000 are represented on the horizontal axis, while those for 2010 are depicted on the vertical axis. This coordinate system allows for a clearer observation of the magnitude of the indicator values and enables a comparative analysis of their growth between the two years. It is important to note that enhancements to traditional spatial analysis methods significantly contribute to a more accurate representation of urban spatial patterns and their evolution (Gu et al. 2009). The advancements in traditional spatial differentiation indicators and their implications are detailed in Table 5.7.

Table 5.7 Presents enhancements to conventional spatial differentiation metrics and their implications

Traditional method	Improvement	Model	Additional notes	Implications for improvement
Variance index	Absolute variance Index	$I_{d1} = \frac{1}{2}\sum_{i=1}^{n}\left\lvert \frac{X_i}{\sum_{i=1}^{n} X_i} - \frac{1}{N} \right\rvert$ (3)	I_{d1} denotes the absolute index of differential, N denotes the number of neighborhoods, and Xi denotes the value of the indicator for neighborhood 0.1	The differential index is classified into two types: relative differential and absolute differential. Absolute differential pertains to the extent of differential in relation to the average distribution pattern, while relative differential pertains to the extent of differential in relation to the distribution pattern of the residential population. This classification allows for a more nuanced understanding of socio-spatial differentiation within the population. Furthermore, it facilitates the visualization of the magnitude of the indicator values, as well as the comparison of these values across two distinct years
	Relative differential Index	$I_{d2} = \frac{1}{2}\sum_{i=1}^{n}\left\lvert \frac{X_i}{\sum_{i=1}^{n} X_i} - \frac{Y_i}{\sum_{i=1}^{n} Y_i} \right\rvert$ (4)	I_{d2} denotes the relative index of variation, Xi denotes the index value of neighborhood i, and Yi denotes the total population of neighborhood 0.1	The differential index is classified into two types: relative differential and absolute differential. Absolute differential pertains to the extent of differential in relation to the average distribution pattern, while relative differential pertains to the extent of differential in relation to the distribution pattern of the residential population. This classification allows for a more nuanced understanding of socio-spatial differentiation within the population. Furthermore, it facilitates the visualization of the magnitude of the indicator values, as well as the comparison of these values across two distinct years

(continued)

Table 5.7 (continued)

Traditional method	Improvement	Model	Additional notes	Implications for improvement
Tabulate results	Display the value of the indicator in the right-angled coordinate system	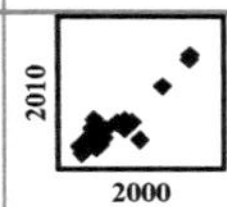	The horizontal axis illustrates the value of the indicator for the year 2000, while the vertical axis depicts the value of the indicator for the year 2010. Each plotted point corresponds to a specific indicator; points that are situated closer to the origin indicate lower values of the indicator. Points located above the diagonal (dotted line) represent indicators that have experienced an increase over the decade, whereas those below the diagonal indicate a decrease. Furthermore, the distance of the diagonal line from the origin serves as a measure of the magnitude of change over the ten-year period, with greater distances reflecting more substantial changes	

5.3.3 Patterns of Spatial Distribution of Urban Residential Population

The population census in China is conducted based on individuals' places of residence, which means that the census data reflect the characteristics of the residential population within the surveyed cities, as well as the various demographic groups residing in those urban areas. To effectively analyze the quantitative indicators of spatial differentiation in residential patterns, it is essential to utilize the spatial visualization capabilities of Geographic Information Systems (GIS). This approach allows for a comprehensive understanding of the spatial distribution patterns of different residential populations. For illustrative purposes, we will utilize data from the Fifth National Population Census to demonstrate the spatial distribution of the residential population in Beijing (see Fig. 5.8).

5.3.3.1 Spatial Distribution Pattern and Characteristics of the Residential Population

(1) **Residential spatial pattern based on population density**

The population density map of Beijing's neighborhoods, as illustrated in the Fifth National Population Census (2000), reveals a concentric-circle pattern of population density that decreases from the city center to the outskirts. The neighborhoods with the highest population density are predominantly located within the urban core and its immediate vicinity, while those surrounding the county and district governmental offices in the outer suburbs exhibit significantly higher population densities compared to their neighboring areas. A comparative analysis with the Fourth National Census (1990) indicates that, over the past decade, the polarization of population density among neighborhoods has diminished, and the distribution of neighborhoods characterized by extremely high density has expanded outward. This phenomenon of suburbanization is reflected in the evolving spatial dynamics of urban living, as evidenced by the decline in density within the most populated neighborhoods in the central area and the outward migration of very high-density neighborhoods, suggesting that suburbanization has alleviated the overcrowding issues traditionally associated with the inner city.

(2) **Spatial distribution of migrant population**

The neighborhoods with a high concentration of migrant populations are predominantly located in the suburban areas, while the government seat neighborhoods in the outer suburbs also exhibit significant appeal to migrants. An analysis of the data from the year 2000 reveals that neighborhoods characterized by a high or medium migrant population ($\geq$12,000 individuals) are primarily situated in the central and suburban regions, in addition to the government seat towns in the outer suburbs and their adjacent rapidly urbanizing areas. Conversely, neighborhoods with a low migrant

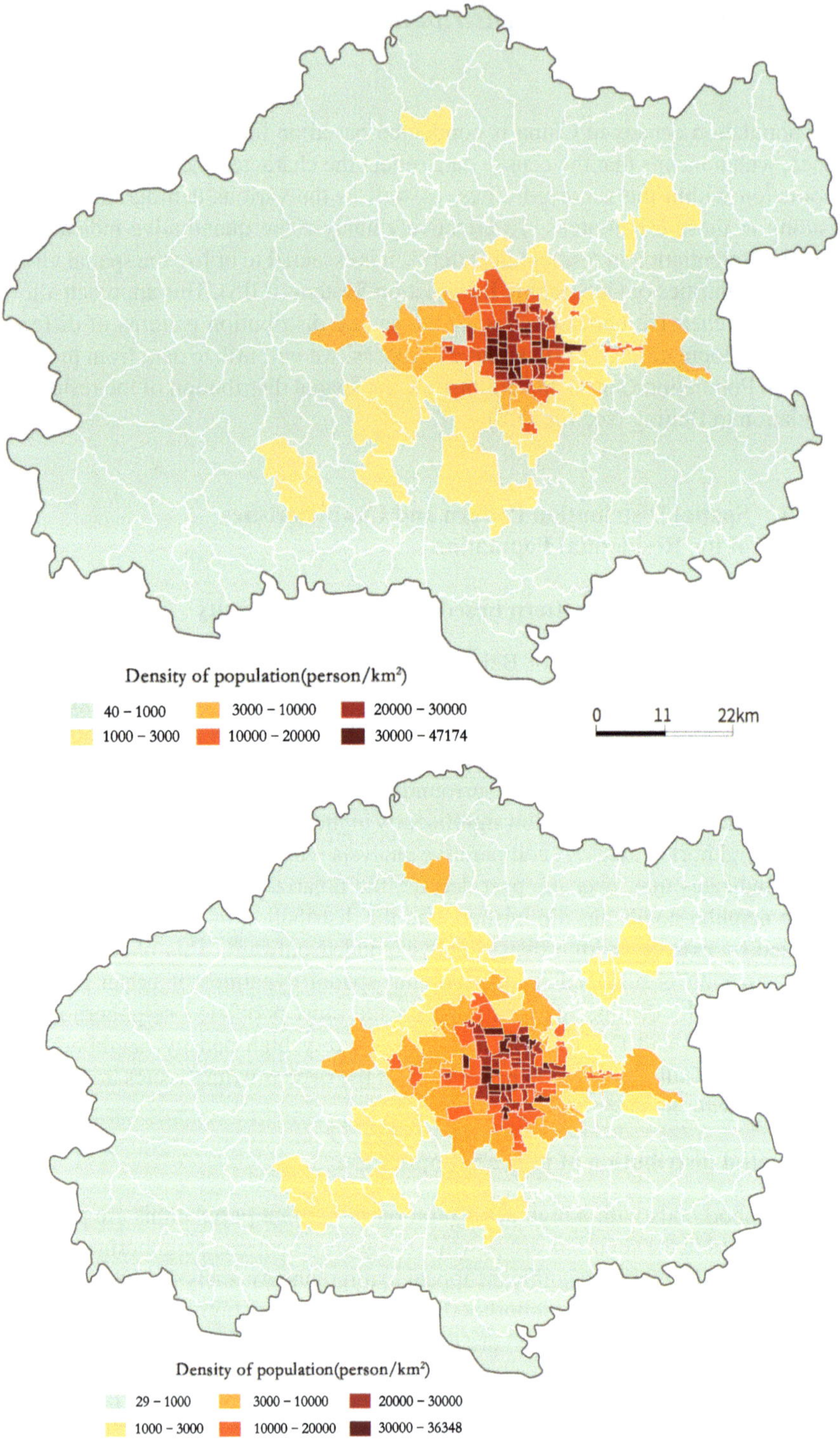

Fig. 5.8 Spatial distribution of residential population by type in Beijing by neighborhood (1990 and 2000). *Note* The first one and the third one are in 1990, while the others are in 2000

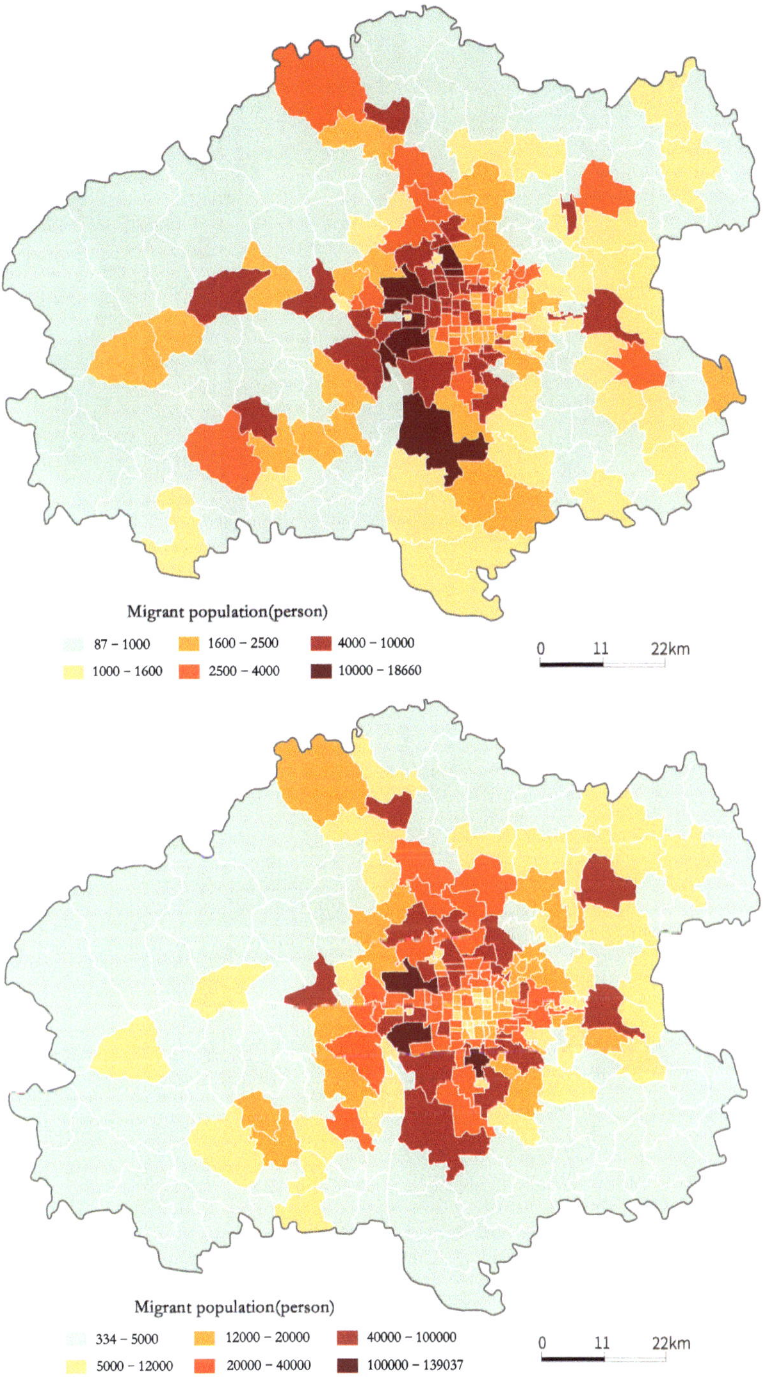

Fig. 5.8 (continued)

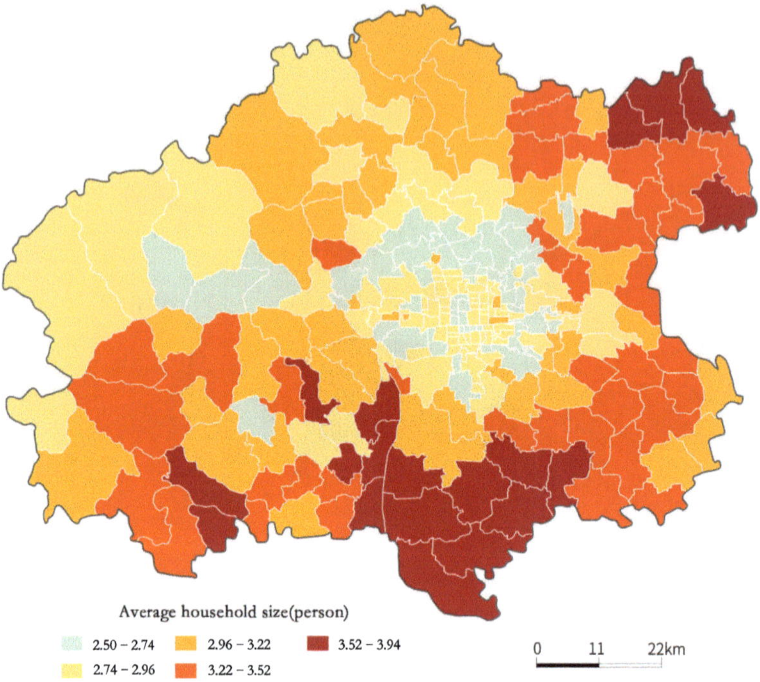

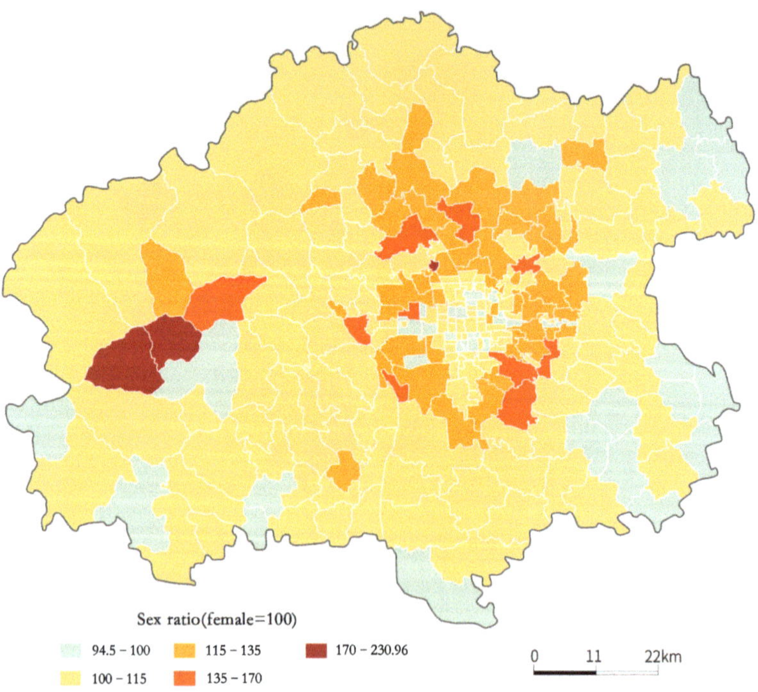

Fig. 5.8 (continued)

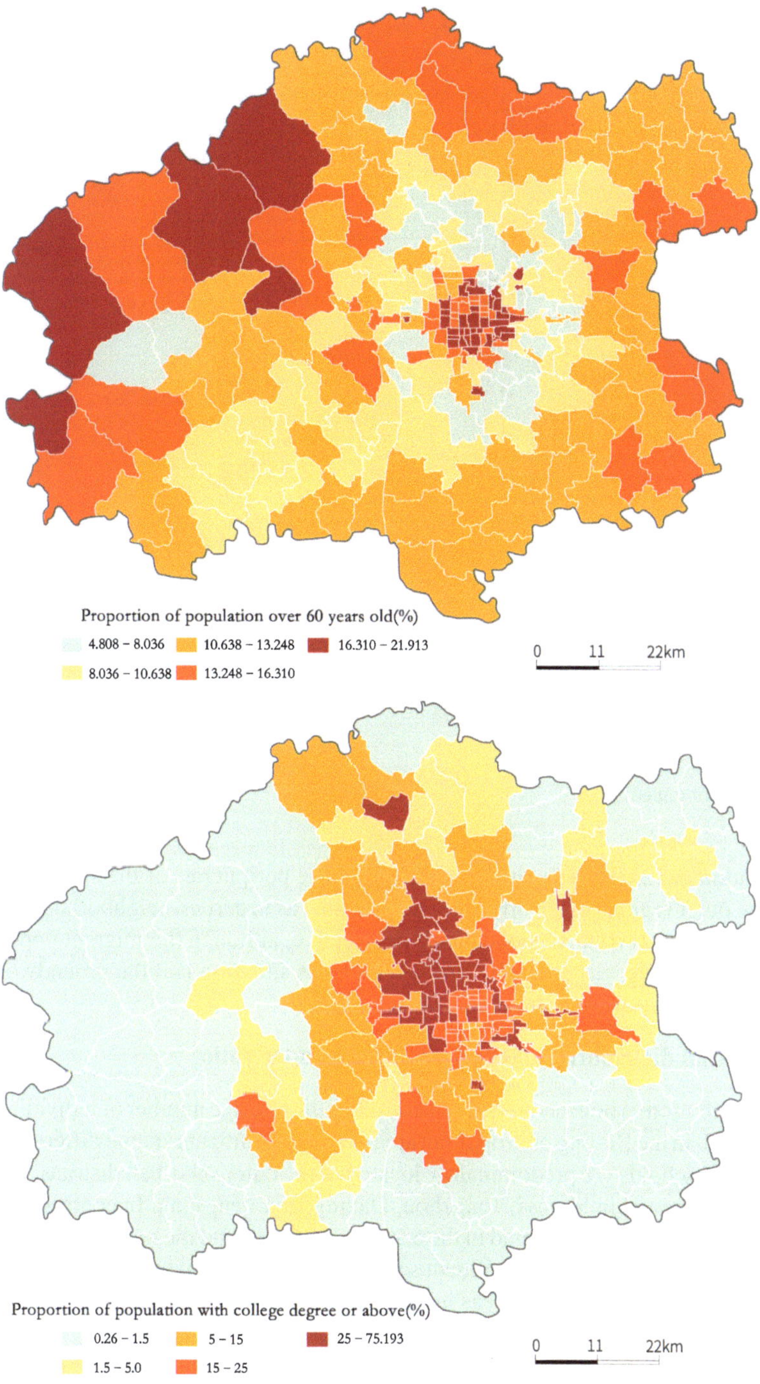

Fig. 5.8 (continued)

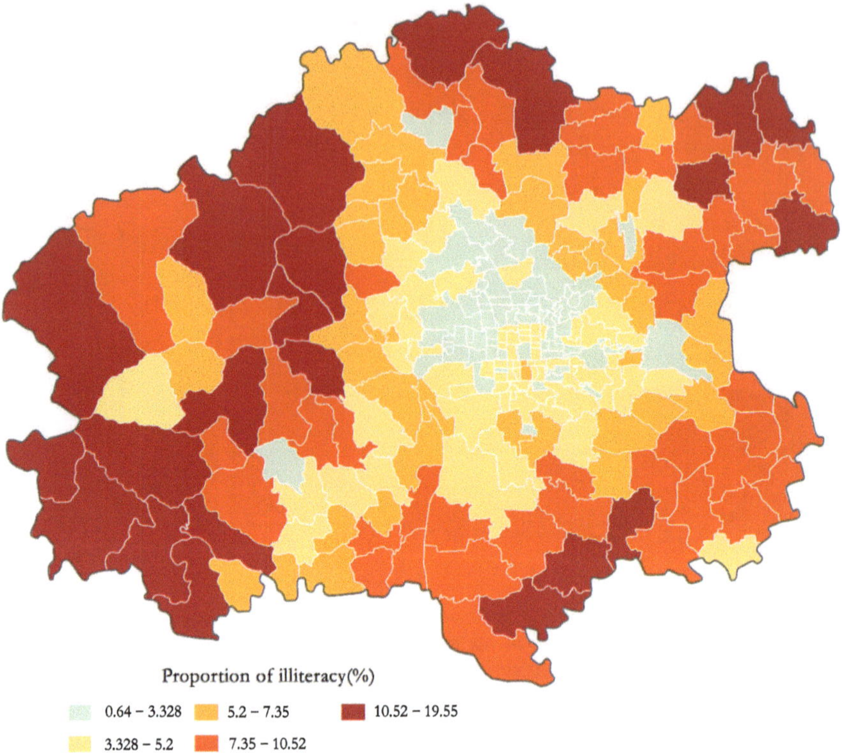

Fig. 5.8 (continued)

population (<12,000 individuals) are found in the peripheries of the suburban areas and the outer regions of six urban districts, as well as in certain neighborhoods within the central area. From an evolutionary perspective, although the migrant population in neighborhoods was generally low in 1990, this situation laid the groundwork for the distribution patterns observed a decade later.

(3) **Spatial distribution of family size and gender ratio**

The distribution of household size, measured by the average number of individuals per household, in the Beijing metropolitan area exhibits significant spatial differentiation. Larger households are predominantly located in the outer suburban districts, particularly concentrated in Shunyi, Tongzhou, Daxing, and Fangshan. In contrast, smaller households are primarily found in the central urban area and the nearby suburbs, with a notable presence in the government seat districts of most outer suburban counties, as well as in the majority of areas in Mentougou and neighborhoods adjacent to the Eighth Ring Road in Changping. This pattern of differentiation indicates that larger household sizes are commonly associated with the outer suburban counties, which correlates with a relatively concentrated rural population in those areas. Conversely, the central urban and nearby suburban regions tend to have smaller household sizes, particularly in the suburbs where most neighborhoods have an average household size of less than 2.74 individuals, a trend linked to the influx of new-generation families purchasing homes or relocating to these areas over the past two decades. Additionally, the gender ratio within the Beijing metropolitan area also demonstrates a degree of spatial variation, with one-sixth of neighborhoods having a higher female population than male, while five-sixths of neighborhoods exhibit a predominance of male residents.

(4) **Spatial distribution of the aging population**

Individuals aged over 60 are primarily concentrated in the central urban area and its adjacent regions, as well as in specific peripheral areas of the metropolitan fringe. This distribution pattern of the aging population within the Beijing metropolitan area suggests that, on one hand, older individuals tend to reside in urban districts and nearby areas where infrastructure conditions are relatively favorable. On the other hand, certain rural areas on the outskirts of the metropolitan region exhibit a higher proportion of elderly residents.

(5) **Spatial distribution of highly educated population**

The population with higher education levels is predominantly concentrated in the Haidian and Chaoyang districts. Conversely, the illiterate population is primarily found in the rural areas surrounding urban centers, with significant concentrations in the Fangshan and Mentougou districts. Within Haidian, areas such as Dongsheng, Tsinghua Garden, Yanyuan, Zhongguancun, Huayuan Road, Beitaipingzhuang, Beixiaoguan, Ganjiaokou, Balizhuang, and Yongding Road exhibit a high proportion of individuals with advanced degrees. Similarly, in Chaoyang, regions such as the Asian Games Village, Xiaoguan, and Heping Street also reflect a significant presence of highly educated individuals. The former district serves as a major hub for higher education institutions, the Chinese Academy of Sciences, and high-tech electronic product markets, while the latter also hosts numerous research institutes affiliated with the Chinese Academy of Sciences. The distribution characteristics of the illiterate population exhibit a pattern that is inversely related to that of the highly educated population. In rural areas, particularly in remote regions, the proportion of illiteracy is relatively elevated.

(6) **Distribution of minorities**

This section focuses on the spatial distribution of fiveminority populations in China: the Manchu, Hui, Mongolian, Zhuang, and Uyghur peoples (see Fig. 5.9). It is observed that the spatial pattern of the Manchu population has historical roots, while the Hui population exhibits a relatively uniform distribution. The spatial concentration of the Zhuang, Mongolian, and Uyghur populations is primarily attributed to the presence of highly educated individuals pursuing studies in Beijing. The Manchu population is predominantly located in the central urban area and its adjacent suburbs, a pattern that can be traced back to the historical settlement of the Manchu people in the inner city. In Beijing, there are 94 neighborhoods with a Hui population exceeding one thousand, and 156 neighborhoods with a population exceeding one hundred, indicating a long-standing coexistence of Hui and Han populations, characterized by a widespread and uniform distribution. However, certain neighborhoods, such as Tongzhou's Yujiawu and Haidian District's Sijiqing (which includes Zizhuyuan), are notable for their relatively high concentrations of Hui residents. The neighborhoods with the largest Uyghur populations are primarily Sijiqing and Ganjiaokou, each with over 300 residents. The distribution patterns of the Mongolian and Zhuang populations show some similarities, with significant concentrations in Sijiqing and its surrounding areas in Haidian District. The reasons for these patterns include: first, the presence of the Central University for Nationalities in Zizhuyuan, which contributes to the relatively prominent minority population in Sijiqing; second, the concentration of higher education institutions in Haidian District correlates with a higher presence of minority populations, suggesting that the concentration of minorities is largely due to individuals pursuing higher education in Beijing; and third, the establishment of a concentrated Uyghur settlement, known as "Xinjiang Village," in Ganjiaokou, which further accentuates the prominence of the Uyghur population in that area.

5.3.3.2 Spatial Distribution Patterns and Characteristics of Employment Status Based on Resident Population

The spatial distribution of various employment statuses, including unemployment and agricultural employment, within the resident population is illustrated in Fig. 5.10.

(1) **Spatial distribution of the unemployed and agricultural workforce**

In Beijing, the unemployed demographic is predominantly located within the eight urban districts and the governmental areas of the more remote suburban counties. The distribution of the agricultural workforce in Beijing exhibits an inverse relationship with the degree of regional urbanization, with a significant concentration in the peripheral regions of the distant suburban counties. While the agricultural population within the eight urban districts is minimal, there remains a presence of agricultural workers in the outskirts of certain towns and districts, such as Chaoyang, Haidian, and Fengta.1 Overall, the largest concentration of the agricultural population within the Kyoto metropolitan area is found in the southern periphery of the metropolitan region.

(2) **Spatial distribution of secondary employment**

The neighborhoods characterized by a significant proportion of individuals engaged in the secondary sector are predominantly located in the districts of Chaoyang, Fengtai, Changping, and Shuny.1 Additionally, certain neighborhoods in Tongzhou, Daxing, and Fangshan, which are in proximity to the eight central districts of the city, also exhibit a notable concentration of secondary sector employment. Conversely, Haidian District does not demonstrate a distinct advantage regarding the proportion

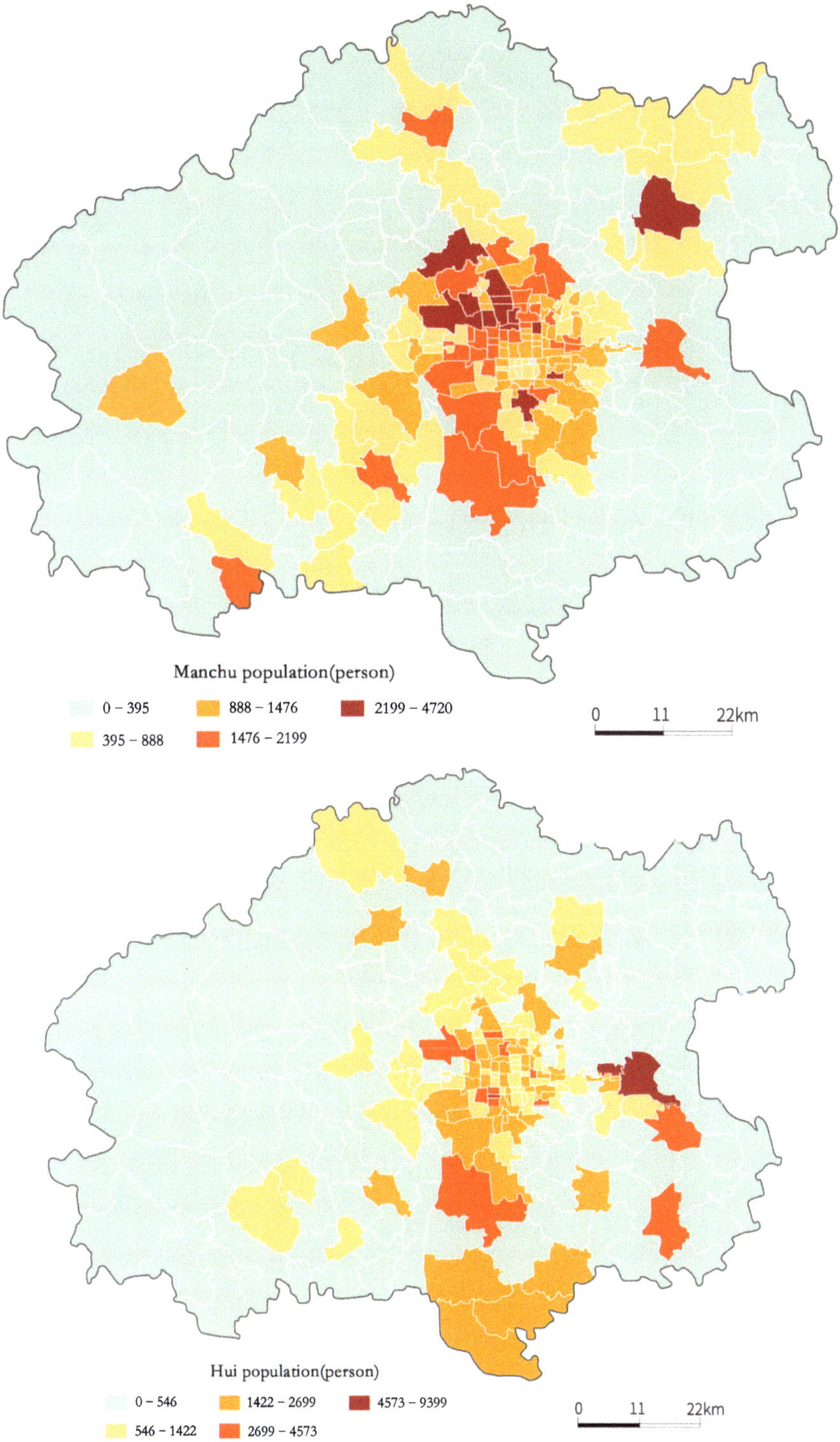

Fig. 5.9 Spatial distribution of ethnic minority status of the resident population in Beijing (2000)

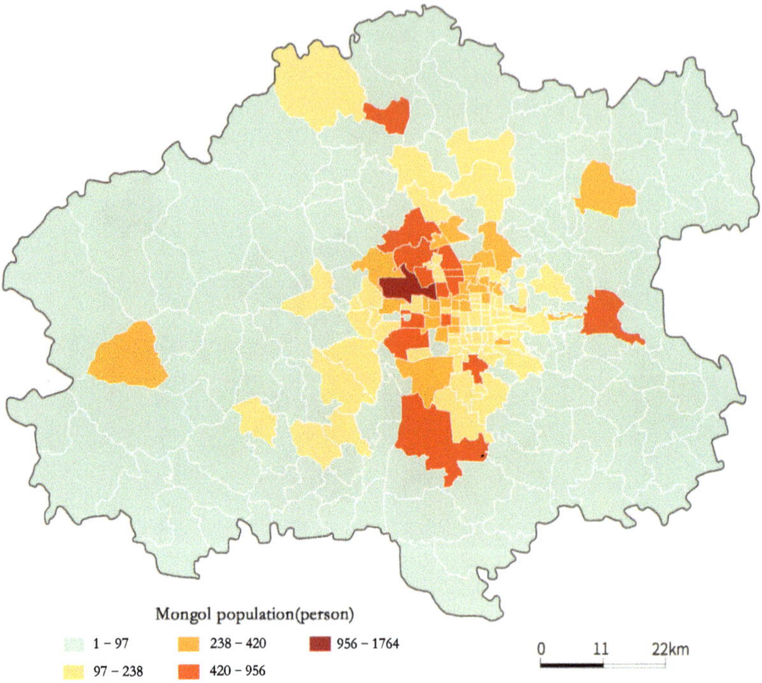

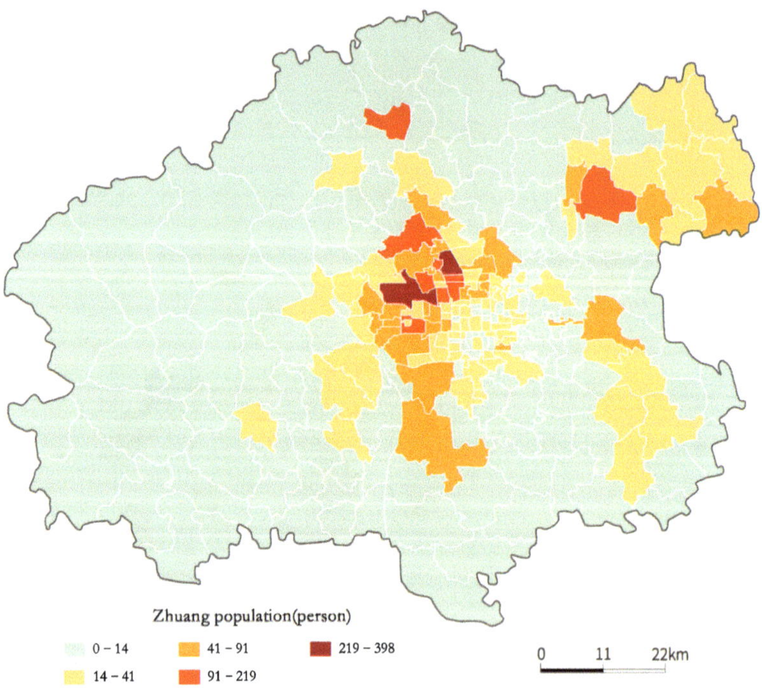

Fig. 5.9 (continued)

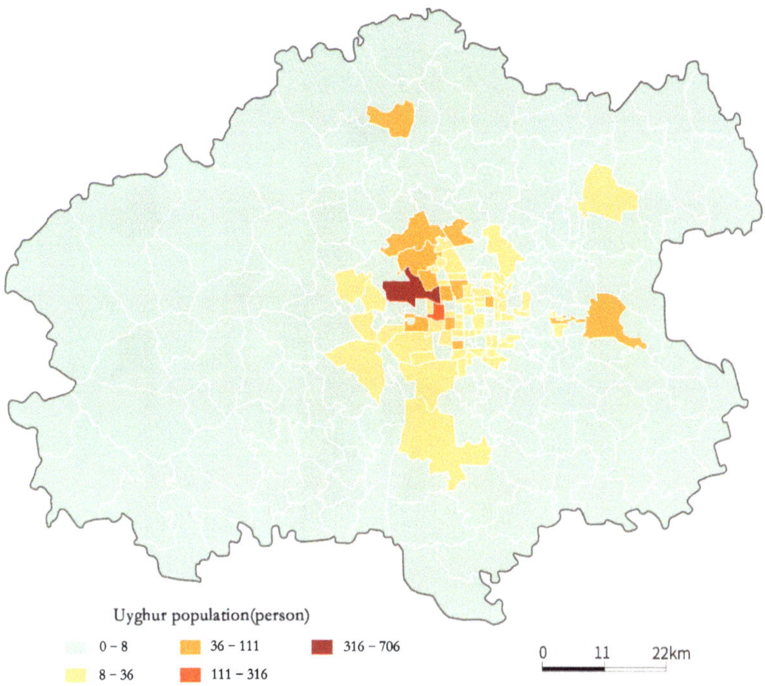

Fig. 5.9 (continued)

of its workforce employed in the secondary sector. Overall, the neighborhoods with a high concentration of secondary sector employment are primarily situated in two principal areas: the majority of Changping and Shunyi in the northern segment of the metropolitan area, and the eastern and southern suburban regions, which include several neighborhoods in Chaoyang and Fengtai.

(3) **Spatial distribution of tertiary employment**

Neighborhoods exhibiting the highest proportions of employment in the tertiary sector are largely concentrated in the central area and its surrounding regions. Notably, Mentougou and Fangshan also show elevated proportions of tertiary sector employment, primarily due to their status as key distribution areas for Beijing's coal mines, which has resulted in a higher concentration of individuals employed in the extractive industry. In the peripheral regions of the metropolitan area, the proportion of the workforce engaged in the tertiary sector remains relatively low. When delineating Beijing into Southern and Northern sectors using Subway Line 1 as a boundary, it becomes evident that the majority of neighborhoods with the highest proportions of tertiary industry employment are situated in the Northern sector, while the Southern sector lags in terms of tertiary industry development.

(4) enlargethispage-24pt**Spatial distribution of the employed population in subcategories of the tertiary sector**

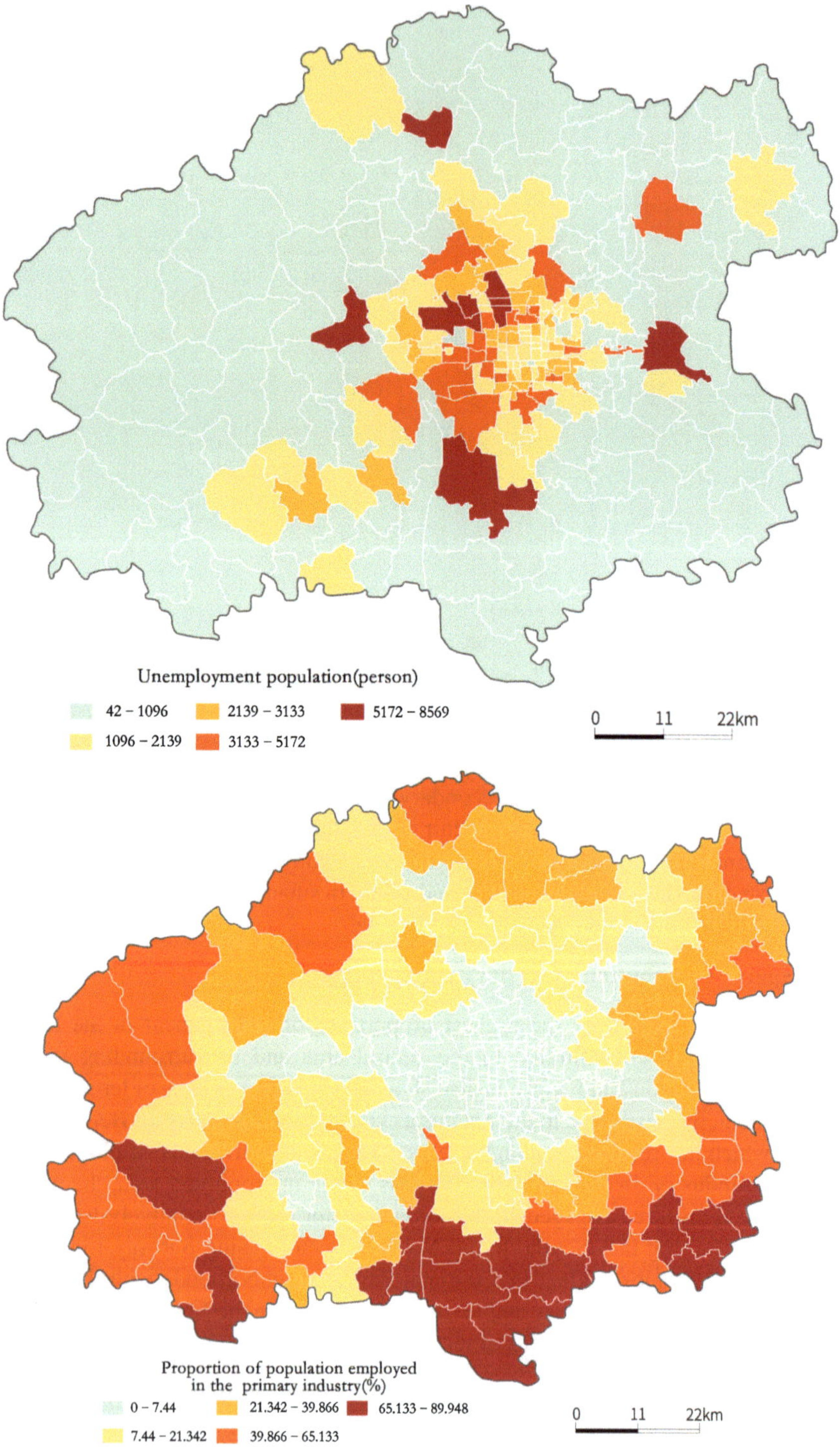

Fig. 5.10 Spatial distribution of employment status of resident population in Beijing (2000)

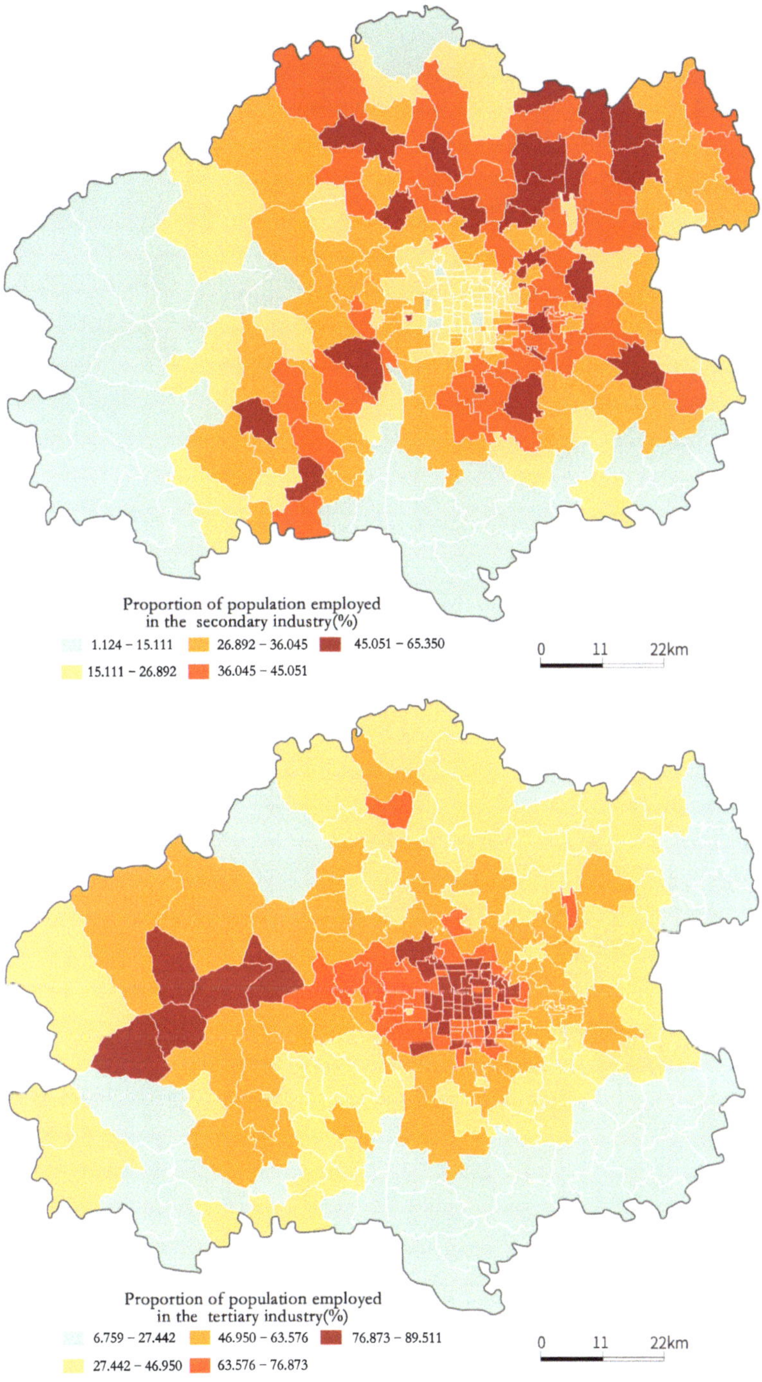

Fig. 5.10 (continued)

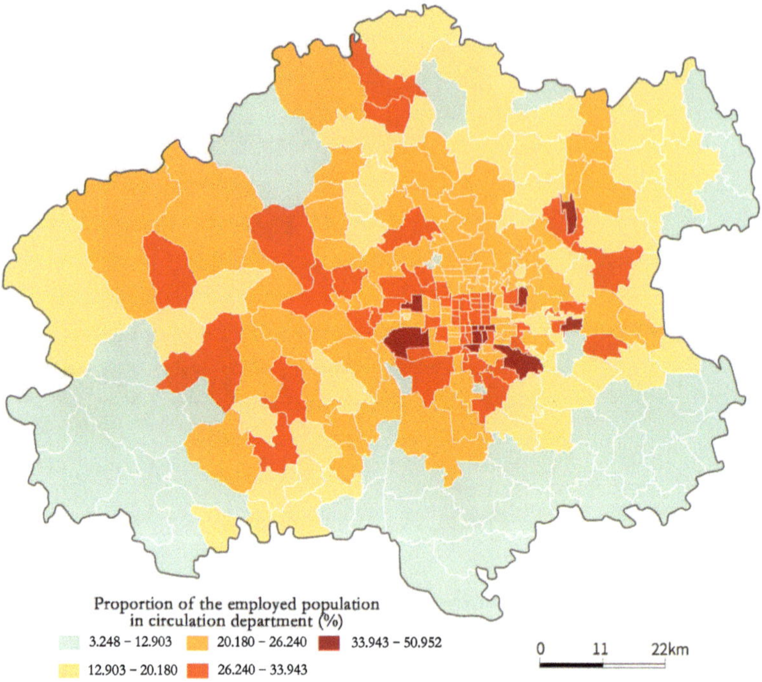

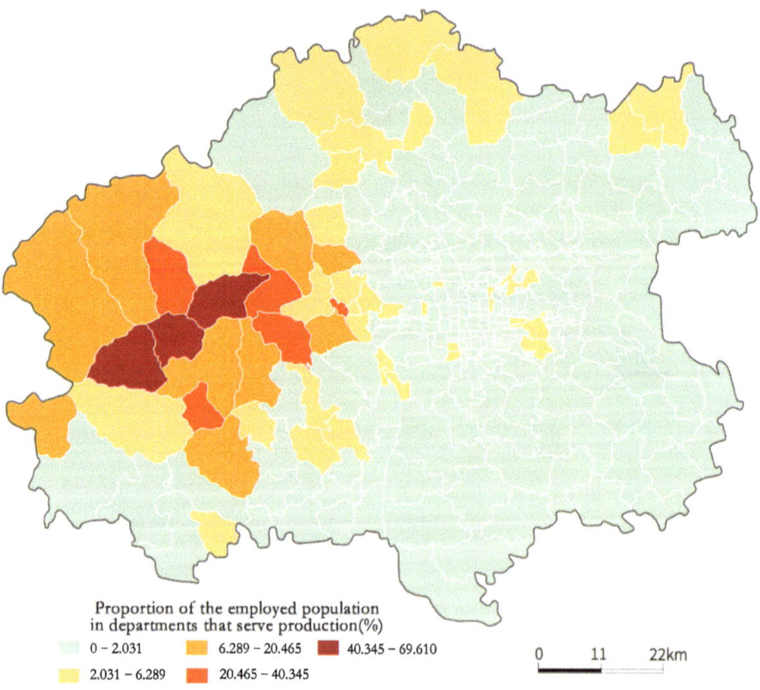

Fig. 5.10 (continued)

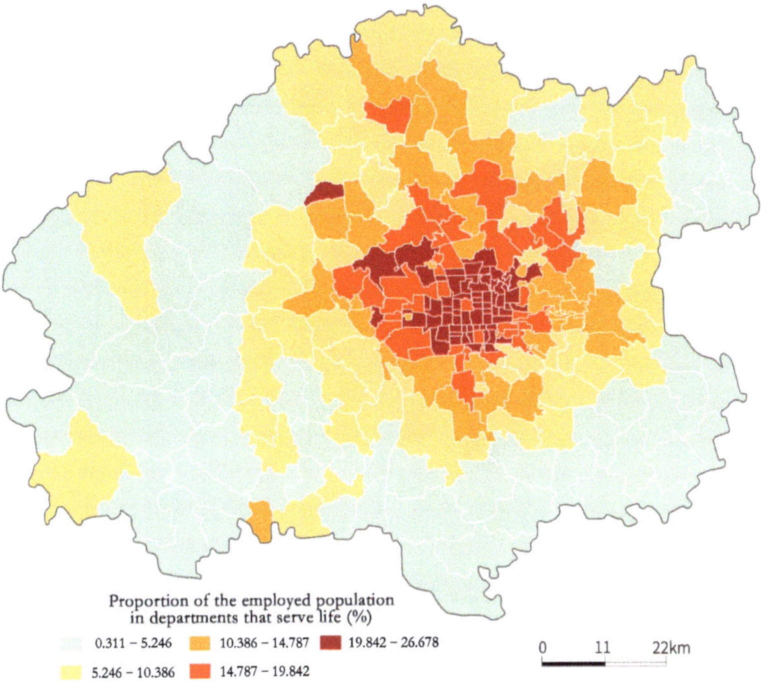

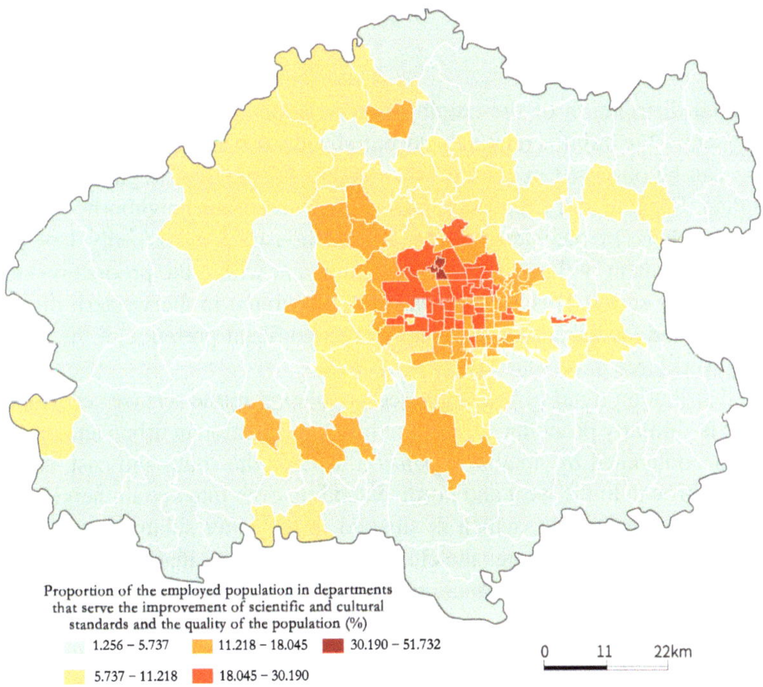

Fig. 5.10 (continued)

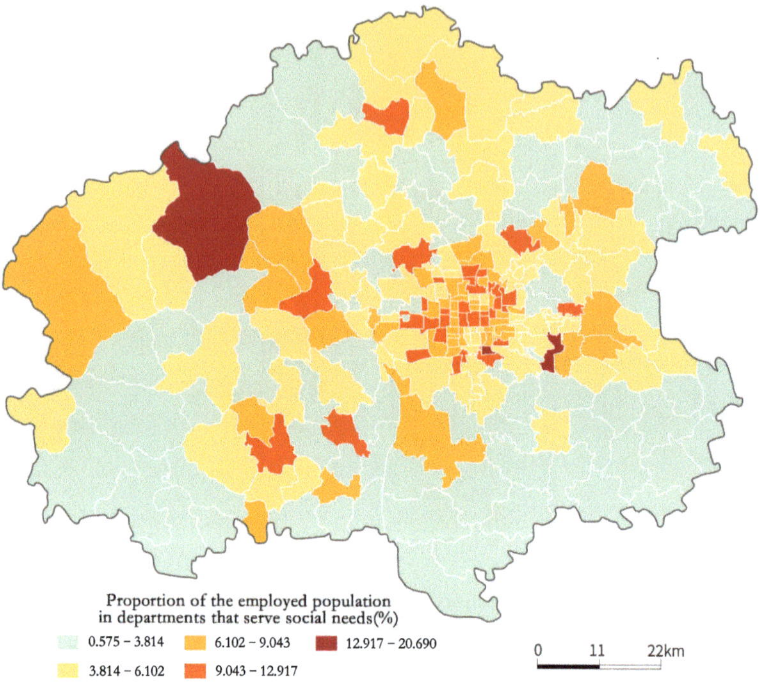

Fig. 5.10 (continued)

The spatial distribution of the employed population within the circulation sector (tertiary sector 1) exhibits a relatively decentralized pattern; however, notable concentrations can be observed in central urban areas and suburban regions, particularly in Haidian, Fengtai, and Chaoyang. Conversely, numerous neighborhoods situated on the periphery of the metropolitan area demonstrate a markedly low proportion of employment within this sector. Employment within the production services sector (tertiary sector 2) is predominantly concentrated in the western districts of Mentougou and Fangshan, attributable to the significant presence of the extractive industry workforce in these areas.

The distribution trend of the employed population in the service sector (tertiary sector 3) is similarly pronounced, with a high concentration in urban and peri-urban locales, accompanied by notable expansion towards the south and east, as well as significant growth to the west and north, extending even into certain neighborhoods of the Changping District, which is situated in the outer suburban regions. This distribution pattern underscores the clustering of industries that are intensive in life services, such as finance, insurance, real estate, and social services, and is closely aligned with the developmental trajectory of urbanized areas.

The distribution of the employed population in sectors aimed at enhancing scientific and cultural standards, as well as the overall quality of the populace (tertiary sector 4), is particularly straightforward to interpret. These sectors are predominantly concentrated in university-dense areas of Haidian District, as well as in adjacent neighborhoods within Xicheng and Dongcheng, and in the vicinity of the Asian Games Village and Datun Road in Chaoyang District. To a lesser extent, there is also representation in the six peripheral suburban districts and counties. Furthermore, regarding sectors that cater to public needs (tertiary sector 5), areas with relatively high employment shares are dispersed across certain neighborhoods in Xicheng and Dongcheng, as well as in suburban and peri-urban districts and counties where government offices are situated.

5.3.3.3 Spatial Distribution Patterns and Characteristics of the Housing Situation of Resident Population

The spatial distribution of housing conditions among the resident population of Beijing is delineated by various indicators, including monthly rental rates, costs associated with purchasing and constructing homes, the acquisition of affordable housing, and rental circumstances (see Fig. 5.11).

In terms of per capita housing space, neighborhoods within Beijing that exhibit lower per capita housing availability are predominantly situated in the central urban areas, whereas the extensive regions beyond these central areas typically offer larger housing spaces. Regarding the distribution of households incurring monthly rents of RMB 1,500 or more, a significant concentration of such households is observed in the urban core and its adjacent suburbs. Conversely, the distribution of households that have expended over 200,000 yuan on home purchase and construction is more widespread than that of households paying monthly rents exceeding 1,500 yuan. Notably, certain distant suburban locales, such as Huilongguan and Beiqijia in Changping, Houshayu in Shunyi, and Yizhuang in Daxing, exhibit higher concentrations of this demographic. The majority of areas with a proportion exceeding 7% arc located outside the urban core.

With respect to the acquisition of affordable housing, neighborhoods with elevated proportions are sporadically located in both proximal and distal suburban areas, reflecting the spatial distribution of affordable housing initiatives. Noteworthy examples include Yizhuang, Huilongguan, and Wangjing, which are recognized for their significant affordable housing developments in Beijing.

The distribution of households renting public housing and those renting commercial housing reveals a concentric circle pattern, exhibiting an approximate complementary relationship. The proportion of households renting public housing diminishes with increasing distance from the city center, with the highest concentrations found in the central urban area, followed by the inner suburban regions, the outer suburban areas, and the lowest proportions at the metropolitan periphery. Similarly, the distribution of households renting commercial housing adheres to a concentric circle model, with the lowest proportions in the central urban area and the metropolitan outskirts, while the highest proportions are located in the inner peri-urban areas, with elevated shares also present in the outer peri-urban regions and the inner edge of the metropolitan area. It is noteworthy that certain areas at the

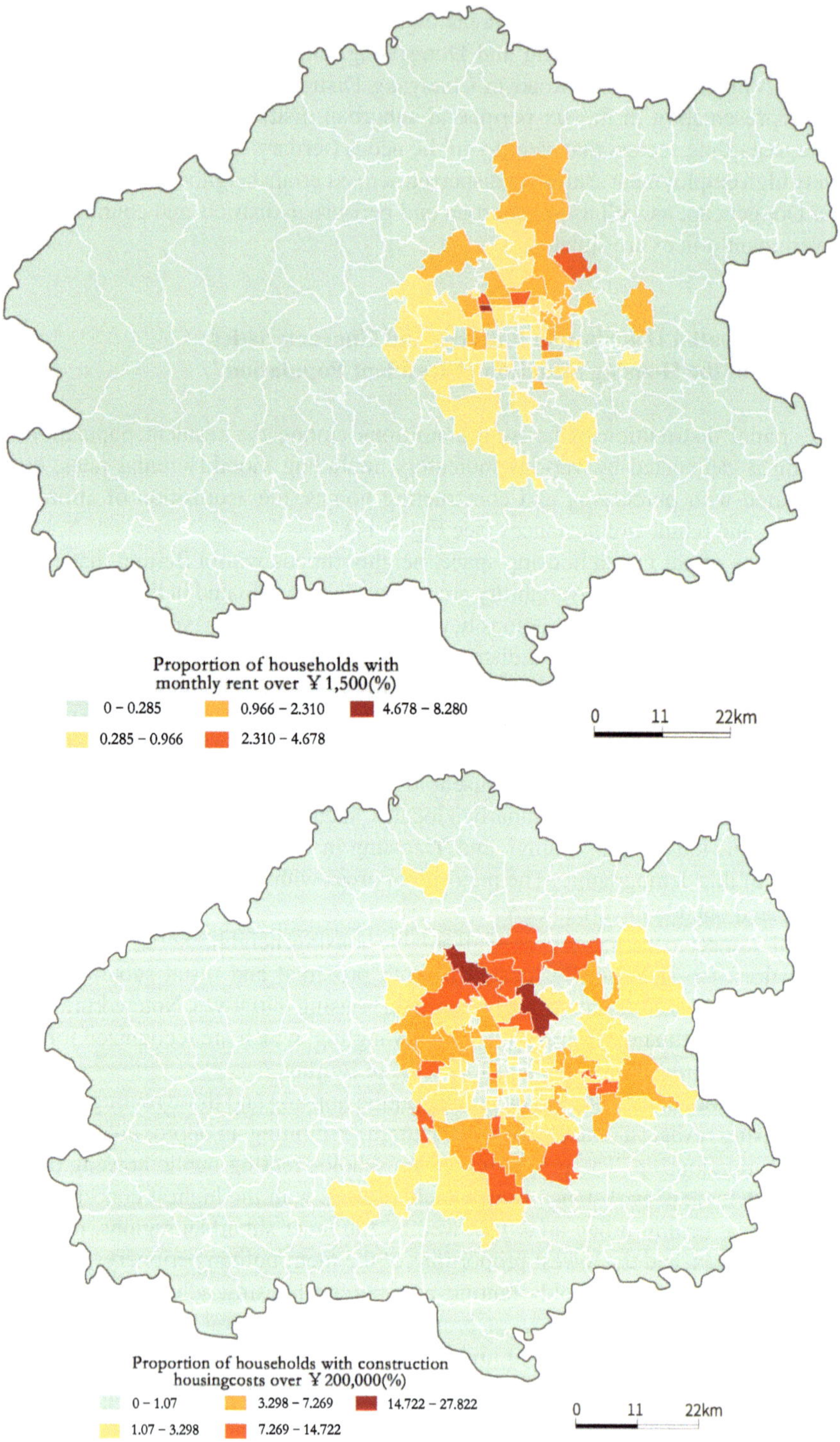

Fig. 5.11 Spatial distribution of housing for the resident population in Beijing (2000)

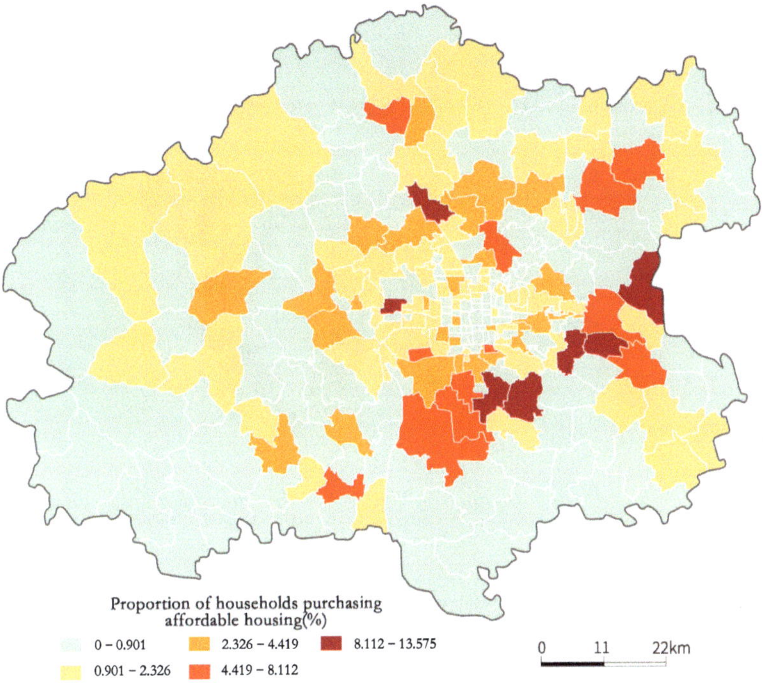

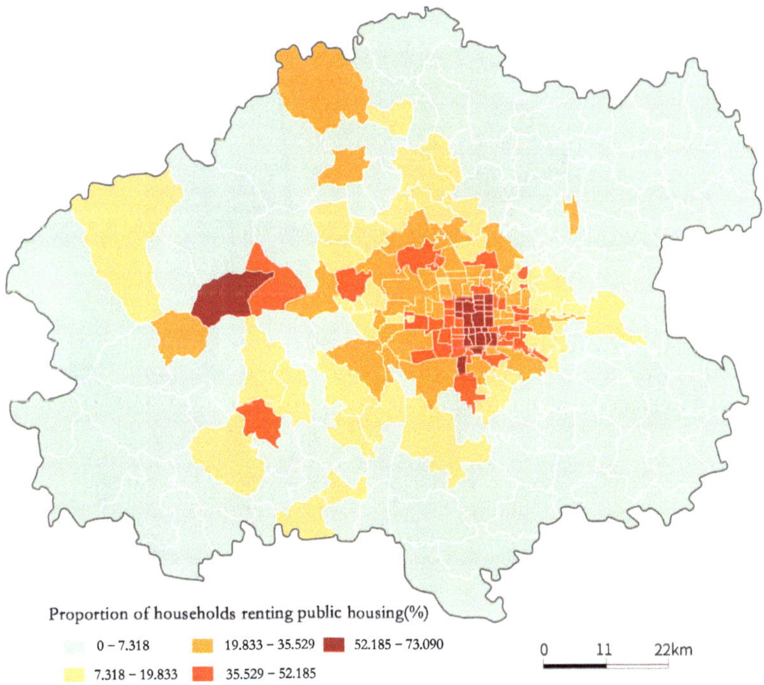

Fig. 5.11 (continued)

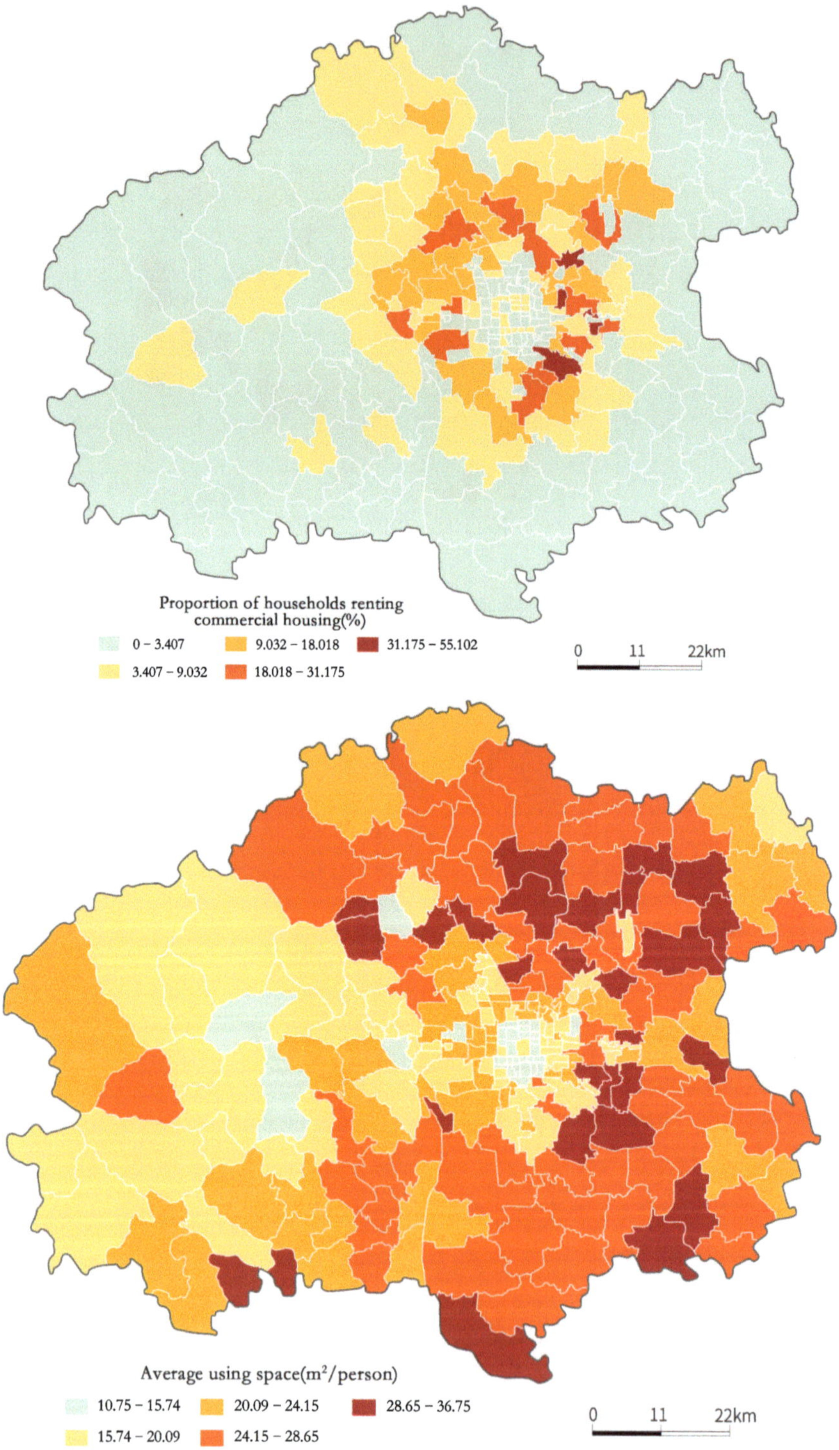

Fig. 5.11 (continued)

metropolitan periphery, where industrialization and urbanization have historically progressed more rapidly, demonstrate a relatively high proportion of public housing, whereas the proportion of commercial housing is comparatively elevated in the more distant suburban and county government districts.

5.3.4 *Restructuring of Differentiation of Urban Residential Space in the Late Period of Planned Economy and the Early Period of Market Transformation (1982–2000)*

5.3.4.1 Restructuring Statistical Characteristics of Differentiation of Residential Population Based on Social Indicators

Table 5.8 presents statistical characteristics of various social indicators for the Beijing metropolitan area, encompassing extreme deviation, maximum value, mean, and standard deviation. The standard deviation serves as a measure of the extent of deviation from the mean, with a reduction in standard deviation indicating a decline in overall sample variation. Conversely, extreme deviation reflects the polarization among individual samples, with an increase in extreme deviation signifying heightened sample polarization.

The findings indicate that from 1982 to 2000, several indicators reflecting the overall trend of diminishing disparities among neighborhoods include population density, household size, the Manchu and Hui populations, the proportion of illiterate individuals, the unemployed population, and the proportions of the employed population in both primary and secondary industries, as well as in sectors that enhance scientific and cultural standards, population quality, and social and public needs across the three industrial sectors. Additionally, the proportion of the employed population in the tertiary sector (specifically tertiary 2, tertiary 4, and tertiary 5) has also shown a similar trend. Conversely, disparities in social indicators such as gender ratio, proportion of the elderly population, foreign population, individuals with university education or higher, the Mongolian and Zhuang populations, and the share of the employed population in the tertiary sector, particularly in the circulation sector and sectors serving living needs (tertiary 3, tertiary 1, and tertiary 3), have exhibited an increasing trend.

The variability of the sample as a whole, as indicated by the standard deviation of the indicators over the 18-year period, does not consistently align with the polarization observed among individual samples, as evidenced by the extreme deviation. A notable example is population density, where the overall disparity has decreased while individual polarization has increased. This suggests that over the past 18 years, although residential density in the metropolitan area has evolved in a more balanced manner, the differences between the extremes of geographic areas' attractiveness for habitation have intensified. This trend aligns with the characteristics of urban sprawl

Table 5.8 Differences in statistical characteristics of social indicators in the Beijing metropolitan area, 1982 and 2000

Indicators	Year 2000				Year 1982			
	Range	Maximum	Mean value	Standard deviation	Range	Maximum	Mean value	Standard deviation
Population densities (persons per sq km)	62,186.8	62,215.9	8902.6	11,644.3	57,035.9	57,080.1	7497.6	13,363.4
Sex ratio (female equal to 100)	136.5	231.0	110.8	15.2	82.0	160.8	101.6	11.2
Average number of persons per household	1.4	3.9	3.0	0.3	5.2	8.2	3.8	0.4
Proportion of population over 60 years old/%	17.1	21.9	12.5	3.5	12.1	17.2	12.4	2.5
Number of migrants in the neighborhood/person	138,703.0	139,037.0	18,760.1	21,973.4	8025.0	8047.0	839.5	1111.7
Manchu population/person	4720.0	4720.0	831.8	804.9	7185.0	7185.0	483.1	848.3
Hui population/person	9399.0	9399.0	957.4	1278.6	13,323.0	13,323.0	959.2	1755.1
Mongol population /person	1763.0	1764.0	146.5	184.5	919.0	919.0	48.5	107.6
Zhuang population/person	398.0	398.0	25.5	43.9	206.0	206.0	9.1	20.0
Uyghur population/person	706.0	706.0	12.7	51.5	-	-	-	-
Proportion of illiterate people/%	18.9	19.6	5.7	3.2	31.1	33.9	16.8	6.6
Population with University Degree or Above/%	74.9	75.2	13.5	13.8	63.4	63.5	4.5	8.1
Employed population in primary industry/%	89.9	89.9	17.1	23.3	87.7	87.8	41.2	32.5

(continued)

Table 5.8 (continued)

Indicators	Year 2000				Year 1982			
	Range	Maximum	Mean value	Standard deviation	Range	Maximum	Mean value	Standard deviation
Employed population in secondary industry/%	64.2	65.3	29.1	12.4	64.2	65.6	32.3	17.3
Tertiary Industry Employment/%	82.8	89.5	54.6	22.1	76.8	80.5	26.5	20.3
Tertiary 1/%	47.7	51.0	21.9	8.1	64.5	65.4	8.9	7.9
Tertiary 2/%	69.6	69.6	3.1	8.7	67.8	67.8	3.1	9.2
Tertiary 3/%	26.4	26.7	13.0	7.5	8.3	8.3	1.7	2.0
Tertiary 4/%	50.5	51.7	10.0	7.9	71.2	72.6	9.4	11.2
Tertiary 5/%	20.1	20.7	5.8	3.2	18.8	19.2	3.4	3.4
Unemployed population/person	8527.0	8569.0	1698.0	1509.2	77,183.0	77,574.0	7918.3	7912.2
Managerial personnel (government agencies, party, and enterprises)/%	16.0	16.3	4.8	3.3	11.6	12.1	3.1	2.2
Professionals and technicians/%	48.5	50.5	14.1	10.5	53.1	57.0	11.3	8.8
Office personnel/%	22.3	22.8	9.1	5.4	12.8	13.0	3.1	2.7
Commercial and services personnel/%	50.5	51.5	22.0	9.4	20.1	21.8	9.2	5.7

(continued)

Table 5.8 (continued)

Indicators	Year 2000				Year 1982			
	Range	Maximum	Mean value	Standard deviation	Range	Maximum	Mean value	Standard deviation
Agricultural occupational personnel/%	88.6	88.6	18.0	23.7	80.2	80.4	36.4	29.0
Industry and transportation personnel/%	69.3	74.6	32.4	13.5	57.5	68.6	37.0	14.4
Housing cost over $200,000/%	27.8	27.8	2.0	3.6	–	–	–	–
Number of households purchasing commercial housing/%	27.9	27.9	2.4	4.1	–	–	–	–
Number of households purchasing affordable housing	13.6	13.6	1.6	2.2	–	–	–	–
Number of households renting public housing	73.1	73.1	21.0	20.7	-	-	-	-
Number of Households Renting Commercial Housing	55.1	55.1	5.8	8.4	–	–	–	–
Per capita housing area/m^2	26.0	36.8	22.2	5.5	–	–	–	–

and the development of the real estate sector in Beijing. Furthermore, the increase in overall disparities in the employed population within the distribution sector, coupled with a decrease in individual polarization, reflects a general rise in specialization within that sector.

5.3.4.2 Restructuring of Systematic Differentiation of Residential Population

This study calculates the information entropy of 35 indicators that characterize the living environment within the Beijing metropolitan area. It is noteworthy that the third national census did not provide data on housing indicators; therefore, changes in relevant indices from 1982 to 2000 are inferred from non-housing indicators.

The results of the calculations indicate that the information entropy values for the majority of indicators range between 4 and 5 bits for both 1982 and 2000. A comparative analysis reveals that, with the exception of the agricultural occupational population, the information entropy for all indicators has increased. This suggests that the residential spatial system of the metropolitan area has evolved towards greater complexity, while the spatial system associated with the agricultural occupational population has become simpler. This transformation aligns with the broader trends of industrial structural adjustment and evolution observed in the metropolitan area over the past two decades, characterized by a decline in the agricultural workforce within urban settings and a gradual replacement of agriculture in peri-urban areas by secondary and tertiary industries, leading to an overall contraction of agricultural activities.

Furthermore, changes in information entropy can serve as an indicator of the degree of balance within system distribution. Specifically, a higher information entropy value, derived from spatial data, suggests a reduced spatial disparity and a more equitable spatial distribution. The findings indicate that the spatial distribution of most social indicators has progressed towards a more balanced state over the 18-year period, with the exception of the agricultural occupational population, which has exhibited a tendency towards concentration in the outer suburban regions. By establishing a threshold of 0.5 bits of information entropy change as a standard, it is evident that significant alterations in spatial distribution from 1982 to 2000 are observed in indicators such as population density, the population of ethnic minorities, individuals with university degrees or higher, and the employed population in secondary and tertiary industries. Notably, the populations of the Manchu and Mongolian ethnic groups, as well as the employed workforce in tertiary industries that cater to production, daily life, and social public needs (specifically tertiary industries 2, 3, and 5), have all experienced an increase in information entropy exceeding 0.7 bits. This indicates that the spatial differentiation of these social indicators has undergone a substantial reconfiguration process.

5.3.4.3 Restructuring of Absolute Spatial Differentiation of Social Indicators of Residential Population

The parameters derived from various methodologies exhibit distinct emphases and levels of detail. Information entropy, lacking a reference object, serves to illustrate the disparities among different entities as a macroscopic statistical measure. It primarily emphasizes overarching patterns, thereby providing insights into the general evolution and restructuring of the system; however, it may not adequately capture the nuances of specific differences. Conversely, the differential index incorporates a reference object, which enables it to emphasize finer micro-level details. Consequently, it is essential to compute the differential index for the social indicators of Beijing (refer to Table 5.9).

The absolute differential index, which is derived from spatial data, serves as a metric for assessing spatial differential in relation to a perfectly uniform distribution. The outcomes of this index exhibit a general trend that aligns with those observed in information entropy calculations, albeit with some minor discrepancies. Notably, with the exception of certain indicators—namely, the elderly population, sex ratio, average household size, and agricultural population—the absolute differential index for the majority of other indicators from 1982 to 2000 demonstrates a declining trend. This suggests a reduction in spatial differential for most social indicators.

Despite the overall increase in the aging population, the extent and trajectory of aging vary across different regions. An analysis of the relationship between the proportion of the elderly population and distance within the Beijing metropolitan area reveals a marked intensification of spatial differentiation in aging levels over the past 18 years. Specifically, aging levels in older urban areas have significantly increased, while many neighborhoods in the inner suburbs continue to experience a decline in aging levels. This phenomenon can be attributed to the relatively youthful demographic structure of the inner suburbs, which is influenced by a concentration of migrants and a substantial number of residents who have relocated to suburban areas.

Furthermore, the spatial differentiation of the overall aging level within the metropolitan area has intensified. Similar trends are observed in the spatial differentiation of the sex ratio and average household size, which reflects family size, although the degree of change in these indicators is comparatively modest. The observed results can be elucidated by examining the relationship between the sex ratio and distance: within a 5–25 km radius from the city center, there has been a significant increase in the sex ratio of neighborhoods over the past 18 years, while other spatial areas have remained relatively stable, thereby reinforcing overall spatial differentiation. A comparable explanation applies to the variations in average household size. The fluctuations in the absolute differentiation index concerning the agricultural population are linked to the relative concentration of agricultural activities. An analysis of housing attribute indicators in the Beijing metropolitan area in 2000 reveals that the absolute differentiation index exceeded 0.5, indicating pronounced spatial differentiation characteristics in housing within Beijing.

Table 5.9 Residential spatial differentiation index and its changes in the Beijing metropolitan area (1982–2000)

Indicators	Absolute differential index			Relative differential index			Information entropy		
	Year 2000	Year 1982	Difference in value	Year 2000	Year 1982	Difference in value	Year 2000	Year 1982	Difference in value
Total population	0.298	0.321	−0.023	0.000	0.000	0.000	5.205	4.915	0.290
Population density	0.554	0.657	−0.103	0.446	0.534	−0.088	4.693	4.096	0.597
Sex ratio	0.044	0.035	0.009	0.302	0.319	−0.017	5.472	5.231	0.241
Household size	0.044	0.034	0.010	0.324	0.335	−0.011	5.475	5.231	0.244
Elderly population	0.330	0.308	0.022	0.124	0.089	0.035	5.179	4.946	0.233
Migrants	0.414	0.457	−0.043	0.181	0.187	−0.006	4.950	4.624	0.326
Manchu population	0.376	0.539	−0.163	0.186	0.359	−0.173	5.044	4.290	0.754
Hui population	0.462	0.560	−0.098	0.319	0.342	−0.024	4.786	4.237	0.549
Mongol population	0.417	0.617	−0.201	0.203	0.433	−0.230	4.921	3.984	0.937
Zhuang population	0.476	0.584	−0.108	0.324	0.427	−0.103	4.679	4.063	0.615
Uyghur population	0.621	–	–	0.506	–	–	3.792	–	–
Illiterate population	0.232	0.240	−0.008	0.204	0.176	0.028	5.299	5.026	0.273
Population with university degree or above	0.513	0.619	−0.106	0.292	0.417	−0.124	4.672	4.074	0.597
Employed in primary industry	0.508	0.368	0.141	0.635	0.515	0.120	4.665	4.760	−0.095
Employed in secondary industry	0.322	0.469	−0.148	0.162	0.186	−0.023	5.137	4.583	0.553
Tertiary employed population	0.379	0.501	−0.121	0.114	0.245	−0.131	5.056	4.525	0.530
Tertiary 1 employed population	0.355	0.525	−0.170	0.105	0.259	−0.154	5.089	4.489	0.599
Tertiary 2 employed population	0.450	0.580	−0.129	0.398	0.558	−0.160	4.746	3.896	0.851
Tertiary 3 employed population	0.408	0.597	−0.189	0.168	0.353	−0.186	4.989	4.262	0.726
Tertiary 4 employed population	0.458	0.540	−0.082	0.223	0.314	−0.091	4.851	4.400	0.451

(continued)

Table 5.9 (continued)

Indicators	Absolute differential index			Relative differential index			Information entropy		
	Year 2000	Year 1982	Difference in value	Year 2000	Year 1982	Difference in value	Year 2000	Year 1982	Difference in value
Tertiary 5 employed population	0.412	0.572	−0.160	0.184	0.335	−0.151	4.997	4.257	0.740
Unemployed population	0.333	0.334	−0.001	0.075	0.064	0.011	5.149	4.893	0.257
Managerial personnel (government agencies, party, and enterprises)	0.429	0.519	−0.090	0.229	0.242	−0.013	4.942	4.453	0.488
Professionals and technicians	0.449	0.490	−0.041	0.203	0.234	−0.031	4.886	4.545	0.341
Office personnel	0.423	0.548	−0.125	0.164	0.272	−0.108	4.977	4.383	0.594
Commercial and services personnel	0.365	0.489	−0.124	0.116	0.208	−0.092	5.061	4.580	0.481
Agricultural occupational personnel	0.506	0.365	0.140	0.634	0.512	0.122	4.678	4.777	−0.098
Industry and transportation personnel	0.292	0.430	−0.137	0.173	0.145	0.029	5.192	4.692	0.500
Number of Households Renting $1,500 or More per Month	0.628	–	–	0.471	–	–	4.189	–	–
Households paying more than $200,000 for housing	0.597	–	–	0.461	–	–	4.253	–	–
Number of households purchasing commercial housing	0.607	-	-	0.511	–	–	4.113	-	-
Number of households purchasing affordable housing	0.577	–	–	0.416	–	–	4.421	–	–
Number of households renting public housing	0.501	–	–	0.298	–	–	4.790	–	–

(continued)

Table 5.9 (continued)

Indicators	Absolute differential index			Relative differential index			Information entropy		
	Year 2000	Year 1982	Difference in value	Year 2000	Year 1982	Difference in value	Year 2000	Year 1982	Difference in value
Number of households renting commercial housing	0.542	–	–	0.412	–	–	4.512	–	–
Per capita housing area	0.100	–	–	0.349	–	–	5.450	–	–

5.3.4.4 Restructuring of Relative Spatial Differentiation of Social Indicators of Residential Population

This section involves the calculation of the relative differential index (refer to Table 5.9) in relation to the overall distribution of the residential population. A decrease in the relative differential index of a particular social indicator over time suggests an increasing alignment of its spatial distribution pattern with that of the total population. Analysis of the computed results reveals that from 1982 to 2000, the relative diversity indices for certain social indicators, including the elderly population, illiterate population, agricultural population, non-working population, and the population engaged in agricultural production and transportation, have exhibited an upward trend. Conversely, the relative diversity indices for other indicators have demonstrated a downward trajectory, indicating an enhancement in the congruence between the spatial distribution patterns of these social indicators and the overall population distribution.

Notably, population density is inherently correlated with the general population distribution. The spatial patterns of migrant populations, ethnic minority groups, and individuals with a college degree or higher can be interpreted as a consequence of the distribution and growth of the total population. Furthermore, the alignment between the secondary and tertiary industries and the overall population distribution has improved, particularly in relation to various service sectors that have emerged over the past 18 years, as these services predominantly cater to urban populations, thereby increasing their consistency with the overall population distribution.

In contrast, the elderly and illiterate populations have progressively diverged from the overall population distribution pattern over the past 18 years, attributable to their distinct distribution characteristics. A descending order arrangement of the total population, elderly population, and illiterate population within each neighborhood elucidates the discrepancies among them. For instance, neighborhoods in the historic district characterized by a high elderly population do not exhibit a significant total population, while outer suburban neighborhoods with a notable illiterate population similarly lack a substantial total population. A comparable rationale applies to the observed changes in the relative differential index of the agricultural population, as the relative concentration of agricultural activities does not correlate with the total population distribution. Additionally, the variation in the relative differential index of the non-working population indicates that areas with high population density do not necessarily coincide with areas of high non-working population density.

Previous research has assessed the relative differential index of housing indicators utilizing neighborhood data from Shanghai, which can be juxtaposed with data from Beijing. In the year 2000, the relative diversity indices for households acquiring commercial housing, affordable housing, renting publicly owned housing, and renting commercial housing in Shanghai were recorded at 0.249, 0.356, 0.209, and 0.549, respectively. In contrast, the spatial diversity associated with commercial housing, as well as the spatial differentiation of affordable housing and public housing tenants in Beijing, is markedly greater than that observed in Shangha.1 However, the spatial differentiation of rental commercial housing in Beijing is lower than that in Shangha.1 This indicates that, relative to Shanghai, the alignment between commercial housing development and the residential population in Beijing is considerably less effective. Conversely, the rental housing market in Beijing appears to be more developed, exhibiting a slightly improved spatial alignment with the population compared to Shangha.1

5.3.4.5 Restructuring of Status of Residential Space: Groupment and Miscibility

The segregation index serves as a useful metric for assessing the spatial dynamics between specific demographic groups and their interactions with other populations. It provides insights into the degree of clustering and integration among different population segments; however, its applicability is limited to certain demographic indicators. From a total of 35 indicators, we selected those pertaining to the elderly population, migrant population, minority population, illiterate individuals, individuals with a college degree or higher, the unemployed population, as well as various occupational and industrial groups, in order to compute the segregation index and analyze its variations for these selected indicators (see Table 5.10).

The findings of the analysis reveal a distinct trend. Over the past 18 years, the segregation index for all demographic indicators, with the exception of the elderly population, migrant population, illiterate population, individuals employed in primary industries, and those engaged in agricultural occupations, has exhibited a decline. This suggests that, overall, the demographic composition of the Beijing metropolitan area is becoming increasingly heterogeneous. However, the elderly, migrant, illiterate, and agricultural populations are experiencing a relative decrease in integration. In terms of absolute values recorded in 2000, the segregation index was highest for the agricultural population, with a score of 0.638 for those employed in primary industries and 0.637 for individuals in agricultural occupations, followed by the Uyghur population at 0.506. All other demographic indicators fell below 0.4, with the majority registering below 0.3. This trend indicates a gradual disappearance of small agricultural plots within the urbanized areas of Beijing, with agricultural land becoming increasingly concentrated in the city's outer suburbs. This observation aligns with the expansion of Beijing's urban social landscape, while the Uyghur population continues to demonstrate a significant degree of community cohesion, corroborating the findings of the *Beijing Social Area Study* (Feng and Zhou 2003a). It is important to note that, aside from the Uyghur population—due to the absence of comparative data from 1982—the segregation indices for the Manchu, Hui, Mongol, and Zhuang populations have all decreased. Consequently, there is an overall increase in the integration of minority and Han populations, reflecting the characteristic of multi-ethnic integration prevalent in major cities across China.

Table 5.10 Segregation index of residential space in the Beijing metropolitan area and its changes (1982–2000)

Indicators	Year 2000	Year 1982	Difference in value	Indicators	Year 2000	Year 1982	Difference in value
Elderly population	0.142	0.101	0.041	Tertiary 1 Employed Population	0.107	0.279	−0.173
Migrants	0.287	0.191	0.097	Tertiary 2 Employed population	0.398	0.565	−0.167
Manchu population	0.189	0.363	−0.174	Tertiary 3 Employed population	0.169	0.359	−0.190
Hui population	0.325	0.350	−0.026	Tertiary 4 Employed population	0.225	0.339	−0.115
Mongol population	0.204	0.434	−0.230	Tertiary 5 Employed population	0.185	0.346	−0.161
Zhuang population	0.324	0.427	−0.103	Unemployed population	0.078	0.079	−0.001
Uyghur population	0.506	–	–	Managerial personnel (government agencies, party, and enterprises)	0.230	0.249	−0.019
Illiterate population	0.213	0.204	0.009	Professionals and technicians	0.205	0.256	−0.052
Population with college degree or above	0.358	0.441	−0.083	Office personnel	0.165	0.280	−0.115
Employed in primary industry	0.638	0.599	0.039	Commercial and services personnel	0.118	0.224	−0.107
Employed in secondary industry	0.165	0.248	−0.083	Agricultural occupational personnel	0.637	0.583	0.053
Employed in tertiaryIndustry	0.118	0.308	−0.190	Industry and transportation personnel	0.176	0.196	−0.020

5.3.5 Restructuring of Urban Residential Space Differentiation Under Market Economic Conditions (2000–2010)

5.3.5.1 Statistical Characteristics of Social Indicators of the Residential Population

Table 5.11 gives the statistical characteristics of several social indicators in Beijing, including the range, maximum value, average value and standard deviation. The standard deviation can reflect the degree of dispersion of a set of values. A decrease in the standard deviation indicates that the overall difference in the sample has decreased. The range is the difference between the maximum and minimum values in a set of data, representing the polarization between individual samples. An increase in the range indicates that the samples have become more polarized.

Table 5.11 illustrates that between 2000 and 2010, there was a decline in the range, standard deviation, maximum, and mean values of four variables: sex ratio, illiterate population, employment in the primary industry (agricultural sector), and the cadre and staff of government and institutions. This trend suggests a reduction in the issue of an imbalanced sex ratio in Beijing, as well as a decrease in disparities among neighborhoods. The variations in the size of the illiterate population, employment in the primary industry, and the cadre and staff of government and institutions across neighborhoods have also diminished, primarily attributable to a reduction in the size of these three demographic groups over the past decade. Furthermore, the disparities in employment within the tertiary industry, particularly in tertiary industry sectors 2 (departments serving production) and 3 (departments serving life), have also lessened across neighborhoods. This change is largely due to an increase in the number of individuals employed in the tertiary industry, leading to a more equitable distribution. Notably, while the overall differences in population density indicators and the proportion of individuals engaged in production and transportation have increased, individual polarization has decreased. This indicates that, although the disparity in population density among neighborhoods has grown, the extremes of this disparity have lessened. Conversely, polarization among the Hui population, as well as the proportions of individuals employed in tertiary industry sectors 2 and 3, has increased, despite an overall reduction in differences.

5.3.5.2 Restructuring of Spatial Differentiation of Residential Population Based on Social Indicators

(1) **System differentiation**

Information entropy serves to elucidate two dimensions of information. Firstly, it provides insight into the complexity of the residential spatial system; a higher value of information entropy signifies greater complexity within the system. Conversely, it also indicates the level of equilibrium in the spatial distribution of variables; a lower information entropy value corresponds to a diminished degree of balance in this distribution. As illustrated in Table 5.12 and Fig. 5.12, the period from 2000 to 2010 witnessed an increase in the information entropy of the majority of indicators, suggesting a progressive complexity in Beijing's overall residential spatial system. However, a decline in information entropy was observed for several indicators, including those related to employment in the primary and secondary industries, the tertiary industry (specifically the circulation sector), commercial services, agriculture, production and transportation, as well as the purchasing of affordable housing. This trend indicates a growing concentration in the spatial distribution of Beijing's primary and secondary industries, commercial services, and affordable

Table 5.11 Differences in the statistical characteristics of social indicators in Beijing in 2000 and 2010

Indicators	Year 2000				Year 2010			
	Range	Maximum	Mean values	Standard deviation	Range	Maximum	Mean values	Standard deviation
Population densities (persons per sq km)	62,186.8	62,215.9	8902.6	11,644.3	49,375.9 ↓	49,401.2 ↓	10,728.9	11,708.6
Sex ratio (female equal to 100)	136.5	231	110.8	15.2	68.7 ↓	152.4 ↓	107.5 ↓	11.5 ↓
Average number of persons per household	1.4	3.9	3.3	0.3	1.6	3.4 ↓	2.5 ↓	0.3
Percentage of population over 60 years old / %	17.1	21.9	12.5	3.5	22.4	24.6	9.6 ↓	3.9
Number of migrants in the neighborhood/person	138,703.0	139,037.0	18,760.1	21,973.4	197,353.0	197,420.0	27,417.9	33,394.0
Mongol population / person	1763.0	1764.0	146.5	184.5	2311.0	2312.0	292.3	340.0
Hui population /person	9399.0	9399.0	957.4	1278.6	10,929.0	10,929.0	994.2	1245.5 ↓
Uyghur population / person	706.0	706.0	12.7	51.5	725.0	725.0	34.7	88.2
Hmong population / person	–	–	–	–	629.0	629.0	50.3	78.3
Zhuang population / person	398.0	398.0	25.5	43.9	775.0	775.0	56.6	85.8
The Korean nationality population /person					6699.0	6699.0	143.5	458.8
Manchu population / person	4720.0	4720.0	831.8	804.9	6117.0	6122.0	1085.0	997.9

(continued)

Table 5.11 (continued)

Indicators	Year 2000				Year 2010			
	Range	Maximum	Mean values	Standard deviation	Range	Maximum	Mean values	Standard deviation
Tujia population /person	–	–	–	–	1463.0	1463.0	92.6	152.6
Proportion of illiterate people/%	18.9	19.6	5.7	3.2	16.2 ↓	16.6 ↓	2.3 ↓	2.1 ↓
Population with University Degree or Above/%	74.9	75.2	13.5	13.8	80.3	81.3	28.2	17.3
Employed population in primary industry/%	89.9	89.9	17.1	23.3	71.7 ↓	71.7 ↓	6.3 ↓	12.7 ↓
Employed population in secondary industry/%	64.2	65.3	29.1	12.4	66.6	68.5	22.7 ↓	13.0
Tertiary Industry Employment/%	82.8	89.5	54.6	22.1	74.5 ↓	95.5	70.5	19.3 ↓
Tertiary 1/%	47.7	51.0	21.9	8.1	60.3	65.1	30.0	10.1
Tertiary 2/%	69.6	69.6	3.1	8.7	82.6	82.9	3.4	6.9 ↓
Tertiary 3/%	26.4	26.7	13.0	7.5	32.4	32.7	13.7	7.0 ↓
Tertiary 4/%	50.5	51.7	10.0	7.9	68.2	68.8	16.0	10.3
Tertiary 5/%	20.1	20.7	5.8	3.2	70.0	70.9	7.4	8.6
Managerial personnel (government agencies, party, and enterprises)/%	16.0	16.3	4.8	3.3	12.2 ↓	12.2 ↓	2.9 ↓	2.1 ↓

(continued)

Table 5.11 (continued)

Indicators	Year 2000				Year 2010			
	Range	Maximum	Mean values	Standard deviation	Range	Maximum	Mean values	Standard deviation
Professionals and technicians/%	48.5	50.5	14.1	10.5	50.2	51.6	18.8	11.1
Office personnel/%	22.3	22.8	9.1	5.4	32.6	33.5	14.9	7.7
Commercial and services personnel/%	50.5	51.5	22.0	9.4	58.0	67.5	31.7	10.2
Agricultural occupational personnel/%	88.6	88.6	18.0	23.7	71.4 ↓	71.4 ↓	7.8 ↓	14.9 ↓
Industry and transportation personnel/%	69.3	74.6	32.4	13.5	69.0 ↓	73.6 ↓	23.9 ↓	14.9
Housing cost over $200,000/%	27.8	27.8	2.0	3.6	–	–	–	–
Managerial personnel (government agencies, party, and enterprises)/%	27.9	27.9	2.4	4.1	73.6	73.6	13.7	15.7
Proportion of affordable housing purchased/%	13.6	13.6	1.6	2.2	47.9	47.9	3.5	6.7
Proportion of second-hand housing purchased/%	–	–	–	–	14.6	14.6	2.6	2.7

(continued)

Table 5.11 (continued)

Indicators	Year 2000				Year 2010			
	Range	Maximum	Mean values	Standard deviation	Range	Maximum	Mean values	Standard deviation
Proportion of former public housing purchased/%	–	–	–	–	96.3	96.3	17.7	20.5
Proportion of low-rent housing rented/%	–	–	–	–	13.4	13.4	1.3	1.8
Proportion of Other housing rented/%	–	–	–	–	87.9	87.9	30.2	18.3
Proportion of households renting public housing/%	73.1	73.1	21.0	20.7	–	–	–	–
Proportion of households renting commercial housing/%	55.1	55.1	5.8	8.4	–	–	–	–
Average housing space per capita/m^2	26.0	36.8	22.2	5.5	72.9	83.2	29.7	9.1

Note (1) It is worth pointing out that when calculating the spatial differentiation indicators of the Residential population from 2000 to 2010, the scope and number of neighborhood units were processed on the basis of the 2010 data. When calculating the relevant indicators from 1982 to 2000, the scope and number of neighborhood units were processed on the basis of the 2000 data. Therefore, the spatial differentiation values of various indicators calculated in 2000 are slightly different in the two calculations. This difference is caused by the slight difference in the sample size involved in the calculation, and does not affect the analysis conclusions of the relevant characteristics. The same applies below
(2) "↓" indicates that the statistical value in 2010 has decreased compared with that in 2000

housing over the past decade. In comparison to the period from 1982 to 2000, there has been a notable increase in the number of indicators exhibiting decreased information entropy, thereby highlighting a discernible trend of agglomeration among a greater number of indicators.

The variable exhibiting the most substantial increase in information entropy is the acquisition of commercial housing, which experienced an increase of 0.584. This is followed by the population within the tertiary industry 2 sector, which serves production, with an increase of 0.429. The increases in information entropy for the other indicators remain below 0.3. These findings suggest that the evolution of Beijing's real estate market over the past decade has resulted in a notable spatial reconfiguration of housing, leading to a progressively more equitable spatial distribution of commercial housing.

(2) **Absolute spatial differentiation**

Calculating the absolute differentiation index can reflect a kind of spatial differentiation of each indicator relative to an absolutely even distribution, that is, the spatial differentiation relative to the number of blocks. Indicators with an increase in the absolute differentiation index include total population, Uyghur population, employed in primary industry, employed in secondary industry, tertiary industry 1 (circulation department) population, commercial population, agricultural population, production and transportation population, average housing space per capita, purchasing affordable housing. The relative concentration of agriculture and the suburbanization of industry have led to a more concentrated spatial distribution, resulting in an increase in spatial differentiation. Judging from the absolute differential index values, both the absolute differential index values of housing sources and housing costs are above 0.4, indicating that the spatial differentiation of housing is more obvious.

(3) **Relative spatial differentiation**

Calculating the relative differentiation index can reflect the spatial differentiation of each indicator relative to the population distribution of the neighborhood. Since the census is based on the resident population, the relative differentiation index actually reflects the spatial matching of the spatial distribution of each indicator with the residential population. If the relative differentiation index of an indicator decreases over a period of time, it indicates that its consistency with the spatial distribution pattern of the residential population has increased. Compared to the absolute differential index, the relative differential index has increased for more indicators, including the sex ratio, migrants, Uyghur population, elderly population, non-working population, primary industry employment, secondary industry employment, tertiary industry 1 (circulation department) population, tertiary industry 5 (departments that serve social needs) population, managerial personnel (government agencies, party, and enterprises), office personnel, commercial service personnel, agricultural population, production and transportation population, number of households purchasing affordable housing, number of households renting public housing, number of households renting commercial housing, average housing space per capita, most of the

Table 5.12 Differential index of residential space and its changes in Beijing (2000–2010)

Indicators	Absolute differential index			Relative differential index			Information entropy		
	Year 2010	Year 2000	Difference in value	Year 2010	Year 2000	Difference in value	Year 2010	Year 2000	Difference in value
Total population	0.305	0.298	0.007	0.000	0.000	0.000	5.215	5.205	0.010
Population density	0.466	0.554	−0.088	0.405	0.446	−0.041	4.908	4.693	0.215
Sex ratio	0.042	0.044	−0.002	0.311	0.302	0.009	5.500	5.472	0.028
Household size	0.042	0.044	−0.002	0.326	0.324	0.002	5.499	5.475	0.024
Elderly population	0.159	0.330	−0.171	0.373	0.124	0.249	5.469	5.279	0.190
Migrants	0.396	0.414	−0.018	0.184	0.181	0.003	4.978	4.950	0.028
Mongol population	0.386	0.417	−0.031	0.152	0.203	−0.051	5.018	4.921	0.097
Hui population	0.429	0.462	−0.033	0.309	0.319	−0.010	4.906	4.786	0.120
Uyghur population	0.651	0.621	0.030	0.537	0.506	0.031	3.976	3.792	0.184
Miao population	0.448	–	–	0.253	–	–	4.805	–	–
Zhuang population	0.441	0.476	−0.035	0.256	0.324	−0.068	4.829	4.679	0.150
The Korean nationality population	0.517	–	–	0.362	–	–	4.321	–	–
Manchu population	0.329	0.376	−0.047	0.131	0.186	−0.055	5.150	5.044	0.106
Tujia population	0.457	–	–	0.249	–	–	4.767	–	–
Illiterate population	0.246	0.232	0.014	0.249	0.204	0.045	5.320	5.299	0.021
Population with university degree or above	0.421	0.513	−0.092	0.200	0.292	−0.092	4.960	4.672	0.288
Primary industry employment population	0.652	0.508	0.144	0.739	0.635	0.104	4.188	4.665	−0.477
Secondary industry employment population	0.354	0.322	0.032	0.256	0.162	0.094	5.072	5.137	−0.065
Tertiary industry employment population	0.358	0.379	−0.021	0.095	0.114	−0.019	5.109	5.056	0.053
Tertiary industry 1 population	0.365	0.355	0.010	0.140	0.105	0.035	5.052	5.089	−0.037

(continued)

Table 5.12 (continued)

Indicators	Absolute differential index			Relative differential index			Information entropy		
	Year 2010	Year 2000	Difference in value	Year 2010	Year 2000	Difference in value	Year 2010	Year 2000	Difference in value
Tertiary industry 2 population	0.324	0.450	−0.126	0.185	0.298	–0.113	5.175	4.746	0.429
Tertiary industry 3 population	0.388	0.408	−0.020	0.153	0.168	–0.015	5.055	4.989	0.066
Tertiary industry 4 population	0.418	0.458	−0.040	0.197	0.223	−0.026	4.969	4.851	0.118
Tertiary industry 5 population	0.297	0.412	−0.115	0.202	0.184	0.018	5.212	4.997	0.215
Unemployed population	0.307	0.333	−0.026	0.138	0.075	0.063	5.196	5.149	0.047
Managerial personnel (government agencies, party, and enterprises)	0.400	0.429	−0.029	0.251	0.229	0.022	5.004	4.942	0.062
Professionals and technicians	0.412	0.449	−0.037	0.186	0.203	–0.017	4.987	4.886	0.101
Office personnel	0.396	0.423	−0.027	0.171	0.164	0.007	5.034	4.977	0.057
Commercial and service personnel	0.366	0.365	0.001	0.142	0.116	0.026	5.051	5.061	–0.010
Agricultural professionals	0.615	0.506	0.109	0.714	0.634	0.080	4.318	4.678	–0.360
Production and transportation personnel	0.355	0.292	0.063	0.311	0.173	0.138	5.088	5.192	−0.104
Number of households purchasing commercial housing	0.524	0.607	−0.083	0.351	0.511	–0.160	4.697	4.113	0.584
Number of households purchasing affordable housing	0.649	0.577	0.072	0.516	0.416	0.100	3.983	4.421	−0.438
Number of households purchasing second-hand housing	0.470	–	–	0.315	–	–	4.822	–	–
Number of households purchasing former public housing	0.508	–	–	0.421	–	–	4.731	–	–

(continued)

Table 5.12 (continued)

Indicators	Absolute differential index			Relative differential index			Information entropy		
	Year 2010	Year 2000	Difference in value	Year 2010	Year 2000	Difference in value	Year 2010	Year 2000	Difference in value
Number of households renting low-cost housing	0.544	–	–	0.405	–	–	4.514	–	–
Number of households renting other housing	0.389	–	–	0.207	–	–	4.994	–	–
Number of Households Renting Public Housing	–	0.501	–	–	0.298	–	–	4.790	–
Number of households renting commercial housing	–	0.542	–	–	0.412	–	–	4.512	–
Average housing space per capita	0.111	0.100	0.011	0.334	0.349	−0.015	5.461	5.450	0.011

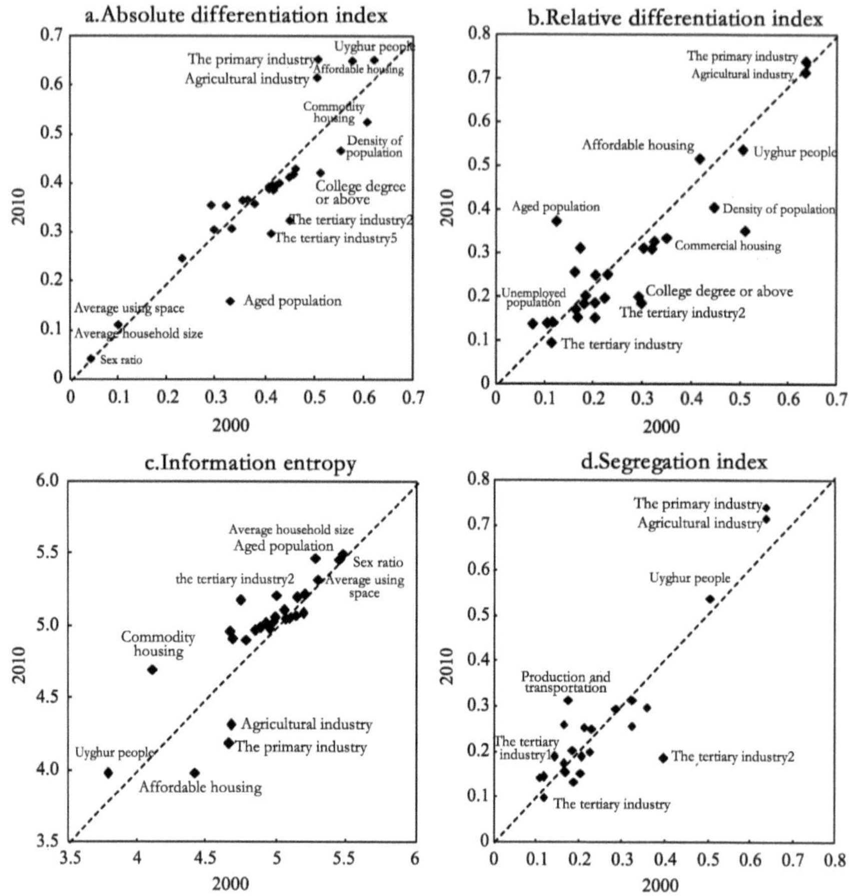

Fig. 5.12 Beijing's social spatial differentiation index (2000–2010)

relative spatial differences in these indicators are increasing, especially the population of various industries and occupations, indicating that the spatial differentiation of occupations is becoming more pronounced. The above changes in the relative differential index also indicate that the consistency of the spatial distribution of these indicators with the spatial matching of the Residential population is deteriorating. In other words, urban problems such as the separation of workplace and residence, including the employed population, the commute of migrants, and the distribution of empty-nest elderly people, are becoming more and more prominent.

(4) **Residential segregation**

Calculating the segregation index can be used to reflect the spatial relationship between a certain type of population and the rest of the population. The segregation index and its change were calculated for the elderly population, migrant population, minority population, population with university degree or above, and various industry and occupation populations. The calculation results (Table 5.13) show that the segregation indices that have increased over the past 10 years include migrants, the Uyghur population, the elderly population, the illiterate population, employed in primary industry, employed in secondary industry, tertiary 1 population, tertiary 5 population, managerial personnel (government agencies, party, and enterprises), office personnel, commercial and services population, occupational population in agriculture, production and transportation population, the most significant increases were in employed in primary industry and production and transportation population. The remaining increases were in minorities including Mongolian nationality, population with university degree or above, and employed in tertiary industry. In terms of segregation index value, employed in primary industry had the highest segregation index, while the Uyghur nationality, the Korean nationality, and production and transportation population also had relatively high segregation indexes. It is worth comparing and analyzing the situation with that of Beijing from 1982 to 2000. the segregation index values for the elderly population, migrants, Uyghur population, illiterate population, and agricultural population were positive from 2000 to 2010, continuing the trend from 1982 to 2000. In other words, the gregariousness of these indicators continued to increase while their intermingling with other populations weakened. This is consistent with the trend in the 18 years prior to 2000. the difference is that from 2000 to 2010, some new indicators have joined the ranks of these indicators, such as employed in secondary industry, employed in some types of tertiary industry (circulation department, departments that serve social needs), managerial personnel (government agencies, party, and enterprises), commercial and services personnel, and production and transportation personnel. Their gregariousness is also becoming stronger while their promiscuity with other populations is weakening. In particular, the segregation index difference between production and transportation workers and those employed in secondary industry is large and worthy of attention, which actually reflects the industrial space evolution trend of Beijing's manufacturing and logistics industry, which tends to concentrate development after 2000.

In summary, the evolution of residential space differentiation in Beijing from 2000 to 2010 can be characterized as follows: ① The gender ratio imbalance has eased, the number of illiterate people, employed in primary industry, and managerial personnel (government agencies, party, and enterprises) has decreased, and neighborhood differences have also decreased. Employed in tertiary Industry, especially tertiary industry 1 and tertiary industry 2 populations, have increased and are more evenly distributed across neighborhoods. ② Beijing's residential space system as a whole is becoming more complex, with employed in primary industry, employed in secondary industry, employed in commercial services, and affordable housing showing a gradual trend of spatial concentration. Commercial housing has undergone significant spatial restructuring, and its spatial distribution is becoming more balanced. ③ The concentration trend in the spatial distribution of the employed population in primary industry and the employed population in secondary industry has

Table 5.13 Beijing residential space segregation index and its changes (2000–2010)

Indicators	Year 2010	Year 2000	Difference in value	Indicators	Year 2010	Year 2000	Difference in value
Elderly population	0.188	0.142	0.046	Secondary industry employment population	0.259	0.165	0.094
Migrants	0.295	0.287	0.008	Tertiary industry employment population	0.099	0.118	−0.019
Mongol population	0.152	0.204	−0.052	Tertiary industry 1 population	0.142	0.107	0.035
Hui population	0.313	0.325	−0.012	Tertiary industry 2 population	0.185	0.398	−0.213
Uyghur population	0.537	0.506	0.031	Tertiary industry 3 population	0.154	0.169	−0.015
Miao population	0.253	–	–	Tertiary industry 4 population	0.199	0.225	−0.026
Zhuang population	0.257	0.324	−0.067	Tertiary industry 5 population	0.202	0.185	0.017
The Korean nationality population	0.363	–	–	Managerial personnel (government agencies, party, and enterprises)	0.251	0.230	0.021
Manchu population	0.133	0.189	−0.056	Professionals and technicians	0.188	0.205	−0.017
Tujia population	0.250	–	–	Office personnel	0.173	0.165	0.008
Illiterate population	0.253	0.213	0.040	Commercial and service personnel	0.144	0.118	0.026
Population with University Degree or Above	0.298	0.358	−0.060	Agricultural professionals	0.716	0.637	0.079
Primary industry employment population	0.740	0.638	0.102	Production and transportation personnel	0.314	0.176	0.138

led to an increase in their absolute spatial differentiation, with housing-related variables showing the most prominent absolute spatial differentiation. ④ The relative differential index of most social indicators, including migrants, the elderly, industries and occupations, has increased. The spatial differentiation of occupations is more pronounced, and urban problems such as the separation of workplace and residence among the employed population, commuting among migrants, and empty-nest elderly people have become more prominent. ⑤ The elderly population, migrants, Uyghur population, illiterate population, and agricultural population continue to follow the past trend of increased clustering and decreased mixing. The difference in value of the segregation index for the population engaged in production and transportation and those employed in secondary industry is large, reflecting the evolution of Beijing's industrial space towards concentrated development of manufacturing and logistics.

5.3.6 Restructuring of the Relationship Between Differentiation of Urban Residential Space and the Distance

5.3.6.1 Restructuring of the Relationship Between Differentiation of Residential Space and Distance from 1982 to 2000

The law of the restructuring of the relationship between population density and distance is very clear (Fig. 5.13): Over the past 18 years, the population density in the central city area (within a radius of about 5 km from the city centre) has decreased; the population density in the inner suburbs (within a radius of 5–25 km from the city centre) has increased to a certain extent, especially in the range of 5–12 km from the city centre. The overall spatial difference in population density distribution is decreasing.

The restructuring of the relationship between the proportion of the population with a college degree or above and distance reflects both the popularization and improvement of education levels since the reform and opening up, and the spatial restructuring of the highly educated population, especially in the neighborhoods of Inner suburb districts (within 0–25 km of the city center), where the improvement in the level of highly educated people over the past 18 years is more obvious. The distribution of migrants before and after the reform and opening up differed greatly. In 1982, the proportion of migrants in most neighborhoods was below 5%. In 2000, the proportion of migrants in most neighborhoods was between 10 and 70%. The distribution of migrants actually also had the characteristic of "universalization", which in turn reduced the differentiation between spatial units. Highly educated people and migrants are representative of this type of social spatial differentiation. Despite their clear differentiation, the force of "universality" has reduced social spatial differentiation compared to the situation in the planned economy era.

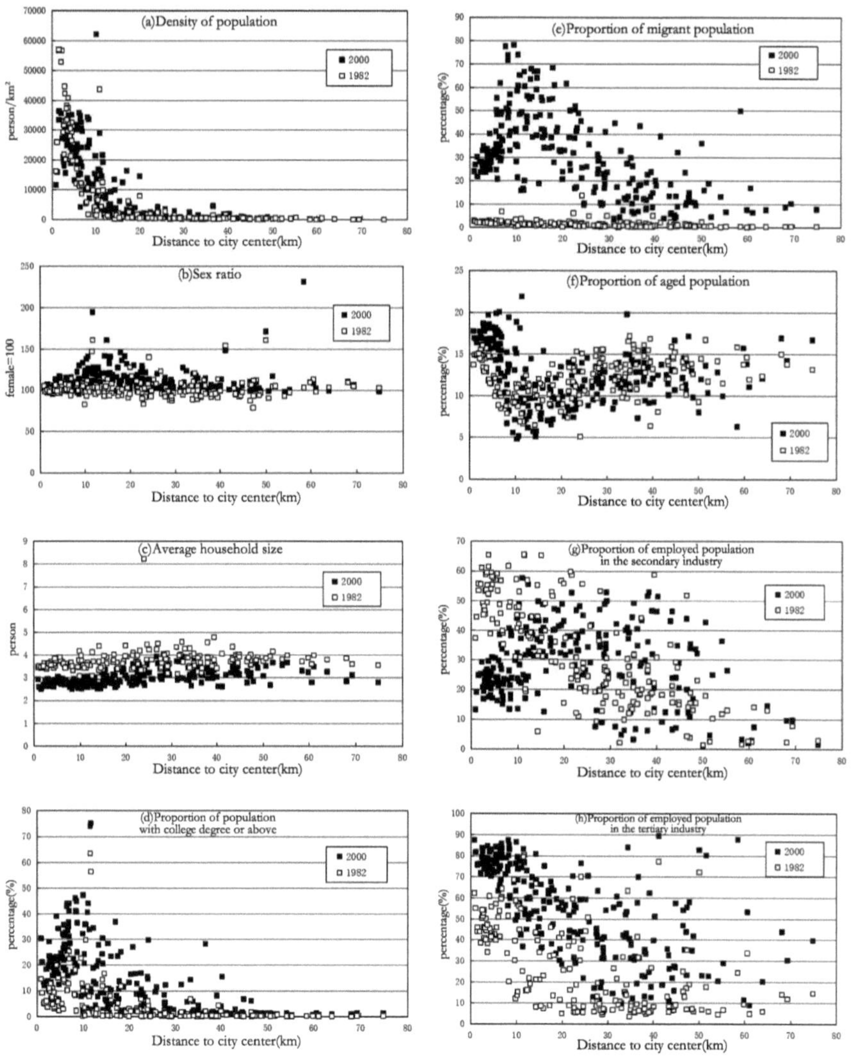

Fig. 5.13 Restructuring of the relationship between major social indicators and distance in Beijing (1982–2000)

The change in the proportion of the elderly population in relation to distance shows an increasing spatial differentiation. In the central urban area and its vicinity, within a distance of 0–8 km from the city centre, there is a relatively obvious change in the level of ageing. In 1982, this distance segment was mostly below 16%, while in 2000 it was mostly above 16%. In the range of 8–12 km from the city centre, the level of ageing in some neighbourhoods has increased, while in a considerable number of neighbourhoods it has decreased. In the range of 12–55 km from the city center, the aging level of neighborhoods as a whole is on the decline. These changes generally reflect two trends: on the one hand, the aging problem is more prominent in the central urban area and its vicinity, as well as in some rural areas in the far suburbs; on the other hand, the age structure in and around the Inner suburb is characterized by a younger population.

The restructuring of the relationship between the proportion of the population employed in secondary industry and distance clearly reproduces the process of Beijing's industrial suburbanization. Within the range of 0–25 km from the city center (approximately the Inner suburb), the proportion of the population employed in secondary industry as a whole shows a clear downward trend. The downward trend is most typical within the range of 0–10 km from the city center. In 1982, the proportion in this distance range was basically above 40%, while in 2000 it basically dropped below 40%. In the range of 10–25 km from the city center, there has also been a certain degree of decline overall. Outside the range of 25 km from the city center, the proportion of employed in secondary industry has shown a clear upward trend. This shows that during the period from 1982 to 2000, Beijing not only began the process of industrial suburbanization, but also completed a considerable part of the process of "outer suburbization", with some of the industry in the Inner suburbs moving to even more distant "suburbs".

Changes in the relationship between the proportion of the employed population working in tertiary industries and distance reflect the development trend of an advanced industrial structure in urban areas. Within a distance of 0–10 km from the city centre, there is a prominent upward trend in the proportion of employed people in tertiary industry. In 1982, the proportion in this distance range was basically below 65%, while in 2000 it was basically above 65%. In areas beyond 10 km from the city centre, the proportion of employed people in tertiary industry has increased to varying degrees. In 2000, the proportion of employed people in tertiary industry in a small number of remote urban neighbourhoods reached as high as 80%, reaching the level of the city centre.

5.3.6.2 Restructuring of the Relationship Between Differentiation of Residential Space and Distance from 2000 to 2010

The relationship between the main social indicators of Beijing's residential population and distance from 2000 to 2010 was constructed using the distance from each neighborhood to the city center and the attribute values of the neighborhood (Fig. 5.14). The main reconstructed features are: ① Population density is showing an increasing trend, with particularly prominent changes in the suburbs. The population density of neighborhoods within 5–25 km of the city center has increased more significantly. ② The sex ratio is trending downward, indicating that the number of women relative to men is increasing. ③ The average household size is decreasing, reflecting the shrinking size of Beijing households. ④ Educational attainment is generally increasing, with the highest levels in the suburbs. Population with University Degree or Above in all distance ranges has increased, with a peak at about 12 km from the city center, which is where several streets in Haidian District with a concentration of colleges and research institutes are located, such as Yanyuan, Tsinghua University, Xueyuan Road, and Zhongguancun. ⑤ The peak of the migrant population has shifted outward. In 2000, the peak of the migrant population was about 10 km from the city center, and in 2010, it was about 20 km from the city center. The proportion of migrants within a radius of 0–15 km has decreased, indicating that migrants are also moving outward, and their spatial distribution is showing signs of suburbanization. ⑥ The proportion of the elderly population has increased, especially in the far suburbs. Specifically, the proportion of the elderly population within 0–15 km of the city center has decreased, the 15–30 km neighborhood has not changed much, and the proportion of the elderly population in neighborhoods beyond 30 km has increased significantly. This indicates that as the labor force goes out to find employment, the problem of rural aging becomes more serious. ⑦ The proportion of the population employed in Secondary Industry has decreased, with a particularly significant decrease in the suburbs, and the peak has shifted outward. The proportion of the population employed in Secondary Industry within 0–20 km of the city center has decreased significantly. ⑧ The proportion of the population employed in Tertiary Industry has generally increased, and spatial differences have narrowed. The proportion of people employed in tertiary industry is extremely high in inner suburbs, and has also increased significantly in outer suburbs.

5.3.7 The Mechanism of Restructuring of Urban Residential Space Differentiation

Social space can be seen as the result of social factors at play in space, or as a process in which society and space interact with one another, a process governed by the "social-spatial dialectic" (Knox and Pinch 2000). In addition, in recent years, Western human geography has emphasized the "cultural and institutional turn" (Clark et al. 2003), focusing on the spatial laws of human phenomena from the cultural and institutional levels, which have provided us with new research perspectives. In fact, the restructuring of residential spatial differentiation in transitional China, represented by Beijing, can be explained from the perspective of institutions, markets, cultural and social change and their interaction with space (Fig. 5.15).

The process of institutional innovation, including the reform of the household registration system and the establishment of a paid land use system, has had a significant impact on the growth and layout of the floating population and industrial activities in urban spaces. Market transformation has posed severe challenges to the traditional urban economic structure and accelerated the transformation of the urban industrial structure towards high-level development. For example, the proportions of the primary industry, secondary industry and tertiary industry in Beijing's GDP have evolved from 6.7, 64.4 and 28.9% in 1982 to 2.5, 32.7 and 64.8% in 2000, and then 0.9, 23.5, 75.6% in 2010 and 0.3, 16.2, 83.5% in 2019 (Beijing Municipal

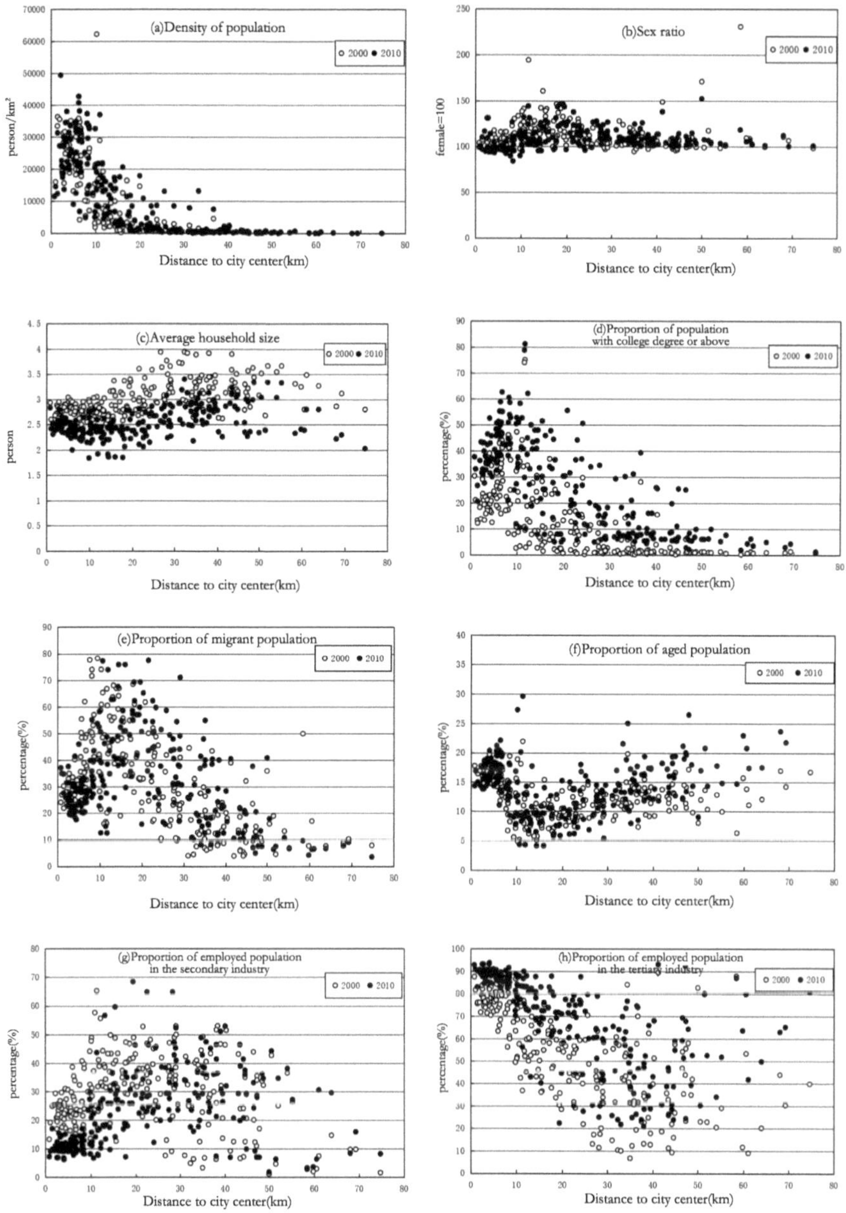

Fig. 5.14 Restructuring of the relationship between major social indicators and distance in Beijing (2000–2010)

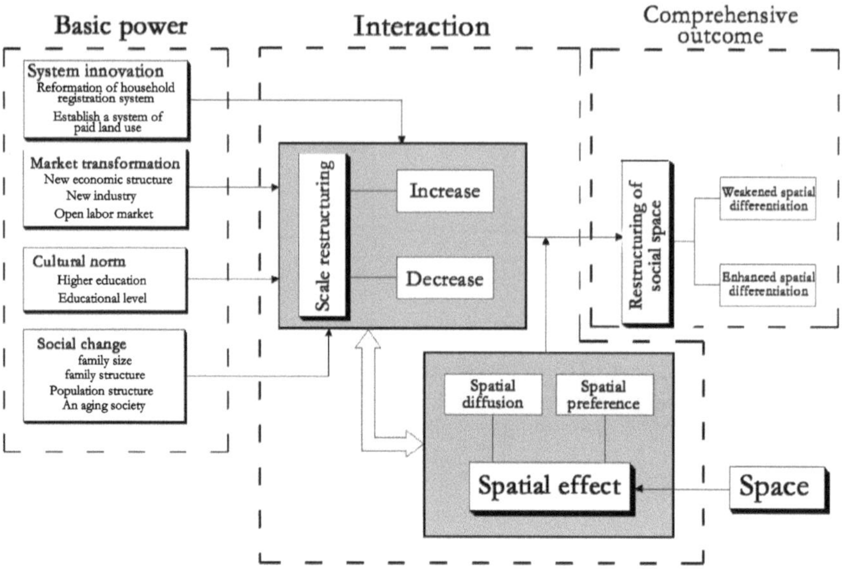

Fig. 5.15 Conceptual model of the mechanism of the restructuring of residential space differentiation in Chinese cities during the transition period

Bureau of Statistics, 2020). It can be seen that the proportion of the primary industry is minimal, the proportion of the secondary industry is gradually decreasing, and the proportion of the tertiary industry is gradually increasing, and its proportion has now reached more than 80%. In fact, the upgrading of the urban industrial structure has promoted the evolution of urban employment space, and the transformation of the industrial structure at the spatial level within cities has promoted the development of differentiation of the employed population in urban space. For example, according to the data on the employment status of the residential population shown in the national census, from 1982 to 2000, the proportion of the employed population in the circulation department and the Departments that serve life in the central area and Inner suburb of Beijing increased rapidly. The proportion of the employed population in the tertiary industry 3 in the central area increased by 17 percentage points, and the proportion of the tertiary industry 1 increased by 11 percentage points. This is obviously closely related to the process of land replacement and the transformation of economic functions in the central area. The labor market, which is open to the whole country, has led to a sharp increase in the number of migrants in Beijing. For example, according to census data, the number of migrants in Beijing increased from 170,000 in 1982 to 2.57 million in 2000, an increase of 14 times in the 18 years from 1982 to 2000, and then to 7.045 million in 2010, an increase of 1.74 times in the 10 years from 2000 to 2010. According to the latest statistics, the permanent migrant population in Beijing had reached 7.456 million by the end of 2019 (Beijing Municipal Bureau of Statistics, 2020). The development of higher education at the cultural level has not only increased the proportion of highly educated people, but also raised the average level of education among ordinary residents. From the perspective of demographic, family and social structural changes, family size is generally shrinking, and the aging of the population is becoming increasingly prominent. In short, the underlying institutional, market and cultural drivers have led to a restructuring of the size of the social elements of the residential population. This is reflected in the growth of many elements, such as migrants, employed in tertiary industry, population with university degree or above, and the elderly population in Beijing, as well as the contraction of some elements, such as family size and the agricultural population.

In addition, social factors of different natures reflect different spatial trends. In other words, from a spatial perspective, there are two effects: decentralization and agglomeration. On the one hand, the development of individual organizational spatial units has a spatial popularization effect, showing an overall decentralization trend. On the other hand, some social factors have specific spatial preferences or tend to agglomerate in specific spaces for the sake of saving operating costs. In fact, during the process of scale restructuring and spatial restructuring of social elements, they have a comprehensive intertwined effect: elements that have experienced growth in scale can have a spatial diffusion effect, such as Employed in tertiary Industry, Population with University Degree or Above, and Distribution of Floating Population in Beijing. They can also have a spatial preference effect, such as the increasing elderly population deciding on the location of their residence based on the convenience of facilities such as medical care and services. Social elements that have experienced a reduction in scale can also correspond to the above two spatial effects. for example, household size corresponds to spatial diffusion, while the agricultural population corresponds to spatial preference, because it is necessary to save land costs as much as possible and replace expensive land with the more land-intensive tertiary industry.

The combined effect of the social factors of "scale restructuring" and "spatial effect" is the restructuring of urban social space, which in some cases shows a trend of weakening spatial differentiation, while in others it shows a trend of enhancement. The spatial differentiation of most social indicators in Beijing is showing a trend of weakening, with most types of population becoming more mixed, but the spatial differentiation of a few indicators is increasing, and the clustering of a few types of population is increasing. The underlying driving forces for these patterns can be explained in terms of institutional, market, social and cultural changes. The intertwined effects of the restructuring process can be seen in terms of scale and space, and the results of the restructuring can be seen in terms of the strengthening and weakening of social spatial differentiation.

5.3.8 Conclusion and Discussion

All types of urban residential populations have obvious and distinctive spatial differentiation characteristics, such as residential population density, migrants, family size, the elderly population, the population with University Degree or Above, and the minority population. Various indicators of the employed population and housing conditions show the spatial differentiation patterns of major economic and social factors within cities against the background of suburbanization and urban industrial transformation. By calculating the information entropy, absolute differentiation index, relative differentiation index, isolation index and other indicators, the restructuring characteristics of residential space divergence in Chinese cities during the transition period can be quantitatively measured. In terms of empirical research on Beijing, with the exception of a few indicators such as the elderly population, sex ratio, average household size, and agricultural population, the spatial differentiation of most social indicators in Beijing decreased between 1982 and 2000. During the same period, the consistency of the distribution patterns of migrants, people of various minority nationalities, Population with University Degree or Above, Employed in Secondary Industry, Employed in Tertiary Industry, etc., are becoming more consistent with the overall population distribution pattern, while the elderly population, illiterate population, and agricultural-related population are gradually deviating from the consistency with the overall population distribution pattern. The residential mixability of the urban population has generally increased over the past 18 years, but the elderly population, migrant population, and agricultural-related population have shown a characteristic of weaker residential mixability and stronger group nature compared to other populations. The changes in information entropy actually reflect the fact that, with the exception of a few indicators, urban residential space as a whole has evolved in the direction of greater complexity and heterogeneity. This is completely consistent with the findings of a study on the evolution of Beijing's urban social space structure based on social area analysis (Feng and Zhou 2003a; Feng 2004). With the upgrading and adjustment of the city's industrial structure and the implementation of urban spatial development strategies, the employment structure of the population has also changed accordingly, and this is reflected spatially, reflecting the spatial relationship between employment and residence, the characteristics and problems of the residential and commuting patterns of specific population groups. This trend was obvious in Beijing from 2000 to 2010. Overall, the spatial differentiation of urban residents' occupations has become more pronounced. Indicators such as the elderly population, migrant population, Uyghur population, illiterate population, and agricultural population continue the past trend of increased clustering and decreased mixing. The consistency of the spatial distribution of these indicators with the spatial matching of the Residential population is deteriorating, reflecting more prominent urban issues including the separation of workplace and residence among the employed population, the commute of migrants, and the left-behind or "empty-nest" elderly.

It is worth emphasizing that after 2000, China's urban real estate market experienced rapid development. A large number of villa districts, Affordable Housing, and policy-guaranteed housing were built in the suburbs, along with a large number of ordinary commercial housing developed and constructed in the suburbs. This has greatly increased the complexity of housing in suburban areas, and there is a clear differentiation of housing types. This background will inevitably have a significant and lasting impact on the differentiation of China's urban residential space from 2000 to 2010, and will inevitably be reflected in the data of the sixth national census.

Therefore, using the sixth national census data broken down by street block to study the differentiation and historical evolution of urban residential space is expected to yield interesting results. An empirical study of the differentiation of residential space in Beijing from 2000 to 2010 supports the above conjecture. The results of the study show that from 2000 to 2010, the complexity of the urban housing system increased, and commercial housing became more widely distributed, but its spatial differentiation also increased. In the past ten years, the rapid development of the real estate market has led to a significant spatial restructuring of urban housing, and a significant increase in housing spatial differentiation. The absolute differential index of various types of housing, such as low-cost housing, commercial housing, Affordable Housing, and former public housing, is relatively high, indicating that all types of housing are relatively far from an even distribution and show a certain trend of spatial concentration. Among them, the information entropy of the number of households purchasing commercial housing has increased significantly in 10 years, and the statistical indicators such as the Mean value, maximum, range and standard deviation of the number of households purchasing commercial housing in each neighborhood have also increased significantly, indicating that while the number of commercial housing units in each neighborhood is rapidly increasing, the differences between neighborhoods are gradually widening, further exacerbating the differentiation of urban residential space.

The research methods for studying the differentiation of residential space in Chinese cities, as well as the spatial scale at which the data is used, require further discussion. Empirical research in Beijing shows that analysis based on the differentiation index and segregation index gives us more details about the restructuring of urban residential space beyond the analysis of social areas in the city. These details can be more intuitively understood from the restructuring of the relationship between urban residential spatial differentiation and distance. The explanation of the restructuring of urban residential spatial differentiation needs to seek the underlying driving forces from the perspective of institutional, market and cultural changes, and an explanatory framework needs to be established with the help of social-spatial dialectics.

Some scholars believe that compared to data at the neighborhood committee level, the block is not suitable as the fundamental unit for dividing social space or residential space, mainly because the degree of social spatial differentiation based on the block is lower than that of the neighborhood committee (Li and Wu 2006). We do not agree with this view. In fact, the degree of differentiation is a relative concept, and different spatial scales are not comparable. Changes in the degree of social spatial differentiation should be judged based on comparisons between the same spatial scale at different points in time. In addition, the choice of research data type also needs to consider the operability of its source. For Chinese census data, the fundamental unit that is publicly available is the neighborhood rather than the neighborhood committee. Moreover, using neighborhood committee data has the drawback of making it difficult to accurately restore urban maps at the neighborhood committee level and their changing relationships. Although we do not oppose the use of data at the neighborhood committee level, this section of the study further shows that the "neighborhood scale" is an operational spatial scale that can be used to show the characteristics of urban social spatial differentiation or residential space differentiation in China.

5.4 Habitability of Urban Communities and Their Differentiation

5.4.1 Habitability of Urban Communities

Urban communities in China, commonly referred to as "urban communities," are primarily formed by the aggregation of diverse groups engaged in various non-agricultural activities within a specific area. These communities are characterized by high population density, diverse social structures, concentrated political and economic functions, significant lifestyle differences, vibrant leisure and recreational activities, and well-developed public service facilities. As China's socioeconomic and socio-spatial development continues to advance, urban residents' concerns have shifted beyond basic survival needs such as food and clothing. Increasingly, they are paying attention to factors such as safety, transportation convenience, environmental civility, and social friendliness—giving rise to the concept of "habitability."

Today, building a "habitable city" has become a key urban development goal in China and serves as an important criterion for assessing a city's competitiveness, attractiveness, and development potential. A habitable city refers to one that is well-suited for human survival, living, and development. It embodies a clean and beautiful natural environment, a harmonious and civilized social ecology, a safe and convenient infrastructure, a comfortable and efficient living environment, convenient transportation, comprehensive public services, a strong economy, and a high degree of harmony between humanity and nature (Zhang 2016).

The habitability of urban communities focuses on identifying what kind of community is most suitable for residents to live in. It encompasses aspects such as comfort, safety, the quality and availability of public service facilities, and the overall happiness and satisfaction of residents. Community habitability directly impacts the quality of urban life: communities with high habitability tend to attract more residents, foster stronger community ties, and instill a deep sense of pride and belonging among inhabitants.

Several factors influence community habitability. These include the quality of public spaces where residents can freely interact, urban planning strategies, public policies, and broader national decisions. Higher levels of community habitability are typically associated with better human environments, improved public service facility provision, and greater opportunities for social exchange and community vitality. As a result, enhancing the habitability of urban communities plays a crucial role in fostering social cohesion and promoting the overall economic and social development of China's cities.

5.4.2 Evaluation of the Habitability of Chinese Urban Communities

5.4.2.1 Indicator System for Evaluating the Habitability of Chinese Urban Communities

As the fundamental unit of urban construction and habitable city development, urban communities serve as the primary social environment where residents live, develop, and engage in daily activities. They also function as crucial platforms for the government to manage and serve residents, as well as to conduct social work. Urban communities exhibit a range of characteristics, including social, economic, territorial, natural, and environmental attributes. In China, the primary function of urban communities is residential, with extended social functions encompassing economic activities, socialization, social regulation, social welfare protection, and social participation. The indicator system for evaluating urban community Habitability is primarily established based on these five functions, as detailed in Table 5.14.

5.4.2.2 Evaluation of Habitability of Chinese Urban Communities

Against the backdrop of China's rapid urbanization, the study of urban community habitability has become an essential pathway for promoting sustainable societal development. Consequently, numerous scholars have investigated the Habitability of Chinese urban communities from diverse perspectives, including public spaces, urban transportation, and human settlements, while incorporating regional characteristics into their analyses.

The habitability of urban community public spaces constitutes a critical aspect of overall urban community Habitability. These public spaces, which encompass residents' living and recreational areas, reflect the historical development, social civilization, and cultural level of urban communities. The extent of public space directly influences residents' spatial freedom and activity. As a core element of urban planning, urban community public spaces are intrinsically linked to residential and other structures within communities, making them a vital component of urban spatial organization. Scholars typically evaluate the public habitability of Chinese urban communities based on factors such as convenience for daily activities, suitability for living, and ecological health. For instance, a study on community Habitability in Changsha revealed significant disparities among districts and counties, exhibiting a "core-periphery" spatial distribution pattern. Issues such as job-housing imbalances, oversized neighborhoods, insufficient elderly care facilities, and uneven distribution of medical, sports, and other public service facilities were identified (Zhou 2018). Rational allocation of public service facilities within urban communities is essential for achieving equity and justice, as well as fostering harmonious development among humans, nature, and society.

Table 5.14 Community habitability evaluation indicator

Modules	Submodule	Number	Indicator	Unit	Property
Physical space rationalization	Site selection	1	Compliance rate of construction sites with disaster prevention and mitigation standards	%	Constraint
		2	Average housing construction area per capita in the community	m^2	Constraint
	Wind environment	3	Compliance rate of building ventilation	%	Constraint
		4	Wind speed in the pedestrian area around the building	m/s	Constraint
	Light environment	5	Compliance rate of building sunshine	%	Constraint
	Thermal environment	6	Average outdoor heat island level in the residential area	°C	Constraint
	Sound environment	7	Compliance rate of urban noise control indicators	%	Constraint
Perfect service facilities	Administrative offices	8	Proportion of households within a 15-min walk of a community service center	%	Constraint
	Scientific research and education	9	Proportion of households within a 10-min walk of a Primary school	%	Constraint
		10	Proportion of households within a 20-min walk of a Secondary school	%	Constraint
	Medical care	11	Proportion of households within a 15-min walk of a health care service station	%	Constraint

(continued)

Table 5.14 (continued)

Modules	Submodule	Number	Indicator	Unit	Property
	Culture and entertainment	12	Proportion of households within 20 min walk of sports facilities	%	Constraint
	Commerce and finance	13	Proportion of households within 10 min walk of commercial facilities	%	Constraint
Infrastructure support	Transportation facilities	14	Ratio of community parking spaces to car ownership	%	Constraint
		15	Coverage of residential parking lots and garages within a service radius	%	Constraint
		16	Proportion of households within a 10-min walk of a bus stop (including rail stations)	%	Constraint
	Municipal facilities	17	Centralized sewage treatment rate	%	Constraint
		18	Harmless treatment rate of domestic waste	%	Constraint
	Disaster insurance and safety	19	Emergency shelter area per capita	m^2	Constraint
	Intelligence and wisdom	20	Proportion of intelligent homes	%	Constraint
Beautiful environment and landscape	Environmental quality	21	Community environmental function area compliance rate	%	Constraint
	Squares and parks	22	Public green space area per capita	m^2	Constraint
	Landscape greenways	23	Local plant index	–	Constraint
		24	Outdoor permeable surface area ratio	%	Constraint
	Sculptures and small works of art	25	Community landscape satisfaction	%	Guiding

(continued)

Table 5.14 (continued)

Modules	Submodule	Number	Indicator	Unit	Property
Neighborhood care and harmony	Age-appropriate facilities	26	Service coverage rate of day care centers for the elderly	%	Constraint
		27	Number of social welfare beds per 100 elderly people	per	Constraint
	Playground	28	Residents' satisfaction with community public service facilities	%	Guiding
	Virtual space	29	Participation rate of residents in the community network virtual platform	%	Guiding
Efficient governance services	Self-government	30	Residents' satisfaction with the work of creating a civilized community	%	Guiding
		31	Property management coverage rate	%	Guiding
	Neighborhood organization	32	Complaint rate of community civil cases	per 1,000 households	Guiding
	Security organization	33	Complaint rate of community criminal cases	per 1,000 households	Guiding
	Volunteer work	34	Participation rate of residents in community activities	%	Guiding

The habitability of urban social spaces in Chinese communities must adopt a people-centered approach, addressing issues of social justice. Initiatives such as planning and constructing green spaces and plazas are crucial for enhancing urban Habitability (Wu 2010). Comprehensive improvements in the quality of community public service facilities and the vigorous development of urban community public space resources are imperative for advancing the construction and planning of habitable urban communities in China. Only through these measures can we better align urban development with the goal of creating habitable and sustainable urban environments.

The Habitability of community transportation is a critical component of research on urban Habitability in China, as it embodies the right of residents to freely choose their preferred modes of transportation. To enhance residents' sense of well-being, it is essential to meet their transportation needs as comprehensively as possible during their daily commutes. Evaluating the Habitability of urban transportation not only captures residents' subjective perceptions of the transportation environment but also directly reflects whether the local transportation conditions are conducive to a high quality of life. Consequently, achieving habitable transportation is a vital step in building habitable cities. For instance, a study on the Habitability of urban community transportation in Xi'an revealed several challenges. The current public transportation system often lacks a people-oriented design, leading to low subjective ratings of commuting convenience in large residential communities and streets adjacent to the central business district (CBD). Additionally, the inconvenience of transfers within the rapid rail transit system has resulted in poor Habitability and low resident satisfaction in many areas. In contrast, residents tend to express higher satisfaction with commuting in communities that offer comprehensive functional amenities (Bi 2017). These findings underscore the importance of prioritizing user-centric transportation planning to improve urban Habitability and resident satisfaction.

From the perspective of human settlements, higher urban Habitability signifies the presence of a human and natural environment that is conducive to comfortable living. In other words, a habitable city is characterized by a favorable or superior human settlement environment, with the quality of this environment being directly linked to the quality of life of urban residents (Lü 2012). For instance, a study on the Habitability of the urban human settlement environment in Dalian revealed that while the overall evaluation of urban community Habitability is relatively positive, significant regional disparities exist within the city. High-value habitable areas are predominantly concentrated in scenic regions and areas housing commercial activity centers and administrative hubs (Liu 2009).

Overall, when examining Chinese cities as the research focus, the China City Competitiveness Report indicates that the most habitable cities in China are primarily distributed in coastal areas. This distribution is largely attributed to factors such as the relatively advanced economic development of coastal regions, environments suitable for human habitation, and convenient transportation infrastructure. A comparison of communities across the country reveals notable differences in Habitability, which are closely tied to the level of urban development and geographic location. Communities in first-tier cities such as Beijing, Shanghai, and Guangzhou exhibit significantly higher Habitability than those in second- and third-tier cities. Specifically, first-tier cities generally outperform their counterparts in both hard conditions, such as environmental quality, and soft conditions, including property management services and surrounding supporting facilities.

Within individual Chinese cities, disparities in community Habitability also exist. Research indicates that the urban community environment often exhibits a spatial distribution characterized by concentric circles and multi-core structures. These structures radiate outward from large commercial centers in a hierarchical manner. Communities closer to these commercial hubs tend to be more habitable, while those farther away experience lower Habitability and reduced resident satisfaction with their living environments (Ni 2018).

5.4.3 Differentiation of Habitability of Urban Communities in China

The differentiation of urban community Habitability is a prominent research topic in urban geography and a critical issue in contemporary Chinese society. Currently, China's urbanization process has entered a phase of steady advancement, with the focus of urban construction gradually shifting toward the development of habitable cities. As an integral component of urban public space, urban communities directly influence residents' daily lives and the healthy, sustainable development of cities. Consequently, Chinese scholars have conducted extensive theoretical and practical research on the differentiation of urban community Habitability, examining it from perspectives such as residential differentiation, spatial disparities in the allocation of public service facilities, and variations in urban activity spaces.

5.4.3.1 Residential Differentiation in Urban Communities

Amid rapid urban–rural integration, class differentiation within urban communities has become increasingly pronounced, leading to a gradual diversification of urban communities. The exploration of residential differentiation aims to identify the spatio-temporal differences and characteristics of residential communities by comparing key elements across different periods and locations.

The phenomenon of residential differentiation in urban communities is both an inevitable outcome of urban development and a reflection of the subjective preferences of residents from different social classes. Spatially, this differentiation manifests in factors such as housing types, natural landscapes, human environments, and supporting facilities. Socially, it is primarily reflected in residents' personal circumstances, including disposable income, education level, occupation type, and consumption habits. Research conducted by Chinese scholars on residential differentiation in urban communities can be broadly categorized into two main areas. The first area focuses on single-year data, using first- and second-tier urban communities in China as case studies to examine the spatial differentiation of housing prices. For instance, some scholars have conducted micro-level studies of five major urban communities in Chengdu, revealing that high-income individuals tend to cluster in polycentric spatial patterns within the main urban area, while low-income residents are predominantly distributed at the city's periphery. Retired individuals and the elderly, on the other hand, are often concentrated in circular patterns within older urban districts. Addressing the housing challenges faced by vulnerable groups, optimizing industrial layouts, and leveraging the strengths of various sectors are essential steps to mitigate residential spatial differentiation and promote social equity and justice (Wang 2019). The second area involves selecting indicators that reflect housing or community environment characteristics, constructing an index system, and analyzing the spatial changes in residential differentiation. For example, some scholars employed web crawler technology to gather data on 1,397 residential quarters in Wuhan, selecting indicators such as building age, greenery coverage, and the availability of supporting infrastructure to study residential differentiation. The findings revealed a spatial pattern in which residential quarters in central areas are densely surrounded by supporting facilities, while those in urban fringe areas suffer from a lack of such amenities (Gong 2018).

5.4.3.2 Spatial Differentiation in the Configuration of Public Service Facilities

Studying the equitable allocation of public service facilities in urban communities is essential for meeting the needs of urban human settlements and constructing habitable cities. The spatial differentiation of public service facilities directly influences residents' quality of life. The level of public service facility provision reflects the status of a community to a certain extent, serving as an indicator of the differentiation in urban community Habitability. High-end residential areas typically feature comprehensive and highly accessible public service facilities, with well-allocated community resources. For example, some scholars analyzed Shahekou District in Dalian using network analysis tools and found that public service facilities are predominantly concentrated around commercial areas. Over 70% of residents in residential buildings have access to six types of public service facilities: shopping and dining, daily services, sports and leisure, medical facilities, elderly care facilities, and science, education, and cultural amenities. Among these, the distribution of facilities, the obstruction caused by land features, and the degree of community enclosure significantly impact residents' access to public services (Han et al. 2019).

5.4.3.3 Differentiation of Activity Space

Differentiation of activity space is a critical dimension in the study of urban community Habitability. The scope of residents' social interactions and the characteristics of their spatial activities are shaped by various factors, including individual social attributes, family social attributes, and community public space attributes. The social communication spaces where residents engage in interactions are primarily categorized into four types: community roads, community squares, community green spaces, and community open spaces. Significant differences exist in the choice of activity spaces among individuals of different classes, ages, genders, and occupations. Factors such as accessibility, size, and safety of these spaces significantly influence residents' preferences. The spatial differentiation, category attributes, and scope of public activity spaces directly affect residents' usage patterns and, indirectly, the Habitability of urban communities. However, different user groups often create conflicts in space utilization due to varying needs, with the most common conflicts being related to time, space, and noise (Hu 2019).

To address these issues, it is essential to study the differentiation of urban community activity spaces. For instance, some scholars, drawing on the theory of everyday life, examined the Pingjiang Historic District to explore the differentiation characteristics, current status, and mechanisms of its everyday life spaces. They found that residents' everyday residential spaces, consumption spaces, and leisure spaces exhibit patterns of hollowing out, marginalization, and fragmentation. The spatial structure of everyday and non-everyday life spaces demonstrates a semi-network structure of "decentralization" and a "heavy east, light west" forked structure. The study emphasizes the need to view the positive and negative impacts of living space differentiation dialectically and proposes restructuring the living space system to ensure residents' daily needs are met, thereby enhancing their sense of happiness and satisfaction (Zhou 2019).

5.4.4 The Driving Mechanism of Differentiation of Habitability of Urban Communities in China

Differentiation of urban communities are an inevitable outcome of China's accelerating urbanization process. Currently, Differentiation of urban communities has become a focal point of research in disciplines such as human geography, urban studies, social sciences, and economics. Over the long-term development of cities, the complexity and diversity of Habitability differences in communities have become increasingly pronounced. Therefore, exploring the spatial differentiation characteristics and driving mechanisms of urban community Habitability holds significant practical and theoretical importance.

5.4.4.1 Historical and Cultural Factors

Historical and cultural factors play a crucial role in shaping residents' perceptions of urban communities. Scholars argue that during China's urban development, the traditional historical and cultural heritage embedded in naturally formed traditional communities and streets has had a lasting impact on the psychology and physiology of future residents. This influence persists over time, meaning that the quality of historical legacies related to community living environments, quality of life, and human settlement environments shapes residents' psychological understanding of urban community environments (Wu and Cui 1999).

5.4.4.2 Market Mechanism

The market mechanism plays a decisive role in determining the scale and nature of urban communities in China, with market factors increasingly becoming one of the primary drivers of social space restructuring. Under the rapid development of the market economy, the differentiation of the housing market, driven by the need to meet residents' demands and market requirements, is a key force behind the differentiation of residential communities. The hierarchical differentiation of housing demand leads to diversification in supply. Differences in residential location preferences among households from various social groups inevitably result in the segregation and differentiation of residential spaces (Qin and Zhou 2011). In large cities, urban planning concepts directly influence the spatial differentiation and structural reorganization of urban communities. Due to China's social transformation and systemic reforms, the role of market mechanisms in the formation, development, and transformation of communities has continued to grow. This has led to changes in the location structure of old urban areas, with older communities declining and new communities emerging, alongside the development of commercial communities (Li and Lan 2012). As new communities become more functionally complete and larger in scale, their Habitability tends to improve.

5.4.4.3 Government Decision-Making

Government decision-making also significantly influences the spatial differentiation characteristics of urban community Habitability. The Habitability of urban communities is closely tied to national and local government policies. Government decision-making has a dual impact: on one hand, it can enhance community Habitability; on the other, it may inhibit community development to some extent. First, government policies and decisions affect urban functional zoning, which in turn influences the spatial distribution of communities at different levels. Second, government planning impacts the greenery, accessibility, and density of urban communities, thereby affecting their overall quality. Government functions exert a mandatory effect on urban spatial management and regulation. Initiatives such as the elimination of shantytowns, the restructuring of hazardous buildings, and the redistribution of urban land use help integrate residents from different social classes, contributing to the fairness and equity of urban communities (Wang et al. 2000).

5.4.4.4 Level of Economic Development

The level of economic development is a fundamental driver of differences in the Habitability of urban communities. First, these differences arise from the spatial interaction between the level of urban socioeconomic development and residents' social income levels. Socioeconomic conditions provide essential economic support for urban community Habitability and serve as a financial foundation for enhancing public service infrastructure. Second, the spatial needs of residents from different social classes vary significantly, and their behavioral preferences play a crucial role in shaping the Habitability differences among communities. High-income individuals tend to favor mid-to-high-end communities characterized by robust infrastructure, eco-friendly environments, and excellent security. In contrast, low-income residents generally opt for communities located on the outskirts of commercial centers, within the range of their disposable income. These preferences influence the spatial distribution of Habitability in urban communities, leading to the formation of a concentric spatial pattern. In this pattern, high-end, mid-range, and low-end residential areas radiate outward from the central commercial district, with Habitability gradually decreasing as distance from the center increases.

5.5 Migration and Changes in Residents' Living Space

5.5.1 The Meaning of Migration

With the continuous development of the social economy, people's housing needs are evolving, leading them to relocate to new residential areas and resulting in changes to their residential locations. This behavior of changing residential locations is referred to as "migration" (Hou and Wang 2014). In China, however, there is ambiguity in the statistical methods and scope of indicators related to the concept of "migration," as well as a lack of clear definition. This issue stems from the relatively strict household registration system in China, where the movement of residents in residential spaces often does not align with changes in population and household registration status. As a result, migration in China exhibits complex internal characteristics.

Internationally, migration is a comprehensive concept that integrates three-dimensional indicators of time, space, and displacement. For example, *the Population Handbook* published by the Population Reference Bureau defines migration as "the movement of people across a certain boundary for a specific settlement purpose." William Peterson provides a more specific definition, describing migration as "the geographical displacement of a population at a certain time, resulting in a change in their permanent residence" (Peterson 1984).

Migration can be categorized into active migration and passive migration. Active migration refers to the process in which individuals relocate their residence with a certain degree of autonomy in choosing their living location. They seek to purchase or rent homes that best meet their needs. Subjective factors influencing active migration include personal attributes such as age, education level, economic status, marital status, and occupational status. Objective factors include the characteristics of the current residential space, availability of supporting facilities, and transportation conditions. In contrast, passive migration occurs when individuals are unable to choose their residence according to their preferences due to external constraints. For example, this type of migration may result from government-led urban renovation and transformation projects or housing allocations by organizations to their employees (Chai et al. 2000).

5.5.2 *Characteristics of Migration of Urban Residents in China*

5.5.2.1 Characteristics of Migrants

(1) **Age and Gender**

The age of migrants is predominantly concentrated between 15 and 60 years old, with a significant proportion being young and middle-aged individuals (Zhou 1996; Liu and Feng 2014). Teenagers around the age of 15 typically migrate with their parents, often driven by parents seeking better educational resources or job changes. Some young people may also drop out of school after completing junior secondary education and take up lower-level jobs. Young adults aged 20–30, who are entering the workforce and reaching marriageable age, exhibit a high demand for housing. Middle-aged individuals, whose careers and financial situations have stabilized, often seek to improve their living conditions, especially with the implementation of the two-child policy, which has increased family size. The young elderly around 60 years old, who are at retirement age, may relocate due to changes in their work situation or to live with their children. In contrast, populations under 15 and over 60 years old are less mobile, as they are less capable of migrating and often reluctant to adapt to new environments due to their attachment to existing ones.

In terms of gender, males tend to be more mobile and migrate over wider ranges compared to females. Females are more likely to migrate to large cities but are also more constrained by family responsibilities (Lin and Zhu 2014).

(2) **Education level and occupation**

Among migrants, individuals with secondary or higher education levels are more active in migration, while those with only elementary education are less likely to migrate. Occupation and education level are closely correlated. Individuals with secondary education (including junior secondary, senior secondary, and specialized secondary education) are primarily engaged in commerce and service-related jobs, such as sales, couriers, chefs, and auto mechanics. These occupations require certain professional skills but not high educational qualifications, and they are characterized by high mobility, moderate income, and a high degree of migration. Tertiary-level jobs, such as office personnel, administrative staff, and teachers, are relatively stable with considerable income, but individuals may still migrate in pursuit of higher earnings. Those with primary education are typically low-level laborers who work long hours for low wages, face high barriers to relocation, and are the least active in migration.

The economic situation of migrants is closely tied to their education level and occupation. Residents with higher economic status have both the willingness and ability to change their living environment and engage in migration. In contrast, those with fewer financial resources are more focused on meeting daily expenses and are less likely to actively migrate.

(3) **Marital status and family structure**

Unmarried and divorced individuals are more active in migration, while married and widowed individuals are less likely to migrate due to the need to consider the preferences and actions of family members. This suggests that marital relationships tend to restrain migration, whereas changes in marital status (such as divorce or remarriage) increase the likelihood of migration.

Family size also influences migration patterns, consistent with marital status. Smaller families are more likely to migrate. Two-generation households with three or four members, as well as single-generation households, are more influenced by work and education factors, making it easier for them to reach a consensus on migration and thus exhibiting higher migration activity. In contrast, larger families, such as two-generation households with five members, three-generation households, and four-generation households, face more constraints. These include fewer suitable housing options and varying adaptability of family members to new environments, making migration less likely. However, due to limited per capita living space or changes in work and marital status, some family members may choose to migrate together to reduce relocation costs (Chai et al. 2000).

5.5.2.2 Periodic Characteristics of Migration

(1) **Planned economy period (1949–1978)**

During the planned economy period, the Danweilization system was the dominant mechanism for regulating urban socio-spatial resources. This system was characterized by the allocation of Danweilization benefits and the use of land without compensation. Residential housing was a by-product of urban construction and economic development, with the housing model heavily influenced by Danweilization housing. The government planned residential land based on the principle of proximity, arranging housing near Danweilization. The distribution of Danweilization in urban social space was highly uniform, with housing conditions determined by the socio-economic status of the respective *Danwei* (Fang 2014). Overall, residential spaces exhibited relative homogeneity, forming a stable, equal, and uniform "honeycomb-like" spatial structure composed of numerous Danweilization housing groups.

During this period, residents were highly dependent on their *Danwe.1* Housing supply within the same *Danwei* was limited, and residents had to wait for housing allocations based on regulations and employee rank. The only way to improve housing conditions was to achieve a higher status within the *Danwei* (Wang and Feng 2013). Due to institutional constraints, migration behavior during this stage was relatively limited, with most migration occurring within the same residential area as residents moved between different housing types. The primary manifestation of migration was changes in living conditions—such as the size, orientation, and floor of residences—as employees' status levels changed.

(2) **Transition period: from early reform and opening up to market transition (1978–1998)**

As the welfare housing system became increasingly unsustainable, housing construction efforts began in 1978, marking the start of the reform of the housing allocation system. This reform was implemented in stages, beginning with the "3–3 system" housing sales plan in 1982, followed by the rent increase subsidy and rent-sale combination policy in 1988, and culminating in the full commercialization of housing in 1998. Policies such as the implementation of housing provident funds and the sale of public housing enhanced residents' autonomy in choosing their homes and promoted small-scale migration.

This period represented a transitional stage in housing reform. While the *Danwei* still significantly influenced residents' housing choices, the shift from free to paid land use and the joint investment in housing by the state, *Danwei*, and individuals began to loosen these constraints. However, residents' housing choices—such as location, size, and supporting facilities—remained limited by the nature of their *Danwe.1*

(3) **Markct cconomy pcriod (post-1998)**

After 1998, the market economy period began. With the growth of the non-public sector and the deepening of housing system reforms, the allocation of housing in kind ended, and a monetized housing system was fully implemented. The *Danwei* community further disintegrated, and urban residents, with more diverse occupations and incomes, gained greater autonomy in choosing their residences. Migration became more frequent and widespread.

Since the housing reform, residential migration has been primarily influenced by real estate market conditions, housing demand, and economic factors. The rapid development of the real estate market and the growth of the second-hand housing market provided residents with a wider range of housing options. Advances in information and communication technology further broke down barriers to accessing housing information, enabling residents to find ideal homes more easily. Housing prices varied significantly based on location, quality, supporting facilities, and the ecological environment, allowing residents to choose housing according to their preferences and financial capabilities. This greatly facilitated migration.

Simultaneously, driven by government policies and developers, housing prices in central urban areas rose, prompting urban residential spaces to expand toward the suburbs. Social housing emerged, and lower-income groups began migrating to suburban areas. This shift not only alleviated population pressure and functional overconcentration in central urban areas but also addressed resettlement and compensation issues arising from rapid urban development. The direction of migration began to reflect residents' socio-economic status, leading to a trend of homogeneous agglomeration and heterogeneous differentiation. Additionally, increasingly convenient transportation conditions extended the distance of migration.

5.5.2.3 Characteristics of Spatial Differentiation of Migration

(1) **Spatial characteristics of migration within cities**

First, in medium and large cities, the number of people moving out of the city center exceeds the number moving in. Rapid urbanization, urban renewal, and the renovation and demolition of old urban areas have forced many residents to relocate. Due to high housing prices in the city center, low economic compensation for demolition, or inadequate living conditions in resettlement housing, only a small proportion of residents move back to the city center. The primary reason for migration to urban centers is the availability of a large number of jobs. Additionally, high-density residential areas developed for profit provide housing options for migrants. Among these migrants, the majority are non-registered populations who seek jobs and settle in the area by renting housing.

Second, the number of people migrating into the inner suburbs is greater than the number migrating out, and migration activity in these areas is relatively high. The tight supply of residential resources, high housing prices, rapid price increases, and job saturation in urban centers force residents to seek new homes and even jobs, leading to frequent migration. Urban residents with weaker economic foundations and floating populations have become the main groups migrating to inner suburbs. The relatively low housing prices, abundant housing supply, and the development of industrial suburbanization in inner suburbs provide employment opportunities and make these areas attractive destinations for migration. Although supporting facilities in the suburbs may not develop as quickly, most residents commute to work in the city during the day and return home at night, which generally meets their daily needs. Furthermore, improving transportation conditions and infrastructure in the suburbs have attracted a large number of floating populations and economically disadvantaged residents to migrate to the urban periphery. At the same time, environmental problems caused by overdevelopment in central urban areas and changing consumption concepts have led some residents to move from the city center to quieter, ecologically superior suburban residential areas. The primary direction of migration for inner suburban residents is toward the nearby urban center, driven by the pursuit of high-quality educational resources and job opportunities.

Finally, migration to and from outer suburbs is relatively limited, and migration activity is less pronounced. The outer suburbs of large cities are the main areas of rural urbanization. Economic development and regional construction in these areas lag behind those of inner and near suburbs, making them less attractive to urban residents from other areas. As a result, the out-migration rate in outer suburbs is slightly higher than the in-migration rate. The main reason for out-migration is the pursuit of better educational resources and jobs, while in-migration is primarily driven by the relocation of surrounding farmers and job changes.

(2) **Spatial characteristics of inter-provincial migration**

Since the Reform and Opening Up, the scale of inter-provincial migration in China has continued to expand. The inflow and outflow of population exhibit a concentrated and "multipolarization" distribution pattern, with the center of gravity of inter-provincial migration shifting eastward and northward. Overall, the direction of population migration reflects a trend from the central and western regions to the eastern regions. The inflow of inter-provincial migrants is predominantly concentrated in the eastern coastal areas, particularly in Guangdong, Shanghai, Zhejiang, Jiangsu, and Beijing, which are part of the three major metropolitan regions: the Pearl River Delta, the Yangtze River Delta, and the Beijing-Tianjin-Hebei region. Areas with a large inflow of population from other provinces typically experience a relatively small outflow of population. Conversely, areas with a large outflow of population are mainly concentrated in the central and western regions, with Sichuan, Anhui, Hunan, Henan, and Jiangxi being the provinces with the largest scale of population outflow (Liu et al. 2010; Liu and Feng 2014; Yang and Ning 2015).

The spatial differentiation of inter-provincial migration among urban residents in China is primarily influenced by the natural environment and socio-economic development (Wang et al. 2012). The eastern coastal region, characterized by relatively flat terrain, a pleasant climate, and a superior living environment, has experienced rapid urbanization and economic development. Cities in this region, such as Beijing (the national capital), Shanghai (a global financial center), Guangdong (a major economic province), and Zhejiang (known for its thriving private economy), possess strong comprehensive strength and competitiveness. These cities offer abundant job opportunities for diverse groups, making them highly attractive to urban residents and encouraging inter-provincial migration to these areas. In contrast, some provinces in the central and western regions have lower livability, slower urbanization, relatively sluggish economic development, and high population densities, leading to large-scale emigration.

5.5.3 The Impact of Migration on Residents' Living Space

Chinese society is undergoing a transitional period marked by significant changes in the social system, lifestyles, and value systems of urban residents. The social status and economic levels of urban residents are increasingly diverging. Residents' priorities have shifted from pursuing "economic wealth" to seeking "physical and mental comfort" at a spiritual level. Additionally, with the gradual rise in investment and consumption awareness, the frequency of residential migration within cities has increased. This migration reflects a continuous adjustment between residents' actual living conditions and their expectations (Liu et al. 2008) and significantly influences urban residents' lifestyles.

5.5.3.1 Changes in the Job-Housing Relationship

With the complete end of the planned economy's housing distribution system, residents' autonomy in choosing their homes has greatly increased. As migration becomes more common, the separation of workplace and residence has become more pronounced. Rapid urban development and constant adjustments to the urban spatial structure have led to high-density land development and soaring housing prices in city centers, prompting residents to move to peripheral areas. However, most residents continue to work in the city center, resulting in significant changes in the job-housing relationship.

First, there are changes in the job-housing relationship across different areas. The central city offers abundant employment opportunities, but high housing prices mean most residents do not live near their workplaces, leading to a prominent separation of workplace and residence. Inner suburbs, with more housing options, increasingly complete supporting facilities, and affordable prices, attract low- and middle-income groups and young people. These groups often initially rent housing near their workplaces in the city center and later migrate to the inner suburbs to purchase homes. This active migration has exacerbated the separation of workplace and residence. However, the suburbanization of industries has also created more job opportunities in the inner suburbs, allowing some residents to work closer to home, reducing the separation compared to the city center. In the outer suburbs, dominated by primary and secondary industries, migration is less active due to transportation and income constraints. Most residents work near their homes, resulting in the lowest level of job-housing separation. Additionally, residents who previously lived in Danweilization housing, located close to their workplaces in the old city, have experienced increased separation due to housing reform policies, demolitions, relocations, and job changes.

Second, changes in the job-housing relationship caused by migration directly affect commuting times. As urban residential and work locations become more dispersed, traffic congestion in the city center during peak hours has worsened. Residents who have migrated to the inner suburbs face longer commuting distances. Whether migration occurs within the city center or from the center to the periphery, the proportion of medium-length commutes has increased, while short commutes have decreased. Most residents now need to leave earlier, and the time allocated for commuting on weekdays has increased compared to pre-migration periods. However, with continuous improvements in transportation infrastructure and the availability of more transportation options, long commutes have become more manageable (Feng and Zhou 2004).

Commuting times affect residents differently based on income levels. Lower-income groups, constrained by housing prices, often migrate to the suburbs, resulting in longer commuting distances and reliance on public transportation, which increases their commuting time. In contrast, higher-income groups have more flexibility in choosing where to live and how to travel, making their commuting times less affected.

5.5.3.2 Changes in Travel Patterns

As urban spaces continue to expand into the suburbs, some city residents are increasingly moving to suburban areas. However, the degree of matching between transportation facilities and residential locations varies, and combined with differences in residents' economic strength and reasons for migration, this leads to variations in their choice of transportation and tolerance for travel time and cost. Consequently, residents' travel patterns have changed before and after migration.

Before migration, the proximity of workplaces and residences meant that residents primarily relied on non-motorized travel methods such as walking or cycling, with limited use of cars. After migration, improvements in living standards and urban infrastructure, along with rising car ownership, have significantly increased the use of public transportation and private cars. Accordingly, non-motorized travel methods have decreased.

For long-distance migration within the city, changes in travel patterns differ between active and passive migration. Residents who actively migrate from the city center to the inner suburbs for better housing conditions, or from the inner suburbs to the city center for better education and job opportunities, often have stronger financial resources and the ability to upgrade their transportation. Before migration, they primarily walked or used public transportation, but after migration, they tend to rely more on private cars. In contrast, residents who are involuntarily relocated from the city center to the suburbs due to demolition or other reasons often have lower economic levels and fewer transportation options, relying more on public transportation after relocation (Luo 2017).

5.5.3.3 Changes in Shopping Behavior

Shopping behavior, as one of the key daily activities of residents, is significantly influenced by changes in residential space.

First, migration has altered shopping locations. Before migration, residents primarily purchased food and daily necessities at markets and supermarkets, while clothes and appliances were bought at shopping malls or specialty stores. With urban upgrading and transformation, the distance between residential areas and traditional markets has increased, while the number of small supermarkets has grown, and medium-to-large supermarkets now offer a wider range of products. After migration, most residences are farther from urban commercial centers. On weekdays, residents mainly shop at small supermarkets near their homes for food and urgent daily necessities. On weekends, they visit large supermarkets and shopping malls near their residential areas for general items, while high-end clothing and valuables are purchased at large commercial centers in the city (Li et al. 2013).

Second, migration has changed shopping methods. While the frequency of shopping remains relatively unchanged, residents adapt their shopping habits to meet daily needs (Zhang and Chai 2009). After migration, reduced accessibility to shopping locations and the unfamiliarity of the new residential environment, combined with the rise of e-commerce and express delivery services, have led to an increase in online shopping and a decrease in physical store visits. Online shopping, once a supplementary method for non-essential items, has become a primary shopping channel.

5.5.3.4 Changes in Leisure Space

First, changes in recreational spaces. Urban recreational spaces include public spaces like parks, squares, bars, cafes, sports venues, and cultural centers, as well as natural spaces such as beaches, waterfronts, and forests. Long-distance migration has a greater impact on leisure activities. Before migration, residents lived closer to urban commercial centers, where recreational resources were abundant. Bars, restaurants, and cinemas were popular leisure venues for young people. After migration, suburban recreational facilities are more scattered and less accessible. However, the increase in public activity spaces within residential areas and the improved natural environment have made communities and surrounding areas the main leisure destinations. On weekends, the availability of diverse travel options reduces the impact of migration on leisure venue choices.

Second, changes in leisure patterns. Before migration, residents were less homebound and preferred activities like shopping, dining out, and watching movies after work. On weekends, suburban outings were popular. After migration, improved living conditions and longer commuting times have made residents more inclined to spend leisure time at home. Weekday leisure activities now include walking around the neighborhood, watching TV, and surfing the internet, while weekends are reserved for visiting shopping malls, meeting friends, or taking short trips.

5.5.3.5 Changes in Social Activities

Social activities are based on stable social relationships within a certain range. Migration gradually separates residents from their original social circles, increasing the spatial distance from neighbors and making the new environment unfamiliar. Both time and space constraints have significantly limited social activities, leading many residents to rely more on family or online social platforms for daily interactions. Migration has weakened traditional in-person socializing, with residents maintaining connections through phone calls, WeChat, and other digital means. The frequency of socializing has also decreased due to the inhibiting effect of distance and the time needed to adapt to the new environment (He 2017).

In addition, migration significantly impacts the social activities of the elderly. After retirement, the elderly have more leisure time and an increased need for physical and psychological care and companionship. However, due to family separation or the busy schedules of family members, the elderly often rely heavily on their social networks. After migration, the elderly face challenges in maintaining their original social networks due to limited mobility and unfamiliarity with social tools. Combined with their relatively weaker ability to adapt to new environments and integrate into new social circles, the time the elderly spend going out has significantly decreased, the variety of social activities has diminished, and their overall social satisfaction tends to be low (Li and Chai 2013).

5.6 The Interaction Among Residential, Working and Commuting

When studying urban residential space, it is essential to consider not only where residents live but also where they work and the means of transportation connecting the two. Therefore, the interactions among residence (place), workplace, and commuting are integral and important components of urban residential space.

The mainstream commuting patterns of Chinese urban residents have evolved through three stages of development. The first stage was characterized by short-distance commuting, primarily by foot or bicycle. The second stage saw medium- and short-distance commuting, mainly by bicycle or public transport. The current stage is marked by the diversified development of commuting patterns, with centripetal long-distance commuting—equally reliant on motor vehicles and public transport—becoming the dominant pattern. The choice of residence, workplace, and commuting methods, times, and distances among Chinese urban residents has been heavily influenced by socio-economic development and broader societal changes. These interactions at the individual behavior level exhibit periodic characteristics. At the level of urban social space, the trend has shifted from the "integration of workplace and residence" to the "separation of workplace and residence." Currently, the interactions between residence, work, and commuting among Chinese urban residents show significant differences due to the combined effects of multiple factors.

5.6.1 *The Period of Traditional Residential Space: Short-Distance Commuting Within the Compound of Danwei*

The planned economy period was also the era of traditional residential space dominated by the *Danwei* compound. During this time, the residence, work, and commuting patterns of Chinese urban residents were characterized by short-distance commuting, with walking and cycling as the primary means of transportation. Most *Danwei* compounds exhibited a spatial distribution where workplaces were located close to residences. As a result, the commuting patterns of urban residents mainly included intra-*Danwei* commuting within walking distance, inter-*Danwei* commuting, and reverse commuting from the city center to suburban *Danwei* (Ta et al. 2016). During this period, the *Danwei* served as the fundamental unit of urban socio-spatial organization in China. The *Danwei* courtyard was its physical manifestation, creating a spatial morphology that was "small but complete" in both physical and social terms. Under the unique historical and socio-economic conditions of the time, employment for urban residents was centrally arranged by the state through labor and personnel departments. Residents did not face the employment pressures that emerged after the transition to a market economy, but they were also restricted in terms of job flexibility and mobility. Many spent their entire careers working in the same *Danwe.1* The stability and security of employment fostered a community of interests among *Danwei* members, who collectively worked to improve living conditions and enhance welfare benefits.

Additionally, during the planned economy period, enterprises focused on meeting production targets rather than pursuing profits. This led to the diversion of production funds to build ancillary living quarters and provide living benefits within the *Danwei*, becoming an inevitable development trend (Liu and Chai 2012). In this socio-historical context, characterized by a free land use system, the primary behavior of *Danwei* was to build residential areas as close as possible to workplaces and occupy more land to reserve space for future development. Studies have shown (Xu 2011) that the average distance between workplaces and residences during this period was less than 3 km. Residents enjoyed the convenience of living near their workplaces, with walking or cycling as the primary modes of transportation, resulting in short commuting distances and times.

However, in dual-income families where spouses worked for different companies, one spouse often had to commute between the residential area and the workplace, resulting in a longer commuting distance compared to the intra-*Danwei* commute of the other spouse. Another type of commuting involved native city residents and employees of smaller *Danwei* (without living quarters) living in state-allocated housing, who often engaged in reverse commuting from the city center to the inner suburbs.

Overall, during this stage, the interaction of residence, work, and transportation commuting among Chinese urban residents formed a work-life balance centered around the *Danwe.1*

5.6.2 Period of Transformation of Residential Space: Gradually Breaking Through the Limitations of Danwei Space for Short-To-Medium Distance Commuting

After the Reform and Opening Up and before 1998, the housing system reform began, and the influence of Danweilization residential space gradually diminished. This marked a transitional period in the development of residential space for Chinese urban residents. During this time, the distance between residents' places of residence and workplaces showed a consistent upward trend. The proportion of short-distance commuting decreased, while short-to-medium distance commuting increased. During this period, the planned economy transitioned to a market economy, and the previously closed spatial interfaces of *Danwei* spaces began to open, gradually integrating into the broader urban spaces. The transformation of *Danwei* communities into urban communities meant that the functional roles of the various components of *Danwei* spaces also changed. The increased openness allowed production and living areas, which were originally oriented toward internal *Danwei* services, to assume more diverse urban functions. This shift also meant that *Danwei* residents transformed into urban residents, with more possibilities in terms of living and working arrangements. The previous work-life balance, dominated by short-distance commuting within the *Danwei*, was gradually disrupted. Intra-*Danwei* commuting decreased as a proportion of total commuting, while inter-*Danwei* commuting and commuting from the inner suburbs to the city center or newly established employment centers increased.

With the advancement of land market reform, the value of land became more apparent, leading to differences in land value within urban built-up areas. Land value and rent were highest in the city center and gradually decreased toward the urban fringe. This situation influenced the development and construction of residential areas, with urban peripheries such as the inner suburbs, where land prices were relatively low, becoming preferred locations for new housing. Until the termination of the welfare housing system in 1998, Chinese residents still enjoyed the benefits of being allocated housing by their workplaces or purchasing housing at prices far below market rates. The key difference from the planned economy period was that housing allocated by workplaces was no longer necessarily located near the workplace but might be situated in the inner suburbs, a certain distance away. This led to a general trend of outward migration of residential locations, while workplaces remained primarily in the original *Danwei* compounds in the city center. As a result, the separation of workplace and residence for Chinese urban residents gradually increased, leading to longer commuting distances and times. For example, in Guangzhou, the average commuting distance increased from 2.54 km in 1984 to 5.41 km in 1998, while the average commuting time rose from 26 to 34 min (Mao et al. 2000). Despite these increases, commutes during this period could still be classified as short-to-medium distance. Limited by the socio-economic conditions of the time, the primary modes of transportation for urban residents during this period were bicycles and above-ground public transport.

5.6.3 *The Period of Market-Oriented Residential Space: Diversified Commuting, Mainly Long-Distance Commuting*

The Notice of the State Council on Further Deepening the Reform of the Urban Housing System and Accelerating Housing Construction marked the end of the welfare housing system and initiated the monetization of housing distribution and the marketization of the real estate sector. Therefore, the period after 1998 can be regarded as the era of market-oriented residential space development for Chinese urban residents. During this period, the separation of workplace and residence among urban residents increased significantly, leading to the emergence of long-distance and even ultra-long-distance commuting. The phenomenon of "arriving home in the early hours of the morning and leaving in the early hours of the morning" highlights the immense pressure faced by long-distance commuters.

With the further transformation of the economy and rapid urbanization, the built environments of large and mega-cities such as Beijing, Shanghai, Shenzhen, and Guangzhou underwent dramatic changes. The rapid population growth exacerbated the scarcity of spatial resources, leading to the continuous outward expansion of urban built-up areas and the formation of a "spreading pancake" urban spatial development pattern. Most Chinese cities retained a monocentric urban structure, with soaring land prices in the city center. Economic development and industrial restructuring, characterized by the trend of "retreating secondary industries and advancing tertiary industries" in the city center, strengthened its function as a central business district (CBD). This further generated agglomeration economies, attracting a large number of enterprises and creating a positive feedback loop between job growth and land value appreciation.

The transformation of urban functions toward employment areas under market forces led to the decline of residential areas in the city center. The separation of workplace and residence became an inevitable consequence of urban expansion. On the one hand, the high spatial value of the city center maximized land utilization, resulting in crowded residential spaces and compressed public service facilities, negatively affecting residents' quality of life. On the other hand, the rapid urban expansion promoted the construction of new urban areas with relatively low land prices, lower plot ratios, and higher greening rates, offering a better living environment compared to the "old, run-down, small" buildings in the city center. Rising economic levels enabled residents to afford private cars, and improved urban infrastructure provided convenient conditions for car-based transportation. These factors, combined with the "crowding out effect" of the city center, triggered the spontaneous out-migration of residents to suburban areas.

However, the improvement in residential functions did not coincide with equivalent economic development in the suburbs. The relative lag in industrial development meant that outer urban areas, represented by new districts, were unable to provide sufficient high-quality jobs. Migrated residents still needed to seek employment opportunities in the central city. At the same time, the influx of floating populations into cities, unable to afford high housing rents in the city center, settled in suburban areas or urban villages with relatively low housing prices and rents, becoming the backbone of long-distance commuting. For example, in Tiantongyuan, a well-known suburban "sleeping city" in Beijing, the average commuting distance for full-time employees is 7.96 km, with an average commuting time of 38.5 min. Among them, 25% of respondents had a commuting distance of over 13.9 km, with an average commuting time exceeding one hour. The overall commuting pattern is an uneven radial flow centered on Tiantongyuan, with the most common direction being from the suburbs to the city center, forming a centripetal flow of long-distance commuting (Fu et al. 2018).

The functional differences between the urban center and suburbs are particularly pronounced in large and mega-cities. Residents in first-tier cities such as Beijing, Shanghai, Guangzhou, and Shenzhen often work far from home for higher pay or closer to home for lower pay. The proportion of high-income earners commuting more than 10 km to work is 1.4–2.1 times that of low-income earners, while the proportion of low-income earners commuting less than 3 km is 1.8–2.1 times that of high-income earners.[1] To balance low housing prices with better job opportunities, residents of large and mega-cities often endure painful long-distance commutes. Beijing residents face the most challenging commutes, with an average distance of 13.2 km and an average time of 56 min.[2]

If commuting reliability is judged by "the amount of planning time needed to ensure arriving on time per kilometer,"[3]high-traffic, long-distance commuting exacerbates urban congestion during rush hours, making ground transportation highly unreliable. This unreliability often causes commuting anxiety, affecting commuters' physical and mental health and reducing their life satisfaction and well-being. Although subway commuting offers relatively reliable travel times, route detours, stops, and transfers can lengthen the journey. Residents relying on public transport often spend more time commuting than those using private vehicles (Li and Li 2007). Commuting problems not only consume urban residents' time for daily life and work but also increase personal travel costs and the externalities of urban socio-spatial organization.

Overall, population mobility has transformed urban regional functions, altering the employment-residential space relationship. The original compact spatial pattern with a balanced division of jobs and residences has been disrupted. The "separation of workplace and residence" and the resulting long-distance commuting have become a new urban challenge. Long-distance commuting from the suburbs to the city center has become the dominant flow, supplemented by reverse commuting from the city center to the suburbs and lateral commuting between suburban new towns. Commuting methods have also become increasingly diverse.

---

[1] Baidu Maps, 2018 China Urban Transportation Report, 2019.

[2] Aurora Big Data, 2018 China Urban Commuting Research Report, 2018.

[3] DiDi Chuxing, "2017 Analysis Report on Traffic Reliability in Major Chinese Cities," 2018.

5.6.4 *The Interactive Relationship Between Work and Residence and Its Mechanism*

To better measure the interaction between residents' residence, workplace, and commuting, scholars typically conduct in-depth research from the perspectives of residents' employment choices and employees' residential choices. Commonly used indicators include the "resident balance index," "employee balance index," "job-housing ratio," "employment accessibility," and the "separation of workplace and residence index," often analyzed using advanced mathematical and statistical methods. Traditional studies of the job-housing relationship rely on census and questionnaire data. However, in recent years, new types of location-based big data, such as public transport card data and Weibo positioning data (Long et al. 2012), have enabled scholars to obtain more accurate information on work-life commuting patterns.

Work-life imbalance manifests differently across spatial units. Geospatial units near typical employment centers usually have more jobs than homes, while units with large residential areas often have fewer jobs than homes (Zhao and Cao 2018). For example, in Beijing, Huilongguan and Tiantongyuan are large suburban residential areas where less than 20% of residents work nearby, making them typical "sleeper towns." In contrast, areas with prominent workplace attributes, such as Jianwai Street, have less than 5% of employees living there, illustrating another form of work-life imbalance (Zheng et al. 2015). A study on the separation of workplace and residence in Beijing found that as the ratio of workplace to residence increases, commuting time first decreases and then increases, forming a U-shaped structure (Wei et al. 2017). Similarly, an empirical study in Shanghai revealed that in areas where workplaces outweigh residences, employees commute longer distances, while residents have shorter commutes. Conversely, in areas where residences outweigh workplaces, residents commute longer distances, while employees have shorter commutes. A more balanced job-housing relationship results in less commuter traffic, whereas a severe separation of workplace and residence leads to greater commuter traffic (Song et al. 2019). Promoting a balance between work and residence can reduce exhaust emissions and environmental pollution to some extent (Chai et al. 2011) and provide more employment opportunities, fostering economies of scale (Anas et al. 1998).

Commuting patterns at the city level are shaped by policy and socio-economic forces, as well as the collective choices of individual residents. In a market-oriented employment environment, urban residents have greater autonomy in choosing where to work and live. One of the main influencing factors is the level of education. Within the same spatial unit, highly skilled labor (and corresponding industries) tends to have a lower employment balance index (a lower proportion of employed people living in the same spatial unit). This is primarily because the supply of jobs meeting the needs of highly skilled workers is relatively limited, requiring them to search over larger spatial areas for suitable opportunities. To secure higher salaries, highly skilled workers often sacrifice shorter commuting distances and times. This difference in commuting time and distance is reflected in the type of work, typically following the order: company employees > commercial services > public institutions > freelance work (Wei et al. 2017).

Additionally, family structure and the division of labor within households also influence residents' choices of residence and workplace, thereby affecting the spatial distance and commuting time between an individual's home and workplace. For example, families with young children often prioritize reducing commuting time to devote more time to childcare. As a result, parents tend to choose workplaces and residences that are closer to each other. In dual-income families, residential choices must balance the commuting needs of both earners to achieve the best overall family outcome.

5.7 Chapter Summary

From the perspective of residential function, the fundamental unit of urban residential space is the smallest spatial unit that constitutes urban residential space. However, from the perspective of spatial statistics, the fundamental unit of residential space is often larger than the functional unit and needs to align with the fundamental administrative division unit to facilitate research using relevant data, particularly census data. Evidence shows that block (street/township) level data is an important and accessible resource for studying the structure and differentiation of urban residential spaces in Chinese cities. The "neighborhood scale" is a suitable spatial scale for examining the characteristics of residential space differentiation in Chinese cities. Although the spatial scale of village dwellings is more microscopic, data at this level is difficult to obtain. Additionally, a significant drawback is the challenge of accurately restoring urban maps and their changing relationships at the neighborhood committee (village committee) level.

By comparing the empirical research results of Beijing with those of other large cities such as Shanghai and Guangzhou (Tang et al. 2016; Zhou et al. 2016a), it becomes evident that the evolution of urban residential spatial structure from 2000 to 2010 exhibits both commonalities and specificities across these cities. In terms of principal factors, the commonalities lie in the relatively stable composition of the main components influencing residential spatial structure. Most factors, such as intellectuals, agricultural populations, and population density, have a lasting impact. However, specificities emerge in the role of certain populations. For instance, in Shanghai, the migrant population is no longer a primary component, while the agricultural population has become a core factor. In contrast, in Beijing and Guangzhou, the social status of migrants has risen, and their distribution has expanded from the city center to the suburbs. Additionally, in Beijing, the white-collar population overlaps significantly with the registered permanent population, while the blue-collar population is predominantly composed of migrants. In terms of residential space types, commonalities include the emergence of distinct social classes, such as intellectuals and agricultural populations, as separate groups. With the development of the market economy, social stratification has become more pronounced. For example,

Beijing and Shanghai exhibit a clear division between white-collar and blue-collar workers, while Guangzhou shows a division between middle-income and low-income groups. A notable characteristic is that Beijing and Guangzhou have high-density, crowded old cities with dense populations, whereas Shanghai has a relatively lower residential population density. In terms of residential spatial structure, major Chinese cities exhibit a combination of ring, fan, and multi-core models. However, the ring structure is more pronounced in Shanghai, Beijing's fan-shaped radial trend is more prominent, and Guangzhou's multi-core model stands out the most. Overall, the latest characteristics of residential space development in China's major cities include the emergence of social stratification based on occupational differentiation, leading to the formation of corresponding residential areas. The household registration factor, particularly whether residents have obtained local household registration, also plays a significant role in shaping residential space. Additionally, the age factor, especially aging, has influenced the formation of urban residential space in the new era, leaving traces of spatial agglomeration.

From 2000 to 2010, the spatial differentiation of urban residents' occupations became more pronounced, as evidenced by research in Beijing. The elderly population, migrants, Uyghur population, illiterate population, and agricultural population continue to exhibit trends of increasing group living and decreasing mixed living. The consistency between the spatial distribution of these indicators and the residential population's spatial matching has worsened, reflecting growing urban issues such as the separation of employment and residence for the working population, commuting challenges for migrants, and the plight of the elderly left behind or living in empty nests. Analysis based on differentiation and segregation indices reveals more nuanced details about the restructuring of urban residential space beyond traditional social area analysis. The underlying driving forces for this restructuring must be sought at the level of institutional, market, and cultural changes. An explanatory framework for these dynamics can be established with the help of social-spatial dialectics.

The habitability of urban communities is a crucial indicator reflecting the condition of urban residential space. Evaluating community work requires establishing a specialized index system that includes the following modules: the rationality of physical space, the completeness of service facilities, the supporting conditions of infrastructure, the beauty of the environment and landscape, the care and harmony in the neighborhood, and the efficiency of governance services. Additionally, academia has conducted extensive research on the differentiation of urban community habitability in China and its formation mechanisms. The differentiation in habitability primarily include variations in residential space, the configuration of public service facilities, activity space, and leisure space. Key factors influencing differentiation are historical and cultural factors, market mechanisms, government decisions, the level of economic development, and individual behavior.

Migration significantly impacts the development of urban residential space. Chinese scholars have conducted numerous studies on the attributes of Chinese urban migrants, the phased development characteristics of migration, and the spatial differentiation of migration. The impact of migration on urban residents' living spaces is mainly reflected in changes to the job-housing relationship, travel patterns, shopping behavior, leisure space, and social activities. Furthermore, the interactive relationship between residence (place), workplace, and transportation commute is an integral part of urban residential space. The choices of residence, workplace, and commuting methods, times, and distances among Chinese urban residents have been strongly influenced by socio-economic development and broader societal changes. At the individual behavioral level, this interaction exhibits phased characteristics: the traditional residential space period: dominated by short-distance commuting within the *Danwei* compound; the residential space transformation period characterized by short-to-medium-distance commuting, gradually breaking through the spatial constraints of the *Danwei*; the market-oriented residential space period marked by diversified commuting patterns, primarily dominated by long-distance commuting. At the urban spatial level, these changes reflect a development trend from the "integration of workplace and residence" to the "separation of workplace and residence." This shift highlights the evolving dynamics of urban residential space and the increasing complexity of commuting patterns in response to socio-economic transformations.

References

Anas A, Arnott R, Small KA (1998) Urban spatial structure. J Econ Lit 36(3):1426–1464
Beijing Municipal Bureau of Statistics (2020) Beijing statistical yearbook 2020. China Statistics Press, Beijing (in Chinese)
Beijing Municipal Bureau of Statistics (2011) Beijing statistical yearbook 2011. China Statistics Press, Beijing (in Chinese)
Bi J (2017) Research on traffic livability and spatial differentiation in urban areas from a microcirculation perspective. Xi'an International Studies University, Xi'an (in Chinese)
Cao R, Chen J, Bai G (2003) Application of niche theory in China's urban development. Geogr Geo-Inf Sci 19(1):62–65 (in Chinese)
Chai H, Feng J (2014) Social space formation in suburban large scale residential areas based on relocation: a case study of Beijing's Huilongguan community. AR Res Dev 33(5):77–81 (in Chinese)
Chai Y, Xiao Z, Liu Z (2011) Comparative analysis of household daily travel carbon emissions in Beijing based on spatial behavioral constraints. Scientia Geographica Sinica 31(7):843–849 (In Chinese)
Cheng C, Li Q, Sha Z et al (2006) High-resolution remote sensing evaluation method for environmental quality of urban residential units. Adv Earth Sci 21(1):24–30 (in Chinese)
Clark GL, Feldman MP, Gertler MS (2003) The oxford handbook of economic geography. Oxford University Press, Oxford
Chai Y, Hu Z, Wu Z (2000) Characteristics and mechanisms of internal population relocation in Tianjin. Geogr Res 19(4):391–399 (in Chinese)
Coffey WJ, Shearmur RG (2002) Agglomeration and dispersion of high-order service employment in the Monstreal metropolitan region, 1981–96. Urban Stud 39:359–378
Cong X, Liu Z (2015) Analysis of urban spatial growth and morphological evolution in Beijing based on GIS and RS. J Hangzhou Norm Univ (Soc Sci Ed) 5:122–130 (in Chinese)
Dong F, Dong W (2017) Research on school district housing prices and their formation mechanism. Soc Sci Front 1:43–51 (in Chinese)

Dou X (2011) Research on residential spatial structure of Wuhan residents. Central China Normal University, Wuhan (in Chinese)
Dupont V (2004) Socio-spatial differentiation and residential segregation in Delhi: a question of scale? Geoforum 35(2):157–175
Fang C (2014) The evolution and internal logic of urban residential space in China. Academic Monthly 46(1):100–109 (in Chinese)
Feng J (2005a) Spatial perception and image structure of beijing residents. Scientia Geographica Sinica 25(2):142–154 (in Chinese)
Feng J (2004) Spatial restructuring of Chinese cities in the transition period. Science Press, Beijing (in Chinese)
Feng J (2005b) Facing the trend of social spatial differentiation in Beijing. Beijing Plan Rev 2:176–179 (in Chinese)
Feng J (2012) Population studies and urban planning based on geographical thinking. City Plan Rev 36(5):27–37 (In Chinese)
Feng J, Liu Y (2007) Restructuring of urban internal space in transitional China: characteristics, mechanisms, and patterns. Prog Geogr 26(4):93–106 (in Chinese)
Feng J, Zhao N (2016) Social space and networks of the LGBTQ community in a postmodern geographical context: a case study of Beijing. Acta Geogr Sin 71(10):1815–1832 (in Chinese)
Feng J, Zhong Y (2018) Social spatial reconstruction in Beijing (2000–2010). Acta Geogr Sin 73(4):711–737 (in Chinese)
Feng J, Zhong Y (2020) Spatial structure of urban residents' quality of life based on residential environment: a case study of Changzhou. Acta Geogr Sin 75(6):1237–1255 (in Chinese)
Feng J, Zhou Y (2003a) Social spatial structure and its evolution in Beijing metropolitan area (1982–2000). Geogr Res 22(4):465–483 (in Chinese)
Feng J, Zhou Y (2003b) Progress and prospects of research on urban internal spatial structure in China. Prog Geogr 22(3):304–315 (in Chinese)
Feng J, Zhou Y (2004) Residential relocation and related spatial behaviors during suburbanization in Beijing. Geogr Res 23(2):227–242 (in Chinese)
Feng J, Zhou Y (2008) Restructuring of social spatial differentiation in transitional Beijing. Acta Geogr Sin 63(8):829–844 (in Chinese)
Feng J, Chen X, Lan Z (2007) Evolution of shopping behavior spatial structure of Beijing residents. Acta Geogr Sin 62(10):1083–1096 (in Chinese)
Feng J, Huang L, Dong Y et al (2012) Spatio-temporal characteristics and mechanisms of urban crime: a case study of property crime in Beijing's eight districts. Acta Geogr Sin 67(12):1655–1666 (in Chinese)
Feng J, Wu F, Logan J (2008) From homogenous to heterogeneous: the transformation of Beijing's sociospatial structure. Built Environment 34(4):482–497
Friedmann J, Wolff G (1982) World city formation: an agenda for research and action. Int J Urban Reg Res 6:309–344
Fu T, Zhang Y, Chai Y (2018) Impact of commuting patterns on health in suburban residents: a case study of Beijing's Tiantongyuan. Prog Geogr 37(4):547–555 (in Chinese)
Garreau J (1991) Edge city. Doubleday, New York
Gong J (2018) Research on the differentiation characteristics of urban residential spaces based on GIS. Wuhan University, Wuhan (in Chinese)
Gu C (2011) Reflections on transitional development and future cities. City Plan Rev 35(11):23–34, 41. (in Chinese)
Gu C, Wang FH, Liu G (2005) The structure of social space in Beijing in 1998: a socialist City in transition. Urban Geogr 26(2):167–192
Han Z, Li Y, Liu T et al (2019) Spatial differentiation of public service facilities in community life circles: a case study of Shahekou district, Dalian. Prog Geogr 38(11):1701–1711 (in Chinese)
He S, Qian J (2017) From an emerging market to a multifaceted urban society: urban China studies. Urban Stud 54(4):827–846

He S, Liu Y, Wu F et al (2010) Social groups and housing differentiation in China's urban villages. An Instal Interpret 25(5):671–691
He Y (2017) Decision-making mechanism of daily activity-travel behavior in relocated households. Kunming University of Science and Technology, Kunming (In Chinese)
Hou M, Wang M (2014) Review of residential relocation behavior. J Cap Norm Univ (Nat Sci Ed) 3:95–100 (In Chinese)
Hsu YA, Pannell CW (1982) Urbanization and residential spatial structure in Taiwan. Pacific Viewpoint 23:22–52
Hu Y (2019) Conflict and governance of public activity spaces in relocation communities in Xi'an. Xi'an International Studies University, Xi'an (in Chinese)
Huang X, Huang X (2013) Evolution of urban social spatial structure in Changchun during the 20th century. Prog Geogr 32(11):1629–1638 (in Chinese)
Icel J, Sharp G (2013) White residential segregation in U. S. metropolitan areas: conceptual issues, patterns, and trends from the U. S. census, 1980 to 2010. Popul Res Policy Rev 32(5):663–686
Johnston RJ, Gregory D, Pratt G et al (2000) The dictionary of human geography, 4th edn. Blackwell, Malden
Kitchin R, Tate NJ (2000) Conducting research into human geography: theory, methodology and practice. Pearson Education, New York
Knox P, Pinch S (2000) Urban social geography: an introduction, 4th edn. Prentice Hall, Englewood Cliffs
Le C (2005) Le Corbusier complete works, vol. 7 (trans: Niu Y, Cheng C). China Architecture & Building Press, Beijing (in Chinese)
Li C, Zhang X (2015) Evolution of social spatial structure in Hefei during the transition period (1982–2000). Scientia Geographica Sinica 35(12):1542–1550 (in Chinese)
Li D, Lan Z (2012) Reflections on community development in China's urbanization process. J Public Manag 9(1):104–110, 127 (in Chinese)
Li Q, Li X (2007) Commuting characteristics of residents in suburban large-scale residential areas in Beijing. Urban Problems 7:55–59 (In Chinese)
Li S, Siu Y (2001) Residential mobility and urban restructuring under market transition: a study of Guangzhou, China. Prof Geogr 53(2):219–229
Li S, Hou Q, Chen S et al (2010) Work, home, and market: the social transformation of housing space in Guangzhou. China. Urban Geography 31(4):434–452
Li X, Li H, Tian S (2013) Research on recreational systems from the perspective of human settlements. J Liaoning Norm Univ (Nat Sci Ed) 2:269–273 (in Chinese)
Li Z, Wu F (2006) Social spatial differentiation in transitional shanghai. Acta Geogr Sin 61(2):199–211 (in Chinese)
Li Z, Wu F (2008) Tenure-based residential segregation in post-reform Chinese cities: a case study of Shanghai. Trans Inst Br Geogr 33(3):404–419
Li, P, Chai Y (2013) Impact of relocation on daily social networks of elderly Danwei members. Hum Geogr 28(3): 78–84, 6 (in Chinese)
Lin L, Zhu Y (2014) Spatial type selection and influencing factors of floating population's initial migration: a survey study in Fujian Province. Scientia Geographica Sinica 34(5):539–546 (in Chinese)
Lim GC, Lee MH (1993) Housing consumption in Urban China. J R Estate Financ Econ 6(1):89–102
Liu X (2009) Study on differentiation characteristics of community livability in Dalian. Liaoning Normal University, Dalian (in Chinese)
Liu T, Chai Y (2012) Influencing factors of urban danwei system formation in China. Urban Dev Stud 19(7):59–66 (in Chinese)
Liu W, Yan X, Cao X (2008) Review on residential mobility within Western cities. Scientia Geographica Sinica 28(1):131–137 (In Chinese)
Liu Y, Feng J (2014) Characteristics and influencing factors of population migration in China: analysis based on the sixth population census data. Hum Geogr 29(2):129–137 (in Chinese)

Lo CP (1975) Changes in the ecological structure of Hong Kong 1961–1971: a comparative analysis. Environ Plan A 7(8):941–963
Lo CP (1986) The evolution of the ecological structure of Hong Kong: implications for planning and future development. Urban Geogr 7(7):311–335
Lo CP (2005) Decentralization and polarization: contradictory trends in Hong Kong's postcolonial social landscape. Urban Geogr 26(1):36–60
Long Y, Zhang Y, Cui C (2012) Analyzing Beijing's jobs-housing relationships and commuting patterns using smart card data. Acta Geogr Sin 67(10):1339–1352 (in Chinese)
Lü Y (2012) Evaluation of livability in Hangzhou's residential environment. Zhejiang University, Hangzhou (in Chinese)
Luo H (2017) Characteristics of commuting mode choice among urban relocators: case study of Nanjing. In: Proceedings of 2017 China urban transportation planning annual conference, pp 2334–2344 (in Chinese)
Mao D, Li X, Xiao L (2000) An analysis on the characteristics of the trips of Guangzhou inhabitants and development of urban communications. Trop Geogr 20(1):32–37
Ni P (2018) Annual report on urban competitiveness in China (No.17). China Social Sciences Press, Beijing (in Chinese)
Pacione M (2001a) Models of urban land use structure in cities of the developed world. Geography 86(2):97–119
Pacione M (2001b) The internal structure of cities in the third world. Geography 86(3):189–209
Pais J (2017) Intergenerational neighborhood attainment and the legacy of racial residential segregation: a causal mediation analysis. Demography 2:1–30
Peterson, W. (1984). Basic demography. Lanzhou University Population Translation Office, Trans. Gansu People's Publishing House, Lanzhou (in Chinese)
Qin B, Jiao Y (2010) Spatial distribution of housing prices and urban structure evolution in Beijing. Econ Geogr 30(11):1815–1820 (In Chinese)
Qin R, Zhou R (2011) Review on urban community differentiation and typology. Planners 27(S1):216–221 (in Chinese)
Sassen S (1991) The global city. Princeton University Press, Princeton
Schnell I, Benjamini Y (2005) Globalization and the structure of urban social space: the lesson from Tel Aviv. Urban Studies 13:2489–2510
Shevky E, Bell W (1955) Social area analysis. Stanford University Press, Stanford
Shevky E, Williams M (1949) The social areas of Los Angeles. University of California Press, Los Angeles
Song X, Wang Y, Yu Y, et al (2019) Statistical verification of commuting distance on job-housing separation. J Geo-Inf Sci 21(11):1699–1709 (in Chinese)
Sun Y (1992) Formation mechanism and demarcation of metropolitan areas: case study of Beijing. Acta Geogr Sin 47(6):552–560 (in Chinese)
Ta N, Chai Y, Zhang Y et al (2016) Understanding job-housing relationship and commuting pattern in Chinese cities: past, present and future. Transp Res Part D: Transp Environ 52:562–573
Tang Z, Chen S, Wang X et al (2016) Socio-spatial structure and evolution pattern of Shanghai central city in transitional period. Planners 32(6):105–111 (in Chinese)
The Ghent Urban Studies Team (1999) The urban condition: space, community, and self in the contemporary metropolis. 010 Publishers, Rotterdam
Wang G, Pan Z, Lu Y (2012) Changes in regional patterns of China's interprovincial migration and influencing factors: analysis based on 2000 and 2010 census data. Chin J Popul Sci 32(5):2–13, 111 (in Chinese)
Wang H (2006) Development zones and socio-economic spatial polarization in Xi'an. Acta Geogr Sin 61(10):1011–1024 (in Chinese)
Wang S (2019) Residential spatial differentiation in Chengdu main urban area. University of Electronic Science and Technology of China, Chengdu (in Chinese)
Wang F, Gao X, Yan B (2014) Beijing's urban spatial structure based on housing prices. Prog Geogr 33(10):1322–1331 (in Chinese)

Wang Y, Feng J (2013) Reconstruction of suburban residents' relocation behavior from life course perspective. Hum Geogr 27(3):34–41, 50 (in Chinese)
Wang X et al (2000) Research on urban social spatial structure in China. Science Press, Beijing (in Chinese)
Wei H, Zhao H, Xiao T (2017) Job-housing separation and influencing factors in Beijing. Urban Dev Stud 24(4):49–57 (in Chinese)
Wu F (2002) Sociospatial differentiation in urban China: evidence from Shanghai's real estate markets. Environ Plan A 34(9):1591–1615
Wu F (2005) The city of transition and the transition of cities. Urban Geogr 26(2):100–106
Wu F, Li Z (2005) Sociospatial differentiation: processes and spaces in subdistricts of Shanghai. Urban Geogr 26(2):137–166
Wu G (2010) Impact mechanism of mega-events on urban tourism spatial development. Hum Geogr 25(5):137–141 (in Chinese)
Wu Q, Cui G (1999) Characteristics and formation mechanism of residential spatial differentiation in Nanjing. City Plan Rev 23(12):24–25 (in Chinese)
Wu Q, Cheng J, Chen G et al (2014) Socio-spatial differentiation and residential segregation in the Chinese city based on the 2000 community-level census data: a case study of the inner city of Nanjing. Cities 39:109–119
Xu D, Wang Z, Zhu X et al (2009) Social spatial structure of Nanjing in transitional period: factor ecological analysis based on the fifth census data. Geogr Res 28(2):484–498 (in Chinese)
Xu X (2011) Transformation of urban supply space patterns in the evolution of Danwei communities: empirical study of Hefei. Huazhong Arch 1:35–38 (in Chinese)
Xu X, Hu H, Yeh AGO et al (1989) Factor ecological analysis of social spatial structure in Guangzhou. Acta Geogr Sin 44(4):385–399 (in Chinese)
Xu Y, Shi X (2016) Spatial social stratification in urbanization and Chinese mechanisms. Soc Sci Beijing 11:85–94 (in Chinese)
Xuan G, Xu J, Zhao J et al (2006) Analysis of social areas in central Shanghai. Geogr Res 25(3):526–538 (in Chinese)
Yang C, Ning Y (2015) Evolution of interprovincial migration patterns in China and its impact on urbanization. Geogr Res 34(8):1492–1506 (in Chinese)
Yang SG, Wang MY, Wang CL (2015) Socio-spatial restructuring in Shanghai: sorting out where you live by affordability and social status. Cities 47:23–34
Yang Y, Feng Y (2019) Social area analysis and urban planning in Suzhou. City Plan Rev 43(11):90–102, 127 (in Chinese)
Zhang L, Lei J, Zhang X et al (2012) Social area analysis of Urumqi city. Acta Geogr Sin 67(6):817–828 (in Chinese)
Zhang W (2016) Core framework for livable city construction. Geogr Res 35(2):205–213 (in Chinese)
Zhang W, Chai Y (2009) Impact of residential space on household shopping travel decisions. Prog Geogr 28(3):362–369 (In Chinese)
Zhao P, Cao Y (2018) Comparative study of jobs-housing balance using multi-source LBS data: a case study of Beijing. J Peking Univ (Nat Sci Ed) 54(6):157–169 (in Chinese)
Zheng S, Xu Y, Zhang X et al (2015) Construction and spatial disparity of jobs-housing balance index: a case study of Beijing. Tsinghua Univ (Sci & Technol Ed) 55(4):105–113 (in Chinese)
Zhong Y, Feng Y (2017) Residential spatial differentiation of urban migrants: An empirical study of Shenzhen. Prog Geogr 36(1):125–135 (in Chinese)
Zhou C, Xu X (1996) Research progress on urban population relocation in Western countries. Hum Geogr 4:23–27 (in Chinese)
Zhou S, Liu Y, Zhu H (2006) Social area analysis of Guangzhou during transitional period. Acta Geographica Sinica 61(10):1046–1056 (in Chinese)
Zhou C, Bian Y, Zhang G et al (2016a) Spatial differentiation and formation mechanisms of middle-class enclaves in Guangzhou. Acta Geogr Sin 71(12):2089–2102 (in Chinese)

Zhou C, Hu J, Tong X et al (2016b) Tracking study on social spatial structure evolution of Guangzhou. Acta Geogr Sin 71(6):1010–1024 (in Chinese)
Zhou J (2018) Evaluation of community environmental livability based on multi-source data. Shenzhen University, Shenzhen (in Chinese)
Zhou K (2019) Differentiation of daily life spaces in residential historic districts. Suzhou University of Science and Technology, Suzhou (in Chinese)
Zhou S, Deng L, Kwan M et al (2015) Social and spatial differentiation of high and low income groups' out-of-home activities in Guangzhou, China. Cities 45:81–90
Zhou, Y. (1996). Suburbanization of Beijing and its implications. *Scientia Geographica Sinica*, 16(3):198–206 (in Chinese)

Chapter 6
Interaction Space

6.1 Networks of Relationships: Interpersonal Relationships and Social Interactions of Urban Residents

6.1.1 Guanxi and Interpersonal Relationships

In the *Modern Chinese Dictionary*, "*Guanxi*" is defined as the state of things interacting and influencing each other, indicating a connection of some nature between people or between people and things (Dictionary Editorial Office, Institute of Languages, Chinese Academy of Social Sciences 2012). From a psychological perspective, interpersonal relationships are relatively stable psychological connections formed through communication and interaction, reflecting the psychological state of an individual or group as they fulfill their social needs. In this chapter, the term "*Guanxi*" refers to social ties—the sum total of connections between individuals in a society—which differs from the commonly understood concept of "relationship" in Chinese culture (Chen et al. 2004).

Currently, research on *Guanxi* predominantly converges within the domains of sociology, economics, and psychology. In the realm of psychology, it is posited that humans have an inherent need to direct their inner emotions towards certain objects, which subsequently precipitates the formation of interpersonal relationships. As such, scholars in this field have comprehensively expounded upon the nature, significance, and interaction skills of interpersonal relationships from an internal perspective. In contrast, sociology and economics initiate their inquiries from the vantage points of social relationship networks and social capital. Western scholars have probed into the motives and patterns of interpersonal communication from diverse dimensions. For instance, Schutz (1958) proposed the three—dimensional theory of interpersonal relationships. This theory posits that every individual harbors the desire and necessity to establish connections with others, albeit

J. Feng et al., *Urban Living Space in China*, Urban Sustainability,
https://doi.org/10.1007/978-981-95-0595-1_6

with varying degrees of manifestation. Schutz classified these needs into three categories: the need for inclusion, the need for control, and the need for affection. Each of these needs can be translated into motives that give rise to specific behavioral inclinations, ultimately leading to the establishment of particular interpersonal relationships. From an economic perspective, Homans put forth the social exchange theory. This theory contends that human interaction essentially constitutes a process of social exchange, encompassing not only the exchange of material commodities but also non—material entities such as emotions, information, and services (Tan 1986). An individual's perception of their relationships with others is primarily contingent upon their evaluation and experience of the rewards and costs associated with these relationships.

In traditional Chinese thought, human relations have always been highly emphasized. Rooted in traditional Chinese culture, Liang (2005) proposed the concept of the ethical nature of interpersonal relationships within the social structure. He pointed out that life essentially exists based on a variety of relationships, and these relationships are a form of ethics. Ethics pertains to the interactions among people, and through these interactions, relationships are engendered. Regarding the nature of Chinese interpersonal relationships, Zhai (2005) put forward a trinity of local patterns composed of *Renyuan*, *Renqing*, and *Renlun*. He posited that *Renqing*, lie at the core of this pattern. They represent the fundamental psychological and behavioral style of traditional Chinese people, which is grounded in kinship (family). *Renlun* is the institutionalization of this basic pattern, providing a set of principles and norms for the pattern and enabling people to adhere to a certain order in their social interactions. Moreover, *Renyuan* is the manifestation of this pattern. It serves as the framework of this pattern, confining all human relationships within an overarching structure that indicates the ultimate origin without further exploration. Sociologist Fei (2005), in his research on the structure of the Chinese rural regions, developed the concept of the "differential order pattern," which has had far-reaching implications. In this pattern, each family takes its own position as the core. Just as a stone thrown into water creates ripples that spread out in concentric circles, the Chinese rural structure is centered around each family. The size of these circles is determined by the influence of the family at the center. With the self as the origin, the circles expand, becoming increasingly distant and less intense. Everyone has a circle centered on themselves and is also part of the circles centered on those with higher status. Based on this, Yang et al. (2008) classified the relationship network of the differential order pattern into three levels: family members, acquaintances, and strangers. Huang et al. (2004) proposed a theoretical model of *Renqing* and *Mianzi*. By synthesizing the four comprehensive concepts of *Renqing, Guanxi, Mianzi, and Baoda*, they constructed a Chinese theoretical model of social behavior. They categorized Chinese interpersonal relationships into three types: emotional, instrumental, and mixed relationships. The interaction norms for these types follow the law of needs, the law of fairness, and the law of human favor, respectively. Luo (2005, 2006), Luo and Ye (2006) suggest that relationships of Chinese can be classified as familiar guanxi, pseudo-familial guanxi, and weak guanxi (Fig. 6.1). Pseudo—familial guanxi refers not only to family members in the traditional sense but also to sworn brothers and those related by marriage.

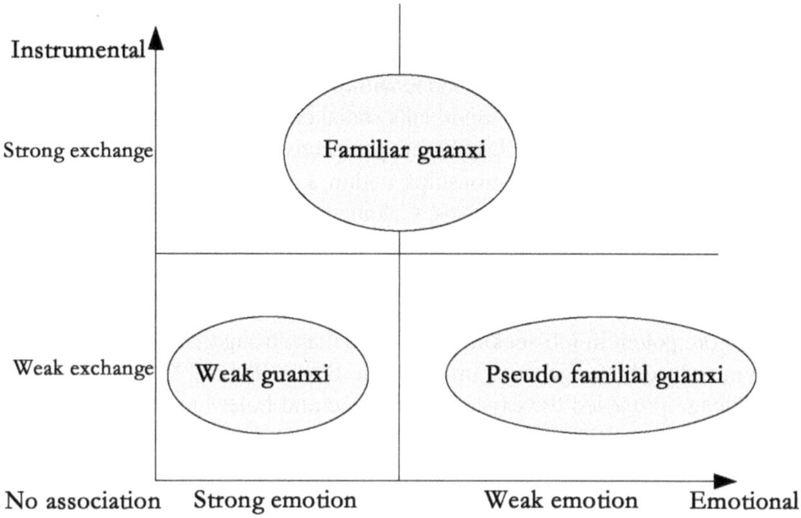

Fig. 6.1 Chinese guanxi dimensions and positional divisions. *Source* Drawn according to the theory by Luo (2005, 2006)

In addition, Wang and Liu (2002) conducted a factor analysis of the data regarding the degree of trust in various social roles. They found that among the four factors, the least trustworthy were strangers and general acquaintances, followed by regular acquaintances and interactors such as leaders, coworkers, and neighbors. Those with a higher level of trust were regarded as friends, including close friends and casual friends. The most trustworthy were relatives, including family members, immediate family, and other kin.

Guanxi function as invisible ligaments that bind individuals together, thereby constructing a complex and extensive relational network within which all individuals are embedded. Every person inhabits diverse groups, organizations, and interpersonal connections with relatives and friends. In this context, the individual can be conceptualized as a node, while the connections between people can be symbolically represented by edges. The configuration and dynamics of these nodes and edges collectively constitute social networks.

6.1.2 Social Network

The social network is a pivotal concept in the study of social structure, particularly within the theory of social resources. Modern sociologists conceive of society as a structure, a discernible network of interconnected institutions, where social connections serve as the foundation of social networks. In this vein, kinship networks, friendship networks, groups, and institutional complexes function as sub—networks that connect individuals at diverse levels and further integrate them into larger—scale social networks (Yang and Wang 2007). Individuals, families, organizations, and institutions can be regarded as nodes within the network. Social networks also determine the magnitude of an individual's social capital. Bourdieu (1983) initially defined social capital from a class-based perspective, positing that social capital represents the aggregate of relationships within a social network that impact the diverse rewards an individual receives. Coleman (1988) defines social capital as a process through which pre—existing interpersonal bonds (such as those grounded in ethnicity and community) can be transformed into economic assets. Granovetter (1973) put forward the theory of strong and weak ties, contending that weak ties might be more potent in job-seeking endeavors than strong ties since they are more likely to introduce heterogeneous information. Burt (1984, 1992) from the University of Chicago proposed the existence of "structural holes" in social networks. In the Chinese context, Bian (1994) suggests that under the planned-economy work-allocation system, personal networks are primarily utilized to acquire information and exert influence on allocation decision-makers rather than merely to gather employment information. Moreover, within the network, it is the influence exerted by strong ties on decision—makers that holds greater significance. The Chinese sociologist Lin (2005) defines social capital as resources embedded in social networks and puts forward the hypothesis of strength of status and strength of relationship. This hypothesis posits that the level of an individual's ascribed status and the presence of weak relationships are positively correlated with the acquisition of social resources. In terms of Chinese empirical analyses, Zhang and Ruan, based on surveys and statistics, analyzed the basic characteristics of Tianjin urban residents' interpersonal networks, including aspects such as the size of the social network, the composition of relationships, the degree of closeness, convergence, and heterogeneity. They also explored the differences between the Tianjin networks and those in the United States, along with the reasons for these differences (Zhang and Ruan 1999; Zhang 1999). Qi (2007), taking a community in Shijiazhuang as an example, studied the social networks of residents in a new-type urban residential neighborhood and proposed the existence of communities that transcend boundaries. It is worth noting that academics have increasingly emphasized the study of social networks from a spatial perspective. For instance, Newman (1972) proposed the classification of spatial territoriality. Through the study of the relationship between people's behavioral activities and the urban environment, Newman recognized that various human behaviors and activities necessitate corresponding territoriality, especially in the residential environment. He then proposed a spatial system composed of private, semi—private, semi—public, and public spaces (Fig. 6.2). Furthermore, some scholars have initiated research on the constraints between space and social interaction, particularly focusing on semi—public and semi—private spaces. The Japanese architect Kisho Kurokawa proposed the concept of "grey space", also known as "rim space". On one hand, it refers to a color; on the other hand, it represents a transitional space between indoor and outdoor areas, which can serve the purpose of integrating indoor and outdoor spaces (Wang et al. 2010).

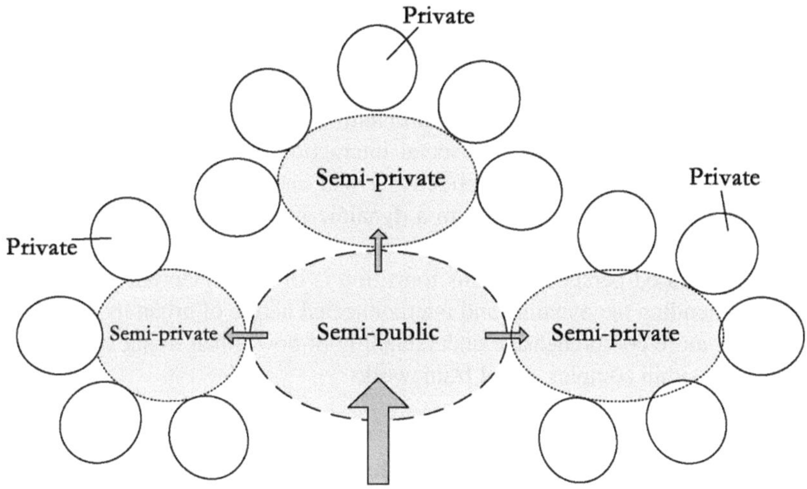

Fig. 6.2 Classification map of spatial territoriality

6.1.3 Research Question and Empirical Research Area

6.1.3.1 Research Question

Currently, amid the rapid development of China's society and economy during the transitional period, the "social spatial segregation" among different social strata in cities is becoming increasingly prominent, and the resultant social problems are emerging continuously (Gu 2007; Li 2010). In urban community life, people's interactions have dwindled to an extremely weak level. The emphasis on individual independence has seemingly led to the disappearance of the closely—knit community. Allowing this situation to persist will significantly impact and impede the integration and harmonious development of urban society. In contrast to traditional rural society, the social and interpersonal networks of urban residents have transcended geographical boundaries, and the neighborhood relationship appears to be characterized by indifference. Frequently, residents in the same neighborhood are separated by a wall or even in close proximity, yet there is no intersection in their respective living spaces.

Notably, urban residents seem to be in a state of contradiction. On one hand, they desire to lead a life of family or personal independence. On the other hand, they sincerely long for a living space imbued with warmth and care. In this paradoxical scenario, what kind of sentiments do Chinese urban residents still hold towards their neighborhoods? What are the characteristics of the specific interpersonal networks within the neighborhood? How does the interaction between the neighborhood interpersonal network and space occur? How does this interaction shape the mature social space of the neighborhood? What role do virtual spaces based on Internet technology play in the formation of neighborhoods, the construction of residents' social networks, and the cultivation of a sense of community belonging?

The aforementioned questions predominantly interpret the living space of urban residents from the perspectives of social interaction and social network connectivity. Answering these questions effectively will undoubtedly propel the research on urban living space from a static to a dynamic perspective, from an individual-centered to a network-connected perspective, and from a physical-space-based to a virtual-space-based perspective. This transition is of great theoretical significance for comprehending the dynamic and interconnected nature of urban living space, as it enables a more comprehensive understanding of how urban living spaces evolve and interact within complex social frameworks.

6.1.3.2 Empirical Research Area and Questionnaire Survey

To further delve into the research questions proposed previously, Huilongguan, one of the most renowned large-scale residential areas in Beijing, was selected as the primary empirical research site. Based on questionnaire and interview surveys regarding the personal social relationship networks and neighborhood relationships of Huilongguan residents, this study analyzes and explores the formation process of the residential community in Huilongguan and the potential micro-mechanisms at the social interaction level. The aim is to gain insights into the social relationships among urban residents and their connection with space.

Western scholars, such as Hawley (1950), have posited that the boundaries of a community should be circumscribed to the extent that residents possess a general understanding of the area's daily life. In the case of "Huilongguan," it can denote either the administrative concept of Huilongguan Town or the cultural settlement of Huilongguan with imprecise boundaries. Huilongguan Town is under the jurisdiction of Changping District in Beijing (Fig. 6.3). Covering an area of 20.7 km^2, according to the sixth national census, it has a population of 325,000 distributed among 124,000 households. The *Huilongguan* Cultural Residential Area, one of the key affordable housing construction areas in Beijing, commenced development in 1998. With a total planned area of 11.27 km^2, a total construction area of 8.5 million m^2, and a designed population of approximately 300,000, it focuses on addressing the housing needs of scientific and educational personnel. Huilongguan stands as one of the largest and relatively independent residential communities in China and even in Asia (Zhang 2002). Predominantly composed of affordable housing, it also includes demolition-relocation housing, replacement housing, group purchase of housing by Danwei, and other housing types, with a prominent housing rental phenomenon. The housing in Huilongguan mainly consists of multi-storey buildings and belongs to the middle-grade residential category. Among the more than 20 living areas in Huilongguan, nearly 80% of the residents are non-locals, making it a typical migrant-populated area.

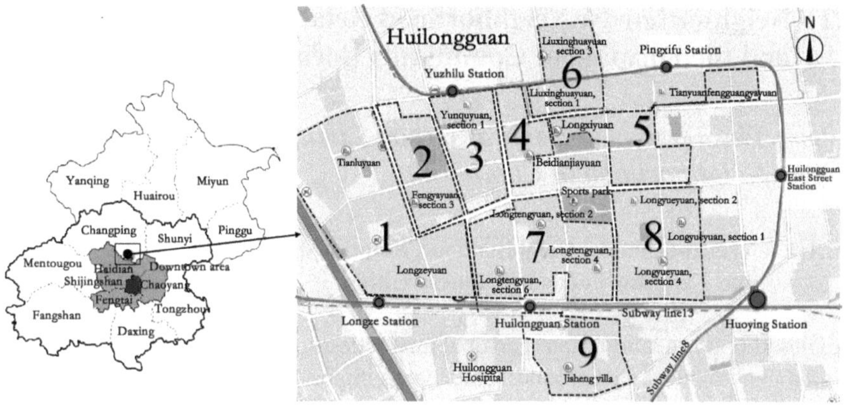

Fig. 6.3 Location map of Huilongguan in the study area and distribution of surveyed districts

In this chapter, the concept of the Huilongguan community primarily refers to the "Huilongguan Cultural Residential Area," which falls within the administrative jurisdiction of Huilongguan Township. The geographical area for questionnaire distribution was mainly concentrated in the region east of the Badaling Expressway, west of Huoying, north of Railway Line 13, and south of Maliandian, constituting the core area of the Huilongguan Cultural Residential Area. From August 2010 to April 2011, the survey team visited Huilongguan over 20 times. Besides conducting numerous resident interviews, the team focused on a questionnaire survey regarding residents' social space. The survey method mainly combined household and outdoor public space investigations. The sampling approach involved a combination of district division (9 districts) and random sampling within each district. A total of 560 questionnaires were distributed, and 550 valid questionnaires were recovered. Based on the establishment time and spatial layout of the neighborhoods, the Huilongguan community was categorized into the West District (with 58 valid questionnaires recovered), Fengyayuan District (59), Yunqiyuan District (53), Longxiyuan District (32), Longjinyuan District (31), Liuxing Garden District (59), Longtengyuan District (69), Longyueyuan District (42), and the South District (26). The number of questionnaires distributed in each area was approximately allocated according to the size of each area. Huilongguan represents the largest and most renowned suburban residential area in Beijing and is among the earliest affordable housing centers established following the housing system reform. With the development of the real estate market, Huilongguan has a high level of housing marketization and a large proportion of rental housing. A significant number of non-local residents, especially "new immigrants," inhabit this area. This not only renders the social composition of the community more complex but also restructures the community's socio-spatial order. Thus, Huilongguan serves as an excellent exemplar of large-scale residential areas that have emerged after the housing reform.

6.2 Neighborhoods: Neighborhood Relationships and the Forming of Community Social Space

6.2.1 The Territorial Basis of Urban Living Space: From Neighborhood to Community

6.2.1.1 Neighborhoods, Neighborhood Relationships and Neighborhoods in Chinese Cities

Neighborhood is a kind of interpersonal interaction space with relatively close residential geographical background, which is an important part of urban living space. Because of their geographical proximity, people in a community have a special kind of social relationship with each other, which is known as the neighborhood relationship. Since ancient times in China, there has been a tradition that "close neighbors are better than distant relatives", and there was also the system of "guilt by association" in the feudal era, which highlights the significance of neighborhood relationship for emotional interaction and social norms from various angles.

Geographers have long been intrigued by the roles of distance, space, and location. However, there remains a lack of consensus regarding the part that proximity plays in either promoting or impeding social interactions. Festinger et al. (1950) conducted a study on the interactions among married engineering students at the Massachusetts Institute of Technology. They discovered that micro—scale distances were of great significance. Interestingly, it seemed that the spatial layout of housing alone could govern the relationships between friends. Nevertheless, the representativeness and generalizability of this finding have faced scrutiny. Some scholars assert that neighborhoods are only crucial during the process when residents move into newly developed housing areas. Others contend that social distance and shared values are the primary factors determining friendships. They also point out that with the shrinking of the world and the weakening of distance constraints brought about by modern technology and mass communication, along with the increase in personal mobility and the prevalence of separation between work and residential areas, people have been liberated from the constraints of neighborhood relationships (Webber 1964).

Neighborhoods have long been a focal point of sociological research in China. A plethora of surveys have yielded specific findings regarding the status quo of urban neighborhoods in the country. These conclusions can be systematically synthesized and categorized into two distinct perspectives.

One perspective posits that in recent years, amidst social transformation and changes in the broader context of housing, the neighborhood relationship, as a form of interpersonal connection, has been gradually waning in significance and has even vanished in certain localities. For instance, an analysis of a 1996 survey conducted by the Beijing Lingdian Company revealed that residents living in apartment buildings exhibited a much lower degree of acquaintance with their neighbors compared to those residing in cottage areas. Specifically, 34.1% of cottage residents reported having a better understanding of their neighbors than those in bungalows. While 34.1% of bungalow residents claimed to know all of their neighbors, the proportion for those in apartment buildings was merely 15%. According to the results of a questionnaire survey on neighborhood relations in Xi'an carried out by Huashang Newspaper in April 2006, 22% of the respondents were completely unfamiliar with their neighbors; 45% only encountered their neighbors occasionally to exchange greetings; 77% knew little about their neighbors' family members; 63% had an average relationship with their neighbors; 14% were satisfied with the state of their neighborhood; and 57.7% did not offer assistance to their neighbors. A questionnaire survey conducted by Jiaxing Daily in July 2007 on neighborly relations in Pinghu indicated that nearly 80% of urban residents could not name more than five neighbors; 92% would not consider turning to their neighbors for help in an emergency; 56.6% had an average relationship with their neighbors; and 77.6% were reluctant to act as "peacemakers" in their neighbors' domestic disputes. Simultaneously, the survey findings demonstrated that neighborhood relationships vary significantly depending on housing types. In new commercial housing communities, interpersonal relations are more aloof compared to old neighborhoods, and the acquaintance-based and semi-acquaintance-based societies that once prevailed in some urban neighborhoods have dissipated. Wang Ying's (2002b) survey of 1,018 residents in 17 new villages in Shanghai showed that neighborhood interactions were more frequent in traditional neighborhoods and socially marginalized rural communities, less so in public housing communities, and least frequent in commercial housing communities. Notably, in high-end commercial housing neighborhoods, residents' social interactions are predominantly centered around relatives and friends residing outside the neighborhood.

Another perspective holds that neighborhoods in urban China still persist today but have taken on new characteristics. An (2001) surveyed the neighborhood relations of 1,500 individuals in Beijing. The findings indicated that 69.8% of the respondents opted for a close relationship with their neighbors, while only 1.7% chose not to interact at all. This suggests that the good moral traditions of neighborliness still prevail among Beijing residents. An also investigated the impacts of urban–rural differentiation, age, and educational attainment. It was discovered that urban residents tend to be more distant from their neighbors compared to rural residents. Regarding the influence of age on neighborhood relations, the older the neighbors are, the closer their relationship; conversely, the younger the neighbors are, the more distant they are from each other. Additionally, a higher level of education is associated with fewer interactions and more distant relations with neighbors. In a hypothetical situation where a neighbor is an elderly widow and the question of willingness to take care of her was posed, 77% of the respondents agreed, while only 6.5% disagreed. In 2006, Sun and Lei (2007) surveyed and analyzed the neighborhood relations of residents in the old urban areas of Beijing. They concluded that, in general, the neighborhood relationships among Beijing urban residents are marked by superficiality and shallowness. The frequency of daily interactions between neighbors is extremely low. Feng et al. (2011) explored the neighborhood relationship between gated communities and their surrounding neighborhoods. Their study pointed out that although there are disparities in social status and the presence of gated enclosures, unlike in countries such as South Africa and Argentina, gated communities in China maintain functional interactions with the surrounding neighborhoods. They do not engender a sense of isolation or discrimination in the surrounding areas. Instead, geographical proximity offers favorable opportunities for mutual understanding, integration, and cooperation among residents of different social classes and diverse living backgrounds.

6.2.1.2 Community and Studies on Urban Community in China

The concept of community was initially proposed by the German sociologist Ferdinand Tönnies at the end of the nineteenth century. He posited that society evolved from a close-knit and humane social grouping (Gemeinschaft), formed by a homogeneous population sharing common factors and values, to a social grouping shaped by contractual relations and rational will. The currently accepted definition of community was put forward by Park (1936), a prominent figure of the Chicago School, and Hillery (1955), a sociologist. They indicated that the essential characteristics of a community should encompass: (1) a defined geographical area; (2) shared relations; and (3) community interaction. Regarding the distinction between community and society, Wu proposed that "Society is an abstract concept depicting the collective life of people. It is a general term for the entire system of complex social relations. In contrast, community is a concrete manifestation of the actual life of people in a specific place, which has a material basis and is observable" (Dong 2008). From a geographical perspective, community represents an overlay of social and regional aspects. In the Chinese context, a community can be defined as a group of interconnected individuals residing within a certain geographical area. Here, residents possess a sense of belonging and identity, and share a social culture and means of interaction. Concerning the definition of the community's scope, Hawley (1950) suggested that the scale of a community should be delimited to the extent that residents have a general acquaintance with the region's daily life. People are familiar with their own neighborhoods, being aware of the daily affairs that are of little concern to those from other areas. In this manner, the boundaries of a community can be more precisely defined.

Is the nature of community relational or territorial? Scholars hold diverse views on whether the most crucial characteristic of a community lies in a specific territory or a common bond. Currently, there are several distinct perspectives regarding the relationship and territory of a community, including the theory of community disappearance, the theory of community survival, and the theory of community emancipation (Cheng and Zhou 1998). The first is the theory of community disappearance. Sociologists such as Tennessee and Turgon posited that modern industry has induced revolutionary transformations in the organization of human society. Under the impact of urbanization and city life, the community, as a grass-roots entity, has started to become nebulous. Residents' sense of belonging to a particular geographical space has a tendency to weaken, and interactions and connections are no longer restricted to the residential space. Wirth (1938) of the Chicago School compared the relationships among urban community residents with those of traditional rural residents. He contended that the relationships among urban residents had transformed from primary to secondary affiliation, and the forces maintaining the community were weakening, leading to the gradual withering away of the community. The second is the theory of community survival, which indicates that communities are still preserved to a certain degree during the process of urbanization and social change. This theory originated from Oscar Lewis's publication *Urbanization without Collapse* in 1952. In this work, he discovered that the lifestyles of Mexican villagers did not experience significant alterations after their relocation to cities, and the sense of human connection and social cooperation remained unchanged. He pointed out that many individuals living in large cities still maintained their small-circle activities. Those within the small circles retained their intimate, mutually supportive, and trusting relationships, while those outside the circles exerted little influence on them. In 1962, Herbert Gans, in his work *The Urban Villager*, depicted the lives of Italian immigrants on Boston's West Side. He posited that the community persisted in the form of an urban village. Thirdly, regarding the community liberation theory, in the 1970s, Fischer (1975) contended that the problems in cities were not engendered by large populations but rather by the presence of subcultural populations within cities. These subcultural groups form small circles. Over an extended period, they engage in interactions with one another and gradually develop norms, values, and lifestyles that are mutually comprehended and embraced by all members. Consequently, two co-existent phenomena are observable in cities: cohesion and integration within groups, and friction and conflict between groups. Wellman (1982) put forward the proposition to re-evaluate the concept of community, emphasizing the necessity for its inhabitants to be liberated from the constraints of place and territory and to establish connections with a broader spectrum of friends. This perspective is termed "community liberation".

In the academic realm, scholars exhibit significant disparities in their research regarding the actual circumstances of communities in China. Wang (2002a) posits that there exists a "community growth point" within Chinese neighborhoods. Fundamentally, the basis for the existence of grassroots communities remains intact. There invariably exist certain unique connections and specific interests between individuals, their families, and other people in a given place. These elements contribute to an individual's sense of placeattachment, identity, and the formation of common interests. The potential for the existence of urban communities is intrinsically linked to human nature. Communities hold crucial value in fulfilling individuals' psychological needs. Moreover, if social capital, such as reciprocity among community members, can be generated, it becomes feasible to foster a sense of community belonging and common interests among members (Feng 2002). Nevertheless, some scholars hold a different view. They contend that although the community as a realm of communal life still holds importance in residents' lives, what is of even greater significance is the "disengaged community" (Wang and Wang 2003). In modern urban societies, the highly developed social division of labor, advanced means of communication, increasingly rational social interactions, and frequent social mobility render closed—knit communities unfeasible. The material and spiritual requirements of community residents are sourced more from outside the community than from within (Shen 2004).

6.2.2 Neighborhood Relationships and Their Constrain Factors

6.2.2.1 The Characteristics of Neighborhood Relationships

(1) **Neighborhood acquaintance and interaction**

The survey reveals that residents demonstrate a high level of unfamiliarity with and distant interactions from their neighbors. The number of neighbors known by residents generally ranges from 7 to 10 (Table 6.1). The median number of neighbors known across the entire sample is 10. Those who know 20 or fewer neighbors account for 78.89% of the sample. When relatives and colleagues are excluded from the list of known neighbors, the degree of unfamiliarity is even more pronounced. As many as 7.89% (a total of 40 residents) do not know any of their neighbors. An in-depth analysis of their socio-economic attributes indicates that most of these individuals are highly educated young people who are occupied with their work and mostly live in rented accommodation, which leads to a dearth of neighborly interaction. Regarding neighborhood acquaintance and visitation (Table 6.2), 36.47% of residents know nothing about their neighbors on the same floor, and 45.30% have no interaction with one another. Neighborly acquaintance largely remains at the level of "casual acquaintance", with many being unaware of their neighbors' names or residences. This stands in sharp contrast to the situation in Beijing's old urban areas, where the majority (84.1%) of people are generally familiar with their neighbors (Sun and Lei 2007). In the interactions of modern urban dwellers, due to their residence in buildings where privacy and security are emphasized, the accelerated pace of urban life, social heterogeneity, and security concerns challenging people's trust in others, the phenomenon of neighbors visiting each other's homes is becoming increasingly rare.

Currently, the primary form and mode of neighborly interaction is interpersonal interaction (Fig. 6.4). The vast majority of neighborhood interactions occur as a result of interpersonal exchanges and daily necessities. Additionally, a certain degree of daily recreational activities and timepassing behaviors are also involved. Many residents exchange greetings out of courtesy or as a form of obligatory social gesture. Or they engage in surface—level and shallow neighborhood interactions that do not touch upon significant matters.

An intriguing phenomenon is that, despite neighborly interaction not being as close-knit as in the traditional town or rural societies, from the respondents' perspective, the willingness to interact with neighbors and the state of neighborly relationships appear to be quite positive. Overall, residents perceive their relationships with

Table 6.1 The number of neighbors in the community known to the residents

Number of neighbors known/person	Sample size/person	Percentage/%
0	40	7.89
1–3	58	11.44
4–6	79	15.58
7–10	123	24.26
11–15	29	5.72
16–20	71	14.00
21–30	38	7.50
31–50	27	5.33
51–100	22	4.34
More than 101	20	3.94
Total	507	100.00

Table 6.2 The level of mutual understanding and visiting among neighbors

Understanding the situation on the same floor	Number of samples	Proportion/%	Visiting or dropping in	Number of samples	Proportion/%
Very well—understood	46	8.83	Frequent interaction	45	8.64
Basically understood	285	54.7	Occasional visits	240	46.06
Completely unaware	190	36.47	No interaction	236	45.3
Total	521	100	Total	521	100

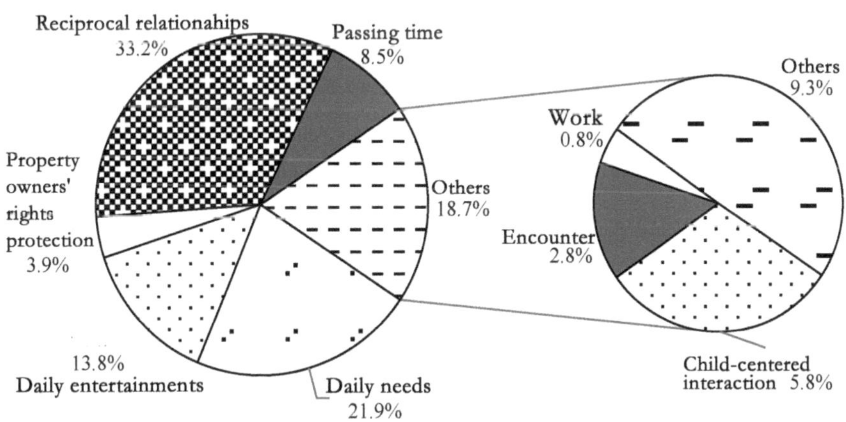

Fig. 6.4 Main forms of neighborhood interaction

neighbors to be harmonious. Only 8.64% consider their neighborly relationships to be indifferent. Three-quarters of them can respond amicably to neighbors' initiatives and exhibit a positive inclination to communicate with neighbors (Table 6.3). Over 80% of the residents have friends within the Huilongguan community, while only 17% do not have many friends in the area. The survey of residents' key-issue discussion networks also indicated that over half of the key relationship network members that residents would consult and turn to for help were located within the Huilongguan area.

(2) **Neighborhood watch and mutual assistance**

In their daily lives, the individuals people turn to for assistance are predominantly friends and relatives, with relatively few instances of mutual aid from neighbors. As per the survey (Table 6.4), when Huilongguan residents encounter difficulties in their daily routines, they initially seek help from friends, followed by relatives, then classmates, colleagues, and property managers, and finally neighbors. In the survey regarding those who are the first choice for seeking help, relatives, rather than friends, account for the highest proportion, and neighbors are less frequently selected. Nevertheless, simultaneously, the neighbors' willingness to offer mutual help remains amicable. However, such mutual help is relatively at a low level and often occurs sporadically. Hypothetically, when a neighbor requests help, only 3% of people state that they would not assist (mainly those living in rented accommodation). Currently, 78% of residents have received help from a neighbor, and 70% have provided help to a neighbor.

(3) **Neighborhood conflicts and tensions**

Based on the questionnaire survey, approximately 83% of the residents in Huilongguan have never experienced any conflicts with their neighbors. The principal causes of observed or encountered neighborly conflicts are a lack of compliance with public ethics and divergent viewpoints. When such conflicts arise, the predominant approach

Table 6.3 Neighborhood relationships and willingness for neighborhood interactions

Neighborhood relationships	Number of samples	Proportion/%	Willingness for neighborhood interaction	Number of samples	Proportion/%
Harmonious	251	48.18	Start chatting with them	395	76.55
Ordinary	225	43.18	Just give perfunctory responses	67	12.98
Cold	45	8.64	Completely ignore	9	1.75
Total	521	100	It depends	45	8.72
			Total	516	100

Table 6.4 Categories of objects residents seek help from

All object types	First object to seek help from		Objects to seek help from	
	Number of samples	Proportion/%	Number of samples	Proportion/%
Relatives	236	46.00	353	68.81
Colleagues and classmates	64	12.48	283	55.17
Friends	108	21.05	365	71.15
Neighbors	22	4.29	168	32.75
Property management	76	14.81	233	45.42
Others	7	1.37	23	4.48
Total	513	–	1425	–

is to take the initiative in mediation. If needed, residents will resort to third-party agencies like property management companies and neighborhood committees for mediation.

(4) **Neighborhood activities and exchanges**

In the survey regarding the participation of Huilongguan residents in neighborhood activities (Table 6.5), 60% of the residents were aware of such activities, with the majority being informed by the neighborhood committee. Approximately 24% of them had actually participated in these activities, and the participants were predominantly elderly individuals who were retired and at home. Around 55% of the residents expressed their willingness to participate, either for amusement or to enhance acquaintance with neighbors. The remaining residents stated that they were reluctant to engage in neighborhood exchange activities. On one hand, they lacked interest, and on the other hand, they did not have the time to do so.

According to diverse triggering events, reasons, and interaction depths, there exist multiple types of ways for residents to become acquainted and interact with one another (Table 6.6). These include activities such as group buying during the initial stage of moving into the community, rights-protection and cultural-sports events in daily life, "scouting-style" diplomacy, and interactions facilitated by the community network. A shared living background, comparable life cycles, and the occurrence of interaction—prompting events enable the sustenance of amicable relationships among neighbors. Huilongguan features a significant number of affordable housing units. Initially, the majority of residents were young individuals who had recently graduated and were engaged in the IT and education industries, sharing the common experience of being "Beijing drifter" workers. This commonality greatly facilitated their communication and interaction. When they first moved in, neighbors collaborated in group-buying activities and building renovations, which provided opportunities for them to get to know and bond with each other. In daily life, numerous residents started to interact as they exercised in the park or neighborhood green

Table 6.5 Participation in and willingness for neighborhood activities

Participation status	Number of samples	Proportion/%	Channels of learning about activities	Number of samples	Proportion/%	Willingness to participate	Number of samples	Proportion/%
Know and participated	127	23.92	Introduced by neighbors	74	18.97	Yes, to get to know neighbors	128	25.45
Know but didn't participate	190	35.78	Notified by the neighborhood committee	199	51.03	Yes, just for fun	152	30.22
Didn't know	214	40.3	Learned online	77	19.74	No, not interested	109	21.67
Others	40	10.26	No, no time	114	22.66	–	–	–
Total	531	100	Total	390	100	Total	503	100

spaces or actively participated in ball games. Some also took part in the community's rights—protection initiatives (Dong 2005), uniting residents under the impetus of the community's common interests and needs to jointly confront external disruptions. Moreover, the regular and daily cultural—sports activities organized by the neighborhood committee, mainly targeting the elderly, allowed some residents to deepen their interactions and foster closer relationships. In Huilongguan, a novel model of neighborhood interaction centered around children's needs has emerged, similar to what is known as "scouting" in the West (Knox and Pinch 2000). The study revealed that due to children's need to play in public spaces, many parents frequently met and gradually became acquainted. As children played together, parents would bring their children to each other's homes, gradually strengthening neighborly relations, particularly among women, and there was a substantial amount of "place—based diplomacy" centered on children. Only 4% of the surveyed residents who were asked about their stance on their children playing with neighbors' children expressed opposition. Furthermore, the community-based Internet, firmly rooted in the physical locale of Huilongguan, has exerted a profound influence on neighborhood interactions. By integrating people's virtual-network interactions with the physical space, it has promoted more favorable neighborhood interactions and enhanced residents' sense of community belonging.

Table 6.6 Table of residents' mutual acquaintance events in neighborhoods

Type	Events and content	Main occurrence places	Level
Group purchasing in the early stage of moving into the community	Group purchasing, decoration	Each resident's house in the community	Shallow, moderate
Rights-protection and cultural-sports activities in daily life	Greeting when meeting	Corridors and community roads	Shallow
	Playing ball, exercising	Sports venues, community green spaces	Shallow, moderate
	Rights - protection events, cultural activities	Outdoor spaces, neighborhood committees	Moderate
Children-led interactions	Sunbathing, playing	Public green spaces, small playgrounds	Moderate
	Children interacting with each other	In a child's home	Moderate
Interactions relying on the community network	Online forum discussions	Virtual network	Shallow, moderate
	Group purchasing, second—hand trading, etc.	Combination of virtual network and physical space	Moderate
	Cultural and sports activities	Combination of virtual network and physical space	Moderate, deep

(5) **Neighborhood preferences and perceptions**

The survey on residents' preferences for neighborhood types shows that most residents prefer a neighborhood that is helpful and courteous, and to a lesser extent, ethical, and able to get along without interfering with each other. How do residents perceive their neighbors and their relationships in the community? Three major questions were designed in the questionnaire to ask residents about their perceptions of close, average, and lukewarm neighborly relations in different spatial scales: neighborhood interactions in small residential area, their own neighborhood relationship, and neighborhood relationship of society, in order to investigate whether there are differences between spatial and neighborhood perceptions. Table 6.7 shows that there is a clear correlation between residents' perceptions of neighborhood relationships at different scales.

A specific analysis of neighborhood relations within each scale range (Fig. 6.5) reveals that residents' perceptions of neighborhood relations across different scales are rather confounding. Notably, the larger the scale range, the more evident the sense of coldness in these relations. Modern residents tend to consider their immediate neighborhoods to be more harmonious. However, when it comes to a broader and more distant scale perception, as the scale range expands, the feeling of coldness becomes more pronounced, and the perception of neighborhoods becomes increasingly perplexing. The primary reasons for this phenomenon are as follows. On one

Table 6.7 Correlations among three neighborhood relationships perceived by different scale

	Parameter	Perception of neighborhood interaction in the community	Perception of one's own neighborhood relationships	Perception of social neighborhood relationships
Perception of neighborhood interaction in small residential area	Pearson correlation	1.000	0.355**	0.254**
	Significance (two-tailed)		0.000	0.000
	N	544	519	520
Perception of one's own neighborhood relationships	Pearson correlation	0.355**	1.000	0.246**
	Significance (two-tailed)	0.000		0.000
	N	519	521	507
Perception of social neighborhood relationships	Pearson correlation	0.254**	0.246**	1.000
	Significance (two-tailed)	0.000	0.000	–
	N	520	507	521

** Significantly correlated at the 0.01 level (two-tailed)

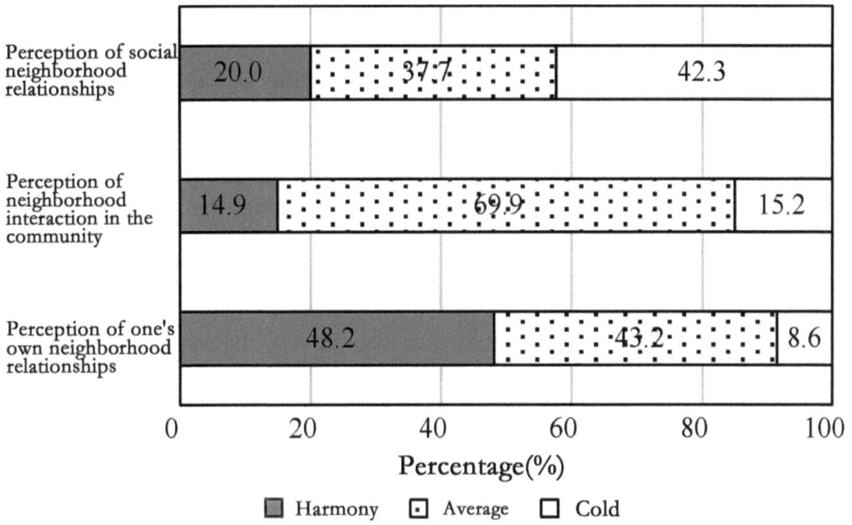

Fig. 6.5 Neighborhood perceptions

hand, residents have the impression that their own neighborhoods are more harmonious, yet they are also influenced by social propaganda and public opinion. On the other hand, through various media channels and case analyses, residents have come to perceive that neighborhood relations in modern society are more complex and relatively colder compared to the past. This disparity between perception and reality renders the exploration results of neighborhood relations in the intermediate transition scale more ambiguous, with the majority of people opting for the "average" assessment.

6.2.2.2 Neighborhood Sense of Belonging and Influencing Factors

(1) Neighborhood responsibility and sense of belonging

Western scholars contend that neighborhood bonds ought to establish a state of favorable interaction (Sugden 1986). The growth of positively interacting neighborhood connections gives rise to the accumulation of social capital. This not only meets the social requirements of an individual but also spurs the entire community to foster a more pleasant living environment (Hanifan 1916). Neighborhoods boasting a high degree of social capital are often regarded as ideal places for raising children, characterized by clean public spaces, amicable residents, and relatively secure streets (Jacobs 1961).

When residents inhabit neighborhoods where neighbors are concerned about the safety of others and willing to offer assistance to those in distress, the neighborhoods tend to be safer. In the survey, when presented with the scenario of witnessing a thief stealing in their neighborhood, 57% of the residents opted to call the police, 17%

chose to inform the building superintendent, and 15% decided to confront and stop the thief, with the latter group mainly consisting of young and middle-aged men under 40 years of age. Overall, the proportion of residents who turned a blind eye to the thief was merely 2%. To some extent, this indicates that the residents possess a robust sense of community responsibility.

When surveyed regarding residents' perspectives on the notion that "The community is our home and its construction depends on all of us" 87.5% of the respondents either agreed or somewhat agreed, whereas a mere 3.4% disagreed or somewhat disagreed (Fig. 6.6). In the event that the collective interests of the community are at stake, 77.2% of the residents indicated that they would or might take part in joint activities organized by fellow community residents to convey the issues to government authorities or submit a joint petition. Conversely, only 6.3% stated that they would not. The majority of residents, predominantly those residing outside the city center and those living in rented accommodations, expressed their willingness or potential willingness to participate in such joint activities for the purpose of reflecting problems to the government and writing a joint petition. Overall, it is evident that residents exhibit a strong sense of community participation and a heightened awareness of safeguarding their rights within the community.

The survey results indicate that 70% of the residents perceive their interaction with neighbors as being of an average level, while 15% feel that it is close and another 15% consider it unfamiliar. Regarding the residents' feelings about being referred to as "Huilongguan people," 16% reported feeling comfortable and proud, 4% expressed discomfort and a sense of discrimination, and 56% stated that their feelings were ordinary and not particularly intense. In the survey question "If you were to move, would you choose to remain in your current neighborhood?", it was discovered that 19.4 and 52.0% of the residents respectively indicated a strong desire and a fairly high willingness to stay in their current neighborhood. Only 8.2% of the residents said they would have little or no attachment, and the remaining 20.4% of the residents showed an indifferent attitude, stating that they did not care. This collectively demonstrates that, on the whole, residents have a relatively strong sense of attachment to the community.

Overall, the neighborhoods where the residents of Huilongguan live possess the traits of a typical urban neighborhood. Although neighborly interactions are relatively lacking in familiarity, neighborhoods still hold a significant position in the residents' lives. As time goes by, residents gradually become integrated into the new community, and their perception of the neighborhood is favorable. They yearn for a neighborly relationship that is founded on warmth and mutual care.

(2) **Neighborhood influencing factors**

The study delved into the correlations among respondents' personal attributes (including gender, age, and socio-economic attributes), the duration of their residence, the type of housing they occupy, their experiences of and willingness for neighborhood mutual assistance, as well as their participation in neighborhood exchange activities. Among these factors, the positive correlation between the education level and the monthly income was remarkably evident. Consequently, the education level

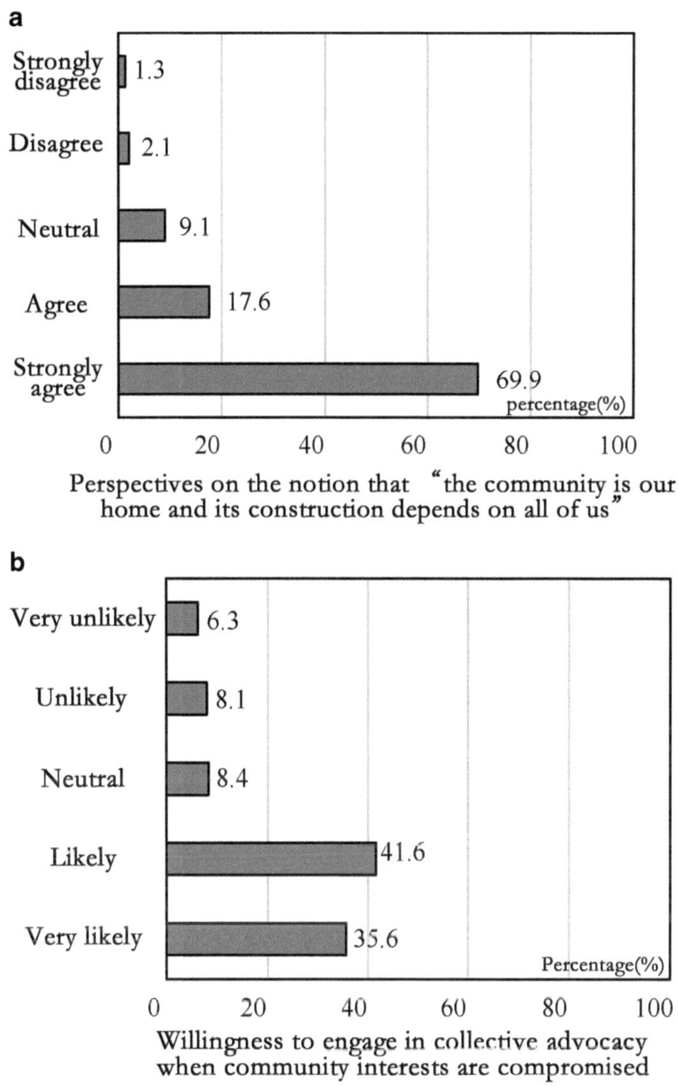

Fig. 6.6 Community participation and awareness of rights

was selected to represent the economic attributes of the respondents. Simultaneously, correlation coefficients were employed to explore the relationships between the basic attributes of the residents and their neighborhood relations. The neighborhood relations were gauged by the number of neighbors the residents knew. The findings of this analysis are presented in Table 6.8.

Table 6.8 Main factors affecting neighborhood relationships

Neighborhood relationships	Correlation test	Male	Age	Beijing Hukou	Party member	Unmarried	Education level
	Pearson correlation	0.041	0.072	0.056	−0.054	−0.047	−0.121**
	Significance (two-tailed)	0.334	0.091	0.189	0.206	0.270	0.005
	N	550	550	550	550	550	550
	Correlation test	Length of residence	Rental housing	Willingness for neighborhood mutual assistance	Experience of neighborhood mutual assistance	Participation in neighborhood activities	Number of acquainted neighbors
	Pearson correlation	0.184**	0.054	0.099*	0.112**	0.077	1.000
	Significance (two-tailed)	0	0.205	0.020	0.008	0.072	–
	N	550	550	550	550	550	550

* Significantly correlated at the 0.05 level (two-tailed). ** Significantly correlated at the 0.01 level (two-tailed)

The analysis results reveal that there is no substantial association between gender, age, household registration (hukou), political affiliation, and marital status and neighborhood relations. Notably, the finding that gender has no significant influence on the variations in neighborhood relations is somewhat unexpected. This can be attributed to the fact that urban women have entered the job market on an equal basis, and the prevalence of numerous dual-income couples has diluted the traditional pattern where neighborhood relations were predominantly shaped by women. There exists a significant negative correlation between education level and neighborhood relations. Specifically, individuals with higher education (and thus relatively higher monthly incomes) tend to know fewer neighbors. Regarding residence, a significant positive relationship is observed between the length of residence in the neighborhood and neighborhood relations. That is to say, the longer one has lived in the neighborhood, the more neighbors one is likely to know. Whether a person rents a house or owns a home does not have a significant impact on neighborhood relations. Furthermore, those who have had experiences of neighborhood mutual assistance or possess a stronger personal willingness to offer mutual help tend to enjoy better neighborhood relations and know more neighbors. The frequency of participation in neighborhood activities is not significantly correlated with neighborhood relations. Presumably, this is because the nature of neighborhood activities often makes the group of interacting individuals more fixed, and as a result, it does not exert a significant influence on residents' awareness of their neighbors.

Through questionnaires and interviews, it has been identified that the primary factors contributing to social indifference in neighborhoods include the rapid pace of modern life, high work pressure, the privatization of housing, and the scarcity of public space (Fig. 6.7a). This same set of characteristics was also evident when respondents were asked to select the most influential factors (Fig. 6.7b). Some individuals held the opinion that modern neighborhoods are not actually characterized by indifference in the sense of being unhelpful (such as rejecting requests for assistance), but rather that people simply lack the time for social interaction. Further in-depth interviews lent support to this perspective. Neighborhood interactions are indeed restricted by a lack of motivation, time, and suitable venues. Many residents who originate from rural areas perceive that modern individuals have become more guarded and wary, and are reluctant to engage in interactions with others easily. Simultaneously, the rapid tempo of urban living and the work-related stress make it arduous for residents to allocate time for building relationships with their neighbors. Moreover, the absence of sufficient public interaction areas in the architectural design of residential buildings is another significant contributing factor.

(3) **Constraining mechanisms of neighborhood relationships**

The influencing and constraining factors of neighborhood relations are categorized into four aspects: time, intention, event, and space. Based on this classification, a conceptual model of neighborhood relationships constraints, namely the TIES model (Fig. 6.8), is constructed.

The first factor is time. An analysis of the 24-h activity paths of residents in Huilongguan reveals that on weekdays, residents' working hours and commuting

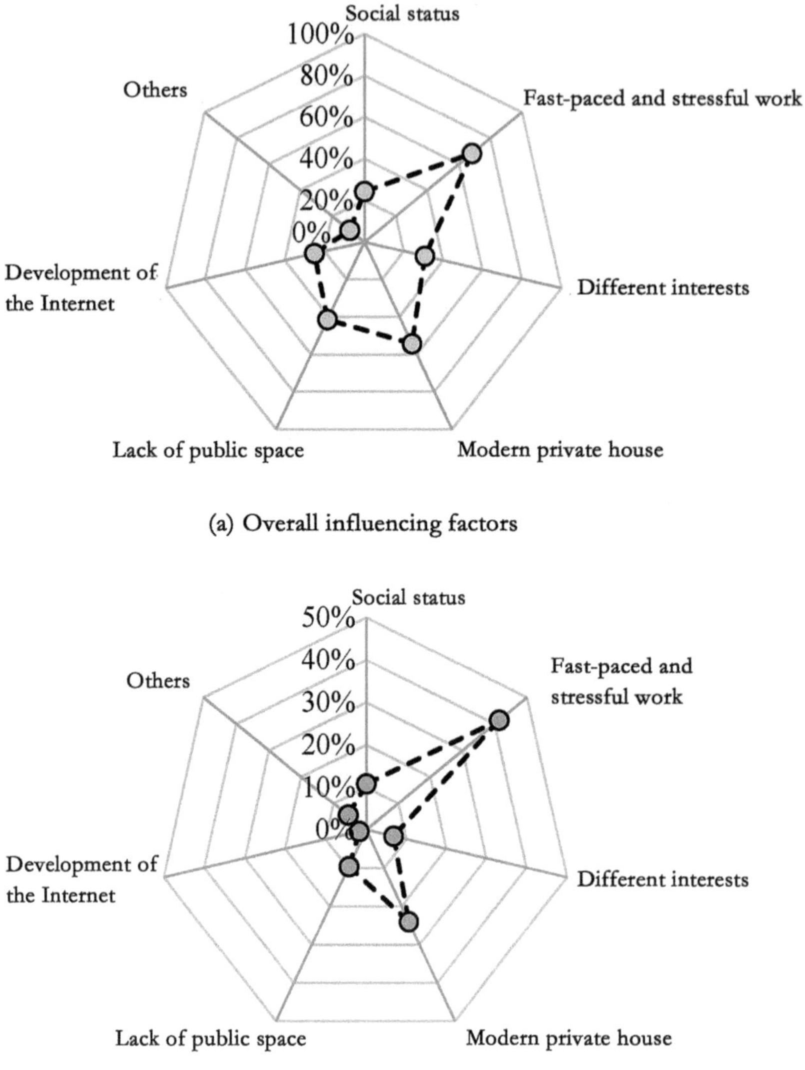

(a) Overall influencing factors

(b) Chief influencing factors

Fig. 6.7 Major factors affecting neighborhood indifference

time are the most prominent aspects. Due to the inconsistency in the time spent away from home and at home, it is more challenging for them to encounter others in the vicinity of their homes. Interactions with others mainly occur during the time periods of outdoor exercise and outdoor activities near their homes. Time constraints are also evident in other aspects. For instance, they are manifested through capacity limitations. The effective duration of commuting time varies among residents because of

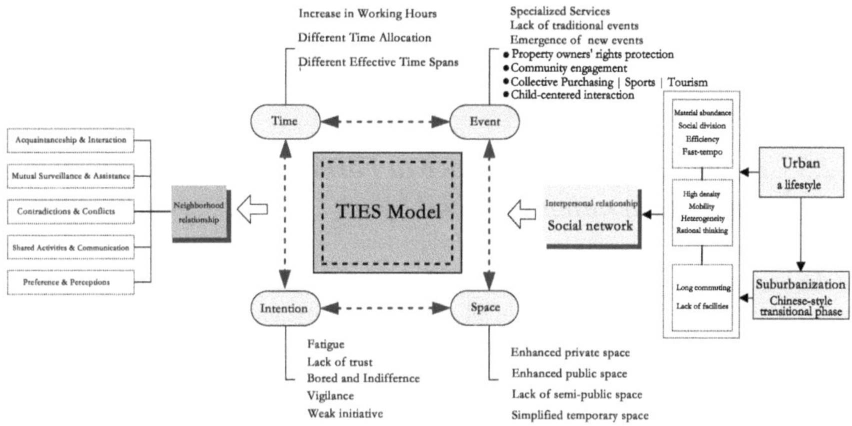

Fig. 6.8 Conceptual model of neighborhood relationships constraints

differences in commuting vehicle, which are influenced by economic conditions and other factors. Secondly, there is the factor of intention. The rapid pace and high stress of urban life, along with the extended commuting time caused by suburbanization, and the increased fatigue of residents, all reduce their inclination to engage in interpersonal interactions. The trust crisis and a sense of defensiveness also have an impact on their willingness to interact. The third factor is the event factor. The heterogeneity and mobility of the urban population have led to a shortage of topics that are of common interest to the residents, which in turn affects the interaction within the community. Some new events in urban life, such as community rights protection, neighborhood group purchasing, community cultural and sports activities, and children's interactions, play a crucial role in maintaining neighborhood relations. In particular, there are child-centered interactions among residents. For example, when parents with children meet, matters related to their children become a common topic of conversation, which leads to their acquaintance and growing familiarity. Or, when interactions between children trigger interactions between their parents, such scenarios are relatively common in neighborhoods. Finally, there is the spatial factor. Public space and semi-public space have a significant influence on neighborhood relations and the perception of the community. They provide a spatial medium for neighborhood interactions.

Neighborhood relations, which represent a distinct social-relation network formed by the spatial proximity and clustering of residents, are restricted by time, intention, events, and physical space. In large suburban communities, the sense of neighborliness and community characteristic of traditional societies has been significantly affected. Certainly, these four constraints interact with one another and are not entirely discrete. More crucially, underlying these constraints are urban living patterns and the profound imprints of suburbanization and urbanization. In particular, the pronounced separation between work and life and the timeconsuming commuting associated with suburban residence, along with the lack of facilities within large suburban settlements, heighten residents' sense of insecurity and mold the unique interpersonal relationships and social networks of suburban communities.

6.2.3 From Neighborhood Space to Formation of Community and Regeneration of Social Space

In the context of suburban neighborhood space, through the interactions among the various actors involved, a certain level of both quantity and quality of neighborhood relationships are established. This development endows the neighborhood space with the characteristics and functions of a community. As a result, it effectively drives the revitalization and renewal of the social functions within the suburban neighborhood space.

6.2.3.1 From Neighborhood Space to Formation of Community

Huilongguan Residential Area initially came into being as a suburban neighborhood space due to the government's actions in response to the housing system reform. Its purposes included addressing the housing issues of employees from high-tech industries and universities in Haidian District, as well as resettling the demolished and relocated residents from the old urban area. Later on, the market gradually took the leading position, and a large number of "new Beijingers" congregated in Huilongguan, giving rise to the characteristics of a large suburban residential area (Fig. 6.9). Interestingly, although the neighborhood committees in Huilongguan did organize some cultural and sports activities, it was not the administrative factors that shaped the stable social and neighborhood relations there. Instead, it was the spontaneous actions of the residents and the awakening of a shared social consciousness. Cultural and sports activities, human rights protection incidents, and the "scouting"—based interactions among parents have become the organic driving forces in building the neighborhood relations. The impetus for constructing neighborhood relations is organic in nature. Certainly, information technology communication tools also play a crucial role in promoting the establishment and formation of neighborhood relations from the perspective of the virtual space.

Overall, the formation of the common areas in Huilongguan is a direct outcome of the combined efforts led by the government and driven by the market. These efforts have provided a spatial foundation for the community, as communities are first formed in terms of space (Grant 2009). Residents with diverse backgrounds and identities gather together and engage in interactions to different extents, all centered around family life. Public recreational spaces (physical space) and community networks (virtual space) strengthen the connections among residents and give birth to the unique subcultures of Huilongguan. Once the conditions for connection and the formation of culture among community members have been relatively established,

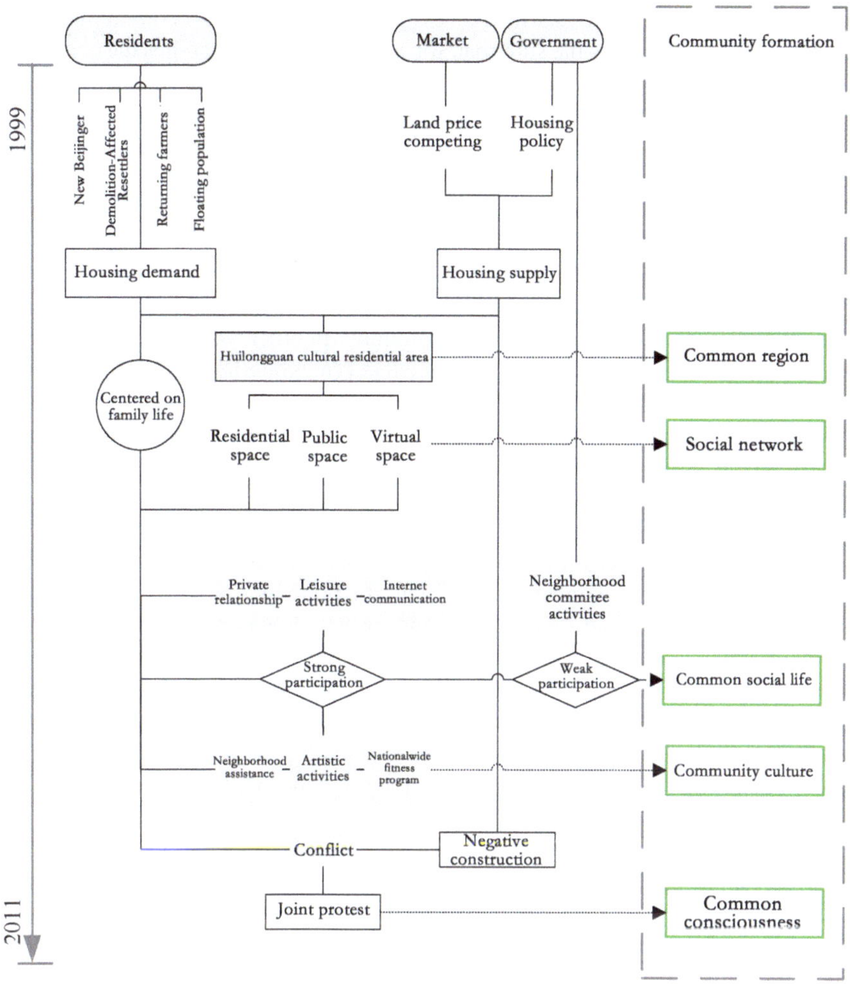

Fig. 6.9 Community formation and its drivers in the Huilongguan residential area

events that are of common interest act as a catalyst for the emergence of a sense of commonality and belonging, eventually leading to the formation of a typical suburban community. It can be concluded that it is the interactive development of residents in their social lives that ultimately forms and creates the community.

6.2.3.2 Regeneration of Social Space of Residential Area

Neighborhood associations allow the residents of large suburban communities to establish interactions that are accompanied by a specific perception, identity, and a sense of belonging to the area. This process gives rise to suburban communities and, to some degree, revitalizes the suburban social space. It also embodies the concept of the dialectical unity between social space and suburban neighborhoods. The suburban space serves as the essential foundation. Due to the close proximity and overlap of living spaces, neighbors engage in interactions with one another, generating certain social connections that ultimately lead to the creation of new suburban communities. Through analysis, it has been discovered that the geographical area is the starting point and fundamental platform of a community, a role that cannot be supplanted by the interconnected virtual network. However, the virtual space holds great significance for the establishment of suburban neighborhood relationships and the development of a sense of belonging to the community. In other words, the geographical spatial element and the social interaction element (neighborhood relationships) of a community remain the most crucial. Without the former, although an online community based on the Internet can still enable interaction, Internet users will lack a sense of attachment to the physical region, and it will be challenging to form a spatial platform for activities and interactions. Without the latter, the geographical spatial area is merely an empty physical shell and cannot be regarded as a community. This comprehensively clarifies the relationship between neighborhoods and the formation of communities, which is in accordance with the principles of socio-spatial dialectics (Soja 1980; Knox and Pinch 2000), and further elaborates on the socio-spatial dialectics at the level of modern suburban communities.

Neighborhoods have indeed reshaped the suburban social space. In the suburbs, apart from the native inhabitants, a significant number of residents are individuals who have migrated from rural areas to cities to embark on new work and life pursuits amidst the backdrop of urbanization. Additionally, some are former residents of the old city who have experienced passive demolition and reconstruction (and are among the main participants in suburbanization). Their original social relationship networks have been challenged by the change of geographical location. The new interpersonal relationship networks, which are constructed with a mindset of mobility and caution, give rise to a fragmented social space. When suburban residents actively engage in rights protection, participate in various activities, and establish good neighborhood relations and a sense of community due to their neighborhood ties, they ultimately develop a sense of identity and belonging to the suburban area. This leads to the cultivation of "placeness" sentiments and, to a certain extent, reshapes the original social space that was based on interpersonal relationship networks. As a result, the suburban community becomes an important carrier of the social space within a metropolis. In essence, it is the presence of a certain quantity and quality of neighborhood relations in Huilongguan's residential area that has facilitated the formation of the Huilongguan community. Precisely because of this, these urban "new immigrants" residing in Huilongguan have been able to establish new social relations. Against the backdrop of losing, destroying, and detaching from their original neighborhood relations and social relationship networks in their places of origin, they have contributed to the creation of a new social space in such a large suburban residential area as Huilongguan. Only in the context of the disruption of the neighborhood and social relationship networks in their original places of residence have these urban "new immigrants" in Huilongguan managed to construct new social relations in this large suburban residential area, thereby contributing to the regeneration of the social space.

6.2.4 Summary and Discussion

This section focuses on Huilongguan, the most renowned large-scale suburban residential area in Beijing, as the research object. It investigates the neighborhood situation, constraints, and its relationship with social space therein. Through the study of the Huilongguan case in Beijing, a series of general features regarding neighborhood relations and interactions among residents in Chinese urban settlements during the era of rapid urbanization can be distilled.

Overall, Huilongguan exhibits the traits of a typical urban neighborhood. Neighborly interactions here are relatively shallow, with Renqing interaction being the primary mode. However, residents demonstrate a high inclination to interact and hold a favorable assessment of the state of neighborhood relations. In daily life, neighborly conflicts are infrequent, and while there are relatively few instances of mutual assistance, there is a strong underlying willingness to offer help. Based on different triggering events, reasons, and interaction depths, neighborhood communication takes various forms. These include group buying during the initial stage of moving into the community, rights-protection and cultural-sports activities in daily life, "children-led" diplomacy, and interactions facilitated by the community network. Precisely these communication channels enable residents to maintain amicable neighborly relations. Regarding neighborhood responsibility and a sense of belonging, Huilongguan residents generally possess a strong sense of responsibility, a high degree of community participation and a keen awareness of rights protection, along with a distinct sense of attachment to the community. Evidently, the neighborhoods in Huilongguan play a crucial role in residents' lives. Furthermore, the results of quantitative analysis indicate that gender, age group, household registration, political affiliation, and marital status have no significant bearing on neighborhood relations. Conversely, there is a significant negative correlation between education level and neighborhood relations. There is also a significant positive correlation between the length of residence in the district and neighborhood relations. Additionally, those who have had experiences of mutual assistance in the neighborhood or have a stronger personal willingness to offer help tend to enjoy better neighborhood relations.

A constraining conceptual model of neighborhood relationships is established to explore the relationship between the formation of new communities and the regeneration of socio-spatial structures in the suburban areas of China's large cities amidst the backdrop of rapid urbanization. Suburban neighborhoods are restricted by factors such as time, intention, events, and physical space. More significantly, underlying these constraints are urban living styles, along with the profound impacts of suburbanization and urbanization, which have molded the unique characteristics of interpersonal relationships and social networks within suburban settlements.

Although government-led and market-driven measures have provided suburban residential areas with the spatial foundation for the development of communities, it is the social life interactions among residents that play a more direct role in community formation. This interaction is, of course, based on the platforms offered by physical spaces (public recreational areas) and virtual spaces (community networks). It also involves the community culture that emerges from the strengthening of connections between residents, as well as the events that trigger a shared sense of common interest and a sense of belonging among them.

Finally, it is important to note that the emergence of suburban communities like Huilongguan is the outcome of both centrifugal diffusion-style suburbanization and centralized urbanization, and it should be regarded as the result of the combined forces of urban spatial development dynamics. The significance of the regeneration of its neighborhood and social space lies in the fact that, as the earliest large-scale suburban residential area established following China's housing system reform in 1998, it has endured the impacts of a high level of housing marketization. A large number of urban "new immigrants" (mainly young people from other places) reside there, and they are confronted with the reconstruction of their social networks and neighborhoods, as well as the reshaping of the social and spatial order within the community. Their social networks and neighborhoods are in the process of being rebuilt, and the social and spatial order of the community is being reconstituted. Returning to the two initial research questions, the empirical research in this section offers preliminary answers to them. For a large and highly marketized suburban residential area such as Huilongguan, the presence of neighborhood relations of a certain quantity and quality is a crucial factor in the regeneration of suburban social space. This further validates the theory of socio-spatial dialectics proposed by Western scholars. Moreover, although suburban neighborhoods that rely on information technology and experience frequent population migrations may give rise to social interactions outside the community, the geographical spatial carriers within the community still play an irreplaceable role in maintaining a sense of belonging to the community and providing an effective platform for activities and interactions. In other words, "geography" remains of great importance and is indispensable in the modern information society. This is precisely why it is essential and necessary for the discipline of geography to conduct studies on communities.

6.3 Social Networks: Restructuring of Social Relations and Socio-Spatial Interactions

6.3.1 Characteristics of Social Networks

6.3.1.1 Attributes of Social Network Characteristics

Relationships can be characterized in terms of their type, intensity, significance, interaction mode, and frequency, as stated by Marsden and Campbell (1984). The comprehensive statistical data regarding the relationship network between members and the respondents within the important issues discussion network of Huilongguan residents are presented in Table 6.9.

(1) **Types of relationships**

Friends, family members, and coworkers make up the network for discussing important issues among the residents of Huilongguan. Friends are the most significant relationships within the important issues discussion network of Huilongguan residents. They are followed by family members including parents, children, and spouses, and then by coworkers. The combined proportion of these three groups reaches 76%. Among them, the proportion of friends is 32.35%, which is the highest. Additionally, friendship is a rather ambiguous psychological concept. Many individuals define a "friend" based on the degree of psychological intimacy. Ordinary classmates, coworkers, friends from one's hometown, or people met accidentally on certain occasions can, due to specific triggering factors, develop a certain level of psychological closeness and thus become friends. They then become the main participants in the important issues discussion network. The proportion of close family members such as parents, children, spouses, and colleagues is also considerable. This indicates that the people with whom residents have core interactions in their daily lives and at work still account for the largest portion in the important issues discussion network. However, when compared with the results of Qi's (2007) survey of relationship networks in a social district, the conclusions differ. Qi discovered that coworkers accounted for approximately 50% of all contacts, whether for conversation, advice-seeking, or assistance. In contrast, because of the large population and the market economy of a big city, and unlike the original residential neighborhoods formed based on unit compounds, the significance of friends and family members like parents, children, and spouses is more pronounced in the relationship types among Huilongguan residents. Meanwhile, the importance of coworker relationships is relatively less significant.

Neighborhoods are notable for their significance and relatively high degree of community integration. Neighbors rank fifth in the network for discussing important issues, surpassing classmates and those from one's hometown. In particular, the group of people from one's hometown alone has a relatively low proportion within this discussion network. The residents of the Huilongguan community are distinct from

Table 6.9 Members in the discussion network of residents' important issues

Objects	Overall			Male			Female		
	Number	Proportion/%	Ranking	Number	Proportion/%	Ranking	Number	Proportion/%	Ranking
Friends	484	32.35	1	304	36.98	1	180	26.71	2
Parents, children, spouses	398	26.60	2	189	22.99	2	209	31.01	1
Colleagues	255	17.05	3	150	18.25	3	105	15.58	3
Siblings	124	8.29	4	59	7.18	4	65	9.64	4
Neighbors	93	6.22	5	39	4.74	6	54	8.01	5
Other relatives	69	4.61	6	39	4.74	5	30	4.45	6
Classmates	29	1.94	7	21	2.56	7	8	1.19	8
Fellow-townsmen	24	1.60	8	13	1.58	8	11	1.63	7
Others	16	1.07	9	8	0.98	9	8	1.19	9
Blank	4	0.27	–	0	0	–	4	0.59	–
Total	1496	100.00	–	822	100	–	674	100	–

the typical migrant workers engaged in low-end industries. There is a relatively higher proportion of high-level intellectuals among them, and many of these individuals hold Beijing household registration. As a result, there are fewer hometown-based groups, and the residents are gradually becoming more integrated into the local community.

In terms of gender, notable differences exist. Men tend to place more emphasis on friendships, whereas women focus more on family relationships, such as those with parents, children, and spouses. When analyzing the spatial distribution of these relationships, a more evident positive correlation is observed between the individuals within women's important relationship discussion network and the spatial distance. Thus, the geographical and spatial attributes of these relationships are more pronounced. Among male respondents, the proportion of friends in their relationship network reaches 36.98%, which is relatively high. In contrast, among female respondents, the most prominent category of members consists of family members like parents, children, and spouses. Friends come second in this regard. Comparatively, the proportion of male respondents who included colleagues and classmates in their important issues discussion network is higher than that of female respondents. On the other hand, the proportion of female respondents who included neighbors in their discussion network is slightly higher than that of male respondents.

(2) **Quality of relationships**

The members of the network have known one another for 10 years or less (Fig. 6.10a), highlighting the significance of "new" contacts who have been acquainted for 5 years or less (Fig. 6.10b). The average duration of the relationships included in the Important Issues Discussion Network exceeds 10 years. Among those surveyed regarding the number of years they have known the members of the Critical Issues Network, 67.2% of them have known each other for 10 years or less, which constitutes a remarkably high proportion of the respondents. When conducting a more detailed breakdown, the highest frequency is observed at the acquaintance duration of 2–3 years, followed by the 4–5 years of acquaintance and the duration close to 10 years of acquaintance.

Regarding gender differences, female respondents exhibited a stronger emotional bond with both new acquaintances and old friends compared to male respondents. The data indicates that a greater proportion of male respondents than female respondents incorporated individuals they had known for less than 10 years into their important issues discussion network, while a higher proportion of female respondents included those they had known for a longer duration in their discussion network. When taking a broader view, men had a higher proportion of people they had known for 2–5 years within their network. However, women had a higher proportion of individuals they had known for 1 year or less and those they had known for more than 5 years compared to men.

(3) **Ways of interaction and contents in relationships**

The primary mode of contact among members of the social network is meeting in person. Based on the survey findings, 60.1% of the network members predominantly

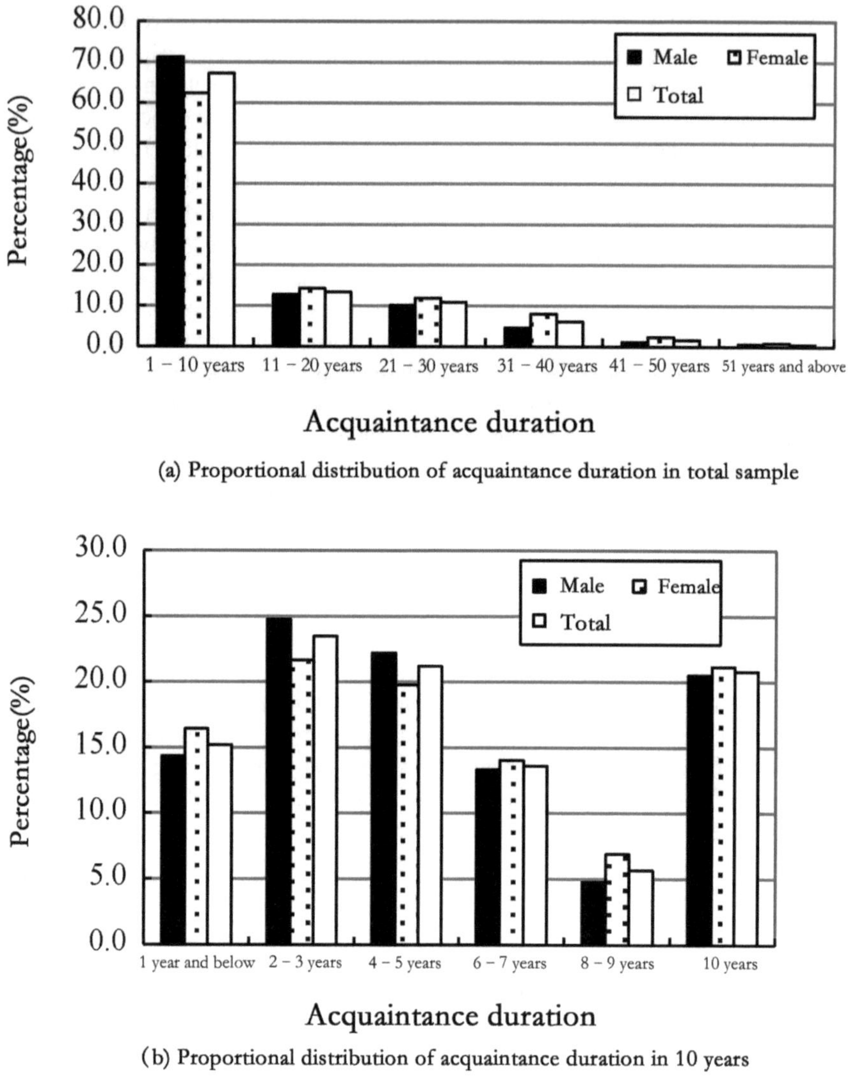

Fig. 6.10 Duration of relationships between residents and their important issues discussion network members

engage in face-to-face interactions, while 47.1% rely mainly on telephone communication. To a lesser degree, communication occurs via the Internet and letters. There is essentially no disparity in the contact methods used by male and female respondents (Table 6.10). According to certain scholars, the social order in post-modern cities has undergone a transformation, with face-to-face communication no longer being the predominant mode of social interaction. Instead, the use of telephones, faxes, and the Internet is favored (Dear and Flusty 1998). Nevertheless, the survey reveals that when it comes to discussing important issues, the residents of Huilongguan community still predominantly utilize face-to-face communication as their main means of interaction.

The primary way of contact among members of the social network is meeting in person. The survey indicates that 60.1% of the network members mainly interact with one another through face-to-face meetings, and this is followed by 47.1% who rely on telephone communication. A smaller proportion of them use the Internet and letters for communication. There is essentially no difference in the contact methods employed by male and female respondents, as shown in Table 6.10. Some scholars, such as Dear and Flusty (1998), have conducted research and posited that the social order in post-modern cities has changed, with face-to-face communication no longer being the dominant mode of social interaction. Instead, tools like the telephone, fax, and the Internet are favored. Nevertheless, the survey reveals that when it comes to discussions on important issues, the residents of the Huilongguan community still predominantly use face-to-face communication as their main means of interaction.

The main categories of important issues discussed among members of the surveyed social networks revolve around work—related official matters, children's development, and family and neighborhood affairs. When it comes to gender differences, the patterns are distinct. Women predominantly discuss children's development. Subsequently, they engage in conversations about work and other business—related topics, neighborhood happenings, and personal relationships. On the other hand, for men, work and other official matters take precedence. Their sharing of news related to TV, entertainment, and travel ranks second. This is followed by discussions about neighborhood affairs and children's development. Overall, men show a greater inclination to emphasize work, while women tend to strive for a balance between family and work affairs.

(4) **Relationship pattern**

There is a distinct differential—order pattern in the relationships within an individual's central network, leading to a polarization in the individuals with whom residents interact. Through an analysis of the order of nominees in the central network of individual respondents and a comparison of the proportions of different relationships (Table 6.11, Fig. 6.11), the following trends are revealed. From the first to the fifth person in the network, the primary contact is typically close family members, such as parents, children, and spouses. Friends come next, followed by colleagues. Notably, starting from the second person, the proportion of friends in the Important Issues Discussion Network is overwhelmingly dominant and on the rise. The proportion of coworker relationships is also increasing, and the proportion of neighbor relationships is steadily growing. However, in the Important Issues Discussion Network, there is a clear pattern of disorderly relationships. As the network expands, the relationships among its members become increasingly distant.

Huilongguan is populated mainly by new arrivals to Beijing. Urban living has led to a significant polarization in the social interactions of its residents. On one hand, the bond with immediate family members has grown stronger. On the other hand, connections with extended relatives and hometowns have weakened compared to

Table 6.10 Interaction situations in discussing issues among social network members

Projects	Overall			Male			Female		
	Number of samples	Proportion/ %	Ranking	Number of samples	Proportion/ %	Ranking	Number of samples	Proportion/ %	Ranking
Work—related matters	680	46.51	1	393	48.64	1	287	43.88	2
Children's development	549	37.55	2	259	32.05	4	290	44.34	1
Matters of relatives and neighbors	543	37.14	3	261	32.3	3	282	43.12	3
News about TV entertainment, tourism, etc	491	33.58	4	265	32.8	2	226	34.56	5
Personal relationships	446	30.51	5	216	26.73	6	230	35.17	4
Social and political events	437	29.89	6	258	31.93	5	179	27.37	7
Marriage and sexuality	349	23.87	7	167	20.67	7	182	27.83	6
Others	75	5.13	8	40	4.95	8	35	5.35	8

Table 6.11 Overview of the order-relationship of the involved objects in the discussion network of residents' important issues

Objects	The first person		The second person		The third person		The fourth person		The fifth person	
	Number of samples	Proportion/%	Number of samples	Proportion/%	Number of samples	Proportion/%	Number of samples	Proportion/%	Number of samples	Proportion/%
Parents, children, spouses	233	50.33	96	23.94	45	15.25	16	8.04	8	6.11
Siblings	44	9.50	38	9.48	19	6.44	13	6.53	9	6.87
Other relatives	15	3.24	20	4.99	18	6.10	10	5.03	6	4.58
Friends	104	22.46	141	35.16	119	40.34	81	40.70	56	42.75
Colleagues	35	7.56	67	16.71	54	18.31	53	26.64	28	21.37
Neighbors	21	4.54	23	5.73	21	7.12	13	6.53	14	10.69
Fellow-townsmen	4	0.86	4	1.00	5	1.70	7	3.52	4	3.05
Others	2	0.43	2	0.50	7	2.37	1	0.50	4	3.05
Blank	5	1.08	10	2.49	7	2.37	5	2.51	2	1.53
Total	463	100.00	401	100.00	295	100.00	199	100.00	131	100.00

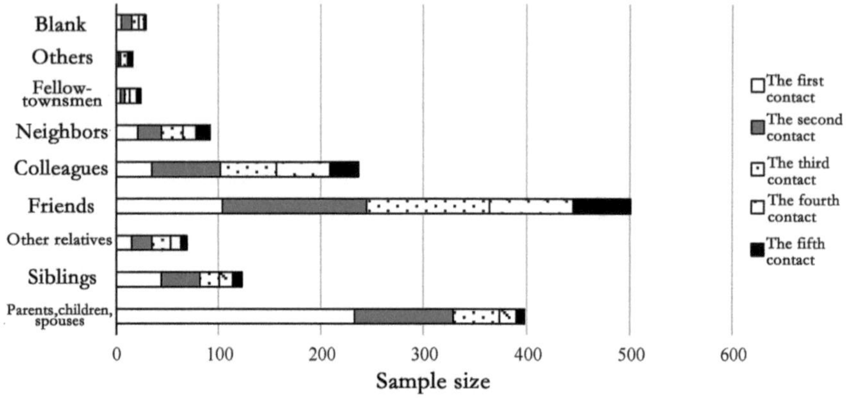

Fig. 6.11 Distribution of relational attributes in different orders within the relational discussion network

those in rural communities. Friends made through hobbies or by chance are playing an increasingly crucial role in residents' lives. The inclusion of neighborly relations also shows that the community is becoming more important in people's daily lives.

The more distant the relationship, the greater the spatial separation and the farther the residential location. As depicted in the spatial distribution of individuals within the relationship discussion network (Fig. 6.12), it is clear that the spatial residence pattern of those engaged in the important-issue discussion network follows a clustering sequence of "within the district > within Huilongguan > within the city > outside the city". The lower the ranking, the more significant the spatial distance between the parties.

The primary mode of communication with each party is in-person meetings, followed by telephone contact. Communication among members of the Important

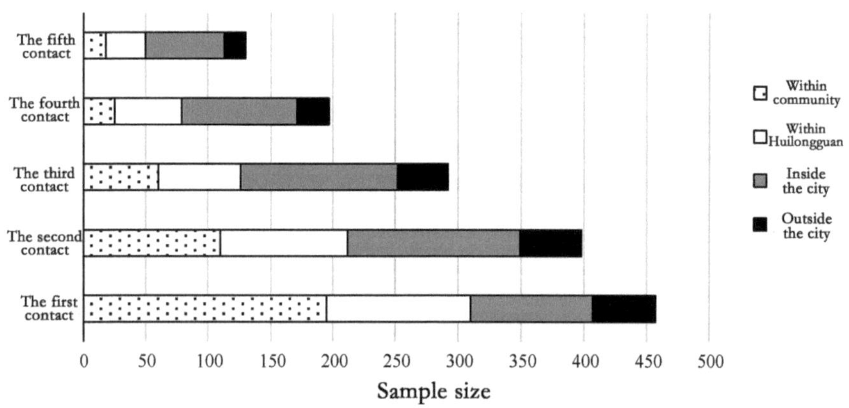

Fig. 6.12 Spatial distribution of individuals within the relationship discussion network

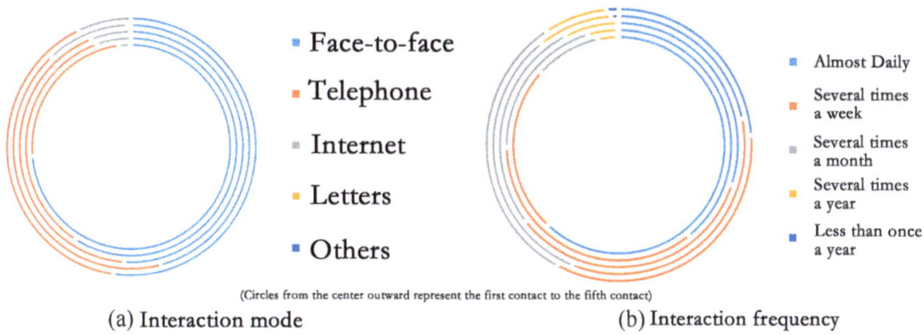

Fig. 6.13 Interaction styles and frequency of each relationship in the relationship discussion network (the circles from the center outwards indicate the first to the fifth person)

Issues Discussion Network remains predominantly face-to-face, with phone calls as the second most common method, while the Internet and letters account for a relatively small proportion. The frequency of contact is relatively high, typically several times a week, and the most frequently contacted individuals are reached almost daily (Fig. 6.13).

Based on the observed changes in the spatial sparseness of different relationship types, it can be posited that a significant correlation exists between the spatial distribution distance of relationships and the degree of relationship proximity. Specifically, the higher the rank of the object relationship, the closer the relationship is, the longer the acquaintance duration, the relatively closer the spatial distribution, the higher the contact frequency, and the contact mode gradually shifts towards in-person meetings.

6.3.1.2 Characteristics of Network of Relationship Membership

The size of the social network measures the richness of the social resources possessed by the respondent. Analyzing social relationship network characteristics can mainly be done from two aspects: network structure and network member similarity.

(1) **Network structure**

Structural characteristics of social networks encompass aspects such as scale (or size), density, homogeneity, and centrality. Interaction characteristics involve relationship multiplicity, interaction frequency and durability, as well as relationship content and direction.

Statistical analysis of the 463 samples indicated that the mean size of community residents' discussion networks was 3.25. In other words, each respondent had an average of 3.25 discussion partners (see Table 6.12). Discussion networks with 5 discussion subjects accounted for the highest proportion, followed by those with 2 and 3 discussion subjects.

The study discovered that the majority of the contacts within the respondents' central network were acquainted with one another. A substantial number of common

Table 6.12 Scale of the discussion network of community residents

	Scale value	Number of samples	Proportion/%
Scale of the discussion network	1	59	12.74
	2	105	22.68
	3	99	21.38
	4	65	14.04
	5	132	28.51
	6	3	0.65
Total		463	100
Average value		3.25	
Standard deviation		1.25	

acquaintances were present among the members of the respondents' social network circles (see Table 6.13). A further examination of the circle of common acquaintances reveals that 44.82% of the respondents assert that they possess a group of individuals they are very familiar with, 26.38% claim to have several such groups, and merely 10.84% indicate having no common acquaintances with a particular member of the community, being acquainted with others only on an individual basis. Regarding gender disparities, women and men exhibit an equal proportion of close relationships. However, the proportion of women having no common acquaintances is higher than that of men.

Table 6.13 The relationship circles of the common acquaintances among discussion network members

Items	Male		Female		Overall	
	Number of samples	Proportion/%	Number of samples	Proportion/%	Number of samples	Proportion/%
No common acquaintances	77	9.66	80	12.29	157	10.84
A group of not—very—familiar common acquaintances	142	17.82	110	16.90	252	17.40
A group of very familiar common acquaintances	357	44.79	292	44.85	649	44.82
Several groups of very familiar common acquaintances	216	27.10	166	25.50	382	26.38
Total	792	–	648	–	1440	–

The average strength of the residents' network for discussing important issues is 1.80. The relationships among network members fall between "average" and "close", indicating a high level of mutual familiarity. To disclose the strength of relationships within the respondents' central networks, the relationships in their important—issue discussion networks were classified into three categories: close, average, and unfamiliar. Respondents with a discussion—network size of less than 2 were excluded from this analysis. Based on valid survey data, the total number of non—directional relationships in the respondents' important—issue discussion networks amounted to 1,704 pairs. The proportion of close relationships among members was 35.5%, while that of unfamiliar relationships was 30.3%. Relationship strength was quantified by assigning a value of 1 to "close" relationships, 2 to "average" relationships, and 3 to "unfamiliar" relationships. The average strength of the residents' important—issues discussion network was computed as 1.80. Notably, a lower score implies a stronger relationship, and conversely, a higher score indicates a weaker relationship.

(2) **Similarity of network members**

Similarity is commonly manifested as the percentage of respondents who belong to the same group in a specific aspect relative to all members of the discussion network. It denotes the degree to which the central individual within a social network shares resemblances with other network members in terms of diverse social characteristics. The similarity among members of the discussion networks regarding important issues is predominantly depicted and described with respect to aspects such as gender, age, educational attainment, and income disparities, among others.

It has been contended that individuals exhibiting similar characteristics are more prone to establishing relationships. This is attributable to the fact that individuals with shared similarities are inclined to cultivate common interests, thereby facilitating the ease with which they can comprehend and support one another (Marsden 1988). Additionally, it has been posited that individuals of disparate statuses are more likely to offer mutual support, engender complementarities among themselves, and engage in the exchange of goods and services (Wellman 1990).

There exists a high level of similarity between the respondents and the members of the social network with respect to gender, age, political affiliation, educational background, and income (Table 6.14). Gender similarity is defined as the proportion of members who share the same sex. The survey results indicate that the overall gender similarity stands at 63.2%. Regarding age, 65.31% of the social network members have an age difference of within 5 years from one another. Moreover, the proportion of discussion network members who are 5 years older than the respondents is higher than that of those who are 5 years younger. In terms of political affiliation, the respondents exhibit similarities with the social network members, with the majority of them being non-party members. Concerning educational attainment, the proportion of respondents having the same educational level reaches 65.91%, and the similarity in educational attainment is notably pronounced. As for income differences, 45.25% of the respondents have incomes that fall within the same level.

Regarding gender differences, in terms of gender, age, political affiliation, educational attainment, and income disparities, the members of men's intimate networks

Table 6.14 Similarities in age, political affiliation, education level and income differences between members of the relationship network and the respondents

Category	Range	Male		Female		Overall	
		Number of samples	Proportion/%	Number of samples	Proportion/%	Number of samples	Proportion/%
Age	Within plus or minus 5 years	553	67.27	424	62.91	977	65.31
Political affiliation	The same	491	59.73	386	57.27	877	58.62
Education level	The same	546	66.42	440	65.28	986	65.91
Income difference	Within the same level*	370	49.14	244	39.1	614	45.25

The specific income levels are: (1) below 500 yuan; (2) 501–1000 yuan; (3) 1001–2000 yuan; (4) 2001–4000 yuan; (5) 4001–10,000 yuan; (6) above 10,000 yuan

exhibit greater similarity compared to those of women's. Specifically, in terms of gender, men dominate the intimate networks of male respondents, accounting for 67.5%, whereas women make up the majority (57.9%) in the intimate networks of female respondents. The slightly lower similarity proportion among women can be attributed to the influence of women's marriage and family life. When it comes to age, the age distribution of network members is more concentrated and homogeneous for men, while it is more dispersed for women. In terms of political affiliation, men show greater similarity than women, yet the proportion of those who are unsure about their political status is also higher among men compared to women. Among male residents, there is a more pronounced similarity in the educational levels of the members within the important-issue discussion networks. In terms of educational attainment, the proportion of respondents who engage in discussions with those having a lower educational level is higher than that of those who discuss with individuals having a higher educational level. Concerning income differences, the similarity is more significant among men. Overall, male respondents have higher monthly incomes than their network members, while female respondents, on average, have lower monthly incomes than their network members.

In summary, with respect to gender, age, political affiliation, educational attainment, and income disparities, the social attributes of the discussion network members and the respondents tend to be more homogeneous and less heterogeneous. Moreover, when making a comparison, men demonstrate a higher degree of homogeneity than women.

6.3.1.3 Networks of Relationships and Spatial Distribution

The spatial distribution of the network members of community residents is predominantly centered around the Huilongguan area. Based on the statistical analysis, the majority of the network members primarily reside in Huilongguan (including the adjacent neighborhoods), with a sample size of 778, which accounts for 52.8% of the total. Among them, the proportion of those living within the immediate neighborhood constitutes 27.8%. Additionally, the number of samples from within the city (excluding Huilongguan) is 512, representing 34.8% of the total. The fewest number of individuals are distributed outside the area. In terms of gender disparities, male and female respondents generally exhibit similar patterns. However, the polarization of spatial distance is more pronounced among females than among males. A slightly larger proportion of male network members are located in Huilongguan (but not in their own neighborhoods) and within the city, whereas a relatively higher proportion of female network members are situated in their own neighborhoods and outside the city. This phenomenon can be attributed to the fact that women tend to place greater emphasis on maintaining relationships and are more attuned to the factor of proximity.

The more distant the relationship, the greater the spatial separation, and the farther the places of residence are from each other. Evidently, from the sequence and spatial distribution of the nominated individuals, as the degree of relatedness of the nominated relationships becomes more remote, their spatial residence patterns follow an evolutionary trajectory: starting from within the district, then extending to within Huilongguan, and finally to locations within the city, with the spatial distances between the objects progressively increasing. The spatial distribution of specific relationship types has distinct characteristics. Parents, children, and spouses are predominantly located within the district, with Huilongguan and other areas following as secondary locations. Friends and colleagues are mainly distributed within the city and in Huilongguan (excluding the district). Siblings and other relatives are dispersed across all four spatial dimensions. Neighbors are primarily situated within the district and in Huilongguan. The pattern of the disparity between the nomination order and the intimacy of the relationship indicates that there is a significant correlation between the spatial distance of the relatives' places of residence and the degree of closeness of their relationships.

Individuals residing abroad predominantly maintain parental relationships, and there is a relatively large number of friends within the same age group. In the context of the discussion network, 12.4% of the individuals with whom relationships are established are currently living overseas. The types of relationships mainly involve parents, children, spouses, siblings, and other relatives. When analyzed from the perspective of age, the majority of those living abroad are parents. Simultaneously, the proportion of friendships is notably higher, particularly among those in the same age group (Fig. 6.14).

Across different spatial scales, the predominant types of relationships vary, giving rise to the spatial distribution patterns of relationship networks (as depicted in Figs. 6.15 and 6.16). Within the area of Huilongguan city, the primary relationship

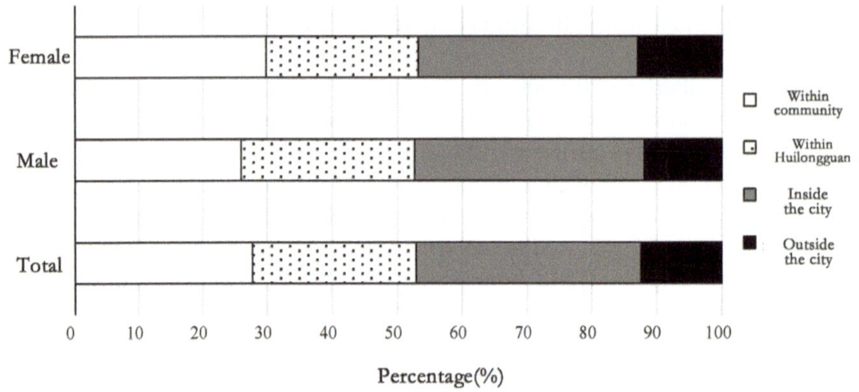

Fig. 6.14 Residential distribution of network members

types include relatives such as parents, children, and spouses. These relationships are concentrated in the local neighborhoods of Huilongguan. Following these are friendships, along with a certain proportion of coworker and neighbor relationships. In the areas within the city but outside of Huilongguan, the network members are predominantly friends, with colleagues being the second most common type. When considering the distribution of relationships across different spatial scopes and the associated characteristics of these relationships, the analysis reveals that parental, sibling, and friendship relationships are the most significant among them.

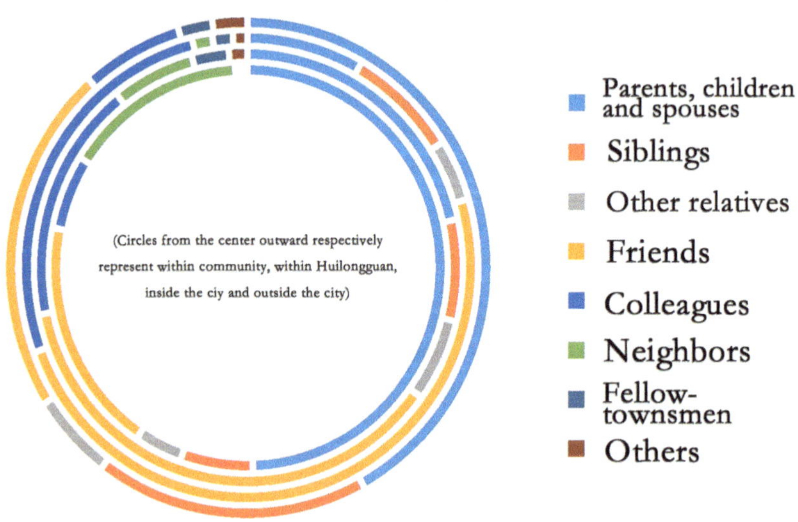

Fig. 6.15 Spatial distribution of network members

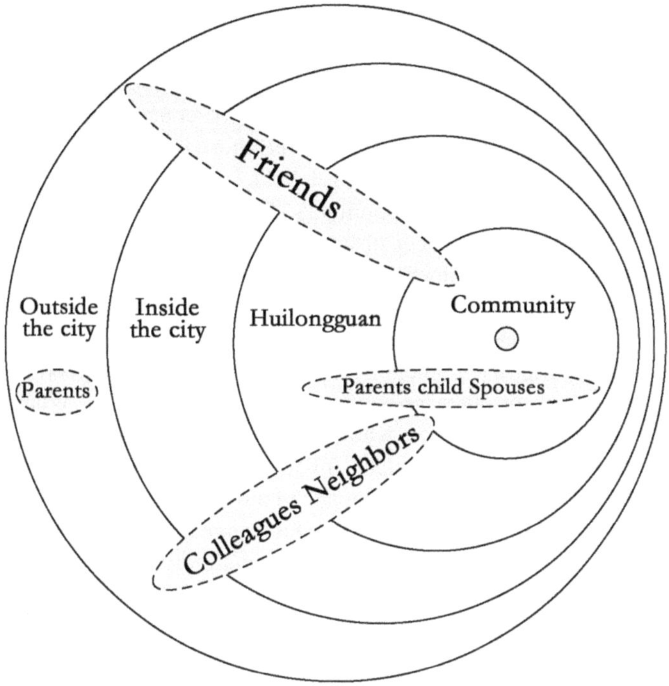

Fig. 6.16 Main types of relationships distributed at different spatial scales

6.3.2 Social Space Based on Interpersonal Networks

The study has discovered that for the residents of Huilongguan, an urban suburban community, on one hand, their relationships with their closest family members are becoming increasingly intimate. On the other hand, they exhibit a strong reliance on friends whom they have encountered either through shared hobbies and interests or by serendipity. The relationship categories of the personas within the Huilongguan residents' Important Issues Discussion Network are predominantly composed of friends, followed by family members such as parents, children, and spouses, and then by coworkers. Qualitative interviews have further revealed that the family remains the primary source of support for individuals when significant events take place. This is particularly true in metropolitan areas, where the growing detachment of small families from the broader society has rendered the family even more crucial. Moreover, any family, when considered as a social unit, also serves as a component of the territorial community. Simultaneously, the significance of friendships forged through "shared interests" and "work" is underscored. When urban dwellers seek advice on important matters, the number of like-minded and close friends formed through "interest connections" and "work connections" often outweighs that of family members and immediate kin. However, the order of importance does not imply a complete substitution or exclusion of family relationships. In large cities, individuals are more inclined to engage in discussions about important issues with a homogeneous group of people. This might be attributed to the fact that people of the same age and background tend to share more common concerns, making it easier to identify with topics and evoke empathy, and thus more likely to elicit a resonance.

The inclusion of neighborhood object relations in the residents' important issues discussion network suggests that a community atmosphere has gradually taken shape within the neighborhood. It has been observed that for the respondents who incorporate their neighbors into this important issues discussion network, the neighborhood holds a relatively high level of significance in their lives. The majority of the corresponding research respondents are housewives and retired elderly individuals, and among them, there are also some who have connections with people from their hometown, coworkers, or other types of relationships interwoven. When analyzing the relationships between neighbors who are simply connected by their shared neighborhood, it is evident that the individuals involved generally have a longer residence duration within the community. They engage in more frequent interactions with one another in their daily lives, and there is a greater degree of overlap in their utilization of public spaces. For them, the life within the neighborhood constitutes a crucial aspect of their overall lives. They maintain more amicable relationships with their neighbors, and those who communicate with each other are capable of further deepening their interactions. During the interaction process, the main conversations and the topics covered are relatively common and on the surface. For instance, they often involve matters at home (especially regarding children), one's own physical state, hobbies, as well as news and events within the community. The presence of green leisure areas, fitness and exercise zones in the neighborhood, along with the parks and public spaces in the community, offers the essential spatial environment for these activities and interactions. This not only facilitates interpersonal exchanges, enabling the formation of relatively stable relationships, but also plays a role in mediating social relations to some degree. Neighborhoods established due to geographical proximity contribute, to a certain extent, to the social network of the residents. This also suggests that the respondents have gradually assimilated into the local community.

As the objects are sequenced in order, relationships tend to become more distant, and there exists a pattern of suboptimal ordering. The mean size of the community residents' discussion network concerning important issues is 3.25, and with a relationship strength of 1.80, it implies a high level of mutual familiarity among the members. The lower the ranking of a relationship, the shorter the duration of acquaintance and the less frequent the interactions. The important issues under discussion predominantly center around work, children's development, as well as friends and neighbors. The mode of contact gradually transitions from in-person meetings to telephone communication, and the circle of mutual acquaintance is progressively narrowing. The members of the discussion network have, for the most part, known each other for approximately 10 years, yet the significance of "new" acquaintances who have known each other for 5 years or less is quite noticeable. There is a substantial degree of similarity between the members of the discussion network and the respondents in terms of social attributes such as gender, age, educational attainment, and political affiliation. In other words, there is a high level of homogeneity and a low level of heterogeneity.

The separation in geographical space and the territorial characteristics of social settings carry significant implications for the individual social networks. There exists a correlation between the clustering pattern of residential locations and the intimacy of relationships. As such, spatial distance serves as a crucial avenue for the study of social bonds. Spatial proximity or clustering contributes to the reinforcement of social networks. For example, face-to-face interactions remain of great importance for mutual engagement, and even relatively loose networks can be strengthened by the factor of spatial proximity. Conversely, spatial distance can exert a negative influence on the maintenance of pre-established relationships. As Fei (2005) noted, "It is not easy to maintain the normal relationship of a couple in a state of separation. A father and son who spend time together day after day and a father and son who are separated by thousands of miles of clouds and mountains are not different in terms of their social status, but the difference in their real-life relationship can be very great." To a certain degree, the network of relationships is still restricted by spatial factors. In terms of directionality, the more distant the relationships are, the greater the spatial separation tends to be, and the farther apart the residents live from one another. Evidently, from the statistics table of the concentration trend, the spatial residence pattern of the nominated relationships follows an evolutionary path of "within the neighborhood > within Huilongguan > in the city > outside the city", with the spatial distances between the subjects gradually increasing (Fig. 6.17). When combined with the observation that relationships are becoming more distant, it can be hypothesized that there is a discernible correlation between the spatial distance of relationships and the degree of relationship proximity.

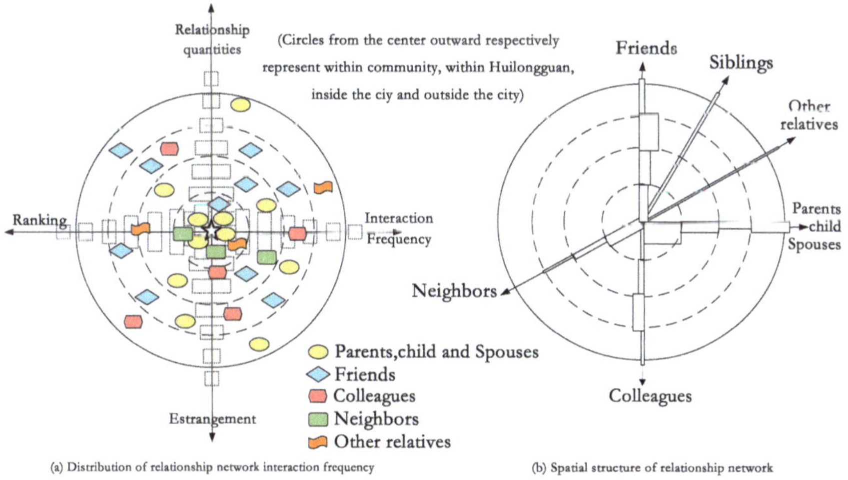

Fig. 6.17 Relationship networks of residents and spatial distribution

Regarding gender differences, notable disparities were observed between male and female respondents. In terms of the similarity of the examined indicators (refer to Table 6.15), male respondents demonstrated a higher level of homogeneity compared to females when it came to the characteristics of network members. Males tend to place greater emphasis on friendships and professional relationships with colleagues, whereas females show a stronger focus on family ties and neighborhood connections. Women exhibit a more intense emotional attachment to their long-standing friends than men do. Additionally, there is a more distinct dichotomy in the duration of acquaintance with their network members among women. At the same time, the polarization of spatial distances is more evident among female respondents than among their male counterparts.

6.3.3 Summary and Discussion

The empirical study grounded in the social relationship network of residents in the Huilongguan residential area of Beijing mirrors, to a certain degree, the characteristics of the social relationship networks of residents in residential areas of large Chinese cities amidst the backdrop of rapid urbanization. This social relationship network manifests the spatial attributes of residents' lives that are centered around social interactions. The social spatial scales that they care about not only encompass the family, the neighborhood, and the community but also extend beyond the community's boundaries to other functional areas within the city. The theoretical significance of this research is as follows: Firstly, since the reform of China's housing system in the late 1990s, spurred by the flourishing urban real estate market, a substantial number of urban residential areas have been constructed in major Chinese cities. The majority of these residential areas are situated in the suburbs and serve as the living spaces for a diverse array of individuals. The residents include traditional urban dwellers who have relocated from the city center or old urban areas due to the redevelopment of old cities, former rural residents who have migrated to the city for employment and home purchases as a result of urbanization, people from rural areas or towns in other regions who are drawn to the city, as well as the general wage earners among the urban household registration population who have bought policy-supported housing, and so on. These complex groups of people are confronted with the challenge of reconstituting the fabric of their social relations within the new residential areas. In other words, they are tasked with reconstructing their social networks in these new neighborhoods. Therefore, after over a decade of community development and construction, carrying out empirical research on large-scale residential areas in Chinese cities such as Huilongguan can be instrumental in uncovering the characteristics of the social relationship network of this kind of community. This can serve as a response to the relevant sociological theories about communities (which are mostly applicable to traditional rural communities). Secondly, from the viewpoint of urban geography, when it comes to the study of the theoretical issue regarding the connection between social relations and space in residential areas,

Table 6.15 Differences in social relationship networks between males and females

Gender	Similarity of network members	Emphasis on relationship types	Emotional connection with old acquaintances	Duration of acquaintance among network members	Distance of residence among network members
Male	Relatively high	Place more emphasis on friendships and colleague relationships	Relatively low	Relatively concentrated, at a moderate—length duration	The proportion of those living within Huilongguan and in the same city is relatively high
Female	Relatively low	Place more emphasis on family and neighborhood relationships	Relatively high	The proportion of those acquainted for relatively short and long durations is relatively high	The proportion of those living within the community and from other places is relatively high

large-scale residential areas like Huilongguan offer abundant and vivid cases for the academic community. They can contribute to addressing the "socio-spatial dialectic," a crucial theory in international urban social geography. Moreover, they can provide a practical basis for the application of this theory within the context of China's rapid urbanization process.

Research findings have demonstrated that there exists a cohesive dialectical relationship between the network of interpersonal social relations in large urban residential areas and space, and they are engaged in an ongoing process of mutual interaction. On one hand, social relations are subject to the constraints of space, and social space, in turn, plays a mediating role in social relations. Through empirical inquiries, it has been discovered that although people's social relationship networks in the city still reflect the connections of non-spatial bonds such as shared interests and kinship, the constraints and impacts of space on the relationships among residents are also remarkably evident in this context. Proximity in location exerts an influence on the interactions among residents and the establishment of their social relationship networks, and the spatial differential pattern of these relationships persists. Physical environments and spatial distances can hinder certain behaviors of residents and facilitate or stifle the development of certain relationships. Space is to some extent the container of social relationships. To some extent, this validates previous theoretical assertions about the "socio-spatial dialectic" (Soja 1980; Knox and Pinch 2000). On the other hand, interpersonal networks play a pivotal role in shaping and influencing space. Bourdieu (1989) posits that space is, in essence, a system of relationships rather than merely a geometric construct of physical space. The distances between spatial entities are interconnected with and interact with the distances in social relationships. Residents are partitioned into distinct social spaces according to the total quantity of capital they hold and the structure of that capital. The social-relations network is an important form of social capital in its own right. As residents engage in living and working within the urban space, they gradually exert an impact on the environment. They adjust and modify it to the fullest extent possible to satisfy their needs and mirror their values, ultimately giving form to and sculpting the urban space in which they reside. Simultaneously, however, the inhabitants themselves gradually acclimatize to their physical surroundings and the people in their vicinity, in a two-way process.

6.4 Virtual Space: Social Interaction Based on Community Internet

6.4.1 Virtual Space: The Impact of the Internet on Interpersonal Relationships

The rapid evolution of modern information and communication technologies (ICT), particularly the Internet, has precipitated a radical transformation of people's lifestyles. As a result, human interactions are no longer restricted to geographical and blood-related ties. Instead, borderless virtual communities have emerged in cyberspace, founded upon the interests, individual desires, and motivations of numerous Internet users. With technological advancements, compact communities have been established, and the high velocity of mobility has rendered people less concerned, both mentally and emotionally, with the restoration of the nominal integrity of cities and neighborhoods (Porterfield and Hall 2003). Individual communities are no longer bound by the territorial limitations of neighborhood and spatial proximity. Since the end of the Second World War, and especially from the 1980s onwards, Western academics have increasingly focused on the study of virtual communities. When comparing virtual social relations to traditional social life, three distinct perspectives have emerged: engulfment, symbiosis, and remodeling. The engulfment theory posits that individuals in real-life settings are increasingly constrained and dissolved by the symbolic world constructed by humans. Consequently, the boundary between the virtual and the real becomes blurred, and the subjectivity dissipates (Ye 1998), ultimately leading to the virtual society supplanting the real one. Symbiosis, on the other hand, implies the co-existence of virtual and real societies, with the virtual society becoming an integral part of daily life. The remodeling theory suggests that Internetbased information has further transformed social structures and processes. Virtual communities have augmented the diversity of communities and social connections (Huang 2010). In general, Internet users within virtual communities exhibit a higher degree of heterogeneity. First, the high accessibility of virtual communities results in an extremely diverse composition of Internet users in terms of social class, age, gender, ethnicity, and religion. Second, in contrast to face-to-face interactions where there is a degree of mutual wariness, online interactions are transient and accidental, and the accountability of individuals within virtual communities is circumscribed. Third, on the Internet, conversation topics are fragmented rather than concentrated on more routine and enduring themes. As a consequence, Internet users lack a sense of trust and closeness towards one another. Although the Internet undeniably surpasses physical space in the ease of information retrieval, there is concern that the pursuit of empathy and support within virtual communities may prove disappointing.

Giddens (2003) has encapsulated two contrasting perspectives regarding the influence of the Internet on interpersonal relationships. One perspective posits that the Internet serves to enhance or complement existing face-to-face interactions. It enables individuals to communicate with family members and friends who are geographically distant. Simultaneously, it facilitates the establishment of new interactions, thereby expanding and enriching people's social networks. Conversely, the other view contends that excessive immersion in online communication will inevitably result in a diminution of face-to-face interactions and interpersonal connections in the real world. This, in turn, can damage relationships and exacerbate social isolation and the phenomenon of "atomization." A randomized survey conducted among American adults revealed a notable correlation. Specifically, the greater the amount of time individuals spent online, the more they became detached from their social environment. Notably, 15% of the respondents reported a decrease in their social activities (Norman and Erbring 2002).

However, through the investigation and research into the relationship network and neighborhood relations of residents in the Huilongguan community, it has been discovered that the community Internet occupies a significant position in the lives of residents and exerts a substantial influence on the overall community's sense of recognition and belonging. Many interviewees frequently and inadvertently mentioned the Huilongguan community network (www.hlgnet.com), as if the Huilongguan community entity manifested itself on the Internet. Therefore, this section will attempt to explore Internetbased social interactions, using the case of Huilongguan as a basis.

6.4.2 The Neighborhoods and Communities Based on Community Internet

On March 9, 2000, the Huilongguan Community Network website (www.hlgnet.com) was officially launched for the first time. It is reported that the founder of the Huilongguan Community Network website was a resident of Longxiyuan in the second phase of Huilongguan. Initially, the community network originated as his personal homepage. Since its first stable version was established on October 7, 2001, the community network has undergone three rounds of revisions. Throughout its development, the Huilongguan Community Website has consistently remained a public welfare website, which was voluntarily established by the neighborhood residents. By the beginning of 2011, the website had amassed over 330,000 registered users. This user base included individuals who had relocated out of the neighborhood, those who intended to move into the neighborhood, as well as those who simply wished to participate in community activities. The website's average daily IP traffic reached 176,000 visits, and the average daily page views (PV) exceeded 700,000. Globally, it ranked 5,848th, and in China, it ranked 677th. Significantly, its number of visits far surpassed those of the renowned large-scale residential areas in Beijing, such as Tiantongyuan and Wangjing communities. As a result, in terms of the number of visits and page views, it became the largest community website in Asia (Table 6.16).

On the community Internet, there are frequent discussions, interactions, as well as buying, selling, and purchasing activities that are closely intertwined with residents'

Table 6.16 Comparison of the traffic of Huilongguan community network with that of Tiantongyuan community network and Wangjing network

Website	Web address	Average daily IP traffic	Average daily PV traffic	Global ranking	Chinese—language ranking
Tiantongyuan community network	www.ttysq.com	6600	22,440	118,484	14,744
Wangjing network	www.wangjing.cn	23,400	191,880	34,913	3731
Huilongguan community network	www.hlgnet.com	175,800	703,200	5848	677

Data source Statistics from the third—party ranking website alexa, web address http://alexa.chinaz.com/; The statistics were made at the beginning of 2011, and the same below

daily lives. As of the year of the survey, the Huilongguan community website features 11 boards. These include Life Tips, Activity Notices, Life Manuals, Good People and Good Deeds, Exposure Platform, Trading Market, News Center, Life Map, etc., offering rich and diverse content. A statistical analysis of the boards browsed by netizens (Table 6.17) reveals that the main focus lies in the discussion and interaction category. This fully reflects the characteristics of a community BBS website. The buying, selling, and purchasing category (encompassing second—hand trading and group buying) follows closely, with a high proportion of visits, indicating its strong connection to residents' daily living requirements.

Table 6.17 Traffic distribution of each section on Huilongguan community network

Category	Sub—website address	Proportion of page views in the recent month/%	Average number of page views per person/time
Discussion and interaction category	bbs.hlgnet.com	54.72	2.15
Buying and selling category	sales.hlgnet.com, jc.hlgnet.com	23.48	14.60
Search and service category	search.hlgnet.com, fuwu.hlgnet.com	8.52	11.70
Community service category	news.hlgnet.com, guide.hlgnet.com, map.hlgnet.com	11.40	7.64
Others	photo.hlgnet.com, etc.	1.88	6.00
Total	–	100.00	–

Data source Compiled from the third—party statistics website alexa

The presence of an owners' association indicates a certain level of regulation within the online community. The website's management and its income—expenditure matters are determined through votes in the elected owners' meeting. The election of the owners' meeting is carried out by registered users, who all possess equal voting rights. In the website's first election in 2002, 30 representatives of the owners' meeting were elected for a one-year term, and another election is held at the end of each term. The website's financial expenditures are reported to the owners' meeting every six months. Regular meetings are convened to deliberate on the website's development and discuss users' suggestions. These include aspects such as website maintenance, equipment replacement, and external donations of the website, all of which are subject to voting procedures. After several years of operation, the website has started generating profits, and all advertising revenues are allocated to website construction and netizens' activities.

In Huilongguan, residents typically exhibit a high level of trust in the community Internet, acknowledging its substantial benefits for neighborhood interactions. An analysis of the questionnaire survey data indicates that 70% of the residents were aware of the community Internet, and the rates of knowledge and usage were significant. The primary means by which residents learned about the existence of the community Internet was through online searches (48%). Family members, relatives, friends, classmates, and neighbors also played a role in spreading awareness, and to some extent, leaflets and posters posted on the community Internet contributed as well. The frequency of community network usage was high, with one—fifth of the residents browsing it daily. Additionally, 22% of the residents had encountered people in real life whom they first got to know online and had established good relationships with. Among these, many were fellow soccer enthusiasts or parents sharing child-rearing experiences.

6.4.3 The Role of Virtual Space in the Social Interaction of Community Residents

6.4.3.1 Virtual Space Provides a Platform for Community Residents to Communicate and Help Each Other

Building on the previous analysis of actual neighborhood relations, we now shift our focus to the perspective of virtual space. Specifically, we select the "Good Deeds and Good People" column on the Huilongguan community website to explore the new forms of neighborhood relations. The "Good Deeds and Good People" column was inaugurated on the Huilongguan community website on April 3, 2002, enabling netizens to post online thank-you letters. By May 1, 2011, a total of 4,630 messages had been accumulated (Table 6.18). Both the senders of these messages (posters) and the beneficiaries (recipients) expressed great satisfaction with this initiative. In this column, both posters and recipients are identified by nicknames. The content of the messages is highly detailed, vividly portraying scenes of amicable neighborly interactions.

Neighborhood residents can enhance the sense of neighborhood identity through mutual assistance. The advent of community networks has enabled residents to avoid the awkwardness of rejection. They can post notices on the Internet, asking those in need to take items they no longer require. When faced with minor, unsolvable problems, they can also post inquiries online, and interested parties will offer unpaid assistance. By integrating the relationship between virtual networks and physical spaces, the gratitude events in the "Good Deeds and Good People" section can be analyzed and categorized into three types (Table 6.19). First, there are "anonymous acquaintances." These are mainly online and transient interactions, exemplified by the gifting of toys or small items for young children. Although some offline meetings may be involved, the contact is relatively preliminary and short-lived. Second, are events initiated by face-to-face interactions in the physical space and further strengthened online, with temporary contacts. In this category, actions such as temporarily giving items to neighbors or returning found money, which start from face-to-face encounters in the real world, are deepened through online communication. Third, there are cases with close contact both in reality and online, where neighbors have developed a friendship-like relationship, often involving assistance during major emergencies. Evidently, there is a strong connection between the online platform and the real space, with each influencing and promoting the other.

Although these acts of neighborhood support are seemingly trivial, their presence has fostered a warm ambiance of assistance and mutual aid within Huilongguan. Numerous residents have been profoundly moved by the help extended to them by their fellow community members, whether these helpers were known or unknown, and whether the assistance was offered online or in person. In response, they have expressed their gratitude on the community Internet and pledged to continue disseminating and promoting the community spirit (Fig. 6.18). Evidently, the community Internet serves as a crucial platform for cultivating and sustaining a positive atmosphere of mutual support within the community.

Table 6.18 Statistics on the number of posts in the "Good Deeds and Good People" column of Huilongguan community network

Year	2002a	2003	2004	2005	2006	2007	2008	2009	2010	2011b	Total
Number of posts	264	331	418	950	869	580	526	396	224	72	4630
Percentage/%	5.70	7.15	9.03	20.52	18.77	12.53	11.36	8.55	4.84	1.55	100.00

2002a: Posts in 2002 started from April 3rd; 2011b: Posts in 2011 were counted until May 1st

Table 6.19 Types of "Good Deeds and Good People" columns based on community internet

	Characteristics	Content	Examples
The first type	Anonymous acquaintance, mainly online and temporary interaction	Sending small items like small gifts, small pets, clothes, etc. anonymously	"I'm at the Parent—Child Hut. My son has been having a bad cough recently. Fashion Xiaoyu 123 not only introduced folk remedies but also gave a pack of sandboard for my son to try. Thank you so much!"—No.4 Train (March 28, 2011) "Thank Lao Jia for sending us a little rabbit. My daughter loves it. Thanks to Jia's family."—June (March 29, 2011) "Thank Lao Xia for finding the accompaniment for the nursery rhyme for my daughter. I had an uncompleted task for days. I posted a message seeking the accompaniment with a try-it-on-the-community-website attitude. I'm so glad. Lao Xia, a netizen, sent me a really great accompaniment. Thank you! You've taken so much trouble! There are so many kind—hearted people in Huilongguan!"—Happykitty (December 13, 2010)
The second type	Face-to-face interaction offline, preliminary expression of gratitude online	Expressing gratitude for acts of kindness in real—life situations	"Thank Mr. Jiang from Dongya Shangbei for helping with moving things. Mr. Jiang from Building 4 of Dongya Shangbei and his girlfriend actively helped us move things. Thanks a lot. Wish you a prosperous life and career."—Siyizhai (March 7, 2011) "I was in a bad situation this morning. I was locked in the bedroom and couldn't get out. In Huilongguan Community Huibao Center, kind-hearted people contacted me and asked their husbands to rescue me. With the joint help of my husband and neighbors, I finally fixed the lock of my room. Through this platform, I especially thank the netizen sister "Lvye Tong Tu Ma Dian" and her husband, and also thank the neighbors of the community. Wish you all a safe and happy life. I also thank the Huilongguan Community website. It makes me feel at home in our Huilongguan neighborhood."—Kaku Automobile Supplies Store (March 11, 2011) "Thank Mr. Chen and Mr. Peng for not pocketing the found items. They returned the lost ID cards and other items to me. I'm very grateful to them. What's more precious is that they found me through various contact numbers. I'm so touched. I feel that Huilongguan Community is really warm and there are so many kind—hearted people. I willalso learn from them and do more good deeds to make Huilongguan Community warmer!"—Momoma (January 22, 2011)

(continued)

Table 6.19 (continued)

	Characteristics	Content	Examples
The third type	Close connection, in—depth interaction, with neighbors having a high—degree of closeness, using major sudden events as the carrier for assistance	Expressing gratitude for deep—level help among neighbors	"Thank the Wolf of the North. When I was on a business trip, he sent my wife, who was about to give birth, to the hospital."—Sanman Duo 1985 (January 2, 2011) "A few days ago, the baby was ill. The baby's father was busy at work. Thank xushi for helping to make an appointment at the community hospital. I also made many phone consultations before. The attitude was always very good, and the answers were detailed. Thanks!"——Little Carrot (March 4, 2011) "Thank the harmonious "Cat Eye" and her husband for sending my little child to the Third Hospital for emergency treatment. At around 5 a.m. on December 21, my little child suddenly fell ill and needed to be sent to the hospital urgently. In a hurry, I called "Cat Eye". Honestly speaking, the husband and wife of "Cat Eye" got up without hesitation and drove my little child to the Third Hospital for emergency treatment."—Autumn Wind (December 21, 2010)

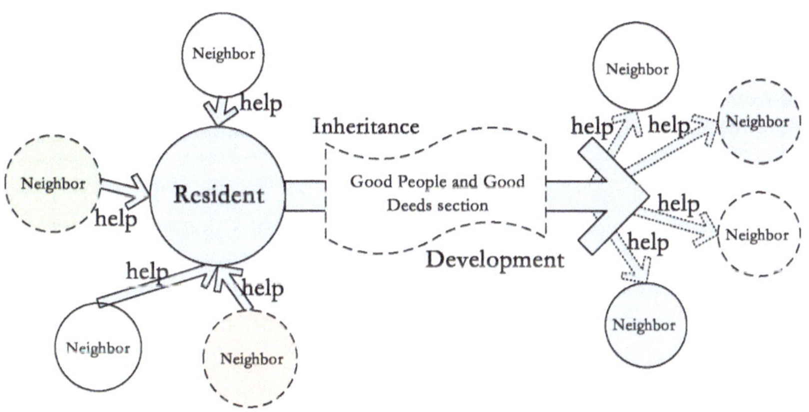

Fig. 6.18 Diagram of neighborhood mutual assistance through community Internet

6.4.3.2 The Features of Temporal and Spatial Dissociation in Virtual Space Create a New Paradigm of Multi-spatial Spectatorship

This book delves into a typical argument centered around the "conflict between mother-in-law and daughter-in-law". The incident originated on the morning of March 24, 2011. A netizen named Lanyunmeigui posted a message on the Huilongguan Community Network. In the post, she described the strained relationship between one of her neighbors and the neighbor's mother-in-law. She also offered her commentary, opining that the neighbor had not managed the situation adeptly and that she herself would have handled it more effectively if faced with the same circumstances. Unexpectedly, the concerned party, namely the neighbor with the screen name Keliy, came across this post on the community website and posted a clarification. In her response, Keliy detailed her own grievances and recounted how her mother-in-law treated others harshly, aiming to present the true facts. Consequently, within a single day, the post attracted nearly 10,000 views and elicited several thousand comments, sparking a significant debate on the community website. The discussion primarily unfolded from two perspectives. One focused on Keliy's mother-in-law and daughter-in-law issues, while the other centered on Lanyunmeigui's act of publicly sharing her neighbor's personal affairs on the Internet.

Keliy posted about the grievances she endured in her interactions with her mother-in-law. Subsequently, a flurry of responses ensued. Some individuals expressed empathy and support for Keliy, others engaged in rational analysis, some offered words of consolation, while a few made derogatory remarks. The back-and-forth exchanges among them created an online rendition of a neighborhood farce. After Keliy's post, the majority of the community network followers demonstrated understanding of her situation. They noted that in contemporary society, the mother-in-law-daughter-in-law dynamic is notoriously challenging to navigate. Many left messages for Keliy, exhorting her to approach the situation rationally and strive for a better life. Conversely, some netizens held a different view from the poster. They contended that Keliy's words betrayed a lack of open-mindedness and that she bore primary responsibility for mismanaging her relationship with her mother-in-law. Moreover, there were onlookers who exhibited indifference. They believed that each family has its own private affairs and that getting involved was pointless, choosing instead to remain as detached spectators.

The reactions of the readers of this article diverge significantly due to Lanyunmeigui's act of gossiping about her real-life neighbor's domestic affairs on the community network. One group of people argued that she was attempting to resolve neighborhood disputes by discussing and seeking solutions with everyone, and thus her actions deserved recognition and praise. Another group opined that since it was a private matter of the neighbor's household, it was inappropriate to casually post it on the Internet. They also suggested that "Guanwang (the Huilongguan community network) is not a completely virtual realm," emphasizing that neighborly matters should not be publicly discussed. Additionally, there was a segment of people who believed that private neighborly affairs simply should not be posted on the Internet at all.

The conclusion of the heated discussion sparked by the mother-in-law-daughter-in-law feud incident was that the protagonist, Keliy, expressed that she was "deeply hurt" and made the final follow-up post, titled "the real last reply." In this post, she hoped to distance herself from the controversy and put an end to any further discussion. Lanyunmeigui's original post was completely deleted, leaving no trace of its origin, while only the subsequent follow-up posts of others awaited deletion. It is reasonable to assume that the friendship between Keliy and Lanyunmeigui will experience a significant setback, and their neighborhood relationship will turn extremely delicate.

Internet-based networks possess characteristics of virtuality, latency, and timespace dissociation, thereby creating new neighborhoods and interaction spaces for community residents. The advent of such networks accentuates anonymity and virtuality. The nature of Internet message—following enables people's reactions to the same event to persist over an extended period. Responses are not necessarily immediate; there is a certain time lag, which reflects a novel mode of interpersonal communication. In the pre-network era, numerous events demanded the synchronization of time and space. However, the network's features of time delay and shared platforms have severely disrupted the original close integration of time and space. Modern—day neighborhood disputes differ from those in the traditional rural society. Most urban residents scarcely have the leisure time or inclination to deliberate on neighborhood conflicts. Conversely, the spatial—temporal fragmentation of social networks allows residents to capitalize on "time fragments" and the unbounded spatial conditions to post and follow up at any time and place. To some extent, this has transformed the way urban community residents utilize time and space for interpersonal interactions. The discussion ignited by the mother-in-law-daughter-in-law conflict on the Huilongguan community network serves as a typical example. It demonstrates that modern neighborhoods and communities are manifesting a new form of social interaction under the influence of new technologies like the Internet.

Residents in real-life inhabit a specific physical space. However, the community network has a unique nature as it is closely associated with the community's geographical area, essentially constituting a semi-virtual community. From the participants in the discussion of the incident, it becomes evident that virtual networks, particularly community networks, are not entirely virtual and anonymous. The virtual network reorders, extends, and broadens the physical space. Simultaneously, it is constrained by the physical space. The roles and spatial domains in the real world impose restrictions and regulations on residents' behaviors within the virtual network (Fig. 6.19). This also enhances residents' sense of responsibility. Moreover, the communication conditions, the speech environment, and the public nature of the network render incidents less manageable. For incidents that escalate, they may even seep back into the residents' real-life physical spaces, ceasing to be completely virtual and anonymous. For those involved, this can exert an influence, or even have a negative impact, on their real-life interpersonal relationships and social interactions.

6.4.3.3 Virtual Space Mediates Community Residents' Group Activities

The Internet hosts a plethora of neighborhood cultural and sports activities. Based on incomplete statistics, the residents of Huilongguan Community have organized dozens of associations, predominantly of a cultural and sports nature. These include the Huilongguan Community Football Association, Tennis Association, Badminton Association, Swimming Team, Mountaineering Association, and the Wild Boar

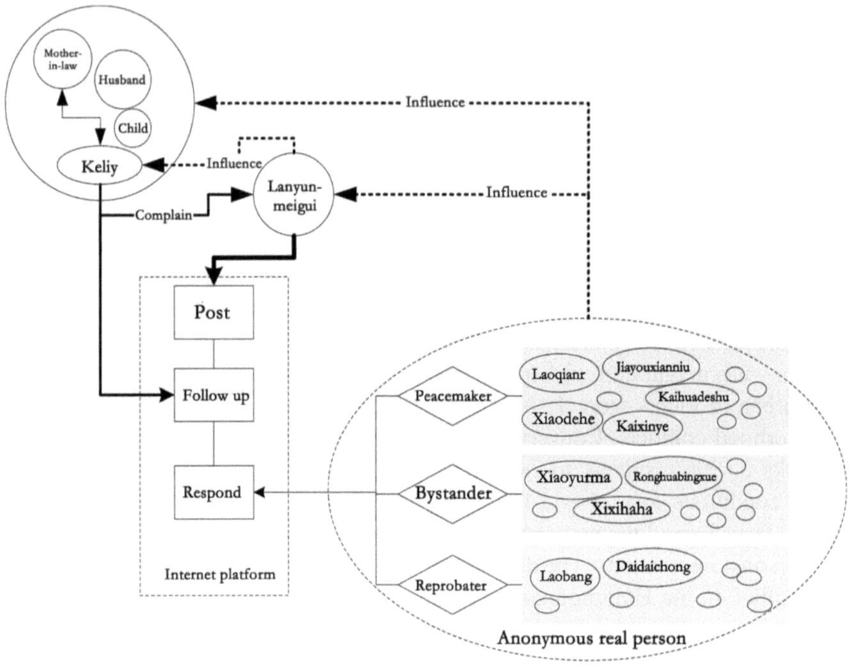

Fig. 6.19 Discussions on the incident of mother-in-law-daughter-in-law conflict

Drama Club, among others. Notably, the Huilongguan Super League (HSL) stands as the largest community football league in the country and has made a significant impact. Since 2002, when the first amateur football team, the Wild Boar Forest Football Team, was founded by the Huilongguan homeowners, football has seen continuous growth among the residents. In 2004, the community organized the first Huilongguan Owners' Football League, setting a precedent in China for the scale, institutionalization, and public-welfare nature of community sports. Renowned as the most popular grassroots league in Beijing, the Hui Super League boasts a level of standardization comparable to that of professional leagues. This event has had a profound influence on neighborhood relations and the overall community's sense of belonging.

The establishment of the Huilongguan Community Network has provided an ideal platform for shared social life. The members of the Huilongguan soccer teams are all netizens who got to know one another via the community network platform. Both the team names and the players' identities are commonly referred to by their online monikers. Names like Fatty Cow, Guoqing, and Gardener frequently surface among the players. These players are homeowners from various local neighborhoods in Huilongguan, each with their own occupations and diverse professional backgrounds. However, through the Hui Super League, they have achieved the aim of making friends and connecting with each other through football. Currently, there are over 20 teams. The annual Hui Super League attracts more than 100 participants and

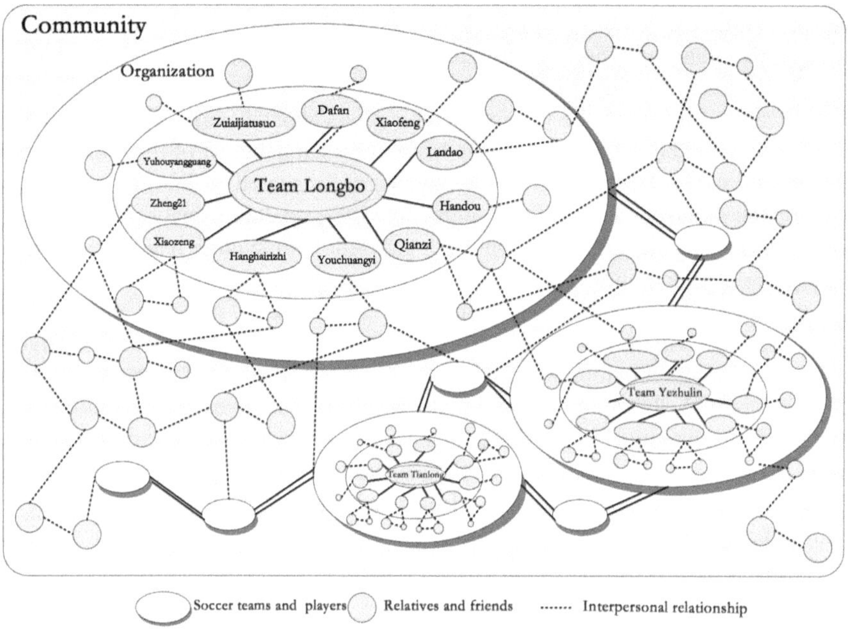

Fig. 6.20 Sports activities of Huilongguan Super League based on community network

10,000 on—site spectators. In 2010, the seventh Hui Super League had nearly 800 registered players, and a total of 210 matches were held throughout the season. The activities in the physical space, in turn, trigger interactions among residents who first met in the virtual space (Fig. 6.20). In 2005, when the Tianlong team vied with the previous champion Boarwood team for a spot in the final, the Tianlong team exerted all their efforts to win the championship. According to the staff's account, "a dozen adults hugged each other and cried, something they had never experienced since starting their careers."[1] Moreover, during each game, many family members bring their babies to the sidelines to watch their fathers play soccer. This offers an excellent opportunity for neighborhood interactions. The convergence of spatial—temporal events and the residents' willingness ultimately contributes to the formation of a harmonious neighborhood community atmosphere.

Physical activities are given added impetus in the virtual space. Post—event, participants or organizers subsequently post summaries of relevant activities and photos of residents' participation scenes on the community network (including activities like voting for soccer babes). Participants respond actively to these postings, attracting the attention of non-participants. All activities taking place in the physical space can be found on the relevant boards of the community network, along with on-site photos. As the scale of these activities expands and habitual patterns are established, with the website serving as the core of the public platform, the online

[1] Cao Xiaofang. The road of HSL. http: //blog.sina.com.cn/s/blog_4a075838010005et.html.2005.7.

medium further magnifies the influence of neighborhood—related activities. This, in turn, heightens the sense of belonging and reliance among community residents.

The Huilongguan Super League represents the "organization" within residents' daily lives and is one of the largest social capitals spontaneously formed and held by Huilongguan homeowners. On one hand, the establishment of the Huilongguan community network constructs an ideal platform for shared social life. It summons residents to partake in cultural and sports activities, gradually enabling these activities to grow in scale and become more organized. On the other hand, the development of related activities in the physical space reciprocally promotes and fortifies interpersonal interactions in the virtual space. Eventually, physical activities are further enhanced in the virtual space. In this process, people's eagerness to engage in cultural and sports activities plays a crucial role. When involved in sports and cultural pursuits, individuals are at ease and joyous, and their motives are straightforward, without being entangled with professional status or background. The players participate in these activities for physical fitness and relaxation (as the HSL advocates, "Let football revert to football, and let the people revert to themselves"), fostering a more harmonious overall atmosphere. People who have previously encountered each other during such events will feel an even closer connection if they meet again within the neighborhood.

6.4.4 Interaction Between Virtual Space and Physical Space

The Internet functions as a potent communication instrument, enabling the transfer of information between individuals who are geographically distant. Nevertheless, the mere flow of information does not inherently foster genuine community. It can merely be stated that the Internet has engendered larger and more efficient social networks, thereby strengthening the bonds between residents and the community. Simultaneously, the virtual space harbors certain negative implications. New acquaintances made online lack a shared social background and a solid communication foundation. Anonymity and the high mobility of online interactions allow people to enter and exit with ease, passing by each other hastily. As a result, aspects such as commitment, trust, and mutual assistance struggle to materialize, and instances of deception and betrayal are not infrequent. This, to some degree, elucidates the indispensability of face-to-face communication. Furthermore, the Internet presents the opportunity to become aware of the homogeneity of the population and geographical constraints, among other factors.

In a virtual community, regardless of their real-life status, every resident enjoys equal rights to participate, express their opinions, and receive timely responses. They also experience the pleasure and satisfaction that come from being respected. Practical and user-friendly community websites offer a platform for homeowners to communicate and exchange ideas, giving rise to a new way of life and bringing about significant changes in the values of community residents. Community Internet exhibits some of the features typical of the Internet, such as anonymity, latency, and the dissociation of time and space. However, it also has a certain territorial entity spatiality, and there is a high degree of correlation between online behaviors and real spatial behaviors, which can even be superimposed. The acquaintance of Internet users is based on the anonymity of network nicknames, endowing it with a certain virtual quality. During the process of organizing activities and offline interactions, residents refer to each other by their nicknames. Many individuals are unaware of the real names and workplaces of those they frequently engage in activities with. The anonymity of the network, on one hand, reveals the virtual nature of the Internet, and on the other hand, it accentuates a sense of trust and intimacy among users.

It is indeed noteworthy that the emergence and utilization of certain special terms vividly showcase the unique cultural connotations of the community. The Huilongguan community network has its own set of distinctive language. Homeowners often use the term "Yezhu (wild boar)" for self-reference, and the community network is titled "Yezhulin (wild boar forest)". "Lanzhu (blue pig)" is used to refer to male homeowners, while "Fenzhu (pink pig)" denotes female homeowners. The term "class leader" is employed to signify a leader or boss. For instance, "Sixth Class Leader" serves as an honorific for Qiang Liu, the founder of the website. "FB" is the pinyin abbreviation of the Chinese word for "corruption", but here it actually refers to the spontaneous gatherings of Huilongguan residents for activities like dining, drinking, and entertainment. "Dragon Coin" is the virtual currency that circulates on the Huilongguan community website. The more frequently one posts online and the more popular those posts are, the greater the number of Dragon Coins one will accumulate. These coins also serve as a symbol of one's seniority and status within the community and can be used in real-life situations to bid for items during various FB auctions.

There exists a superposition of the virtual and real spaces, leading to frequent interactions among residents both online and offline. The community website has emerged as the primary space for homeowners' daily life. When homeowners register as members, they are required to ensure that their IP addresses are consistent with those of the residents on the floor where they reside. As a result, the communication among them is no longer merely a virtual and illusory interaction on the Internet. Instead, it is integrated with their actual daily lives and real individuals. In this regard, it can be asserted that the community network, which is underpinned by the neighborhood relationships in the physical space, has, to a certain degree, transformed into a physical space that harbors rich connotations.

The social life facilitated by the community's Internet has played a distinctly positive role in the further integration and development of the community. It has given rise to a new type of neighborhood relationship and cultivated a strong sense of community belonging among residents. It is precisely these online and offline activities that shape the unique atmosphere of the Huilongguan community. Acts of helping others and mutual assistance are one manifestation, while neighborhood disputes and conflicts are also part of the picture. Additionally, community cultural and sports activities serve to deepen the interpersonal connections among people. The presence of the virtual space encourages residents to gain a deeper understanding of the physical community, gradually coalescing into a sense of belonging. As a novel platform and mode of communication, the content of community online communication complements the physical space and community autonomy, with the former exerting a substantial influence on the latter (Dong 2005). In the virtual space, a wealth of shared topics and common ground emerge among residents, further enhancing their sense of attachment and belonging to the community. For many of them, Huilongguan is not merely a place of residence; it has become their true "home" in a more profound sense.

The Internet, as a new means of communication, has indeed brought about novel forms of community. It enables people to establish social relationships that transcend neighborhoods, cities, regions, and even span across the globe, effectively turning the world into one large community. Nonetheless, this does not rule out the building of social networks and the organization of social activities at the local level. Instead, it exerts an effect of spatial and temporal reorganization known as "relocalization." The community Internet serves as an augmentation to the physical social space, facilitating the construction and development of the real social space (Fig. 6.21). This indicates that while the Internet expands social connections on a global scale, it also has the potential to strengthen local social ties and reshape the spatial and temporal dynamics of local communities. It acts as a bridge that combines the global and the local, enhancing the overall social fabric by integrating different levels of social interactions.

This process of social space construction can generally be categorized into three stages. In the initial stage, the community Internet acts as a platform and medium for neighborhood interactions and activities. It allows individuals to come into contact with one another and establish the first links of connection. During the second stage, the relationships and interactions forged in the virtual space are further solidified through face-to-face exchanges and interactions among people in the physical space. This real-world interaction deepens the bonds that were initially formed online. In

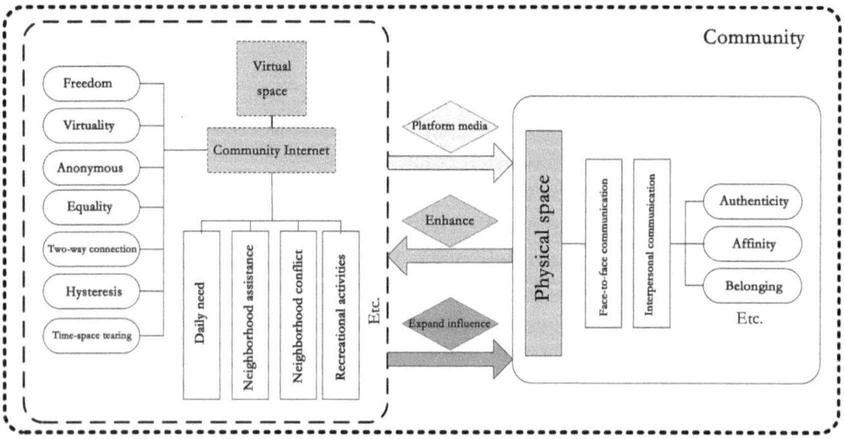

Fig. 6.21 The interaction and superposition mechanism between virtual and physical space

the third stage, through the virtual network, activities are further connected and reinforced, thereby expanding the influence of the activities and behaviors that occur in the physical space. It should be noted that these three stages are not hierarchical but rather mutually reinforcing. The interplay between online and offline interactions, along with the overlap of virtual and physical spaces, creates the possibility for further enhancing the residents' perception of space. This integrated process enriches the social experience of the residents, making the community a more vibrant and connected entity.

6.4.5 Summary

The existence of the Internet, particularly the community Internet, exerts a profoundly significant influence on residents' social interactions and perceptions. In Huilongguan, residents typically exhibit a high level of trust in the community Internet and acknowledge the substantial advantages it brings to neighborhood interactions. The empirical research conducted in Huilongguan demonstrates that the virtual space plays a pivotal role in the social interactions of community residents. It offers a spatial platform for them to communicate and extend mutual assistance. Moreover, the temporal and spatial dissociation characteristics of the virtual space create the necessary conditions for multiple spatial perspectives. The virtual space also assumes the function of a spatial medium for organizing the activities of community resident groups. There is an overlapping phenomenon between the virtual space and the real space, with residents engaging in frequent interactions both online and offline. To this end, it can be asserted that the community network, which is underpinned by neighborhood relations in the physical space, has, to a certain degree, transformed into a physical space laden with rich connotations. From the standpoint of residents' living spaces, the most crucial significance of the Internet lies not in technology or information, but rather in its role as a means of communication. Practical and user-friendly community websites provide a venue for homeowners to communicate and exchange ideas, giving rise to a new way of life and bringing about remarkable changes in the values of community residents. In the virtual community, residents display qualities such as warmth, generosity, compassion, and a sense of responsibility. The community Internet not only inherits the characteristics of the Internet, such as anonymity, latency, and spatio-temporal dissociation, but also exhibits a certain geographical entity spatiality. There is a significant overlap and a high degree of correlation between online behaviors and those in the real space. New communication technologies typified by the Internet have given rise to novel forms of community. Nevertheless, this does not preclude the establishment of local social networks and the organization of social actions. Instead, it exerts a spatial and temporal reorganization effect known as "relocalization." The community Internet serves as a facilitator for the construction of the real social space. The interplay between online and offline interactions, along with the superimposition of the virtual and physical spaces, further enhances the residents' perception of their community space.

6.5 Chapter Summary

This chapter commences from the perspective of social interaction. Taking the Huilongguan residential area in Beijing as a case study, it endeavors to explore the characteristics of the social-relationship network among urban residents. By examining the state of neighborhood relationships, it examines the social-interaction space constructed by relationships within a specific geographical area. It then analyzes the geographical and relational attributes of the community, and probes into the linkages between the social space and various relationships. Ultimately, it deliberates on the influence and function of the community Internet, a modern information platform, on relational linkages, urban communities, and the interaction between virtual and physical spaces.

First, regarding social networks: Overall, the pattern of differentiated sequences of social relations networks of urban residents has grown more intricate. The family still serves as the primary pillar of support. Friendships forged through "interest connection" and "work connection" ties hold prominent significance. Geospatial separation and the territoriality of social facilities exert a substantial influence on personal social networks. Neighborhoods established due to geographical proximity contribute, to a certain degree, to the social networks of residents. The evolutionary trajectory of the spatial residence of nominees indicates a clear correlation between the spatial distance of the relationship's place of residence and the intimacy of the relationship. Specifically, the more distant the relationship, the farther apart the places of residence, and the greater the spatial distance between relationship members. Second, in the context of neighborhood relations: Huilongguan is a quintessential urban residential area, embodying the characteristics of a typical urban neighborhood, where interactions are more often marked by unfamiliarity and shallow interpersonal engagement. Consequently, a TIES model is constructed to circumscribe neighborhood relations, incorporating Time, Intention, Event, and Space factors. More significantly, underlying these constraints are urban lifestyles and the profound imprint of suburbanization. In urban residential communities, residents generally maintain relatively good relationships and engage in active interactions, demonstrating a robust sense of community and identity. The empirical research on Huilongguan reveals the characteristics of the formation process of new urban settlements in China. Here, the establishment of a shared territory, the development of social relations, the emergence of a common social life, the construction of a community culture, and the formation of a collective consciousness jointly constitute the mechanisms for the establishment and sustenance of a "new community." For the formation of an urban community in the true sense, a defined territory, along with the presence of diverse mechanisms and events, is essential. Among these, the relationships forged between residents play the most pivotal role.

Research on interpersonal networks and social space has demonstrated that social relations are circumscribed by space, while social space acts as a medium for social relations. Interpersonal networks play a crucial role in shaping and influencing space. There exists a dialectical and unified relationship between social interpersonal networks and space, with the two engaged in a continuous process of mutual interaction. Studies of neighborhoods and community space have revealed that space furnishes the essential foundation, and proximity reconfigures social space. Through relationship linkages, residents in the neighborhood initiate interactions. They regard the community as their own home, actively safeguard their rights, partake in diverse activities, and thus form harmonious neighborhood relations and a strong community awareness. Eventually, this gives rise to a sense of identity and belonging to the region, fostering a "placeness" sentiment. To a certain degree, this also reshapes the social space initially grounded in interpersonal networks. Research on the community Internet has indicated that its presence does not impede the building of local social networks and the organization of social actions. Instead, it brings about a spatial–temporal reorganization effect termed "relocalization." The community Internet serves as a catalyst for the construction of the real-world social space. The interplay between online and offline activities, along with the superposition of virtual and physical spaces, further enhances residents' perception of the space around them.

References

Bian Y (1994) Social networks and job search processes. Foreign Sociol 4:6–7 (in Chinese)

Bourdieu P (1983) Forms of capital. In: Richardson JG (ed) Handbook of theory and research for the sociology of educations. Greenwood Press, New York

Bourdieu P (1989) Social space and symbolic power. Sociol Theory 7(1):14–25

Burt RS (1984) Network items and the general social survey. Soc Netw 6(4):293–339

Burt RS (1992) Structural holes: the social structure of competition. Harvard University Press, Cambridge

Chen CC, Chen YR, Xin K (2004) Guanxi practices and trust in management: a procedural justice perspective. Organ Sci 15(2):200–209

Cheng Y, Zhou M (1998) Suburban conflicts and management innovation in metropolitan development stages: a case study of Hangzhou. Urban Planning Forum 1:55–59 (in Chinese)

Coleman JS (1988) Social capital in the creation of human capital. Am J Sociol 94(Supplement):95–120

Dear M, Flusty S (1998) Postmodern urbanism. Ann Assoc Am Geogr 88(1):50–72

Dictionary Editing Room of The Institute of Linguistics, Chinese Academy of Social Sciences (2012) Contemporary Chinese dictionary. The Commercial Press, Beijing (in Chinese)

Dong C (2008) Wu Wenzao's Community research theory and its contemporary implications. J Qiqihar Univ (Philos Soc Sci Ed) 4:64–66 (in Chinese)

Dong Y (2005) Huilongguan experiment: the most hopeful place for community autonomy. Community Dev 14(7):10–14 (in Chinese)

Fei X (2005) From the soil: the foundations of Chinese society. Peking University Press, Beijing (in Chinese)

Feng D, Werner B, Zhu H (2011) Gated communities and neighborhood relations in the context of residential suburbanization: a case study of Guangzhou's Lijiang garden. Geogr Res 30(1):61–70 (in Chinese)

Feng G (2002) How modern communities are possible. Zhejiang Acad J 2:5–11 (in Chinese)

Festinger LS, Schacter S, Back K (1950) Social pressures in informal groups. Harper and Row, New York

Fischer CS (1975) Toward a subcultural theory of urbanism. Am J Sociol 80(6):1319–1341
Giddens A (2003) Sociology. (trans: Zhao X et al). Peking University Press, Beijing (in Chinese)
Granovetter M (1973) The strength of weak ties. Am J Sociol 78(6):1360–1380
Grant J (2009) Planning the good community: new urbanism in theory and practice (trans: Ye Q, Ni X). China Architecture & Building Press, Beijing (in Chinese)
Gu C (2007) On building a harmonious society and developing social geography. Hum Geogr 22(3):7–11 (in Chinese)
Hanifan LJ (1916) The rural school community center. Ann Am Acad Polit Soc Sci 67:130–138
Hawley A (1950) Human ecology: a theory of community structure. The Ronald Press, New York
Hillery GA (1955) Definition of community: areas of agreement. Rural Sociol 20(4):111–123
Huang G, Hu G et al (2004) Face: the power game of Chinese people. China Renmin University Press, Beijing (in Chinese)
Huang P (2010) Online communities: we are together. China Astronautic Publishing House, Beijing (in Chinese)
Jacobs J (1961) The death and life of great American cities. Random House, New York
Knox P, Pinch S (2000) Urban social geography: an introduction, 4th edn. Prentice Hall, Englewood Cliffs
Li P (2010) New trends in urban lifestyles: online consumption and leisure—an analysis based on cultural lag theory. Adm Law 6:57–60 (in Chinese)
Liang S (2005) The essence of Chinese culture. Shanghai People's Publishing House, Shanghai (in Chinese)
Lin N (2005) Social capital: A theory of social structure and action (trans: Zhang L). Shanghai People's Publishing House, Shanghai (in Chinese)
Luo J (2005) Social network analysis lecture notes. Social Sciences Academic Press, Beijing (in Chinese)
Luo J (2006) Chinese social networks: personal-centered trust networks. Guanxi Manag Res 3:1–24 (in Chinese)
Luo J, Ye Y (2006) The role of trust in the governance of outsourcing transactions. Learn Explor (2):44–50 (in Chinese)
Marsden PP (1988) Homogeneity in confiding relations. Soc Netw 10(1):57–76
Marsden PP, Campbell K (1984) Measuring tie strength. Soc Forces 63(2):483–501
Newman O (1972) Defensible space. The Macmillan Co, New York
Norman HN, Erbring L (2002) Internet and society: a preliminary report. IT Society 1(1):275–283
Park RE (1936) Human ecology. Am J Sociol 17(1):1–15
Porterfield KB, Hall J (2003) A concise handbook of community planning (trans Zhang X, Pan F). China Architecture & Building Press, Beijing (in Chinese)
Qi X (2007) Towards limited community: a social network analysis of urban residential neighborhoods. Capital Normal University Press, Beijing (in Chinese)
Schutz WC (1958) On categorizing qualitative data in content analysis. Public Opin Q 22(4):503–515
Shen X (2004) Marketization path: The inevitable road for urban community construction. J Grad Stud CentL China Norm Univ 11(1):114–117 (In Chinese)
Soja EW (1980) Socio-spatial dialectic. Ann Assoc Am Geogr 70(6):207–225
Sugden R (1986) The economics of rights. Co-operation and welfare. Basil Blackwell, Oxford
Sun L, Lei T (2007) Investigation and analysis of neighborhood relationships in Beijing old city. Urban Probl 2:56–59 (in Chinese)
Tan G (1986) Homans and Blau's social exchange theory. Soc Sci 10:55–59 (in Chinese)
Wang S, Liu X (2002) Foundation of trust: a rational explanation. Sociol Res 111(3):23–30 (in Chinese)
Wang X (2002a) What is community and what does community do. Zhejiang Acad J 2:20–24 (in Chinese)
Wang Y (2002b) Empirical study on Shanghai urban communities: community types, locational structure and changing trends. Urban Plan Forum 6:33–40 (in Chinese)

Wang X, Wang Z (2003) From "community" to "disembedded community": Community and community construction in the context of modernity. Academic Forum 6:40–43 (in Chinese)
Wang X, Mei M, Luo Y (2010) Utilization of grey space in residential areas: case study of Longhu Lijing Area A. Sichuan Build Sci Res 36(5):241–244 (in Chinese)
Webber MM (1964) The urban place and noplace urban realm. In: Webber MM et al (ed) Explorations into urban Structure. University of Pennsylvania Press, Philadelphia
Wellman B (1982) Studying personal community. In: Marsden PV, Nan L (ed) Social structure and network analysis. Sage publications, London
Wellman W (1990) Different strokes from different folks: community ties and social support. Am J Sociol 96(3):558–588
Wirth L (1938) Urbanism as a way of life. Am J Sociol 44(1):1–24
Yang K, Hwang K, Yang C (2008) Indigenous psychology for Chinese societies. Chongqing University Press, Chongqing (in Chinese)
Yang S, Wang C (2007) Review on driving mechanisms of urban social space evolution in foreign countries and policy implications. Urban Plan Int 22(2):42–50 (in Chinese)
Ye Q (1998) The chaotisation of virtuality and reality: the practical logic of the online world. Soc Res (3):56–57 (in Chinese)
Zhai X (2005) Face, favoritism, and the reproduction of power. Peking University Press, Beijing (in Chinese)
Zhang W (2002) Sales boom in Beijing Huilongguan cultural residential area: the largest affordable housing project in China. China R Estate Inf 7:13 (in Chinese)
Zhang W, Ruan D (1999) Social support networks of urban and rural residents. Sociol Stud 3:12–22 (in Chinese)

Chapter 7
Activity Space

7.1 Activity Space and Literature Review

7.1.1 Concept and Types of Activity Space

Activity space refers to the spatial realm in which individuals carry out most of their activities, encompassing both the geographical range and the behavioral patterns determining specific location choices (Johnston et al. 2000). It can also be understood as the spatial extent residents reach through various production and daily activities in urban areas. Unlike physical urban spaces, the concept of activity space emphasizes individual agency, representing an intangible space formed through residents' use of urban facilities and participation in social organizations (Chai 2000). Human activities occur across various spatial scales, such as household, neighborhood, economic, or urban zones. Larger-scale activity spaces may be discontinuous, often composed of discrete points connected by known routes but separated by regions (Chombart de Lauwe 1952).

Residential activities can be categorized in multiple ways. For instance, activity spaces can be classified by type of activity, with geographers focusing on spaces closely tied to activities such as living, working, commuting, shopping, and leisure. They can also be categorized temporally: long-term scales include life paths and life courses reflecting activities over a lifetime or specific periods, while short-term scales encompass daily 24-h space-time patterns of activity.

Activity space falls within the domain of behavioral geography. Research on residents' activity spaces helps uncover the social characteristics of urban spaces and promotes the transition of urban planning from physical-form-oriented approaches to socio-spatial frameworks.

J. Feng et al., *Urban Living Space in China*, Urban Sustainability,
https://doi.org/10.1007/978-981-95-0595-1_7

7.1.2 Literature Review of Activity Space

Research on activity space advocates understanding urban spaces from the individual subjectivity perspective (Yan et al. 2018). Activity space represents the direct interaction between individuals and their environment, which plays an important role in shaping and defining the scope of their behavioral spaces. It is a key aspect of understanding individual behavior.

Studies on activity space have developed behavioral geography methods focusing on "active individuals" and time geography methods focusing on "passive individuals". The integration of these two approaches enables a comprehensive study of human behavior. Some scholars have proposed activity analysis methods based on daily activity spaces to address and integrate key issues in human behavior research (Chai and Shen 2008). Activity space is closely related to the roles individuals play in society. The individual's space-time activities are the results of a series of activities they expect to engage in. Over time, residents develop familiarity with regions based on their experiences, the locations of their interactions in space, and the degree of cognitive reinforcement of those places (Golledge and Stimson 1997).

Daily activities consist of routine behaviors such as working, returning home, shopping, and recreation (Chapin 1974). These include mobility between home and other locations, regular activity sites (e.g., workplaces or schools), and activities within and beyond these locations (Golledge and Stimson 1997). They can be understood as typically occurring at home, workplaces, consumption place, non-consumption public spaces, and the mobility between them (Zhang 2007). These activities exhibit characteristics such as stability, variability, and flexibility (Carpenter and Jones 1983) and have become increasingly complex, diverse, and personalized due to the continuous restructuring and expansion of urban spaces, the rapid development of transportation and communication technologies, and the growing agency of individuals (Zhou and Yan 2005; Chai and Shen 2006; Zhen et al. 2009; Shen and Chai 2013). Some scholars argue that there is an interdependence between individual activities and spatial structures, with preferences in spatial choices (Rushton 1969). They emphasize the relationship between human agency and activities, with values and psychological processes also becoming important aspects of interpreting activity spaces. Other scholars focus on cognitive processes. In their professional book 'Spatial Behavior: A Geographic Perspective', Geographers Gollege and Stimson (1997) propose the 'anchor point theory', which illustrates the regularity of activity space formation.

Since the 1990s, research on individual activities has shifted from studying spatial behavior to studying behavior within spaces. This shift moves away from universal models and laws unrelated to spatial environments toward highlighting the differences in behavior and environmental contexts, emphasizing the non-universality of behavioral patterns (Chai et al. 2012). The social domains and lifestyles of behavioral subjects emerge according to daily living patterns, and daily activities have their own spatial structures. The social spaces of groups combine with the places characterized by daily activities, forming interactions between group behaviors and spaces. The activity space characteristics of different urban sub-groups, especially disadvantaged and marginalized populations, have attracted particular attention from scholars. Liu et al. (2005) found that poor urban groups exhibit singular spatial activities and fragmented time use, forming narrow daily activity spaces centered around residential areas. Ta and Chai (2017) highlighted the behavioral spatial dilemmas of low-income residents in typical suburban areas of Beijing, using potential and actual activity spaces to illustrate their disadvantages in space-time accessibility and the actual use of urban spaces. Zhao (2007) focused on the identity construction and spatial formation in the daily lives of migrant workers, arguing that their activities contribute to the formation of 'migrant spaces'. Other scholars, from a feminist perspective, have focused on the residential activity spaces of urban women in China (Chai et al. 2003). The daily activity spaces of university students have also entered researchers' vision. Holton (2017) analyzed how UK higher education students use nighttime social activities as a mechanism for establishing identity during their university transition, exploring the interaction between student identity, club activities, and nighttime activity spaces. Tao and Shen (2018) analyzed the interaction between university students' daily activities and urban spaces, finding that suburban students' daily activity spaces are oriented toward urban areas, though the degree of dependence on urban areas varies among groups. Compared to urban students, suburban students have significantly larger daily activity spaces.

With advancements in geographic information technology, activity space research has gained richer data sources and more diverse technical methods. Call detail records (CDR) from smartphones (Xu et al. 2016), GPS data (Shen and Chai 2013), mobile signaling data (Yan et al. 2018), public transit smart card data (SCD) (Liu et al. 2018), and point of interest (POI) data (Xu et al. 2018) provide new approaches for studying the space-time activity characteristics of groups (Cao et al. 2017). By combining geographic information systems (GIS) and big data, differences in activity patterns among groups can be compared, exploring how different groups use urban spaces differently (Kwan 1999). The aforementioned studies on urban sub-group activity spaces have already applied big data methods. However, activity space research based on big data can easily overlook individual agency, which become a significant issue that academia have to reflect on. Therefore, scholars have begun exploring qualitative methods such as small-sample in-depth interviews and small-sample 3D spatial visualization in urban activity space research to compensate for the limitations of relying solely on big data. Kwan (2007) proposed a GIS-based narrative analysis method, combining spatio-temporal activity trajectories with narrative materials such as oral histories, life histories, and biographies from qualitative research to form geo-narratives in a narrow sense. This approach introduced emotional geography analysis within a geospatial technology framework to better showcase individual life experiences. This is a successful example of using 'small data', geo-narrative methods, and 3D spatial visualization for qualitative activity space research. This attempt integrated behavioral and subjective qualitative analysis with time geography methods and GIS, establishing a qualitative GIS research paradigm. At the same time, such research compensates for the shortcomings of traditional quantitative studies that overlook subjective emotions and agency, transforming the neutral individuals in traditional time geography—those without subjective preferences and with shared life experiences and social backgrounds—into vivid and multidimensional individuals (Guan et al. 2013). In recent years, Western qualitative GIS and geo-narrative methods have been introduced to China, promoting the development of qualitative methods in studying Chinese urban activity spaces.

In summary, activity space research can be divided into macro-level studies based on population spatial distribution and large-sample surveys of resident behavior, and micro-level studies influenced by time geography that focus on individual behavior, time and space. Research methods have evolved from traditional statistical data analysis, questionnaire surveys, and activity diary analysis to activity space research using positioning data, GIS, and big data analysis, as well as individual activity space research combining small-sample in-depth interview data with 3D spatial visualization methods and geo-narrative approaches.

7.2 Activity Space Based on Life Paths and Life Courses

7.2.1 Long-Term Behavior and Activity Space Research Based on Life History

Time geography is predicated on micro-individual analyses, but the underlying idea is to establish links between micro-individual situations and macro-contexts and patterns; in other words, to refract macro-problems through micro-analysis (Hägerstrand 1970; Lenntorp 1976). In order to analyze this micro impact, time-geographers have considered long-term and short-term behavior and constructed relationships between long-term life-path choices and everyday behavior (Pred 1981a and b).

The life history of an individual generally reveals his or her long-term institutional roles, which are often influenced by and interact with factors such as his or her family background, social status, and educational background. This causes the individual to be actively or compulsively involved in specific routine or non-routine enterprises and to be bound by an activity bundle. Conversely, this involvement can have the effect of allowing the individual to develop his or her own unique pattern of daily life. This pattern of life influences the choices individuals make about themselves, their families, and their surroundings. Through these choices, they change their past long-term institutional roles. The concept of 'individual', which is the focus of activity space research, is not only the individual based on the person, but also the individual based on the family. The long-term institutional role of each individual in the family often begins with interactions with the daily activities of individuals within the family, and the choice of paths and times of activity is also constrained by the life course opportunities offered by the family as a whole (Pred 1981a and b). Consequently, the family's life path is dialectically related to the daily behaviors of each individual in the family, constrained by the historical stage, geographic context, and socio-cultural system in which it is situated, and influenced by changes in these factors. In particular, decisions based on the family's long-term behaviors, such as relocation and employment, can lead a lock-in effect on the daily activity patterns of the family as a whole (Thrift and Pred 1981).

Therefore, the study of long-term behaviors needs to be conducted within the framework of time and space. It is imperative to consider the contextual and temporal-social nature of 'space-time' as a perspective, to emphasize the interaction between mobility and constraints, and to take into account the combined effects of socio-cultural and spatial processes. The present study explores the nature of activity constraints in the formation of life paths in order to realize a comprehensive understanding of the individual's life course. This study adopts a micro-individual perspective by selecting three families living in the suburbs of Beijing as typical samples to analyze the changes in their life paths and activity space based on a long-time scale.

7.2.2 *Research Samples*

This study selected three representative family samples residing in Beijing's suburban areas. Sample JT1 represents a traditional Chinese extended family structure: The male head was born in 1979 (aged 37 at the time of survey) and the female head in 1981 (aged 35), both employed, with one child, demonstrating relatively favorable economic status (Table 7.1). Sample JT2 constitutes a typical nuclear family: Both spouses were born in 1970 (aged 46), with the male head employed while the female head had resigned from her job and was currently a homemaker, having one daughter and maintaining a strong financial position (Table 7.2). Sample JT3 consists of a male head born in 1982 (aged 34) and a female head born in 1983 (aged 33), both employed professionals without children yet, characterized by moderate household economic conditions (Table 7.3).

7.2.3 *Urban Residents' Activity Spaces Based on Individual Life Paths and Life Courses*

Households and workplaces constitute two critical anchor points in residents' spatio-temporal activity trajectory. Beyond work and domestic responsibilities, most daily activities revolve around these anchors. Consequently, the analysis of individual life paths across the three sample households focuses on their residential and occupational spaces to delineate the evolution of their behavioral activity spaces.

Table 7.1 Life history and occupational trajectory of sample JT1

Years	Life events	Career history	
		Male head	Female head
1979	Husband's birth	Academic pursuits	–
1981	Wife's birth	–	Academic pursuits
1997	Husband enters university	Beijing university of aeronautics and astronautics	–
1999	Wife enters university	–	University of science and technology Beijing
2005	Husband's employment	China patent office	–
2006	Wife's employment	–	Chinese academy of agricultural mechanization
2009	Marriage	–	–
2014	Birth of first child; In-migration of husband's parents	–	–

Table 7.2 Life history and occupational trajectory of sample JT2

Years	Life events	Career history	
		Male head	Female head
1970	Wife's birth	–	Academic pursuits
	Husband enters university	Northwest A&F university	–
1990	Wife enters university	–	China agricultural university
		China agricultural university	–
1994	Wife's employment	–	State-owned enterprise
1997	Marriage	–	–
1998	Birth of first daughter; In-migration of wife's mother	–	–
1999	Husband's employment	Beijing academy of agriculture and forestry Sciences	–
	Out-migration of wife's mother	–	–
2004	Wife's birth	–	Academic pursuits
2015	Wife's resignation	–	–

Table 7.3 Life history and occupational trajectory of sample JT3

Years	Life events	Career history	
		Male head	Female head
1982	Husband's birth	Academic pursuits	–
1983	Wife's birth	–	Academic pursuits
2000	Husband enters university	Beijing institute of technology	–
2001	Wife enters university	–	Tianjin normal university
2004	Husband's employment	Kingsoft corporation	–
2005	Wife's employment	–	A Beijing-based media company
2009	Courtship	–	–
2011	Marriage	–	–
2012		NetEase games	–

7.2.3.1 Extended Families with Children

The changes in the place of residence and employment of the husband and wife in Sample JT1 families are shown in Fig. 7.1. The place of employment of the male parent, Mr. Wang, remained unchanged during the decade, while the place of work of the female parent, Ms. Qiao, changed once. The male parent's place of residence moved five times and the female parent's place of residence moved four times, with three of the moves occurring after the two were married. The first relocation took place in 2010, one year after they started their family, when they purchased a small commercial house in the Yipinge community of Beishatan area; in February 2014, due to the woman's pregnancy and need for care, the man's mother's relocation, the increase in family members, and a change in the woman's workplace prompted the couple to purchase another commercial house in Changyang Bandao community in Fengtai District, and then sell the house they are currently residing in. Because the new house was not yet completed and in view of the renovation factors, the family temporarily rented a turnover in Zhichun Road neighborhood; they officially moved into their new home in 2015.

The life path of Mr. Wang, a male parent, is shown in Fig. 7.2. He lived with his parents since he was a child, and his life path extends vertically upwards with his home as the point of residence, and then he started to go to school, and his residential and occupational paths were separated. in 1997, he came to Beijing because of his admission to the university, and his residential and occupational paths' points of residence were changed and overlapped. 2004, he formally entered the workforce, and his residential and occupational paths' points of residence were changed, and the distance between his residence and his occupation was longer at this time. The distance between work and residence was longer. After one year of work, the first relocation occurs due to the long work-life distance, the residence path changes, and the work-life distance is shortened. After three years of work, she meets the

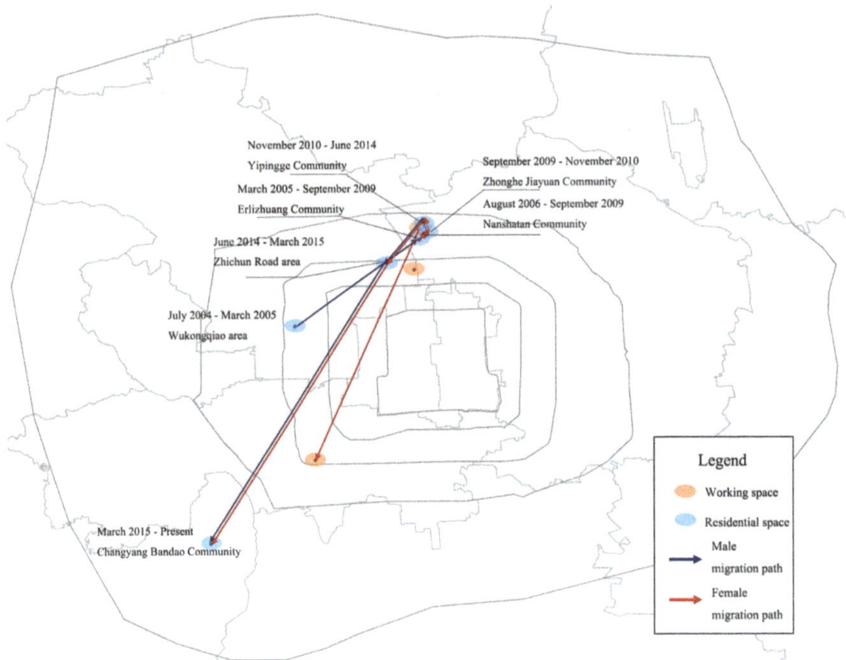

Fig. 7.1 Changes in residential space and employment space for couples in JT1 households

woman and moves for the second time one year after meeting her, with a change in the residence path. Married one year later and promoted in the same period, but no change in residence and workplace, the two paths extend upwards in parallel. The following year, a new home is purchased, the residence path changes, and the work-life distance rises slightly; four years later, due to changes in the family structure, the residence path changes twice, and the work-life distance rises again. In this process, his social role completed the transition from son to husband to father. As can be seen from the transitions in the life paths, there were eight major life events in Mr. Wang's life course, including five relocations, falling in love, getting married, and having a child.

The life path of Ms. Qiao, a female parent, can also be divided into three stages (Fig. 7.3): during the schooling period, the life path extends vertically upwards with the home as the point of residence, and the residential and occupational paths are separated; during the single stage after joining the workforce, the residence and occupational paths' points of residence are changed, and the distance between the job and the home is close to the same at this time; after getting married, the residential path changes three times, and the place of work changes once, and the place of work is moved from the original Beishatan area to Fengtai Science and Technology Park, and the work-life distance also changed. In this process, her social role also changed from daughter to wife to mother. From the transitions in the life path, it can be

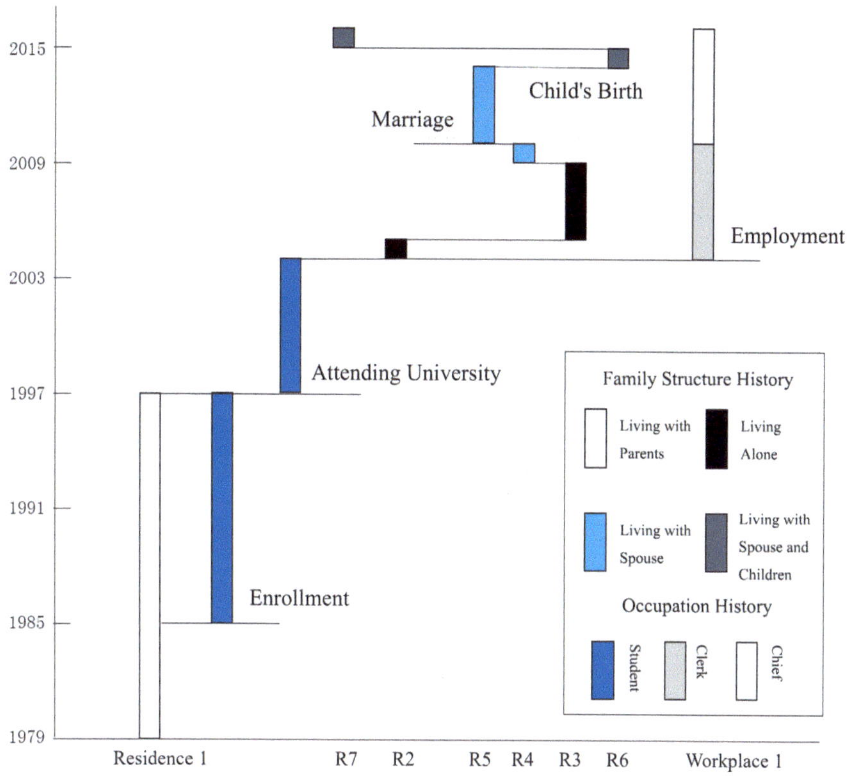

Fig. 7.2 Life path and activity space changes for male parents in JT1 families

seen that there are seven major life events in Ms. Qiao's life course, including four relocations, falling in love, getting married, and having children.

7.2.3.2 Nuclear Families with Children

Changes in the place of residence and employment of the husband and wife in sample JT2 families are shown in Fig. 7.4, with no change in the place of employment of the male parent and two changes in the place of employment of the female parent in the last 20 years. The female parent's place of residence has changed three times, twice after the couple's marriage. The first time they moved was in 1998, when they moved into the housing allocated to them by the female parent's organization after their marriage; the second time they moved was in 2004, when they purchased a commercial house in the Shangdi-Qinghe area and moved into a new house due to the small size of the organization's housing unit and the old type of housing.

The life path of the male parent, Mr. Lu, as shown in Fig. 7.5, was admitted to the Northwest University of Agriculture and Forestry in 1990, and then came to

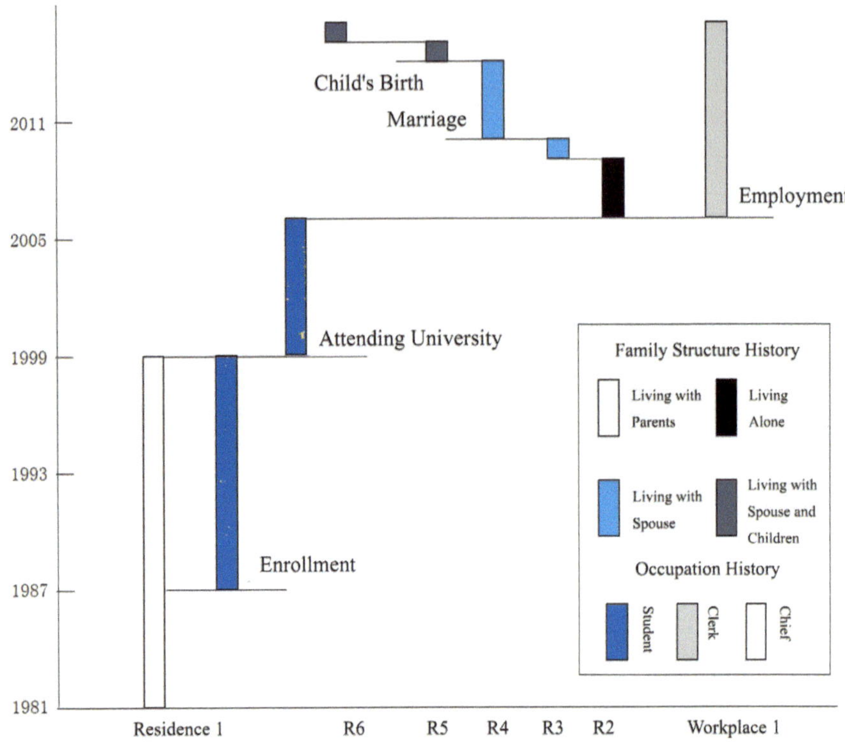

Fig. 7.3 Life path and activity space changes for female parents in JT1 families

Beijing in 1994 as a result of being admitted to the postgraduate program of the China Agricultural University (CAU), where his place of residence and occupational path stationing changed and overlapped. He married his wife, Ms. Zhang, in 1998 and then moved into the housing allocated to the woman's parents' unit, separating his place of residence and occupational path stationing. In 1999, after finishing his studies, he joined the Beijing Academy of Agricultural and Forestry Science and Technology, and his career path changed. After four years of work, he bought a commercial house in Heyun Jiayuan community on Anningzhuang West Road, and his place of residence changed again. In the process, his social role changed from son to husband to father.

The life path of Ms. Zhang, a female parent, is shown in Fig. 7.6. She grew up living with her parents, and during her studies, her life path extended vertically upwards with her home as the station, and her residence and career paths were separated; she was admitted to the China Agricultural University in 1990, and her residence and career paths overlapped; she formally took a job in a state-owned enterprise in 1997, and she moved into her mother's unit-allocated housing, and her residence and career paths were changed to a station, and her career path was close; in 1998, she purchased a commercial house in Heyun Jiayuan community on Anningzhuang West

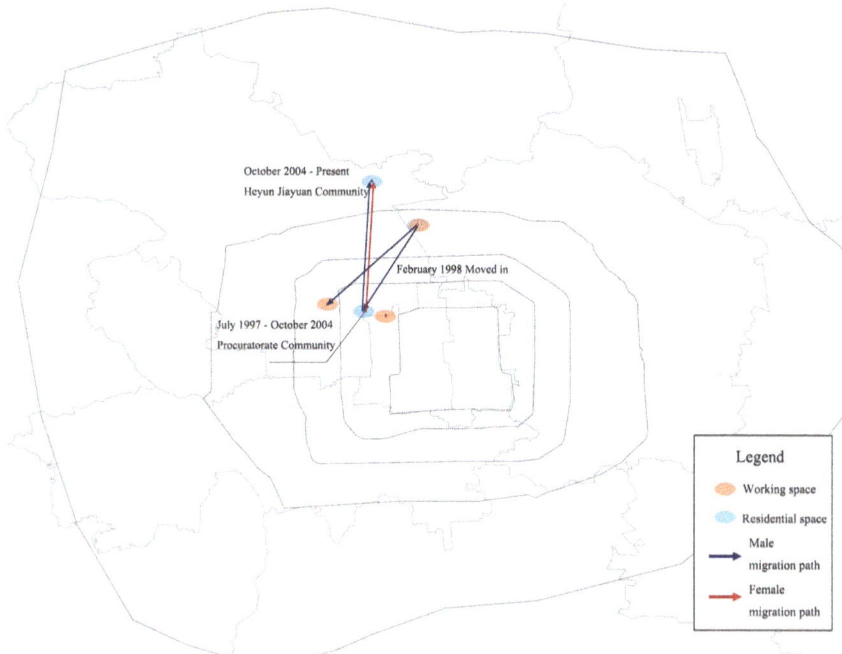

Fig. 7.4 Changes in residential space and employment space for couples in JT2 households

Road. At this time, the distance between work and residence was close; she married Mr. Lu in 1998, and her place of residence remained unchanged; her daughter was born in 2000, and she and Mr. Lu moved to a new home in 2004, with no change in her place of work during this period; she resigned from the unit in 2015 for the sake of her daughter's high school education and her personal career considerations. In the process, her social role has similarly changed from daughter to wife to mother.

7.2.3.3 Nuclear Families Without Children

The changes in the place of residence and employment of the husband and wife in the sample JT3 families are shown in Fig. 7.7. The place of employment of the male parent changed twice, the first time in 2009 following the company's address change, and the second time in 2012 when he jumped from Kingsoft company to NetEase Games company; the place of employment of the female parent changed twice following the company's relocation. The male parent purchased a commercial house on Xiaoying West Road in 2007, and the female parent's place of residence changed more frequently due to the change of place of employment before the marriage, and after the marriage, she moved into the house on Xiaoying West Road and has been living together since then.

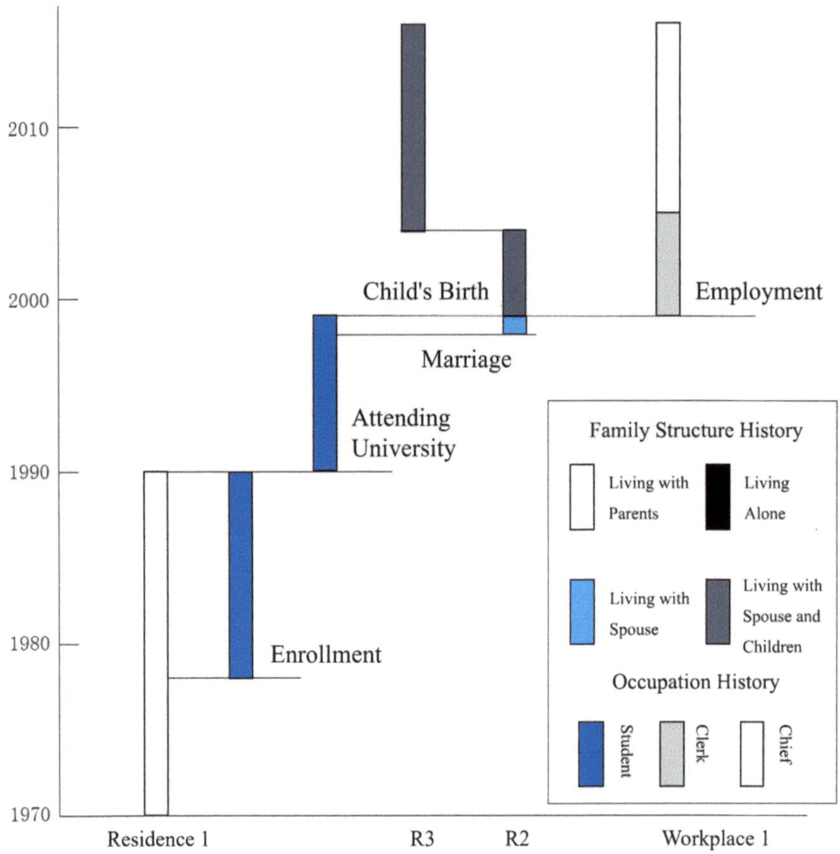

Fig. 7.5 Life path and activity space changes for male parents in JT2 families

The life path of Mr. Liu, a male parent, is shown in Fig. 7.8. In 2000, Mr. Liu came to Beijing because he was admitted to the university, and his place of residence and career path station changed and overlapped. 2004, Mr. Liu formally joined Kingsoft company, and shared a room with his colleagues, and his place of residence and career path station changed, and the distance between his work and residence was relatively close at this time. 2007, Mr. Liu purchased a house on Xiaoying West Road with the financial support of his parents, and the distance between his work and residence increased. In 2009, due to the relocation of the Kingsoft company to the Jinshan Software Building on Xiaoying West Road, Shandi area, Mr. Liu's workplace has also changed. His career path changed following, and the distance between work and residence was close. 2012, Mr. Liu jumped from Kingsoft company to NetEase Games, and his career path changed again, and the distance between work and residence increased again. During this process, his social role completed the transition from son to husband. Mr. Liu showed dissatisfaction with his career status between 2011 and 2012, and thus chose to jump ship, after which his satisfaction

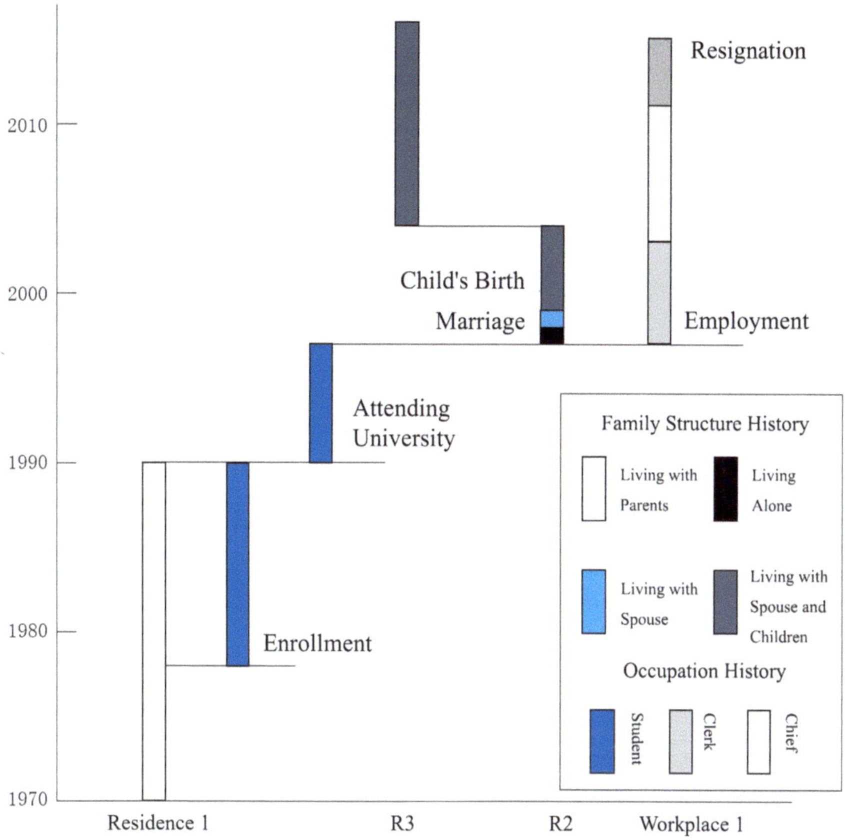

Fig. 7.6 Life path and activity space changes for female parents in JT2 families

level rebounded. As can be seen from the life path transitions, there were four major life events in Mr. Liu's life course, including relocation, career change, and marriage.

The life path of Ms. Fan, a female parent (see Fig. 7.9), can be similarly divided into three stages: during the period of schooling, the life path extends vertically upwards with the home as the point of residence, and the residence path is separated from the career path; during the single stage after joining the workforce, the place of residence changes twice following the point of residence of the career path, and at this time the distance between work and residence is close to the same distance; after getting married, the residence path changes, and the work path remains unchanged and extends upwards. In this process, her social role also changed from daughter to wife. The following section will focus on the changes in Ms. Fan's daily living behaviors before and after marriage.

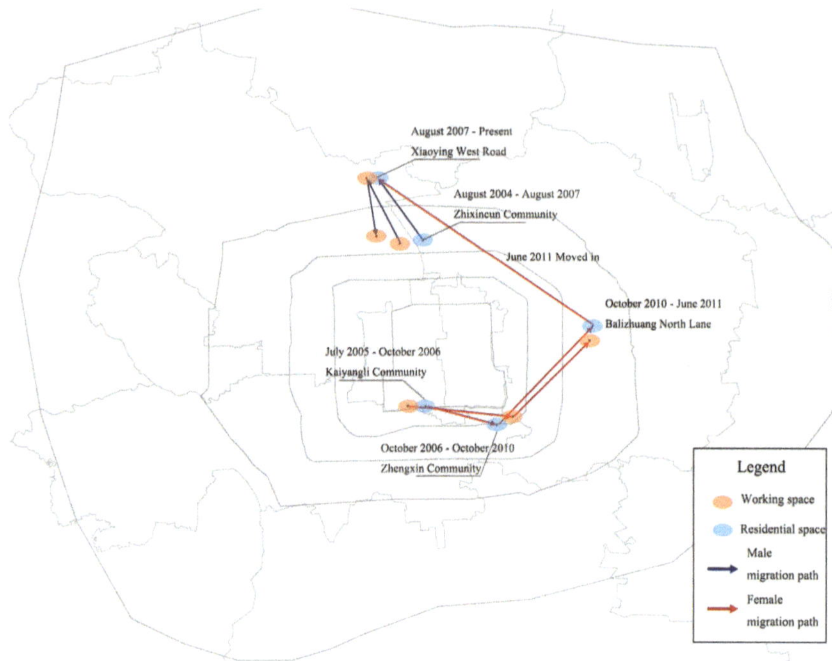

Fig. 7.7 Changes in residential space and employment space for couples in JT3 households

7.2.3.4 Interaction Between Residents' Long-Term Behavior and Urban Space

In an individual's life path and life course, as the most important spatial form of daily activities, the interaction between work and residence space shows the process of change of their behavioral space. By analyzing the individual as a whole, it is possible to grasp the correlation of different life events in a person's life, and to analyze issues such as the mechanism of relocation, the distance between work and residence, and family relationships. In terms of the understanding of space-time constraints, the constraints of different long-term behaviors can vary considerably and are significantly different from those of short-term behaviors. In the case of relocation behavior, capacity constraints, coupling constraints and authority constraints all have a large impact. Changes in the activity space of the three families mentioned above can be classified into three categories: change of residence (JT1, JT2); change of workspace (JT3 Mr. Liu); and change of both residence and workplace (JT3 Ms. Fan).

When the workspace is fixed and the residential space is unstable, individuals spontaneously seek the residential space that best suits their economic conditions and commuting time, for reasons of commuting costs. Relocation behavior is constrained by ability constraints, such as income, and compositional authority constraints, such as the *danwei* system, housing discrimination, and hukou. In sample JT1, for example, the behavioral space of both the male and female parent is one in which the workplace

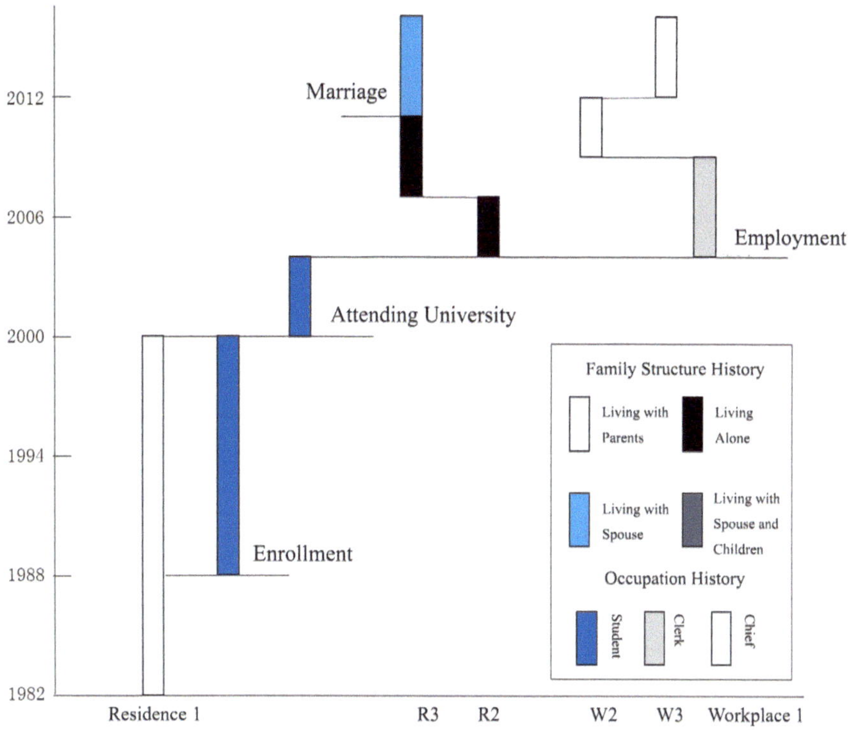

Fig. 7.8 Life path and activity space changes for male parents in JT3 families

is constant and the residential space changes around the workplace. In the early years, the male parent's residential space is not stable and changes with the urban space, and his behavioral decision to move is constrained by his ability to earn personal income.

However, when the residence is more fixed (especially after purchasing a house), the commuting costs associated with a change in the workplace are much smaller than the costs of changing housing. Therefore, this market mechanism-oriented change in the workspace does not affect their residential space changes, but only brings about adjustments in their daily behavioral patterns. For example, the male parent in Sample JT3 experienced two job changes after he purchased his home in the Shangdi area. However, due to the small commuting cost, he chose to change his commuting mode to adapt to the change of his workplace, and his daily activities continued to revolve around his residential space.

When both residential and workspaces are not fixed, unmarried individuals will spontaneously seek a residential space that best suits their economic conditions and commuting time. At this point, the residential space will follow the workspace. This type of job-residence dependency is the most common and economical spatial combination pattern under the effect of the system and the market. For example, when unmarried, the female parent (Ms. Fan) in sample JT3 chooses her place of

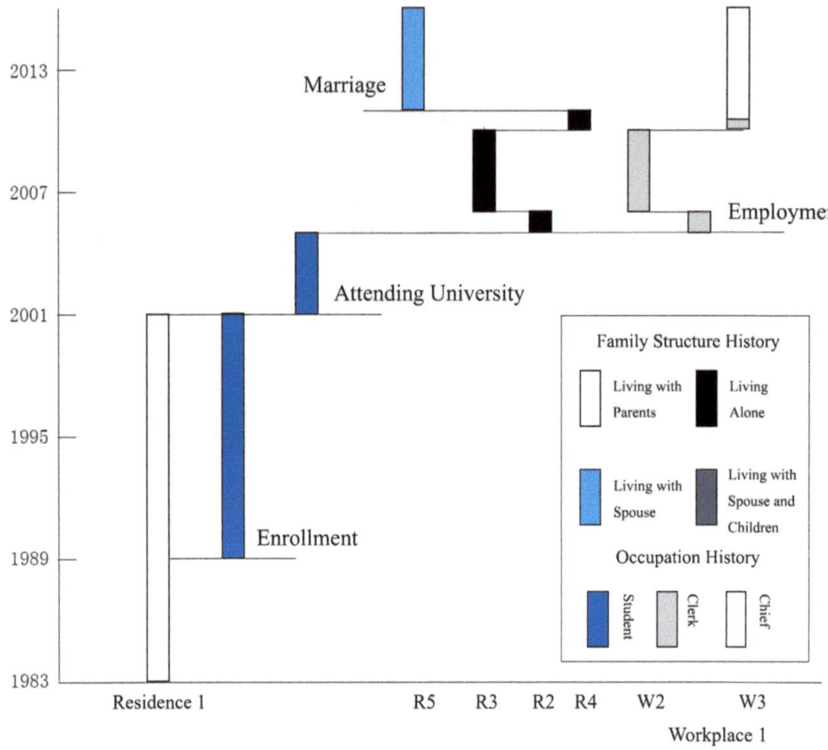

Fig. 7.9 Life path and activity space changes for female parents in JT3 families

residence out of commuting cost considerations. This process constrains their relocation decisions by their ability to meet income and daily behavioral continuity needs, as well as authority constraints on the composition of the workplace time system and other components.

However, when it is necessary to choose housing again after marriage, the relocation decision of the whole family is more influenced by the combination of constraints consisting of family relationships, intergenerational care, and housing facilities necessary for life. For example, the relocation decision of sample JT1 after having a child is made after considering about a number of factors, including the couple's commuting distance and schedule, if the community could meet the family's basic living needs and future educational needs of the children, and the size of the home to accommodate the entire family.

In short, different constraints affect the relocation behavior and activity space changes of the family. The influence of authority constraints is most significant in relocation activities. The constraints themselves change due to spatial evolution, social change, technological advancement, and individual growth, and need to be added to considerations such as historical era and age.

7.2.4 *Spatial Changes in Chinese Urban Residents' Activities from the Perspective of Life History and Long-Term Spatial Behavior*

In the traditional *danwei* system dominated residential space of Chinese urban residents, urban space is a unitized space, a spatialization of the planned economic system in the city (Liu and Chai 2014). Residents have fewer opportunities and possibilities for job changes, housing is allocated by the *danwei*, and only when an individual's title and status have significantly improved can he or she apply to the *danwei* for a larger area of housing. Even then housing adjustments are subject to long waits, queues and cumbersome procedures. Comparatively speaking, in the planned economy, when housing resources were scarce, an individual's living and working space was relatively stable, and it was basically difficult for an individual's and family's life course to have a decision-making impact on changes in the individual's living space.

In the process of China's transition to a market economy, the influence of *danwei* on society, especially on the living space of residents, has gradually dissolved. The market has begun to dominate, and individuals can make decisions based on their economic conditions to improve their living conditions, especially housing and employment conditions, which in turn affects the spatial and temporal utilization and arrangement of work-residence relationships and various related activities.

In this context, the life course and life paths of residents begin to be linked to individual decisions about activity space. Important life events in the life course, such as marriage, the birth of children and co-habitation of the elderly, and other changes in family structure, generate new demand for housing. In a market economy, when urban residents have a certain level of economic strength, they may be inspired by important life events to make new housing choices and relocation decisions. Of course, important life events in the life course will also have an impact on the choice of work (career choice) of Chinese urban residents after the market transformation, which in turn will affect the distance between work and housing and various related activities. At the same time, a family is also a whole, especially in terms of residence, and the adjustment of the residence and employment location of one spouse may lead to the corresponding adjustment of the other spouse.

In Chinese cities after the market transition period, for some families with children, due to the birth of children, the elderly come to live together in order to take care of the third generation. Their household type changes to the extend family. In many cases, two elderly people come together to look after the children and take care of the family, which creates a greater demand for living space in a short period of time. This change in the life cycle tends to stimulate the behavior of residents to buy new and larger homes. At the same time, due to the increase in the cost of living and the pressure of life, residents, especially men, may also face changes in their careers, and these changes in living space will in turn affect the development of various daily activities. This is a feature of the traditional Chinese cultural concept of a close-knit three-generation family relationship that is completely different from that of Western cities.

For the nuclear family with children, both spouses have gone through the transition from being children to husband and wife to playing the role of parents, and the increase in the number of family members will lead to the need for more living space. However, because this type of nuclear family does not live with the elderly, it does not have the same housing and living pressures as the previous type of family. With the implementation of China's two-child policy, some families have welcomed the birth of a second child, and the pressure on living space will further increase. There are also families that do not need the help of an elderly person with one child, but after the birth of a second child have to ask an elderly person to help take care of the child, transforming them into an 'extended family'. This change of roles in life course is also one of the driving forces behind the change in living space.

There is another type of family—nuclear family without children—which compared to the other two types of families, will not experience a large increase in demand for living space in a short period of time. Their demand for living space is relatively stable, and is more likely to be driven by changes in the distance between work and residence as a result of new job-seeking behaviors by one spouse in order to improve his or her job offer, while the employment and residence of the other spouse accommodates that spouse in shaping his or her own decision-making.

Life course theory, as a sociological theory focusing on the interaction between individual life history and social change, can effectively complement the inadequacy of time geography in the analysis of social processes. By reconstructing residents' long-term spatial behavior paths, it is possible to integrate psychosocial factors, multiple social roles and multidimensional social time (Chai et al. 2013), which is the significance of long-term spatial behavior research from the academic level. And by combining the development practice of Chinese cities, the relationship between life course and life path and residents' activity space is a complex systematic relationship. Interestingly, this relationship has always existed in Western cities, but in China, it is the transformation of the economic system and the accompanying social transformation that has contributed to the 'resurrection' of this systemic relationship.

7.3 Activity Space Based on Daily Path and Activity Dairy

The activity dairy records the activities of the residents during the 24 h of the day, including the places where the activities take place and the duration of each activity. The daily path visualizes the spatio-temporal activity trajectories of residents in the form of a space-time coordinate system for 24 h a day.

In contrast to life paths and life courses, daily paths are an effective way of presenting the characteristics of space-time pattern of residents' activities on short time scales. A daily path typically depicts the time and space of different types of activity such as sleeping, household chores, personal affairs, recreation, work, and shopping. From space perspective, it is based on the location of one's own home and involves spatial locations some distance from one's home such as workplaces, shopping destinations, entertainment venues, and other locations where various activities occur. From a time perspective, it lasts from 0:00 to 24:00 and depicts the path of residents' daily activities. Since there is a clear difference between weekdays and weekends in the space-time activity pattern of urban residents, the individual's paths of daily life also show a difference between weekdays and weekends.

The study of the spatial change of residents' daily activities needs a clear clue, and it can draw on the relevant research results of the life course to find the clues for the study of the change of individual life paths. The concepts of path, turning point and duration are used in life course theory to describe the development of individuals in the process of social change, emphasizing the degree of influence of life events on individuals, and arguing that each individual can construct his or her life course through choices and actions under the opportunities and constraints provided by the historical and social environment (Elder 1998). Among them, turning points are the time and direction of changes in individual life events, which help to understand the continuity and rupture of life trajectories and the influence of internal and external factors on the life course, so the study of turning points is particularly important (Liu 2015). And these turning points can be used to act as clues reflecting spatial changes in daily life. The analysis of important life events as turning points in the family not only describes the direction of change of the two states before and after the event, but also dynamically correlates the change of the state of the event with the change of the social environment, the basic characteristics of the individual, and other important events.

The inclusion of behavioral activities and their socio-spatial influences in the event state allows for a dynamic analysis of the changes in the spatial state of residents' daily activities before and after these life events, as well as the differences in the socio-spatial influences on these two states.

7.3.1 *The Influence of Relocation and Geographic Context Change on Residents' Daily Activity Space*

7.3.1.1 Changes in Residential Space and Work-Residence Relationship

Changes in residential space bring about changes and constraints in the geographic context. Urban residence provides households with an abundance of amenities and accessibility to relevant services, as well as the opportunity to minimize combinatorial constraints for individuals within the household to engage in different activities of daily affairs. For example, the changes in male parents' daily living behaviors and activity spaces before and after several moves in samples JT1 and JT3 are influenced by the urban space and work-residence relationship in which they live (Fig. 7.10).

In Sample JT3, prior to moving, male parent Mr. Liu shared a flat with a colleague in the Zhixincun neighborhood, which is close to the company location, where the

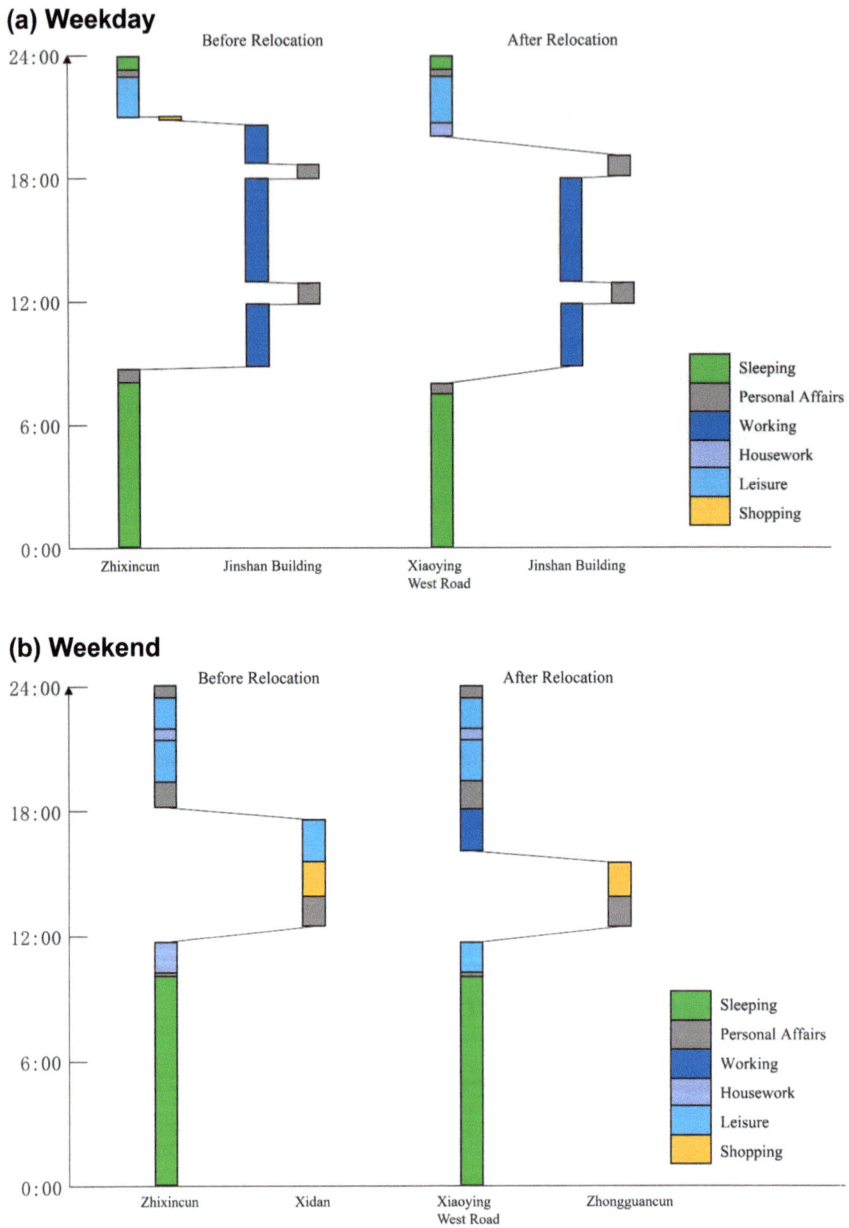

Fig. 7.10 Path of daily activities of the male parent in JT3 families before and after the first relocation

daily activity space overlaps with the workspace, and commuting is dominated by bicycles and public transport. On non-working days, activities are mainly within the home, with a smaller range of activities, and most of the outings are long-distance leisure, shopping and dining activities directed towards the city center.

Considering into account the rising prices of housing in Beijing and problems such as starting a family, in 2007, Mr. Liu in his parents' financial assistance in the form of a mortgage loan to buy a set of commercial housing in Xiaoying West Road, renovation and then formally move in.

'..... *My father had the foresight, those years Beijing housing prices rose quite fast, the family put up the down payment to buy this house, I slowly pay the mortgage every month......*'

After this relocation, Mr. Liu's commuting distance rose significantly, and his commute changed to travel by public transportation such as subway and bus. With the increase in daily commuting time, the type of trips he takes tends to be more multi-purpose single trips, where he completes activities such as dining and shopping during his commute, and his lifestyle shifts to an urban-dependent one. On non-working days, his activities are still dominated by activities within the home.

In 2009, Kingsoft company relocated to Jinshan Building on Xiaoying West Road, Shandi area, and Mr. Liu's workplace changed. At this time, it is only 1 km from Mr. Liu's home to the company, and his commute is mainly on foot. Due to the proximity of work and residence and the more flexible company time system, Mr. Liu is more flexible in his time use, and the proportion of non-work activities outside the home after returning home from work has risen. Due to the insufficient density of community dining facilities, service facilities, leisure facilities and retail commercial facilities, the distance for out-of-home leisure activities has increased significantly, and activities such as dining, leisure and shopping are more inclined to the Shangdi and Zhongguancun areas. On non-working days, in-home activities are still dominated, and out-of-home non-work activities are more oriented towards the direction of Olympic Forest Park (Fig. 7.11).

For Sample JT1, on the other hand, before the first relocation (Fig. 7.12), Mr. Wang lived with others in a dormitory provided by the company located in Wukongqiao area. As a result, there were high restrictions on individual freedom, and Mr. Wang's daily activities basically revolved around his workplace, overlapping with his workspace, with only a few leisure activities in his residence. On weekdays his activity space overlaps with his workspace, his use of time is more regular, and his commute is dominated by public transportation. On non-working days, leisure activities dominate, with a larger scope of activities, basically focusing on the city center and scenic areas in the far suburbs. His lifestyle is characterized by 'concentrated activity distribution and active travel'.

And after the first move, Mr. Wang shared a room with a friend in Erlizhuang area, and the proximity of work and residence brought about a reduction in commuting distance. The increase in accommodation size and conditions increased Mr. Wang's time for activities within his home. His daily activity space is still dominated by his workspace, with a slightly expanded scope, when the commuting mode changes to a combination of bicycle and public transportation. Non-workday leisure activities

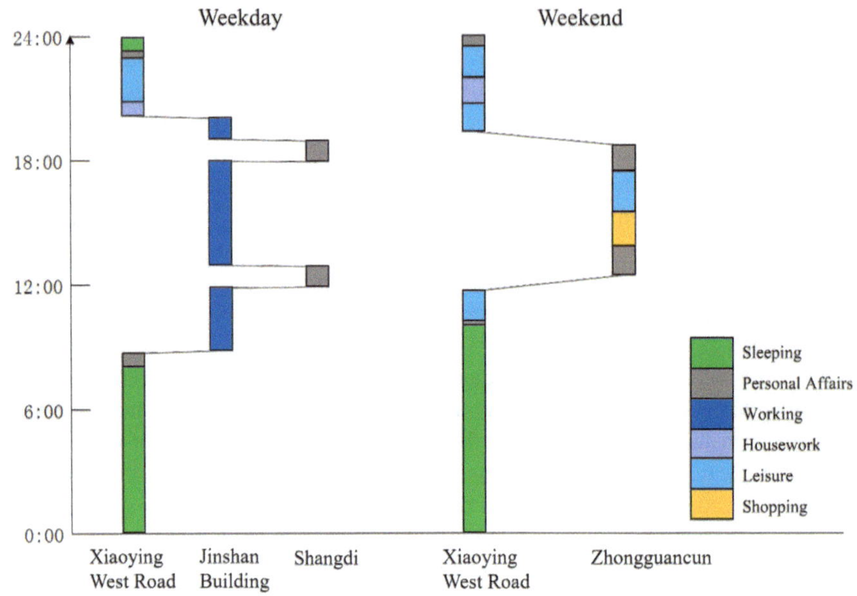

Fig. 7.11 Path of daily activities of male parents in JT3 families after the first change of place of work

remained the same. During this period, Mr. Wang's activity space characteristics can be categorized as urban-dependent, with a higher degree of spatio-temporal activity trajectory fragmentation, multiple short-duration, short-distance outings in the middle of the day outside the home, and more outings in the evenings than other types. On non-working days, there are still more active travel activities, and outings to the city center and areas near home are distributed throughout the day.

The second relocation occurred one year after Mr. Wang and Ms. Qiao met. They shared a house because they were living together. The house is in Zhonghe Jiayuan community, which is close to Ms. Qiao's workplace. Mr. Wang's commuting distance has increased compared to the previous move, and his commuting mode remains a combination of bicycle and public transportation. However, he spends significantly more time at home after work, and most of his daily activities are family-centered and close to home, such as walking to the Olympic Forest Park after meals. Non-working day activities were dominated by leisure and shopping. In terms of timing, mornings are basically dominated by home-based affairs, with a small amount of eating out at lunchtime, afternoon non-work activities are mainly leisure and shopping within 3 km of the home, and there are almost no out-of-home non-work activities in the evenings (Fig. 7.13). If there are long-distance non-work activities, the schedule will be adjusted accordingly.

> ...... sometimes want to go climbing on weekends, then definitely have to get up early...... we participated in a hiking association, usually more than 7:00 a.m. collection...... Mainly go to the outskirts of Beijing, wild mountains, wild Great Wall would like to try...... bring

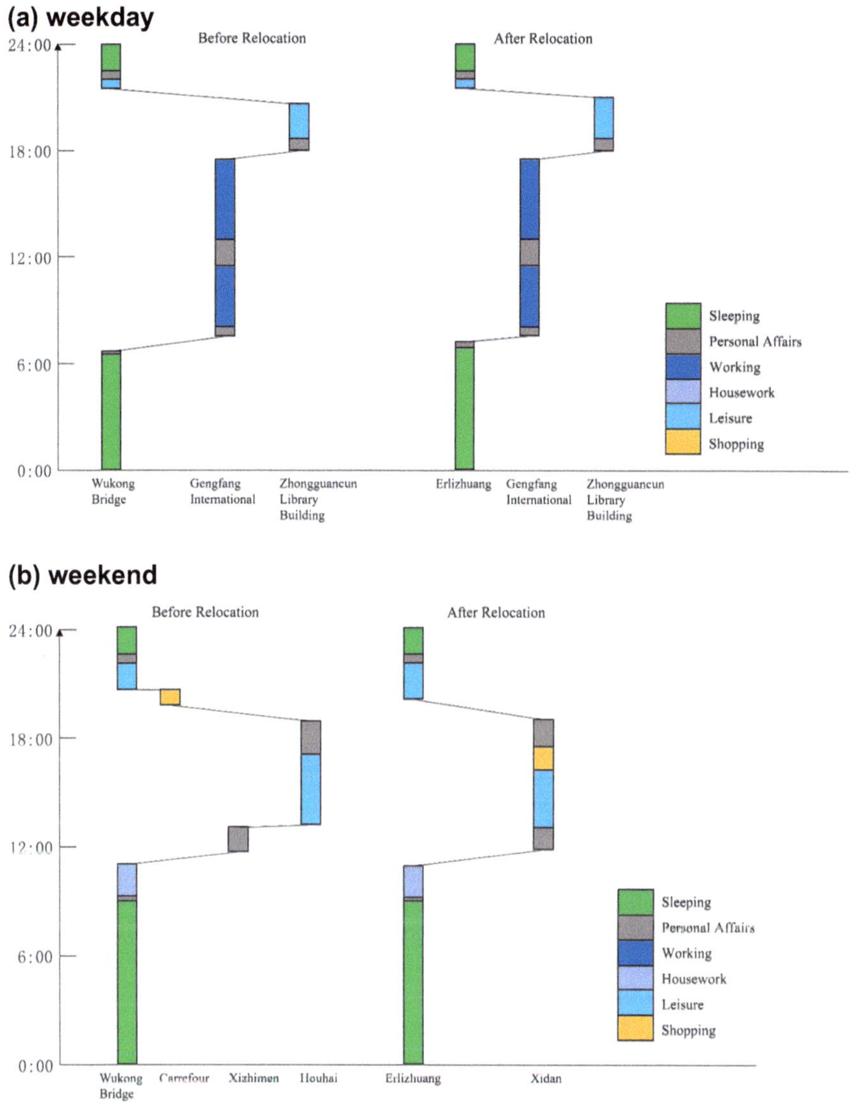

Fig. 7.12 Path of daily activities of male parents in JT1 families before and after the first relocation

food and drink, outside is a whole day...... almost six o'clock in the evening to the city, and then eat and chat together, to the home on the nine o'clock......I can't do much, just take a bath and rest and go to sleep......

(Mr. Wang, male parent of sample JT1 family)

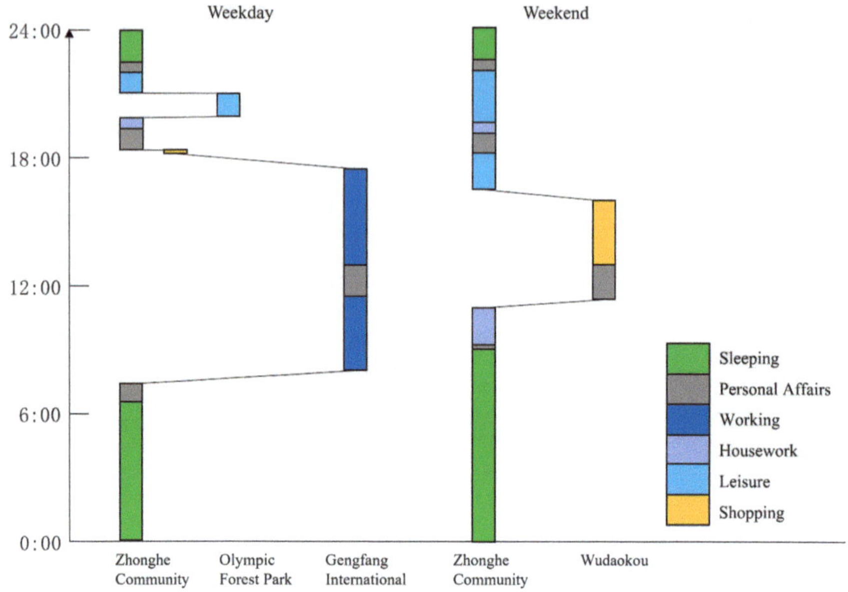

Fig. 7.13 Path of daily activities of male parents in JT1 families after the second relocation

7.3.1.2 Relocation Decision and Household Structure

Urban housing and residence are fundamental to everyday life, and are likewise the result of important life decisions based on household structure and income. When a household chooses a type of car, a place to live for work or study, it is also choosing each member's potential activity participation and mode of travel.

The fifth move in the Sample JT1 household occurred after the birth of Mr. Wang and Ms. Qiao's first son. The family structure changed in February 2014 when Mr. Wang's mother moved in due to Ms. Qiao's pregnancy and need for care. The conflict between the increase in family size and the small space in the old home prompted Mr. and Mrs. Wang to decide to purchase a new home.

The location and size of the new home took into account the change in the workplace of Ms. Qiao, the female parent, and the family's financial ability, and the final choice was the Changyang Bandao community in Fengtai District, which is closer to Ms. Qiao's workplace.

> ...... always have to start from the family, she finished maternity leave, although there is my mother, but also with children, to breastfeeding, then the new house is certainly the closer to her the more convenient...... and we want to buy a new building, do not want to be a small household, then I happened to know about this side of the community, came over to visit a few times, kindergartens, primary schools, junior high schools have supporting, at least children do not have to worry about school, and finally decided to buy here......
>
> (Sample JT1 family male parent Mr Wang)

After officially moving into his new home, Mr. Wang's father moved in at the same time out of the need to care for his children. At this point, Mr. Wang's social identity changed from husband to father. The change in family structure and living space also affected the change in Mr. Wang's daily activity space.

7.3.1.3 Changing of Employment and Commuting

Urban functions and land use conditions determine the configuration of urban facilities in different spatial locations. Traffic road network conditions limit the ability of different spatial locations to connect to the city. The distribution of employment centers determines the distribution pattern of jobs and residences in the city and the general direction of commuting. All of these factors are important geographic contextual factors that residents must face in order to live in the city. These three factors play an important role in constraining the mode of transportation used by individuals, the routes they choose for commuting, and whether they can carry out non-work activities on weekdays, thus affecting the efficiency of commuting and the organization of non-work activities.

For example, Mr. Zhang, the male parent in Sample JT2 household, his work-life distance varies with his place of residence and employment, which in turn changes his commuting time and commuting distance. In turn, work-life distance and work hours constrain their accessible urban spatial opportunities, further limiting their possibilities for non-work activities. Figure 7.14 shows the activity path of Mr. Lu on a particular weekday before and after the move.

Before the relocation, Mr. and Mrs. Lu's family of three lived together in the housing allocated by Ms. Zhang's mother's workplace, and Ms. Zhang's mother lived with them out of the need to take care of her granddaughter. Because of the close proximity of his residence to his workplace, Mr. Lu's commute is dominated by public transportation, with a commuting time of about 20 min, and his daily activities basically revolve around his place of residence. Non-working days are dominated by family trips for shopping, leisure and other activities, with the scope of activities expanding around the place of residence.

And after the relocation, Ms. Zhang's mother moved out. As Mr. Lu's working hours were more flexible, he took on more of the task of picking up and dropping off his daughter. The change of residence increased the commuting distance as well as the commuting time. Picking up and dropping off the child during the commute imposed more time constraints on Mr. Lu, which similarly limited his ability to engage in non-work activities outside the home. His out-of-home non-work activities are significantly reduced, with the only shopping and dining activities occurring after picking up his daughter. In contrast, path-choosing behavior due to changes in the transportation network in the Shangdi area directly affects his travel arrangements.

> This road was not open before, you can only go from the Xiaoying West Road side, the morning especially blocked, but I have to send my child to school well, I have to watch the time, catch no blockage before going out...... Now this side of the road is open you can go directly on the G7 to Xiaojiahe to the third ring, saving more......

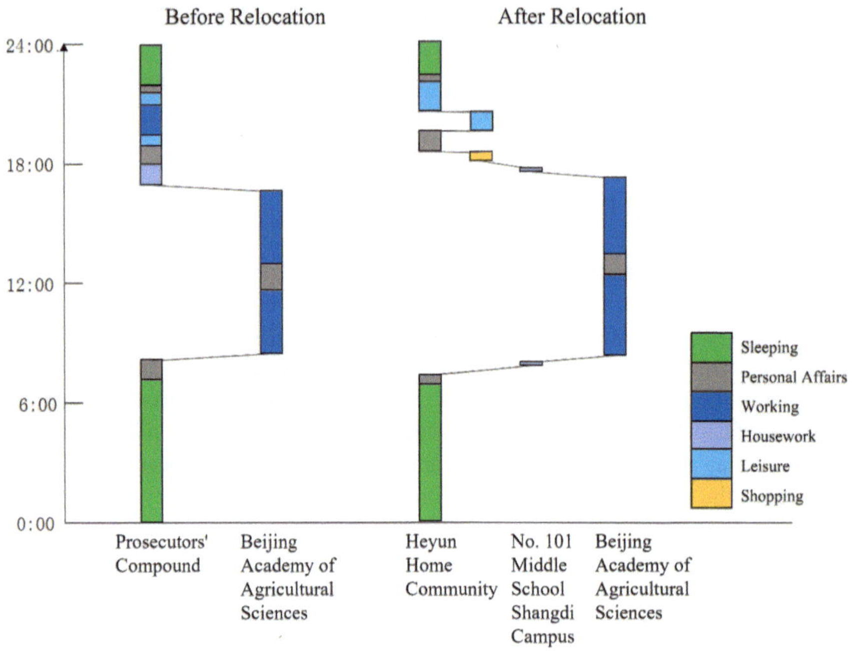

Fig. 7.14 Paths of activity of the male parent in JT2 families on weekdays before and after relocation

(Mr. Lu, male parent of sample JT2 family)

The time system of the work unit and individual career pursuits similarly constrain the individual's commuting activity arrangement and time utilization in a day. Taking Mr. Wang, the male parent of Sample JT1 family, as an example (Fig. 7.15), Mr. Wang was assigned to be in charge of an important project from April to September 2010 by his company. During these five months, Mr. Wang worked overtime on basically all weekdays, and at the busiest time he did not even have a double day off. Due to the work pressure brought about by the promotion and the project arrangement, Mr. Wang arranged more time for work, the daily working hours increased significantly, the time for activities within the home decreased, and there were basically no leisure activities on weekdays. On non-working days, Mr. Wang also arranges a certain amount of time for work activities, and the content of his non-work activities is mainly out shopping and a small amount of recreation, and the rest of the time he has almost no non-work activities outside the home.

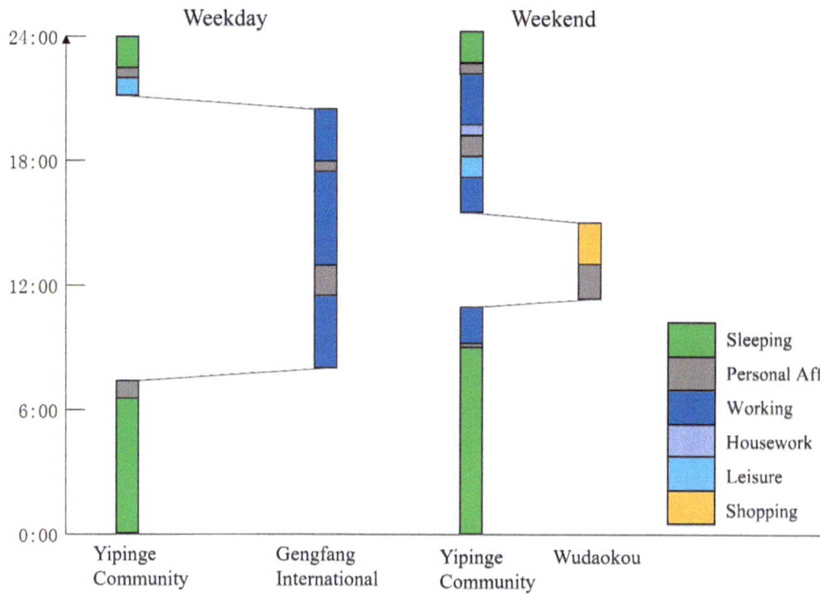

Fig. 7.15 Daily activity path for male parents in JT1 families during overtime

7.3.2 *The Influence of Changing Economic and Social Roles on Residents' Daily Activity Space*

The social nature that human beings possess both endows individuals with different social roles and social relationships, and restricts them to having to take on responsibilities in certain households or other organizations.

7.3.2.1 Changes in Social Status and Economic Conditions

The promotion brought about a change in social status, and the direct effect on the individual's daily behavioral space was not significant. The daily activity paths of male parents in JT1 and JT2 before and after promotion remained largely unchanged.

However, in the long run, the promotion of social status changed the economic conditions of the family, which in turn affected the family's mobility and leisure activities, among others.

The first influencing factor is the impact of changing economic conditions on mobility.

With the rapid growth of private car ownership among urban households in China, small cars have increasingly become an important means of transportation for residents' daily travel. Suburbanization, the growth of small cars and the construction of motorized road systems promote each other, pushing the travel structure of Beijing residents to shift to car travel. Changes in household economic conditions have made it possible for residents to afford to purchase small cars, thus changing the mobility of themselves or their spouses. Owning a car increases residents' employment accessibility and commuting-related mobility, while at the same time leading to corresponding urban problems such as peak hour traffic congestion, which reduces the efficiency of urban traffic operations. The spatio-temporal activity trajectories of households with different car ownership and car driving rights are differentiated.

In sample JT2, for example, the family purchased its first car in 2006 and was driven by the male parent, Mr. Lu. On weekdays, his activity space is slightly contracted compared to when he does not have a car, because he needs to take up the responsibility of picking up and dropping off children in the family while acquiring the right to drive a car, showing a clear tendency to be close to home. On the other hand, on non-working days, use of car expands the scope of leisure activities of the family. Traveling by car often implies long-distance activities, leading to a significant increase in activity space.

> Before the weekend is mainly to go to my parents, incidentally in the supermarket to buy some food, a little farther to take the children to the park sketching...... have a car is a little free, the weekend to go to my home to pick up my father and my mother together out of the clouds to stroll around and eat dinner...... Farther away you can go to Daoxiang Lake ah, Longquan Temple ah......
>
> (Ms. Zhang, female parent of JT2 family)

Also in terms of scheduling, Mr. Lu allocates his daily activities based on car use within a week's time frame. For example, focusing non-work activities on car use trips, while tending to reduce travel activities or make close trips when not using a small car for travel.

The second influencing factor is the effect of changing economic conditions on leisure patterns.

Changes in economic conditions can affect shifts in individual or household leisure styles and the space for leisure activities, e.g., an increase in the number of electronic facilities in the household, such as electronic facilities or the Internet, can similarly change an individual's path of daily activities.

In the case of Mr. Wang in the sample JT1 household (Fig. 7.16), for example, he purchased a desktop computer and installed a broadband network in 2007. Since then, on weekdays, Mr. Wang's place of leisure activities has shifted from the Zhongguancun Library Building to inside his home, where he goes straight home after having dinner with his colleagues near his office, and his leisure activities are mainly browsing the Internet, watching videos and playing games. The activity space shifted from the physical space of matter to the virtual space of the Internet.

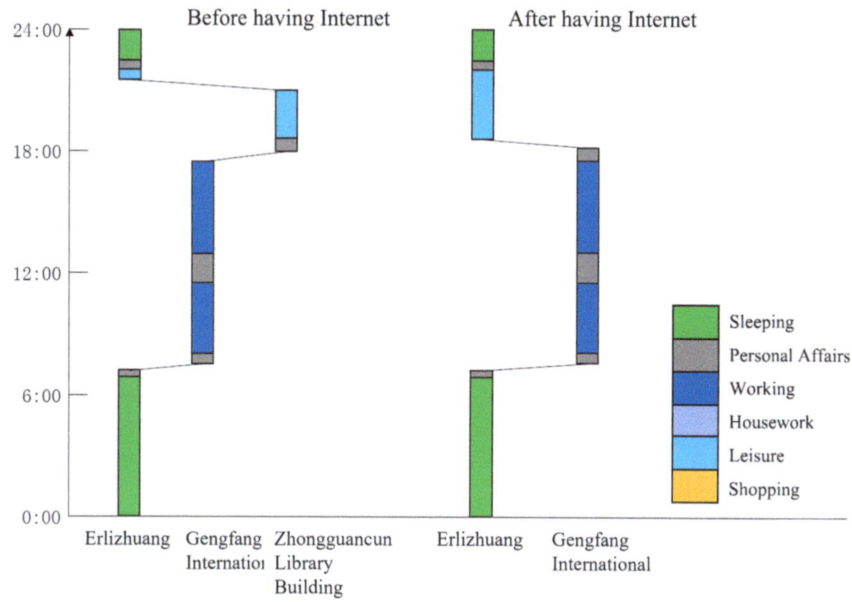

Fig. 7.16 Path of Mr. Wang's daily activities on a working day before and after installing the network

7.3.2.2 Changes in the Social Roles of Individuals in the Family

The social role of the individual in the family changes from that of a child to that of a husband, wife or parent, with an equal increase in responsibilities. This permanent change in roles also affects the planning of daily activities.

First, there is the impact of the transition from single to married.

In sample JT3, Ms. Fan's place of residence changed twice before she got married, following the relocation of her company. Whether she moved from Youanmen area to Fangzhuang area or from Fangzhuang area to Balizhuang area, each time she passively relocated followed the change of workplace, but always maintained a relatively close proximity between work and residence. Her commuting mode is mainly by public transportation, with shorter commuting time and obvious characteristics of close commuting, and the space of daily life overlaps with the space of employment, with locally oriented characteristics. Non-work activities on weekdays are mostly focused on shopping and dining around the workplace; leisure activities on non-workdays have a wider range, mostly departing after 10 a.m. and pointing to shopping centers in the city center (Fig. 7.17).

After getting married, Ms. Fan moved into the commercial house purchased by Mr. Liu, and her commuting distance increased significantly, changing from work-life proximity to long-distance commuting. Her commute is mainly by bus and subway, and her daily commute lasts more than 3 h. Her commute is characterized by leaving

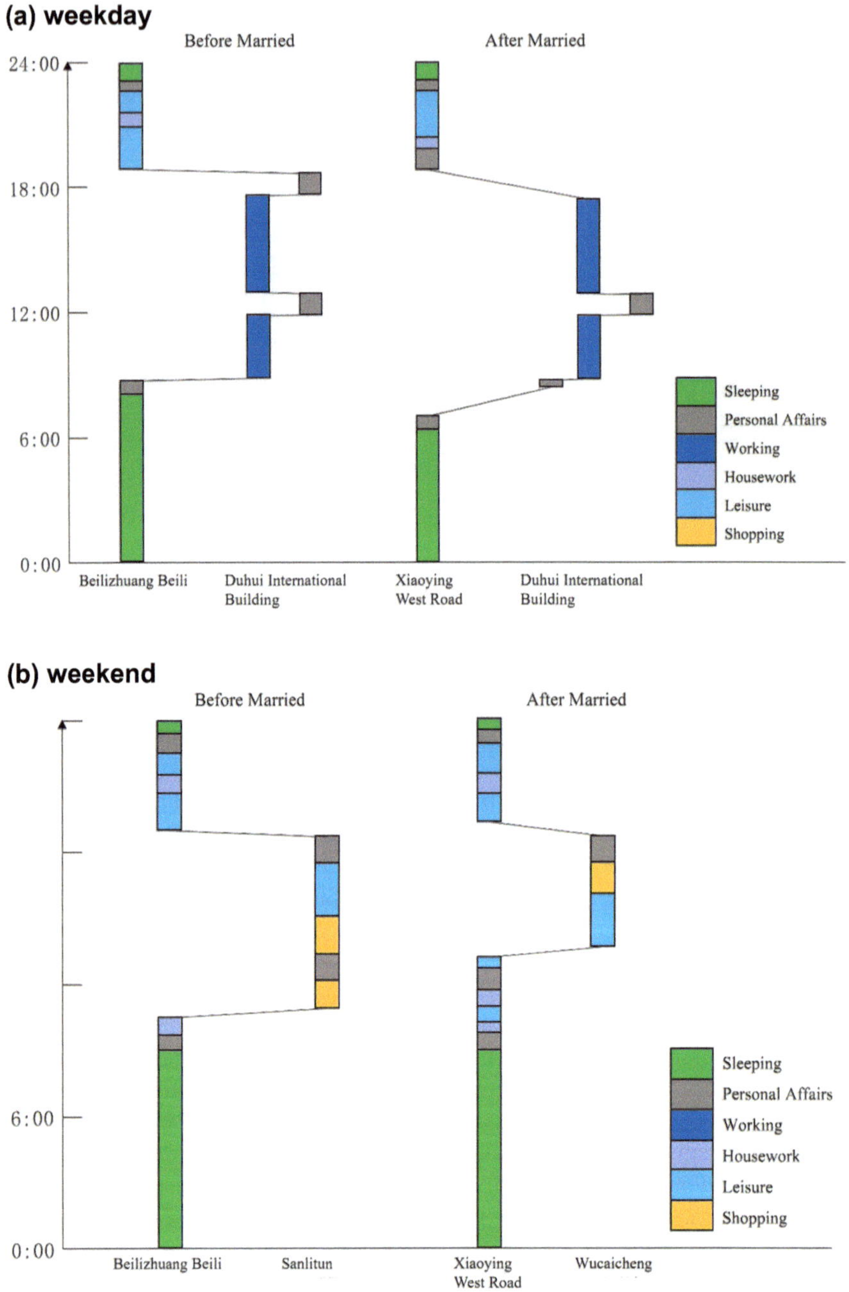

Fig. 7.17 Path of daily activities of female heads of JT3 families before and after marriage

early, returning late, and commuting to the city center by long-distance public transportation. On weekday mornings, she needs to leave as early as 7:00 a.m. She works long hours throughout the day and is at her workplace for almost the entire day. In the evening, she returns home late, and when she arrives home, she focuses on leisure activities within the home and rarely engages in activities outside the home. Her daily life on weekdays was significantly constrained by her work activities, and her non-work activities were dominated by meals on the morning commute and eating out near her workplace, with a small amount of occasional leisure activities, presenting a tendency to be bipartite between being at her workplace during the day and being in the city center with her residence in the evening. The content of non-working day activities is the same as that of her husband, Mr. Liu, with leisure and shopping as the main focus, mostly pointing to shopping centers around her home and distant suburbs.

Second, there is the impact of the change from wife to mother.

Ms. Qiao's social role in the family in the Sample JT1 family has gone through the change from wife to mother, and it is worthwhile to carry out academic observation by taking the changes in her daily life behaviors before and after she got married and gave birth to a child as an example (Fig. 7.18).

Before marriage, due to the proximity of work and residence, the workspace and residential space basically overlap. Ms. Qiao's spatio-temporal activity trajectories are characterized by local pointing on weekdays, and the overlap of employment space and daily life space can be seen in the distinctive characteristics of late departure, early return, and close commuting. Leisure activities during non-working hours are mostly directed to the urban space, with the proximity to the Zhongguancun area as the main focus; rest day activities are dominated by activities within the home, with occasional shopping and entertainment activities directed to the city center.

Ms. Qiao's lifestyle after marriage is one that prioritizes family responsibilities, with the most constricted temporal and spatial paths on weekdays, almost surrounding the home, going home earlier in the evenings and doing less outings after returning home than the male parent.

Her day-to-day activities on weekdays are constrained by both work and family responsibilities, with fewer non-work activities concentrated around the home and dominated by other non-work activities (household chores, personal affairs, socializing, etc.). On her workday, she is also almost predominantly active within the home, with activities outside the home showing a distance-decaying distribution centered on the home in all directions. Non-work activities on weekends were also relatively low, dominated by leisure activities within 12 km, and later away from home, mainly focusing on leisure activities 5–10 km away from home around 12:00 noon and a small amount of shopping activities within 2 km of home in the afternoon. Out-home activities are mainly near home, with a few long-distance trips in the afternoon or evening. Suburban spaces around the home provide essential amenities mainly in the midday and evening on weekdays, and are mainly used for other non-work activities. The urban space provides amenities mainly during the daytime on days off for leisure and shopping activities.

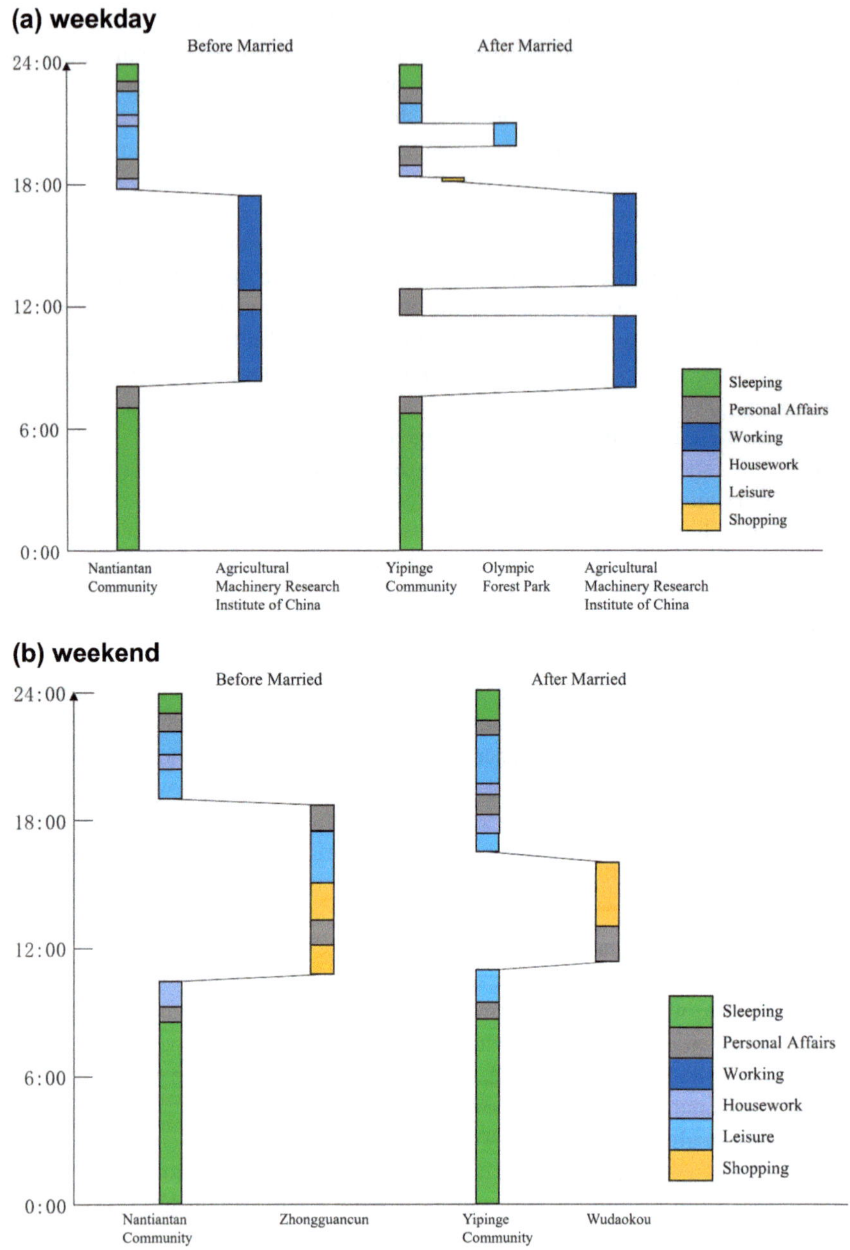

Fig. 7.18 Path of daily activities of female heads of JT1 families before and after marriage

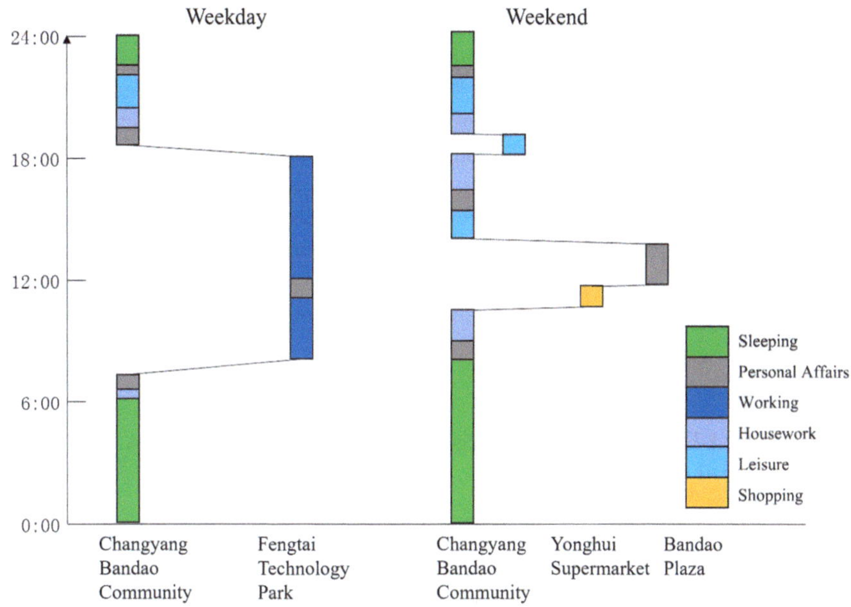

Fig. 7.19 Path of daily activities of female parents in JT1 after giving birth to a child

After the birth of her child, Ms. Qiao's workplace changed to Fengtai Science and Technology Park due to the organization's arrangement. The change in both work and living space had little effect on Ms. Qiao's lifestyle (Fig. 7.19). The increase in the number of family members puts more pressure on Ms. Qiao to raise her children, and her lifestyle continues to prioritize family responsibilities. In order to make it easier for Ms. Qiao to go to work, the couple purchased a car and Ms. Qiao uses it to go to work. Her mobility has increased, and her mode of commuting has changed from public transportation to car travel. At the same time, Ms. Qiao tends to go straight home from work, with a smaller percentage of non-work activities outside the home during her commute. Occasionally, out-of-home non-work activities are undertaken after returning home to compensate for the accessibility constraints during the commute.

7.3.3 The Influence of Household Structure on Resident's Daily Activity Space

Married or not, with or without children, sharing a home with an elderly person, and subjective factors such as the role and status of the individual in the household determine the type of household in which the individual lives. For example, whether it is a single person household, a nuclear family without children, a nuclear family with children, or a extended family with children, and so on.

The characteristics of the type of family in turn determine the resources and help available to individuals in the family and whether they are subject to the dual constraints of work and family. Because it affects the division of responsibilities and the allocation of time in a family, it also has a very important impact on an individual's space for day-to-day activities and employment choices. This section focuses on the differences in temporal and spatial constraints and behavioral patterns between male and female parents in nuclear and extended families, and analyzes the impact of external forces on the division of responsibilities and gender differences in the family.

7.3.3.1 Dual-Earner Families: Division of Family Responsibilities Based on Disposable Time

In dual-earner families, the division of family responsibilities is still based on the disposable time of both spouses. Both husband and wife are subject to the dual influence of employment and family responsibilities, with the less time-constrained members tending to take on more household chores. The scheduling of work and commuting activities between male and female parents can also have a significant impact on the pattern of division of household labor.

For example, in sample JT3 family, due to the longer commuting and working hours of the female parent, Ms. Fan, the division of responsibilities in this family is more inclined to the nontraditional division of labor, whereby the male parent, Mr. Liu, takes on more household chores on weekdays (Fig. 7.20). Conversely, on non-working days, the female parent takes on more domestic activities within the home and spends significantly less time on leisure activities within the home than the male parent.

After Mr. Liu married Ms. Fan in 2011, Ms. Fan moved in with him. At this point, Mr. Liu's social identity completed the transition from son to husband. As Ms. Fan works farther away from home and commutes longer distances, Mr. Liu takes on more domestic activities, and his evening commute pattern shifts, with a decrease in the proportion of non-work activities outside the home that take place after returning home, and some shopping in his evening commute. His in-home activity time has increased significantly, and his daily activities are more family-centered and nearby. His non-workday activities are dominated by leisure and shopping, mostly pointing to shopping centers around his home or to distant suburbs. In terms of time, long-distance non-work activities mainly started around 7:00 am, such as going to the far suburbs to climb mountains and hiking. Afternoon non-work activities are mainly leisure and shopping trips to the neighboring shopping mall (Wucaicheng in Shangdi-Qinghe area or Aeon Mall in Life Science Park).

Due to issues such as competition and promotion within the company, Mr. Liu chose to jump ship to NetEase Games company in 2012. After the second change of workplace, Mr. Liu's commuting pattern returned to what it was after he moved in

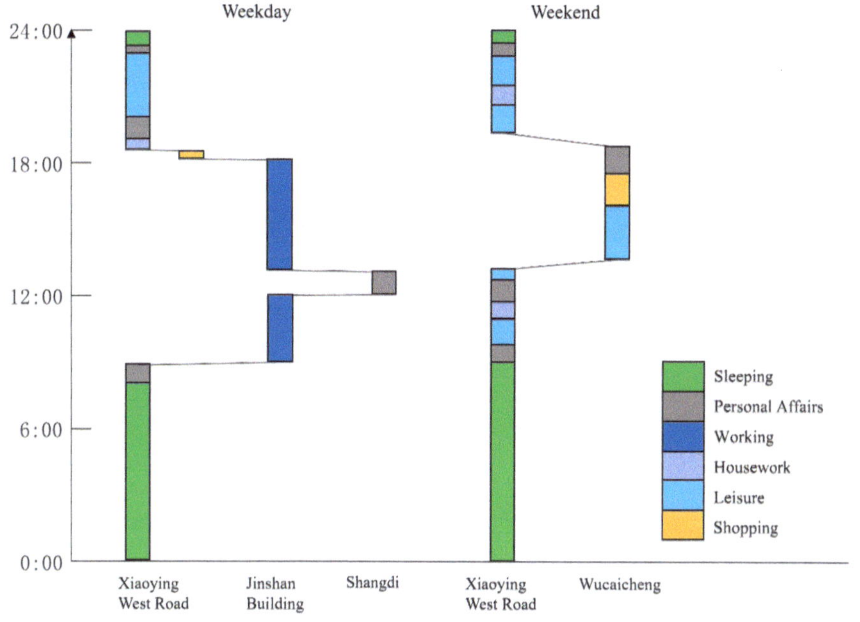

Fig. 7.20 Path of daily activities of male heads of JT3 families after married

2007, with longer commuting distances and commuting by subway and bus. Due to Mr. Liu's flexible work system and Ms. Fan's distance from her workplace, Mr. Liu remains the main bearer of domestic activities in the household, with little change in the division of labor within the household, but he does very little more going out after returning home, and there are almost no non-work activities outside the home in the evenings (Fig. 7.21).

7.3.3.2 Growing Families: More Space–Time Constrained

The family life cycle is usually determined by family size (total number of family members), age, marital status, and the emergence of children in the family and their ages. In particular, the emergence of children can have a significant impact on parents' activity patterns. For dual-income families with children, the conflict between employment pressure and family responsibilities is becoming an important factor affecting the quality of life of people living in large cities. After the birth of children, core family members are subject to more temporal and spatial constraints.

In the case of the sample JT2 family, for example, before marriage, Ms. Zhang's commute was dominated by bicycling and walking due to the close proximity of employment and residence, and the basic overlap of work and residential spaces (Fig. 7.22).The spatio-temporal activity trajectories are characterized by local pointing on weekdays, the employment space overlaps with the space of daily life,

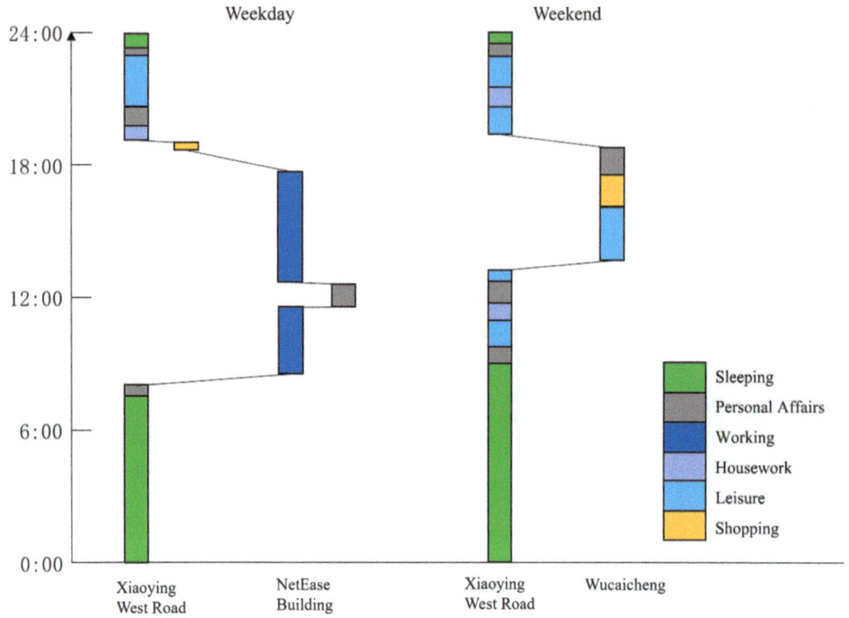

Fig. 7.21 Path of daily activities of male parents in JT3 families after the second change of work location

and it can be seen that there is a clear characteristic of late departure, early return, and close commuting, and most of the leisure activities during non-working hours are in the periphery of the residential space. After marriage, as Mr. Lu is still studying, Ms. Zhang takes on more household chores, the spatio-temporal activity trajectory on weekdays shrinks, and non-working days are also dominated by activities within the home, while activities outside the home are mostly concentrated near the home, with occasional small amounts of long-distance travelling.

After giving birth, Ms. Zhang's lifestyle continued to prioritize family responsibilities due to the pressure of childcare. On weekdays, however, her spatio-temporal activity trajectories are more contracted, with no non-work activities outside the home during her commute, and her home activities dominate when she returns home from work. Domestic activities centered on caring for her daughter became Ms. Zhang's primary activity within the home, and her personal leisure time was greatly squeezed. During this time, Ms. Zhang took on more family responsibilities, and the division of labor between male and female parents tended to be more in line with the traditional division of labor of 'male dominates outside the home and female dominates inside the home', leading to more pronounced gender differences in daily activity patterns.

During the period between moving into her new home and the time before she quit her job, Ms. Zhang's commuting distance increased, her commuting mode changed to a combination of bicycles and public transportation, and her commuting time grew

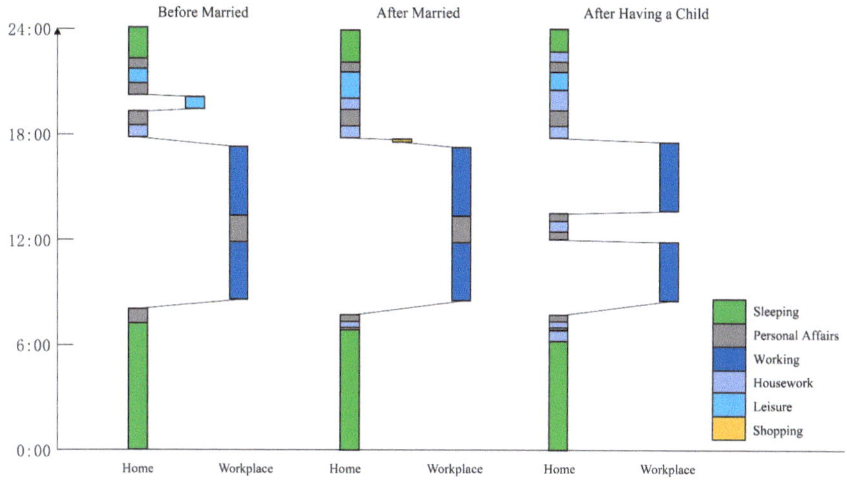

Fig. 7.22 Path of daily activities of female head in JT2 families before and after married and childbearing

and shifted to a long-distance commuting pattern of leaving early and returning late. She has to leave the house earlier in the morning and work longer hours, and works almost the entire day in the city center.

Due to the care needs of Ms. Zhang's father, Ms. Zhang's mother moved out of the household and the couple began to share the task of caring for their daughter. Considering the time system and flexibility of their workplaces, Mr. Lu was responsible for picking up and dropping off their daughter, and Ms. Zhang was responsible for cooking and other household chores, and hardly ever went out for non-work activities outside of the home when they returned home, and most of their daily activities occurred in the neighborhood and community around their home.

In 2015, Ms. Zhang's daughter was admitted to Beijing 101 Middle School, and she resigned from her job at a state-owned enterprise for reasons of personal career development and to take care of her daughter's education. Since then Ms. Zhang has had a more flexible daytime schedule, taking on most of the housework and the task of taking her daughter to school (Fig. 7.23). The task of picking up her daughter from school is still undertaken by Mr. Zhang, whose schedule is also constrained by her daughter's school time system.

> ...... Attendance in that previous company was getting stricter and less effective, and development was not as good as when I first started working...... My daughter was quite competitive and got into a good school, and I wanted to put more effort into her as well...... It just so happens that I have a friend in the education and training, only with the weekend time classes, usually compiled teaching materials, in the network to the students to answer questions, the monthly salary is also not bad, then the trend of early retirement...... Usually at home to write, do housework, the weekend go to Classes, easy and also can take care of the family.......
>
> (Ms. Zhang, female parent of sample JT2 family)

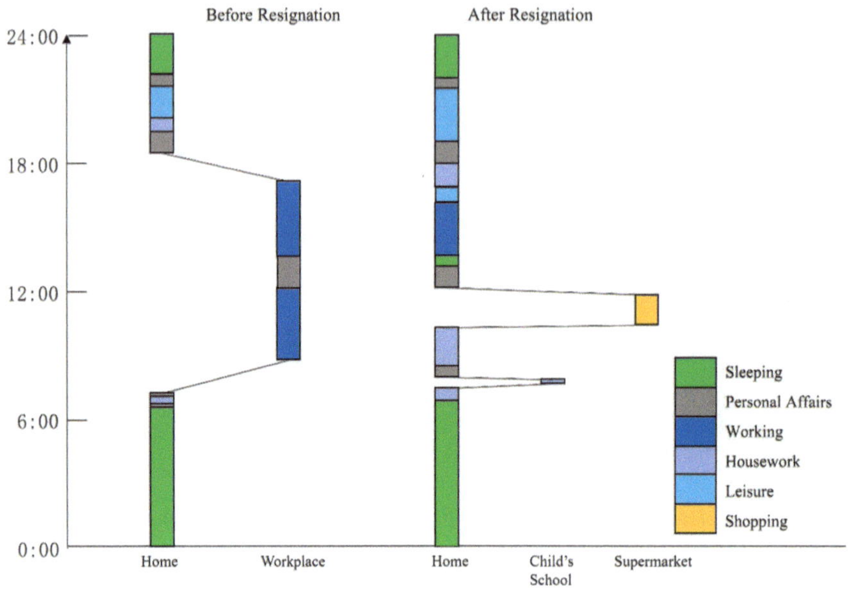

Fig. 7.23 Changing of female parent workday activity paths in JT2 families before and after resignation

7.3.3.3 Extended Families: Space–Time Flexibility Due to the Co-living of Grandparents

Families can also seek help from extended family members (e.g., grandparents, grandparents-in-law, siblings). Involvement of extended family members affects gender differences in the division of family responsibilities and changes the behavioral patterns of male and female parents. In the life course of urban Chinese elder, with the birth of the third generation, due to the busyness of their children, grandparents often take over the responsibility of caring for them in place of 'nannies' before the third generation becomes more independent, resulting in a typical pattern of out-of-home mobility and daily life in a given stage. In extended families, the addition of grandparents during the workday reduces the burden of household chores and childcare and gives parents of both genders more space-time flexibility. This allowed both male and female parents to feel fewer time constraints and to engage in more time flexible activities.

Take the example of the older people in Sample JT1 households (Table 7.4). From the weekly out-of-home activities of the two grandparents, it can be seen that on weekdays, due to the need to accompany and care for their grandchildren, the grandparents' daily trips revolved around life-sustaining activities such as shopping and errands around the community, with the exception of walks within the community; the rest of the time they spent most of the time in the home, with no other out-of-home travelling. Because one person must stay home with the grandchildren at a time, both grandparents go out on their own for activities (except for morning walks with the grandchildren in the community). On weekends, it was often the male parent who drove for family outings, meals, or excursions, thus creating a larger range of activity space. All of the grandparents' daily leisure is based on indoor low-grade leisure, such as reading, practicing, watching television, etc. The grandfather occasionally goes to the community activity center to play cards with other elderly people.

> We are retired and have nothing to do, come over to help take care of the grandson, neighbors, friends know not much, we both usually mainly stay at home to read books, watch TV, generally do not go out...... mainly to look at the child Well, the child is now a little bigger, more things in the house for fear of falling, but also have to carefully stare at the He...... afternoon air is good old companion will push the child out for a walk, I sometimes accompany her to go together, the child is big, she sometimes cannot look at a person...... do not take the child when you read books, practice writing, I am so a hobby, that is happy in it!...... My partner watches TV, mainly to listen to the opera...... On weekends, my son and daughter-in-law are in the house, so I take it easy, and sometimes go out to play cards, in the community activity center, there are several old partners......
>
> (Sample JT1 family male parent Mr. Wang's father)

The co-living of grandparents was clearly helpful for the female parent, but this help appeared to have a greater effect on reducing the time constraints of the male parent, increasing more the likelihood that they would be involved in activities with a more flexible schedule.

It is useful to carry out further analysis using the example of the daily activity paths of male parents in JT1 families before and after their fifth move (Fig. 7.24).

After the move, Mr. Wang's lifestyle changed from work-life proximity to long-distance commuting due to a significant increase in commuting distance, with a total daily commuting time of nearly three hours, and a predominantly subway commuting mode. He leaves home earlier in the morning, works longer hours, spends almost the entire day working in the city center, and returns home later in the evening, with fewer outings after returning home. Non-work activities in the morning are mainly around 8:00 a.m. and are dominated by short dining activities near the office. Midday non work activities are the most frequent, mainly around 12 o'clock, and are mainly eating out near the workplace, with a small amount of leisure activities as well. In the evening, non-work activities are mainly concentrated between 17:00 and 21:00, and are dominated by long-distance leisure and dining activities near home, with few shopping activities on weekdays. On the other hand, non-working days are basically dominated by activities near home, with occasional outings of about half a day to city centers or distant suburbs, dominated by leisure and shopping activities. Whereas on non-workdays the help of older people does little to ease the time–space constraints for both male and female parents, given that older people have care needs of their own, regular activities at home on weekends increase.

Table 7.4 One-week out-of-home mobility patterns of grandparents in JT1

Family member	Day type	Activity	Frequency	Location	Activity companions	Activity duration (min)	Travel companions	Transportation mode	Travel duration (min)
Female	Weekdays	Walking/ Strolling	7	Residential compound	Alone	40	–	Walking	–
		Shopping	5	Convenience vegetable market	Alone	40	Alone	Walking	15
		Shopping	2	Yupeng supermarket	Alone	20	Alone	Bicycle	20
	Weekends	Dining	1	Peninsula plaza	Spouse, children, grandchildren	90	Spouse, children, grandchildren	Walking	20
Male	Weekday	Walking/ Strolling	3	Residential compound	Alone	90	Grandchildren	Walking	-
		Shopping	2	Convenience vegetable market	Alone	40	Alone	Walking	15
		Shopping	1	Yupeng supermarket	Alone	20	Alone	Walking	30
		Personal Care	1	Community vicinity	Alone	180	Alone	Walking	5
		Socializing	1	Community area	Alone	40	Alone	Walking	5
	Weekends	Dining	1	Peninsula plaza	Spouse, children, grandchildren	90	Spouse, children, grandchildren	Walking	20

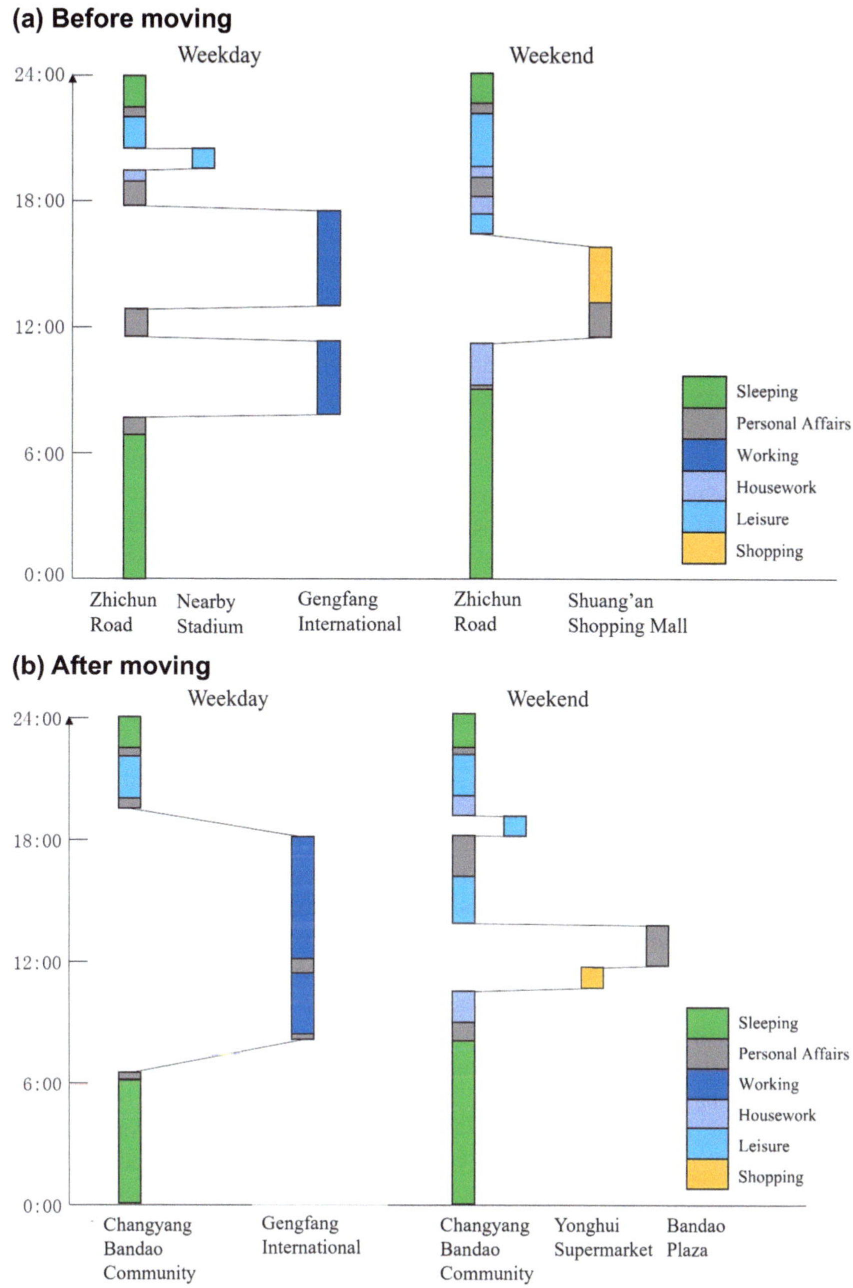

Fig. 7.24 Path of Mr. Wang's daily activities before and after relocation

7.3.4 Activity Space of Chinese Urban Residents in the Interactive Perspective of Life Paths and Short- and Long-Term Behavior

Important events in the life course of urban residents in China since the market transition, including relocation, employment choices, and changes in the social roles of members within the household, affect the routine activity patterns of each family member by altering the geographic context, mobility, and division of responsibilities within the household (Chai and Feng 2016). The life course of urban residents belongs to the long-term spatial behavioral perspective, while the habitual activities of urban residents belong to the short-term spatial behavioral perspective, which suggests that there is some degree of interaction between the long-term and short-term behaviors of residents.

The long-term behaviors of urban residents in China, such as employment and relocation, change the social and physical space that they come into contact within their daily lives through changes in their work or living environments. The social and cultural atmosphere in which individuals live often changes as well. After the change of work arrangement or living space location, individuals often change the time and space utilization of daily activities according to their actual needs. In order to adapt to the new external environment, individuals will often adjust their daily behaviors to make their daily activities harmonize with the rhythm of the microenvironment in which they live. Individuals maintain their own sense of ontological security by completing the routinized activities allowed by the new environment (Giddens 1998). Once the rhythm of life in which the individual adjusts his or her daily behaviors fits with the surrounding environment and is accepted or forced by the individual, this new routine is perpetuated and solidified as a pattern of daily life for a relatively stable period of time until it is altered by new influences.

Changes in residential and employment spaces in urban China bring about changes in the occupational and residential relationships of individual residents in households, as well as changes and constraints in geographical contexts. These factors affect the opportunities for individual residents to minimize combinatorial constraints in engaging in different matters of daily activity. For suburban residents in Chinese cities, residential space is separated from employment space, and they may be able to move to other geographic areas or urban centers for employment by means of small cars and public transportation. And due to the separation of workplace and residence, individual's daily activity space revolves around residence and workplace, and the space of leisure, shopping, and other out-of-home non-work activities expands outward, constituting the residents' composite living space characteristics. The workplace has a strong spatial correlation with out-of-home non-work activities, and is one of the main starting points for out-of-home non-work activities such as shopping, leisure and eating out.

While the improvement in social status does not have a significant direct impact on the path of daily activities of individual residents, the rise in status, such as promotion, brings about a change in the economic conditions of the household, which affects the mobility of the household and the way in which leisure activities are carried out.

The use of automobiles is the main factor that increases the level of mobility and the level of utilization of urban space by residents. An increase in household car ownership can significantly expand the activity space of weekday residents and increase their mobility.

Changes in the social roles of individual residents and changes in family structure in long-term behavior, as well as the scheduling of work and commuting activities of male and female parents in short-term behavior, combine to have a significant impact on the pattern of division of household labor within the family. In terms of commuting time, when the male parent commutes longer distances than the female parent, the family prefers the traditional division of labor. If the female parent works longer hours, the family prefers a non-traditional division of labor.

In terms of family structure, nuclear families without children are more likely to choose a non-traditional model of division of labor, which is generally a type of division of family responsibilities centered on disposable time. In nuclear families, in general, there are more space-time constraints on the activities of family members. The division of household labor among male parent significantly affects the spatial patterns of activity of female parent during the workday. The division of household labor moderates the effect of gender on individual activity patterns. In extended households, intergenerational assistance in urban Chinese households helps married female residents balance employment and family life. In nontraditional division of labor households, gender differences between male and female parents are smaller than in traditional division of labor households. Male parents instead spend more time on housework activities as well as shopping activities, and there are no significant gender differences in the allocation of time between male and female parents for other activities. Moreover, on weekends, females take on more household activities in both types of households.

7.4 Activity Space Based on Home and Pocket of Local Order

The 'Home' is a symbol with both physical space and sociocultural significance in geographic research. The home separates the 'inside' and 'outside' of the family through the physical space constructed on the basis of the house, bringing about a physical spatial division, whereby the inside and the outside of the home become two different spaces and bring about corresponding differences in social relations. Therefore, some scholars have proposed that 'home' is a symbolic place full of binary contradictory concepts, internal and external, private and public, male and female, sacred and secular, etc. are all embodiments of the duality of home space (Feng et al. 2015).

Home space is an independent space separated by walls, a space that contains a combination of relationships between individuals and others, individuals and society, and family and society. It is likewise a system of different spaces and resources that provide a series of restrictions and opportunities for each individual in the space. In the home space, activity occurs as a process in which each individual uses the space within the home (e.g., kitchen, bedroom) and the resources of the home in a certain order. In this process, there are contradictions and combinations between individuals, which require the family to realize internal coordination mechanisms so that the space and resources within the home can be used appropriately.

The home shows the interaction of the individual with other members of the home and with the activities and spaces within the home. The home is 'the reference point of human existence' (Relph 1976). The sense of belonging in the home is demonstrated through boundaries such as walls and gates in opposition to the larger space outside the home. Individuals have a need to interact externally in society, and individuals also travel through transportation in order to cross this boundary into public space. Social space is the place where people interact with each other and with space. There is a relationship between home space and social space that is both antagonistic and inclusive. The home is a key site for understanding society and space (Domosh 1998), so this section hopes to discuss the individual's daily behavioral activities and the ordering of family activities from the perspective of the home.

7.4.1 Accessibility of the Home as an Activity Space

The in-home space, as a typical representative of private space, provides basic protection and security as well as emotional support for the members of the family, and is also the main space for social interaction based on blood relations. The social space outside the home is more open, and is the space where individuals gather with acquaintances or strangers based on relationships such as karmic ties and township ties. In the out-home space, the individual is separated from the family and nuclear relatives and is more expressive of public will. The in- and out-home space are antithetical to each other, and when individuals are frustrated or hurt in the public space, they retreat to the private space within the home to seek solace.

The in-home space has its own private nature, where relatives can naturally gain access to the space, but others need to obtain an invitation or permission from a family member to enter. The ease of access to the in-home space is a direct reflection of the family's interpersonal patterns and the degree of protection of private space. In daily life, who an individual resident chooses as an activity companion directly affects the individual's decision-making about the location of subsequent activities. Based on the existing interviews and activity logs, and with reference to Zhao et al.'s (2013a, b) study, this section classifies individuals' activity companions into four basic types according to whether they are related by blood or marriage and the degree of autonomy (Table 7.5). At the same time, the interviewed residents' willingness to choose was investigated, which in turn was divided into a single sample and a family sample to discuss residents' openness to home space separately.

The meeting places for activities with peers were classified into three categories: one's own home (inviting friends and relatives to visit one's home without having to go out of the home), other people's homes (invited or initiated to visit friends and relatives in other people's homes, etc., which is an out-of-home activity, but requires access to other people's private space), and public spaces (all public spaces excluding one's own home and other people's homes, including recreational and entertainment venues, restaurants, and public places, etc.).

All single samples preferred to meet with their activity companions in public space, and the activity companions were mainly friends and colleagues. Among them, the single male samples' willingness to go to other people's homes and to invite people to their own homes were closer, and they were significantly more open to their homes, inviting friends to their homes on weekends or holidays, while other people's homes were more often the homes of their hometown friends living in the same area of Beijing. The higher degree of dependence on this kind of geo-sustained ties is a result of the reconstruction of the original geo-ties in their hometowns in other places. Due to the exclusion of the household registration system and the land system, these new immigrants to Beijing are in a relatively isolated state in the city, with insufficient social resources, and the friendships established by hometowns from the same place and the period of study have become the most important social resources in the city, and they have built up a social network and become a common whole through the geo-ties. They build up a social network through geographical ties and become a common whole.

Single women who live alone are extremely guarded and have very low willingness to invite others to their homes. They also have a heavier demand for privacy in their homes and will not easily agree to allow others to enter their homes, except for blood relatives, and prefer to choose public spaces for their general social activities.

'Previously there were old folks who cheated money, afraid of being cheated, there is a precautionary mentality...... mainly still and the old family familiar friends to and fro, generally about outside...... my own personality is also more introverted,

Table 7.5 Classification of activity companions

Activity companions	Description	Notes
Solo activities	Choosing to independently complete daily activities without companionship	–
Core kin	Spouses and children	Blood or marital relationships
Extended kin	Parents, grandparents, siblings, and other relatives	Includes lineal and collateral relatives
Friends/colleagues	Non-kin individuals maintaining frequent social contact	Socially constructed relationships

Source Based on Zhao et al. (2013a, b)

less active...... You see the news on microblogging now, there are too many similar cases...... wouldn't dare to let too many people come to the house......'

(Sample B4)

This difference side by side reflects the imperfection of the current social order and the lack of security brought to single women by some focused news. This sense of insecurity prompts them to protect themselves and their home space closely to avoid being victimized by the outside world.

As for the choice of activity locations and activity companions in the household sample, residents' willingness to choose core relatives as activity companions is stronger, and their willingness to choose activity venues in their own homes and public spaces is also stronger. When the activity takes place in one's own home, individuals are more willing to choose core relatives as activity companions, followed by other relatives, and friends and coworkers are the least willing. When the place where the activity occurs is in other people's homes, it is usually invited to enter other people's home space, and the activity that the individual actively chooses is very rare, and is most likely to occur in other relatives' homes. This suggests that residents are currently relatively protective of their homes, and that relatives based on blood and marriage relationships are more likely to be trusted and therefore more likely to enter the home space

'...... I'll go to my parents' home to accompany them, generally do not go to other people's homes, especially colleagues' homes, a lot of only know which piece they live in, specifically in which district do not know...... Occasionally, friends about it to go to their home to eat a meal Get together, but also very little, only a few days of the New Year's holidays...... Now we are still all about the outside restaurant, mainly in other people's homes is inconvenient, a little awkward, after all, with their own homes are not the same......'

(Sample A12)

When the activities take place in public space, individuals on weekdays are more inclined to choose friends and colleagues as their activity companions, which is generally based on karmic relationships and common workspace, and is constrained by work time and space. On non-working days, they tend to choose activities with other family members living with them, which are mostly for leisure or shopping activities. Especially in extended married families with children, activities take place in their own homes and public spaces, with minimal activities in other people's homes. This is probably due to the fact that living with grandparents increases daily interactions in one's own home and replaces activities in other people's homes.

In short, the home space as a behavioral space, family members clearly show the cognitive distinction of blood relationship as 'close' and friends as 'distant'. This is closely related to China's traditional concept of family, which makes it easier to gain trust and maintain solid ties based on blood relations. Individuals are highly protective of their own home space, with a low degree of openness to others, and a low willingness to enter others' home space. This strong sense of protection comes from the lack of security brought about by imperfect public order, and is also a reflection of various factors such as social culture in daily space. With the rapid development of urbanization, and against the background that individual rights and interests and security still need to be further improved and protected, residents have become more and more protective of their individual private space. As a result, there are relatively few opportunities for friends or colleagues who are not related by blood to enter other people's homes. In particular, rapid informatization has brought about the wider dissemination of various types of negative events, and residents will also increasingly shrink the threshold of home space accessibility. For example, Sample A17, after seeing news on Weibo that residents in the same neighborhood had a dispute with a courier that affected his daily life, she chose to change the courier address from her home to the entrance of the neighborhood to avoid similar incidents.

7.4.2 *Home as a Pocket of Local Order*

In real life, people's daily lives often require continuous travel between multiple locations. For example, they have to go from home to work, or from home to the mall, etc. Therefore, people's daily activities are often influenced by a combination of different local orders, which looks like a pocket of local order. The pocket of local order of activities not only affects residents' activities in one day, but also shows regularization and institutionalization when viewed on a weekly or longer time scale (Hägerstrand 1985). Ellegård et al. (2016) extend spatio-temporal activity trajectories to multidimensional paths, arguing that the pocket of place order can be understood using the concept of context in time geography.

7.4.2.1 Home as a Pocket of Local Order

According to Ellegård et al. (2016), the home is a central space and an important node in people's lives. Family members cooperate with each other in both space and time to form the local order of the home. This order influences the daily activities of each family member and maintains the normalization of family life. In Ellegård's study, she created an activity-oriented path representation. She has developed a new way of representing 'paths' from the perspective of the complex contextualization of daily activities, expanding spatio-temporal activity trajectory to multidimensional path. The concept of 'context' in time geography is used to understand the pocket of local order. The context is categorized into three dimensions, everyday context, geographical context, and social context.

The empirical analysis is carried out using the context of the daily activities of Mr. Lu and Ms. Zhang (sample JT2) before and after the female parent resignation as an example. In Figs. 7.25 and 7.26, on the left is the geographic context constructed based on activity paths, which represents the location where the activity occurs, the time of stationing, and the mode of transportation used. In the center is the daily activity context based on daily activities, representing the order in which different

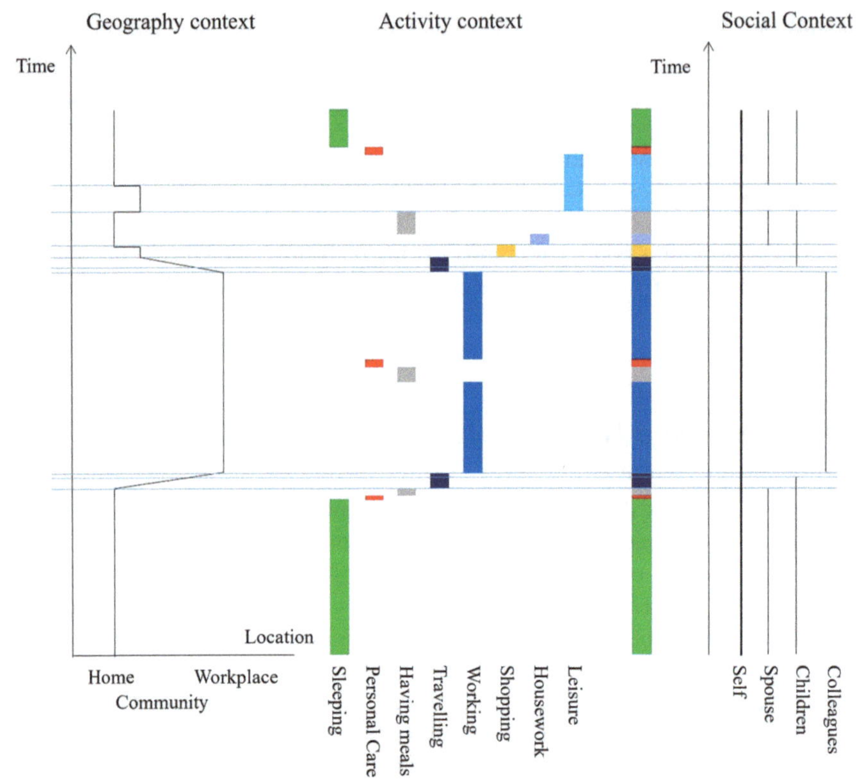

Fig. 7.25 Activity context of male parent in workday before female parent's resignation in sample JT2 household

types of activities occur and the duration of the activities. On the right is its social context, representing activity companions.

Before Ms. Zhang resignation, Mr. Lu was the driver of the family car, and Ms. Zhang used public transport as her main mode of commuting. On the weekday mornings, Ms. Zhang woke up first to wash up, then Mr. Lu woke up. Ms. Zhang started to prepare breakfast for the family; Mr. Lu washed up and woke their daughter up, and then the whole family had breakfast together. Due to the long commuting distance, Ms. Zhang left home first after breakfast and waited for the bus. 10 min later, Mr. Lu left home with their daughter. Mr. Lu sent their daughter to school on time and then went to his workplace, where he arrived earlier than the official start time and read the newspaper before starting work. Half an hour later, Ms. Zhang arrived at her workplace and started her day's work. At noon, both of them took a meal close to their workplace. During working hours, both of their activity companions were colleagues in the workplace. At 16:50, Mr. Lu left the workplace early, and went to school to pick up their daughter, then went to a nearby supermarket with their daughter to buy the necessary ingredients for dinner and arrived home at 17:50. After a short break, their daughter began to finish her homework, and Mr. Lu began to prepare dinner.

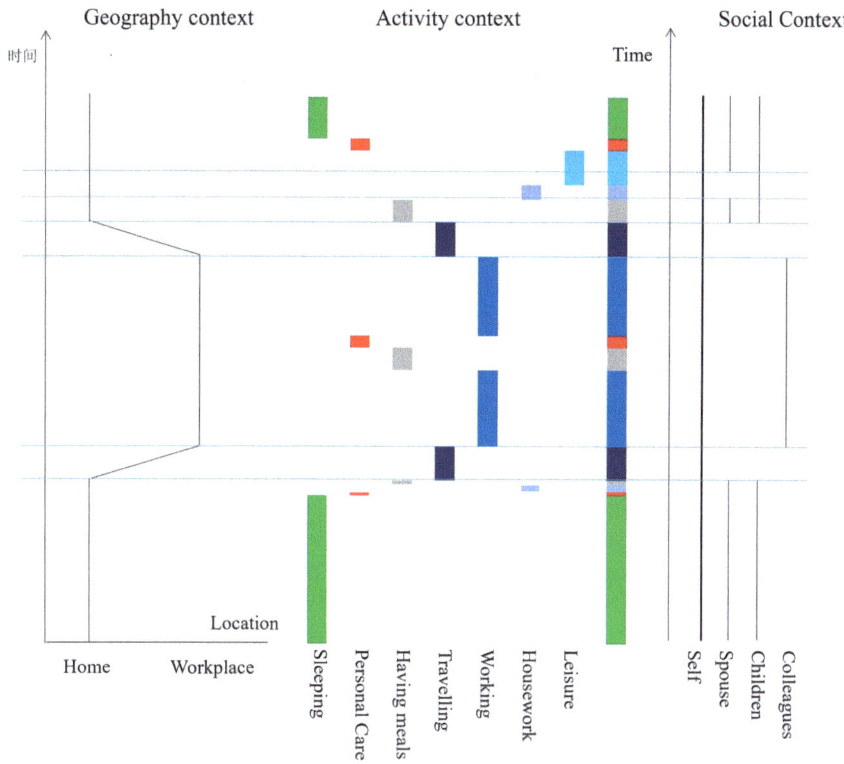

Fig. 7.26 Activity context of female parent in workday before female parent's resignation in sample JT2 household

Ms. Zhang came home from workplace at 17:10, and after taking two buses, she arrived home at 18:30. And then, the whole family had dinner together. After dinner, Mr. Lu went out for a walk, and Ms. Zhang cleaned up the table, washed the dishes, cleaned the house, and began to watch TV dramas. Mr. Lu came back home an hour later, and they watched TV together until 21:40, when Ms. Zhang started to wash up, and 20 min later, Mr. Lu started to wash up as well, after that they both went to sleep. At this point, this family has completed all its activities for one day.

This order has changed after Ms. Zhang resignation (Figs. 7.27 and 7.28). The driving rights of the family car have changed from Mr. Lu's sole use to they taking turns to drive. On a weekday morning, Ms. Zhang woke up 20 min later than before and then woke their daughter up. After that, she prepared breakfast and sent their daughter to school after breakfast with their daughter. After Ms. Zhang and her daughter left home, Mr. Lu woke up, washed himself and started to enjoy breakfast. When Ms. Zhang returned home, Mr. Lu drove out to work and Ms. Zhang began personal care such as washing up. During this period, Mr. Zhang arrived at his workplace to start work. After washing up, Ms. Zhang started washing dishes, tidying up the house and cleaning the house. At 10:30, a friend in the community called

Ms. Zhang went out with her, and they took the supermarket's commuting bus to a supermarket in Xierqi area to buy food ingredients. An hour later, they took a bus home, and when they arrived, Ms. Zhang began to prepare her lunch. After finishing her lunch and cleaning the dishes, Ms. Zhang began to take a lunch break. After an hour, she woke up and began to work, such as preparing lessons for weekend classes and answering questions from students via the Internet. During this period, she also did some leisure activities, mostly of them are browsing the videos of website, etc. At 17:00, Mr. Lu left his workplace and went to the school to pick up their daughter, and then returned home with her. At 17:00, Ms. Zhang began to prepare dinner, and 40 min later, Mr. Lu and his daughter arrived home. Then the family had dinner together. After dinner, Mr. Lu took on the chore of washing dishes and went out for a walk after completing his chores. Ms. Zhang started her leisure activities, which consisted of browsing the Internet and watching TV programs. When Mr. Lu came back home, they watched TV programs together until they washed up and went to bed. And then, they finished the day's personal caring activities in the same way as the previous time use. It can be seen that after resignation, Ms. Zhang has more time flexibility and therefore takes on the main family responsibilities, including dropping off the children, shopping, housework, etc., and the proportion of individual leisure activities rises.

In the former context, the order of the family during the working day is that the female parent is responsible for washing the dishes for breakfast and dinner, while the male parent is responsible for picking up the children, shopping and preparing dinner. In the latter context, the order of the family's workday is changed to one in which the female parent is responsible for breakfast, taking the children to school, shopping, and dinner preparation, while the male parent is responsible for only two tasks: picking up the children and washing the dinner dishes. The fulfillment of the family's local order requires communication and cooperation among the family members in deciding on the division of tasks, places, times and ways of accomplishing the family's activities. This local order also affects the space-time behavior of each individual within the family, forming a regular, patterned, and habitual way of resident behavior.

7.4.2.2 The Local Order of Family Resource Use

Another important development of Professor Ellegård's concept of the 'pocket of local order' is the combination of human activity paths with the space-time allocation of local resources, visualizing how actors use local resources in an orderly way to accomplish their activities (Ellegård et al. 2016).

It is worthwhile to carry out an empirical analysis using the example of the project of picking up and dropping off children within the family.

Before Ms. Zhang resigned from her job, Mr. Lu, as the driver of the family car, had a shorter commute than his wife and thus took on the task of picking up and dropping off children. The activities of picking up and dropping off children are indicated in red (Fig. 7.29). It can be noticed that Mr. Lu needs to complete the activity of picking up children twice a day, and this activity is repeated on weekdays.

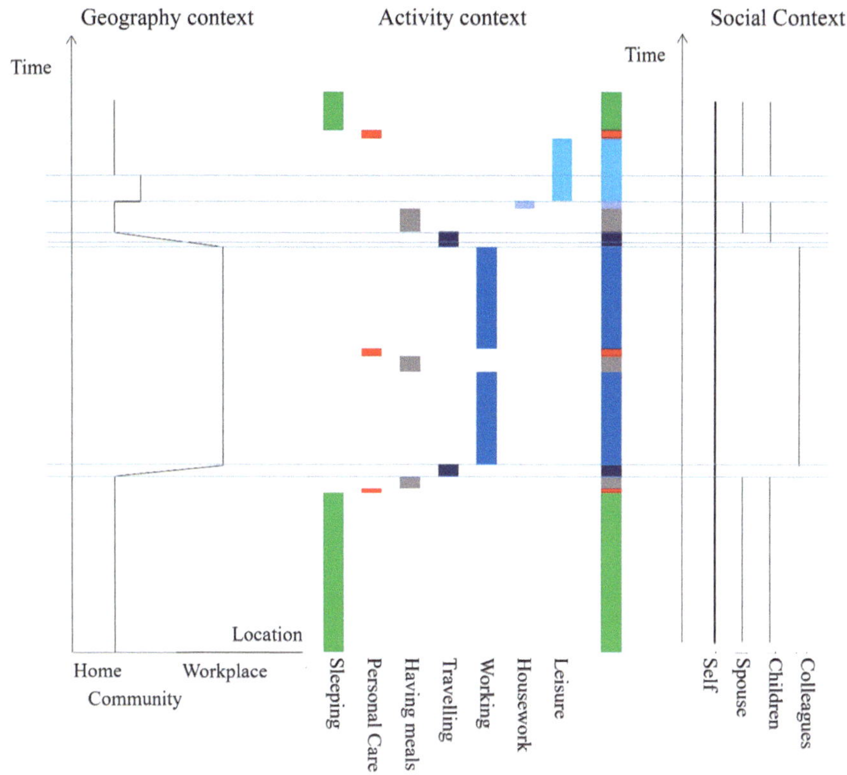

Fig. 7.27 The activity situation of male parents on working days after the female parent resigns in sample JT2 household

Take the activity of picking up children on Thursday afternoon (16:40–18:00) as an example to analyze the basic events that occur during the process of picking up children. The red line represents Mr. Lu's activity path, and the black line represents Mr. Lu's daughter's activity path. Choosing the activity for a Thursday afternoon, Mr. Lu first arrives at their daughter's school after work, picks up their daughter and goes home together, goes to a supermarket around the community to buy vegetables and other daily necessities on the way home, and finally arrives home together. During the completion of the entire activity, five types of basic events occurred, where the number of occurrences of each type of basic event was: stay (occurred 7 times), combine (occurred 1 time), leave (occurred 5 times), arrive (occurred 5 times), and move (occurred 5 times).

However, after Ms. Zhang resigned her job, picking up and dropping off the children become an activity that the couple rotate and also repeat on weekdays.

The blue line represents Ms. Zhang's activity path (Fig. 7.30). And still choosing a Thursday in the morning (6:40–8:00) after breakfast, Ms. Zhang drops her children off at school and returns home. In the afternoon (16:40–18:00) Mr. Lu arrives at his

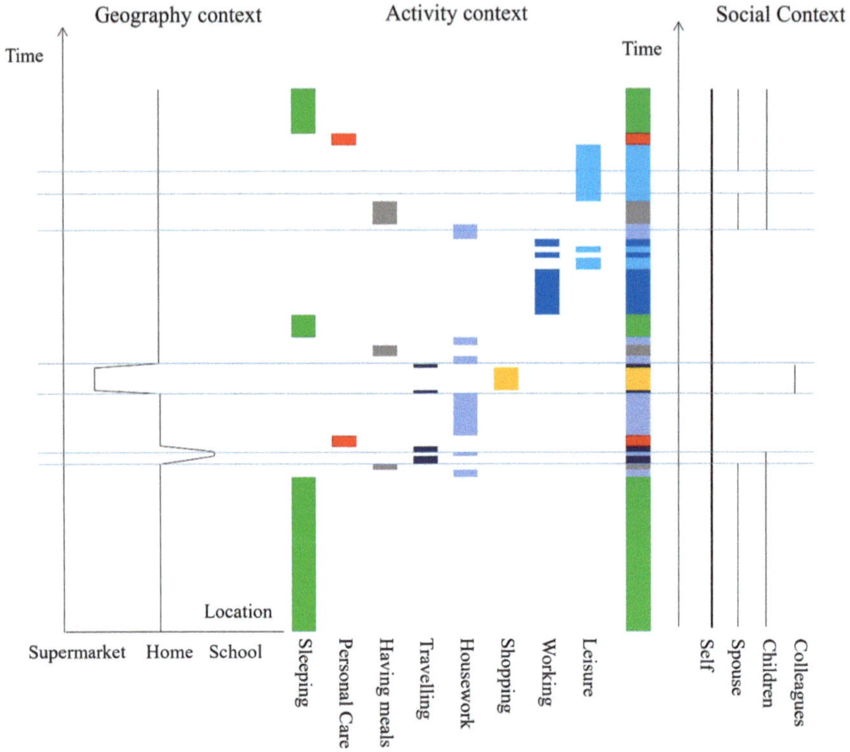

Fig. 7.28 Daily activities of a female parent in workdays after resigning from work in sample JT2 household

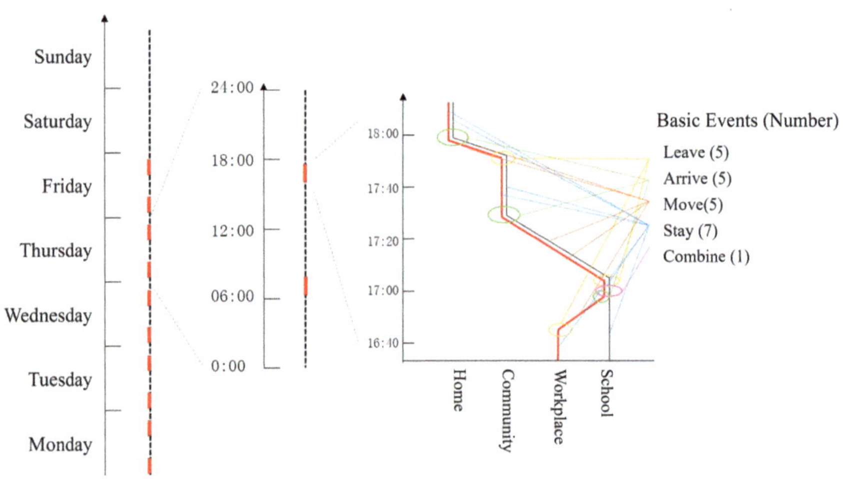

Fig. 7.29 The project of pick-up and drop-off child of female parent before her resignation in sample JT2 household

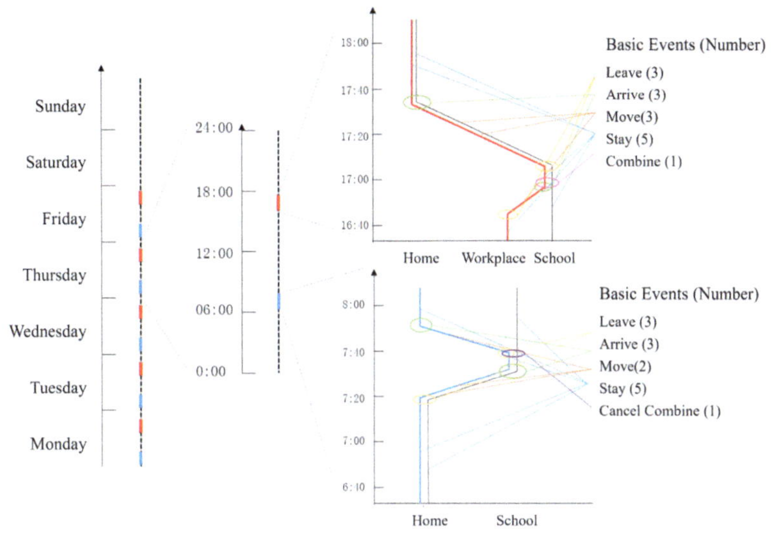

Fig. 7.30 The project of pick-up and drop-off child of female parent after her resignation in sample JT2 household

daughter's school after work and picks up his daughter to return home together. This change in project essentially reflects a change of local order in the family.

7.4.3 *Activity Spaces of Chinese Urban Residents from Household Perspective*

As the spatial carrier of human activities, urban space is itself a complex of urban physical space, economic space and social space. As a matter of fact, urban architectural space and economic space are the spaces artificially created by urban residents on the surface of the natural environment of the city in order to satisfy their basic needs of production and life, and they are the physical space of the city, which is also called the built environment of the city.

In contrast, it is the social and behavioral space formed by human activities on top of the urban physical space. Compared to the physical space of the city, the social and behavioral space of the city is relatively invisible, dynamically changing, and even a virtual space. Behavioral geography suggests that space-time behavior is influenced by subjective factors such as individual perceptions, preferences and choices. Broadly speaking, the collection of urban residents' daily behavioral space constitutes the urban activity space.

Individual behavioral space under the theory of time geography can be represented by the potential activity space under constraints and the actual activity space which can be observed. Individuals' activities and mobility in space-time are inevitably subject to various constraints originating from the individuals themselves, their families, the social environment and so on. And at the same time, the scope of individual behavioral space cannot exceed the scope of the potential activity space under the constraints.

7.4.3.1 Interaction Between Long-Term and Short-Term Behavior of Chinese Urban Residents from Household Perspective

The formation of different types of short- and long-term behaviors varies significantly at different stages of an individual's life course, and is closely related to the social and physical environments in which they are found, as well as to individual and family factors. However, different factors have different constraints on different behaviors at different stages of development. It is through the changes in the influencing factors that lead to the adaptation of other behaviors, resulting in the interaction between individuals' activities.

From the perspective of the family, long-term behaviors lead to changes in short-term behaviors by changing spatial distances, time schedules, and constraints, forcing individuals to choose to change the existing family order or adjusting their time schedules. When an individual's short-term behavior changes, in order to maintain the stability of the internal order of the entire family, other family members will be forced to adjust their own activities or take the initiative to adjust their own activities, thus changing their daily living patterns, and affecting long-term behavior (Fig. 7.31).

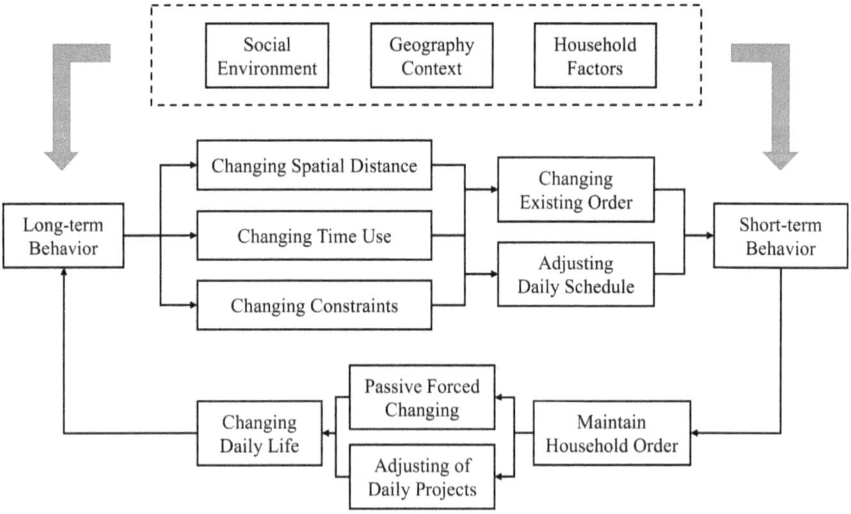

Fig. 7.31 Schematic diagram of the interaction between long-term and short-term behavioural activities from a family perspective

As can be seen from the previous empirical case, family life events (e.g., life cycle changes, social role transitions) and changes in the geographic context in which they are located affect changes in the daily living patterns of each individual resident of the family. This pattern change is influenced by multiple factors and interacts with long-term behavioral activities such as employment and relocation.

Changes in behavioral space and activity space are brought about by the combined effects of workspace and residential space. The daily activities of each individual resident in the household revolve around the two anchor points of his or her workplace and residence. When jobs and residences are close to each other, the short-term behavior of individuals is mostly completed within the community. When jobs and residences are separated, the increase in commuting distance squeezes the individual's time use, which leads to the corresponding adjustment of other short-term behavior, such as shopping. Chinese urban households adjust their short-term behavior in order to harmonize the daily activities of each individual with the rhythm of the micro-environment in which the whole household is located. The changed short-term behavior continues for a relatively stable period of time and solidifies into the new daily living pattern of the family.

The interaction between long-term behavioral activities, such as employment and relocation, and individuals' daily living patterns shows that the changing of workplace and relocation of urban residents lead to changes in daily living patterns through changes in short-term behaviors prompted by environmental factors. Among them, the influence of employment on short-term behaviors changes the daily living patterns of urban residents. For example, Ms. Zhang, the female parents of the JT2 household, changed her daily life from work to family life after resigned her job, and her activity space changed accordingly. The relocation, on the other hand, changes the short-term behaviors associated with the original built environment of urban residents by altering the geographic context, such as a shift in the direction of shopping behaviors as a result of the different community configurations after relocation.

The impact of short-term behavioral changes on the daily living patterns of Chinese urban residents is realized through changes in the short-term behaviors themselves or in the way they are assembled. When the built environment is relatively stable, such changes continue to form a new daily living pattern. When household, built environment and other factors affect the short term behavior of individuals, individuals with cognitive ability adapt to this change by regulating other behavior. Sometimes this is done by regulating other types of short-term behaviors in the same behavioral chain, and sometimes it is necessary to achieve this through long-term behaviors such as employment or relocation.

Individual short-term behaviors may lead to changes in long-term behaviors forced on individuals by his(her)self or by outsiders. Sometimes, when such changes in spatial distance cause changes in travel distance associated with daily activities, individuals make self-determined decisions about whether to stick to their usual behavioral activities and respond to the spatial changes by changing or abandoning the behavior. For example, when Sample JT1 families lived in the Yipinge community, Ms. Qiao, the female parent, usually returned home at noon to prepare her own lunch and take a lunch break due to the close of her workplace and residence. When moving to Changyang Bandao community, Ms. Qiao switched to eating and taking her lunch break at her workplace due to the separation of workplace and residence. At the same time, changes in spatial distance affect the daily schedules of urban residents in China. For example, the squeeze of increased commuting time may sometimes cause certain types of activities to be put on hold, and individuals need to make new daily schedules to fill in the gaps in the use of the remaining time.

7.4.3.2 Interaction Between the Physical and Social Environment and the Activity Space of Chinese Urban Residents

From the perspective of the family, the activities and movement of each individual in an urban family in space-time are subject to various constraints from the individual's self-conditions, the social environment in which he or she lives, and the behavior of other members in the family. At this time, the behavioral space of an individual includes not only the actual activity space of the individual in the family (that is, the spatial extent of the urban space in which individuals carry out their daily activities and mobility between activities in order to satisfy their own needs and in the context of a local order), but also the cognitive space formed through various ways of communication and learning between individuals within and outside the family. The home is an important pocket of order, where the members of the family can carry out stable projects and are isolated from the outside world.

The physical environment includes the physical space structure, physical facilities and its service management, and other physical and virtual space factors that individuals can access or use. Research has focused on the constraints of the physical environment on individual behavior, revealing the influence and mechanism of environmental factors, especially space-time factors, on individual behavior through the study of how individuals complete daily activities in what kind of space (Kwan 2008; Schwanen et al. 2008). The physical environment is the physical carrier that individual spatial behavior occupies and uses. In the situation that both subjective and objective factors allow, the implementation of individual behavior also needs to rely on the physical environment to provide the place where the behavior occurs, facilities and corresponding services, which includes material support such as transportation for activity-based mobility.

As Chinese society and science and technology have evolved, the environment has transitioned from physical space to virtual space, and individuals' behavior has also given rise to virtual behavior. However, the occurrence of virtual behavior requires a certain material environment to support. Although the rules of constraints have been changed, the material environment remains the carrier of behavioral activities and constraints. To illustrate this point, consider the development and application of network technology. On the one hand, the popularization of the network has weakened the space-time constraints to a certain extent, so that some activities can be performed in virtual space. On the other hand, the demand for specific material equipment reinforces the constraints of the physical environment on behaviors related to virtual space. Such as the case of Ms. Zhang, the female parent in Sample

JT2, she currently can work in both virtual space and real space. On weekends, she travels to a tutoring institution to teach her students, while on weekdays, she answers questions and corrects homework via the Internet. However, due to the technological requirements for equipment and network access, Ms. Zhang has to stay at home and cannot go out during this period of time.

The physical environment has historically functioned as a material medium and instrument for individuals' behaviors, thereby constraining their activities. Conversely, alterations in the physical environment engender corresponding alterations in behavior and activities, which, in turn, influence the behavior space. However, such changes also require the cooperation and support of other individuals in the family and their own abilities. The physical environment is undergoing changes; nevertheless, the mechanism of its action on the behavioral space remains in place.

The social environment is comprised of several elements, including the social system environment, traditional culture, behavioral norms and moral code consensus, social and cultural atmosphere, and social interaction network. The relationship between the individual and society is characterized by a dynamic interplay of influence, whereby individual behavior is shaped by the specific social environment and vice versa. This reciprocal relationship is a fundamental aspect of social interaction and is subject to the influence of contextual factors (Valentine and Sadgrove 2012).

The social environment plays an important role in the formation of individual ideology and other subjective concepts, and therefore influences the formation of individual spatial behavior. First, when the social environment changes in urban China, rational families coordinate and adjust their spatial behavior to meet the behavioral norms constructed by the new social environment. In turn, they maintain a sense of security for the whole family through the routinized activities allowed by the new environment. Second, when making individual behavioral decisions, urban residents will make choices based on the social environment, especially the social system, traditional cultural concepts, behavioral norms, and moral code consensus, which have stronger binding power.

From another perspective, the behavior of urban residents in China are also driving changes in physical space. When the demand for a certain behavior increasing and the conditions in the physical space are not available, the residents will change the physical space by appealing to the community, the property, or even the relevant administrative departments. For example, in the Changyang Bandao community where Sample JT1 families live, the issue of activity facilities for the elderly was not initially considered by the community at the beginning. It is because most of the earliest residents of the community were immigrant families who could not afford to buy homes in the urban area, the overall age of the community's residents was predominantly young and middle-age. Subsequently, a number of elderly people moved into the community to care for the third generation and seek intergenerational support, and there was some demand for places and facilities for activities. But the only place in the community where dancing was allowed was piled high with a lot of discarded building materials, and safety was extremely poor. In response to the expressed needs of certain families and residents, the community management committee cleaned and repaired the area. Furthermore, an activity center for the elderly was constructed and some of the elderly usually gather for recreational and social activities such as chess, card games, and chatting.

7.5 Chapter Summary

As China's major metropolitan cities have entered the rapid urbanization stage one after another, urban residential areas have flourished and assumed complex social functions, and the families and residents of these residential areas are carrying out complex and varied behavioral activities. This chapter uses the in-depth interview method in qualitative research and combines interviews with three sample families in Beijing, from a micro research perspective, explores the activity space based on life paths and life courses, daily path and activity dairy, and the pocket of local order. The study also explores the interaction between individual behavior and urban space within the context of rapid urbanization.

In the life path of Chinese urban families and residents, as the most important spatial form of daily activities, the interaction of work-life space shows the process of change of residents' behavioral space. The constraints on residents' long-term behavior are highly variable and distinctly different from those on short-term behavior. The behavior associated with employment and relocation are closely interrelated and jointly influenced by environmental, personal, and family factors, and contribute to changes in the individual's daily activity patterns.

When examining residents' activity spaces from daily path and activity dairy, it is necessary to combine important life events such as changes in residence, workplace, economic and social roles, and family structure in the life course as research clues to explore their impact on individuals' daily path. Changes in residence and workplace bring about changes and constraints in the job and residential relationships and geographic contexts of each individual, affecting their opportunities to minimize the combinatorial constraints of engaging in different daily activities. An individual's rising social status has been shown to influence household mobility and leisure activity patterns by enhancing household economic conditions. Conversely, changes in family structure and shifts in the social roles of individuals within the household directly impact the division of domestic labor and the use of household resources.

In China, residing in urban areas has been shown to exacerbate gender disparities in activity and mobility. However, the family functions as a coordinating mechanism, influencing and restricting individuals' daily activity patterns through different choices of family strategies. In this manner, it helps them to better cope with the pressures they face in their own lives, thus constituting a more complex nested order of activity places. It has been established that fixed resource use, space and time allocation in the family form relatively fixed activity norms and order among family members, which in turn promotes the occurrence of routine activities. This process of activity order formation is also a process of continuous adaptation and interaction between people and space.

References

Cao J, Tu W, Li Q et al (2017) Spatio-temporal characteristics of group activities based on large-scale mobile phone location data. J Geo-Inf Sci 19(4):467–474 (In Chinese)
Carpenter S, Jones PM (1983). Recent advances in travel demand analysis. Alder-shot: Gower
Chai Y (2000) Urban space. Science Press, Beijing (In Chinese)
Chai H, Feng J (2016) Behavioral space of suburban residents in Beijing based on family life course. Prog Geogr 35(12):1506–1516 (In Chinese)
Chai Y, Shen J (2006) Urban space research based on residents' mobility and activity behaviors. Hum Geogr 21(5):108–112, 54. (In Chinese)
Chai Y, Shen J (2008) Research on human spatial behavior based on activity analysis. Sci Geogr Sin 28(5):594–600 (In Chinese)
Chai Y et al (2012) Urban geography: ideas and methods. Science Press, Beijing (In Chinese)
Chai Y, Ta N, Zhang Y (2013) Revisiting time geography with life course theory: long-term spatial behavior perspectives. Hum Geogr 28(2):1–6 (In Chinese)
Chai Y, Weng G, Liu Z (2003) Feminist perspectives on the behavioral space of urban female residents in China. Hum Geogr 18(4):1–4 (In Chinese)
Chapin FS (1974) Human activity patterns in the city. Queen's Q 29 12):463–469
Chombart de Lauwe PH (1952) Parisetl' agglomeration parisienne. Presses Universitaires de France, Paris
Domosh M (1998) Geography and gender: home, again. Prog Hum Geogr 22(2):276–282
Elder GH (1998) The life course and human development. In Richard ML (ed) Handbook of child psychology (volume1: theoretical models of human development). Wiley Sons Inc., New York
Ellegård K, Zhang X, Zhang Y et al (2016) Study of human activities based on nested local orders. Hum Geogr 31(5):25–31 (In Chinese)
Feng D, Li P, Zhu H (2015) Progress and implications of geographical studies on "home" in western literature. Prog Geogr 34(7):809–817 (In Chinese)
Giddens A (1998) Modernity and self-identity. SDX Joint Publishing Company, Beijing (In Chinese)
Golledge R, Stimson R (1997) Spatial behavior: a geographic perspective. The Guilford Press, London
Guan M, Gu Z, Ta N et al (2013) Application of qualitative GIS in spatio-temporal behavior research. Prog Geogr 32(9):1316–1331 (In Chinese)
Hägerstrand T (1970) What about people in regional science. Pap Proc Regional Sci Assoc 24:6–21
Hägerstrand T (1985) Time-geography: focus on the corporeality of man, society and environment. In: Aidas T (ed) The science and praxis of complexity. The United Nations University Press, Tokyo, pp 193–216
Holton M (2017) Examining student's night-time activity spaces: identities, performances and transformations. Geogr Res 55(1):70–79
Johnston RJ, Gregory D, Pratt G et al (2000) The dictionary of human geography, 4th edn. Blackwell, Malden
Kwan MP (1999) Gender, the home-work link, and space-time patterns of non-employment activities. Econ Geogr 75(4):370–394
Kwan MP (2007) Affecting geospatial technologies: toward a feminist politics of emotion. Prof Geogr 59(1):22–34
Kwan MP (2008) From oral histories to visual narratives: re-presenting the post-September 11 experiences of the Muslim women in the USA. Soc Cult Geogr 9(6):653–669

Lenntorp B (1976) Paths in space-time environments: a time-geographic study of movement possibilities of individuals. The Royal University of Lund, Sweden, Londe
Liu W (2015) Life course theory and its application in long-term spatial behavior research. Hum Geogr 30(2):1–6 (In Chinese)
Liu T, Chai Y (2014) Production and reproduction of spatial-social relations in Chinese urban danwei compounds. Nanjing Soc Sci 7:48–55 (In Chinese)
Liu Y, He S, Li Z (2005) Space-time structure of daily activities among urban poor in Nanjing. Chin J Popul Sci S1:85–93 (In Chinese)
Liu L, Yu H, Jin H (2018) Characteristics of weekend outdoor leisure behavior of Beijing residents based on bus smart card data. Ar Res Dev 37(6):54–59 (In Chinese)
Pred A (1981a) Social reproduction and the time-geography of everyday life. Geogr Ann (Ser B: Hum Geogr) 63(1):5–22
Pred A (1981b) Production, family, and free-time projects: a time-geographic perspective on the individual and societal change in nineteenth-century US cities. J Hist Geogr 7(1):3–36
Relph E (1976) Place and placelessness. Pion, London
Rushton G (1969) Analysis of spatial behavior by revealed space preference. Ann Assoc Am Geogr 59:391–400
Schwanen T, Kwan M, Ren F (2008) How fixed is fixed? Gendered rigidity of space-time constraints and geographies of everyday activities. Geoforum 39(6):2109–2121
Shen Y, Chai Y (2013) Daily activity space of suburban mega-community residents in Beijing based on GPS data. Acta Geogr Sin 68(4):506–516 (In Chinese)
Ta N, Chai Y (2017) Behavioral space dilemma of low-income residents in Beijing suburbs based on income group differences. Acta Geogr Sin 72(10):1776–1786 (In Chinese)
Tao Y, Shen Y (2018) Campus differences in students' daily activity spaces under university suburbanization: case study of East China Normal University in Shanghai. World Regional Studies 27(3):134–145 (In Chinese)
Thrift N, Pred A (1981) Time-geography: a new beginning. Prog Hum Geogr 5(2):277–286
Valentine G, Sadgrove J (2012) Lived difference: a narrative account of Spatio-temporal processes of social differentiation. Environ Plan A 44(9):2049–2063
Xu D, Huang Z, Lü L et al (2018) Spatial characteristics of urban leisure tourism based on POI mining: case study of Nanjing. Geogr Geo-Inf Sci 34(1):59–64 (In Chinese)
Xu Y, Shaw SL, Zhao Z et al (2016) Another tale of two cities—understanding human activity space using actively tracked cellphone location data. Ann Assoc Am Geogr 106(2):489–502
Yan Q, Li C, Chen C et al (2018) Activity space characteristics and community differentiation in Changchun based on mobile signaling data. Hum Geogr 33(6):35–43 (In Chinese)
Zhang X (2007) Study on daily life space: formation of Shanghai's urban daily life space. Tongji University, Shanghai (In Chinese)
Zhao Y (2007) Migrant workers: identity construction and spatial formation in daily life. Society 27(6):175–188. (In Chinese)
Zhao L, Hong Z, Yan H (2013a) Reflections on constructing urban public communication spaces. Urban Dev Stud 20(1):80–85 (In Chinese)
Zhao Y, Chai Y, Martin D (2013b) Home space and family relationships: a activity-mobility behavior perspective based on international comparison. Geogr Res 32(6):1068–1076 (In Chinese)
Zhen F, Wei Z, Yang S et al (2009) Impact of information technology on urban residents' travel characteristics: a case study of Nanjing. Geogr Res 28(5):1307–1317 (In Chinese)
Zhou S, Yan X (2005) Relationship between urban spatial structure and transportation demand in Guangzhou. Acta Geogr Sin 60(1):131–142 (In Chinese)

Chapter 8
Leisure Space

8.1 The Process of Formation and Evolution of Urban Leisure Space

8.1.1 Research on Leisure Space and Urban Leisure Space in China

Urban leisure space is a product of the process of urban development, and the demand for leisure drives the development of urban leisure economy, which, to a certain extent, promotes the supply of leisure space. As an important gathering place for human beings, the city naturally becomes an important carrier of leisure in the sense of time and space (Miao 2003). Satisfying the leisure needs of urban residents is a function that cities should have, and it is also a reflection of the level of urban development and the quality of life of citizens (Wei and Li 2007). Leisure demand and leisure space supply develop in constant contradiction and reconciliation, generating a coupling mechanism that derives the evolution of diversified structural patterns and spatial alienation of leisure space (Tao and Li 2005; Feng 2006; Lou and Xu 2011). Leisure economy as well as leisure industry directly affects the layout of urban space, and there is an interdependent and mutually influential relationship between leisure economy and leisure space (Guo et al. 2008). Urban leisure space carries residents' leisure and recreational behavior and is an important part of urban living space.

The distribution of leisure space in Chinese cities varies greatly, while possessing significant characteristics of circle structure (Yang and Zhou 2008; Feng 2010; Lou et al. 2016). The leisure space within the city has a diverse and multilevel structure, and gradually presents the characteristics of a polycentric distribution, including multiple centers such as urban green spaces, city squares, and recreational business district (RBD) (Bao and Gu 2002; Zheng 2005; Zhu et al. 2014). Among the extended and integrated areas in the periphery of the city, the recreation belt around metropolis (ReBAM) is the most typical place for recreational activities (Wu 2001; Zhao and

J. Feng et al., *Urban Living Space in China*, Urban Sustainability,
https://doi.org/10.1007/978-981-95-0595-1_8

Xu 2008; Zhao et al. 2016), and the spatial layout of recreational space is affected by various factors of economy, demand, society, and culture, and the process of the evolution of the development of the urban space is closely related to the economic and social development and the demand for recreation (Guo et al. 2008; Chai et al. 2013; Ma et al. 2019a). The geographic extent of Chinese cities varies widely, and the time course of urban development and evolution is also different, resulting in a wide range of differences in the structure of China's urban spatial systems, and different patterns of leisure spatial structure in different cities. There are evolutionary regularities in the leisure spatial structure patterns of large cities such as Beijing, Shanghai, Guangzhou, and Shenzhen at different times (Liu et al. 2000; Wu and Jia 2002; Bao and Gu 2002; Zhang 2008; Lin 2016; Lou et al. 2016; Xu and Lou 2019).

Chinese urban leisure research is also gradually developing in the direction of "humanization" (Zhang 2012; Li and Li 2010a, b), and the research on urban leisure has gradually shifted from the direction of leisure space, leisure system, and leisure facilities to the exploration of urban residents' leisure behavior. Under the influence of the postmodern view of leisure time and space, academics have begun to study the spatial imagery of leisure behavior and its impact on leisure behavior. Leisure space is the hardware carrier of urban leisure activities, while the leisure activities and behaviors of urban residents, as well as their cognition, perception and emotional expression in the process of leisure belong to the "software" factors. Leisure urbanization needs to analyze the subjectivity of the city as a "place" and the perception, experience, attitude of leisure individuals as well as the emotional expression between them, so as to break the development bottleneck between leisure space and urbanization, and to form leisure towns with harmonious development inside and outside of the city and local characteristics.

8.1.2 Leisure Demand and Leisure Economy Drivers

The idea of leisure has always existed in traditional Chinese culture, and leisure is not only the pursuit of a better life, but also a natural state of life (Liu 2001). Leisure, as one of the needs of human beings, is a higher-order spiritual need above the basic physiological needs. From agricultural rest in the agricultural era, to festivals and celebrations in the feudal era, to urban idylls in the industrial era as well as extensive recreational activities in the information age, the need for leisure has been present throughout the entire course of human development (Ma 2002). Chinese history is permeated with traditional Chinese leisure ideas, whether in Confucianism, Taoism, or Buddhist thought (Lu 2011). Modern and contemporary urban development in China has always been accompanied by the integration and development of traditional leisure thoughts and Western views on leisure.

Because of the similarity and relevance of the concepts of leisure, recreation, free time, entertainment, and tourism, etc., China has derived many different research directions in the field of urban leisure, and even mixed concepts, resulting in the emergence of "urban leisure space", "urban public space", "urban recreation space", "urban free-time space", "urban entertainment space", "urban tourism space", etc. These concepts have different research objects. There are different concepts of urban leisure space, urban leisure space, urban recreation space, urban tourism space, and so on, and these concepts have differences and overlaps in the research objects and similarities in the research scope (Bao and Gu 1998, 2002; Liu and Chai 2001; Wu et al. 2001, 2003; Yu 2003; Qin 2003; Fang and Bu 2003; Bian 2003; Tao and Li 2003, 2005; Tao and Huang 2006; Wu 2005; Feng 2006; Wang 2007; Shi et al. 2008; Bian and Sha 2008; Wang et al. 2009; Hu and Zhong 2012). The identification of the concepts of leisure, recreation, and tourism in some studies has allowed urban space research to have a further clear delineation in terms of research objects and fields of existence (Yu and Ma 2006; Ye 2009), gradually forming the mainstream research direction of urban leisure space, recreation space, and tourism space. The generic expressions of leisure space, recreation space and leisure and recreational space will appear in this chapter when describing and summarizing.

The demand for leisure is the intrinsic driving force for the development of the leisure economy in cities. The psychological needs of urban residents and the gradual increase in leisure time have driven the aggregation of leisure elements within cities and the evolution and reconstruction of urban leisure space. As the gathering centers of various development elements in the region, urban centers and suburbs are firstly developed as the space for human beings to satisfy leisure demand (Miao 2003). The demand for leisure is elastic and is affected more by the external environment and personal internal reasons and so on, and there will be stage-by-stage fluctuations in demand, and the substitutability of the demand for leisure is high compared with other basic life activities. Therefore, the urban leisure support system largely affects the level of elasticity of leisure demand (Wei and Li 2007), and the development of urban leisure economy needs to build a support system for leisure service, management and influence.

The gradual diversification of urban functions and the agglomeration of urban functions have raised the level of urban economy, and at the same time raised the level of residents' leisure demand. Urban residents have generated more leisure demand outside of production and life, and leisure activities have begun to shift from family space to community space, urban centers and urban periphery, and leisure and recreation have become an important extension of urban functions (Yang and Zhou 2008).Nowadays, the leisure demand of Chinese urban residents still focuses on entertainment and recreation, and the public welfare of leisure activities and the demand for leisure activities with rich cultural connotations are still relatively low compared to the level of economic development, and the level of leisure demand needs to be improved (Wang and Wei 2020). Beijing, Shanghai, Guangzhou, Hangzhou, Wuhan, and Chengdu have formed sizable recreation and leisure groups as a way to improve their quality (Xu and Chen 2012; Shan 2019; Jiang 2020). Leisure activities in terms of artistic cultivation and social value enhancement have driven the creation of new urban leisure spaces such as museums, art galleries and art and culture leisure neighborhoods (Xia and Wang 2010; Zhao 2020). However, the activities of home leisure and community leisure are still the carrying space for the main leisure needs of Chinese urban residents.

With China's economic and social development, the leisure time of urban residents shows the differentiated characteristics of different classes, and the space for leisure behaviors also extends with the differentiation of gathering places, showing heterogeneous characteristics (Shi et al. 2008). The demand for leisure and recreation is transforming, especially the development of cities from a single supply of open space gradually transformed into the ability to meet the demand for leisure and recreation of different classes. Different leisure and recreation motives and leisure purposes make residents have different preferences for leisure activities and leisure places, and diversified and differentiated leisure and recreation activities within the city will affect the heterogeneity of the spatial structure of the urban leisure and recreation system (Xiao and Song 2008). In addition, urban leisure demand shows a systematic tendency, and its spatial structure is characterized by complexity, diversity and permeability (Yu et al. 2018). Urban leisure space needs to provide a material foundation for diversified leisure demands and good social leisure values. How to optimize the supply structure of urban leisure products and services, and how to introduce corresponding leisure consumption policies by the management authorities to promote the consumption upgrade of urban residents' leisure demand will still be the issues that need to be faced by China's urban leisure development (Wang and Wei 2020). The development direction of urban leisure needs to be guided by positive and healthy leisure concepts, with personal and social values embodied as the characteristics of leisure demand, and modern leisure aiming at improving the quality of life and living gets developed (Zheng 2005). The differentiation of urban leisure space serves more for the differentiated leisure needs of different classes, thus realizing the systematic construction of urban leisure.

The supply of urban leisure space is the external driving force for the development of urban leisure economy, which is affected by the joint constraints of demand pull and supply thrust, which in turn affects the spatial and temporal layout of urban leisure behavior (Guo et al. 2008). The emergence of China's urban leisure economy is inextricably linked to the improvement of residents' living standards, and the development of leisure public space is closely related to the change of urban functions and the development process of urbanization (Guo et al. 2008; Lin 2016). At the beginning of the reform and opening-up period, China's urban leisure was relatively homogeneous. With the comprehensive deepening of China's reform and opening-up and the upgrading of the consumption structure of urban residents, the function of the city has evolved from being production-oriented to being service-consumption-oriented, the production space has gradually been marginalized and suburbanized, and the manufacturing industry has gradually shifted to the peripheral zone of the city, whereas the space for public leisure and the information space, service space, and consumption space associated with it have tended to move to the center of the city The public leisure space and the information space, service space and consumption space related to it tend to the center of the city and the community, and the industries related to the leisure economy occupy the center of the city, forming a new pattern of urban leisure space.

8.1.3 Reconciling the Contradictions Between Demand and Supply of Urban Leisure Space

The demand for leisure and the supply of leisure space in Chinese cities have always been a process of continuous conflict and coordinated development. China is in the stage of rapid urbanization, and in the process of urban–rural dualistic structure as well as the construction of new urbanization, there is a situation in which the total amount of leisure demand is large while the supply of leisure space and industry in China's urban and rural areas is significantly insufficient (Yu et al. 2018). The sharp increase of urban population causes the tension of urban space resources and the extraordinary utilization of land resources, and the expansion of urban areas directly or indirectly causes the change of open space, which in turn affects the development of leisure activities, and the imbalance between the demand for leisure and the supply will be more prominent (Xiao and Song 2008). The contradiction between supply and demand of leisure will be the inevitable problem facing the development of China's leisure industry for a long period of time in the future.

With the improvement of the level of leisure economic development and the perfection of the leisure industry system in the era of the experience economy, the field of urban planning and urban management has also begun the overall planning and effective management of urban leisure and recreation systems to correct and rectify the mismatches and imbalances that have occurred between the urban leisure supply and the intrinsic leisure demand of local urban residents, so as to maximize the current balance between urban leisure supply and demand as much as possible (Zhao 2013). Leisure demand and space supply in China's major cities show their own differences, with different characteristics in different historical periods. In the process of urban development, because a number of different government functions are associated with the management of urban leisure space, the function of urban leisure space has received fragmented public services, and the leisure space in the city is constantly being changed and encroached upon (Ye et al. 2015). The supply of urban leisure space cannot meet the evolving demand for leisure, which requires increasing the proportion of leisure space in the allocation of urban spatial resource, integrating the government's leisure space management department and strengthening its planning and regulatory power.

In addition, certain dislocation and imbalance have been created in terms of urban leisure space, and the leisure construction of some cities has been influenced by the concept of tourism orientation, which has resulted in the outward orientation of the overall leisure space and the provision of urban leisure and recreational space for external tourists, while neglecting the planning and construction of the daily inner leisure and recreational space for the internal city dwellers (Zhao 2013; Yuan and Wu 2018). At the same time, driven by market economy and profit, the supply of urban commercial leisure and recreational space is detached from the demand of mass leisure consumption, presenting an imbalance in spatial utilization and an intensification of class differentiation (Zhao 2013; Zhao and Chen 2016). In the context of globalization, the supply of leisure space in Chinese cities has also been influenced by Western urban leisure styles, resulting in the emergence of modernity of urban leisure and a pan-entertainment supply of leisure space (Lin 2016). This has provided urban "moderns" with access to new leisure styles (Zhao and Chen 2016), while at the same time there has been a decline of localized leisure in the provision of urban leisure under the concept of leisure modernity.

The concentration of tourism and leisure resources in the central city of Shanghai has been high and expanding in the past 70 years from 1949 to 2015, which belongs to the typical single-core development and regional radiation pattern. From the city center to the suburban counties, the balance of the urban tourism and leisure public space is increasing, showing a trend from low-level concentration to optimized and balanced development, but the speed and strength of the development are not enough to meet the demand (Li 2014; Lin 2016), and the public space for tourism and leisure in each suburb and county is gradually expanding with the growth of population and the development of industrial transfer. The spatial pattern of urban parks in Beijing evolved from 2000 to 2010, from urban center agglomeration to peripheral multi-directional and multi-location agglomeration. The balance ratio indices of various urban parks in various sectors and circles differed greatly, the change of the balance ratio curve was complicated, and the unbalanced distribution among subregions was obvious (Mao et al. 2012). The development level of the public leisure resource system in Guangzhou City from 1990 to 2016 varied greatly over the years, and the level of public leisure services in Guangzhou City has experienced four stages: slow development, steady rise, rapid growth and sustained growth, with obvious differences in the internal influencing factors for the development of the level of public leisure services at each stage (Xu and Chen 2018).

In the context of experience economy, urban space will be dominated by consumption, information, service and leisure, and the original production-oriented urban spatial layout can no longer adapt to the needs of experience economy urban development. The heterogeneity and fragmentation of post-suburban environments, multiculturalism and fast-developing subcultures have become the contemporary urban social landscape (Feng and Zhou 2008). Combined with multi-level and diversified leisure demands, the urban leisure industry will also diversify from scale batch to resident customization (Zheng 2005). The development of urban leisure also requires government regulators to possess the concepts of breakout and reconstruction, the correction of the uneven development of the overall outward-oriented urban leisure supply under the concept of tourism orientation; and the correction of the imbalance of the urban public leisure supply system under the concept of equalization; guidance on the high profit-driven urban commercial leisure supply detached from mass leisure consumption and attention to the decline of local leisure in the urban leisure supply under the concept of modernity (Zhao and Chen 2016), and ultimately the optimization of the urban leisure supply to achieve the balance between urban leisure supply and demand as much as possible.

8.1.4 *The Evolution of Urban Leisure Space in China*

Chinese cities in different historical periods or different stages of the same period, different classes of urban residents have different needs, and there are differences in the supply of urban space, the urban leisure behavioral space generated through the demand for leisure and the urban leisure material space generated through the supply of leisure are constantly coupled and developed, which ultimately affects the formation of the different structures of urban leisure space in China. As shown in Fig. 8.1, the coupling between urban leisure demand and leisure supply forms the pattern of urban leisure space (Feng 2007). In addition, considering the relationship between social system, economic development conditions and life productivity, the structural evolution of urban leisure space is also guided by leisure behavior, and the evolution process of leisure space shows causality and dynamics.

8.1.4.1 Formation and Evolution of Leisure Space Location in Ancient Chinese Cities

The evolution of urban leisure space in China has gone through a development process from closed to open, from simple to complex, from point, line, surface to network, from specialization to massification, and from low level to high level (Feng 2007).

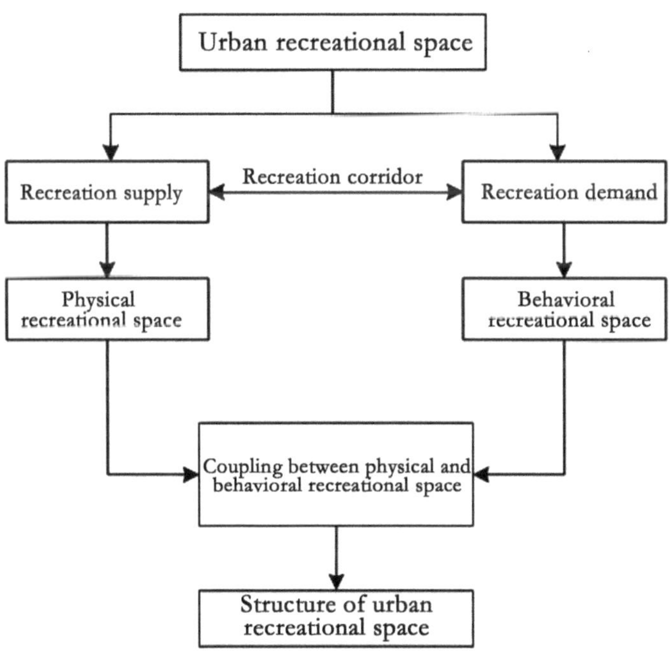

Fig. 8.1 The coupling of Urban Leisure material - behavioral space (*Source* Feng 2007)

In China, in the age of Yangshao culture, about 6000 years ago, the spatial structure of primitive settlements where clans lived together already showed the dawning of leisure and recreational space (Shan 2000). Subsequently, the diversification of urban leisure and recreational spacelayout was also reflected in the concepts of ritual sites and urban architecture during the Yin-Shang periods and the Spring and Autumn and Warring States periods (Jin 2009). Rituals and religious activities have always been indispensable elements in urban spatial layout, from meditation and contemplation during the ancient Greek and Roman periods to the feudal ceremonies and blessing rituals of the Chinese emperors, all of which influenced the layout of ancient urban leisure space.

Ancient Chinese cities were subject to the multiple influences of feudal culture, Confucianism, and imperial power and patriarchy, and urban public leisure space has been greatly restricted. Citizens' leisure and recreational activities were limited by opening hours or spatial layout, and during the Sui and Tang dynasties, a certain amount of trade and leisure space and suburban Buddhist monasteries were developed, and urban residents relied on Buddhist monasteries and suburban scenic areas for their recreational recreation (Chen 2013), but it was not until the middle of Northern Song dynasty that truly open urban leisure space was liberated from the confines of the closed *fang* (neighborhood). The evolution of China's cities from the square and lane system to the street system is the most direct reason for the formation of urban public open space. Specifically manifested in the type of open space from a single to diversified direction, open space has obvious public nature. The shape of open space has been transformed from traditional point to line, and the opening time of street open space has been extended from daytime to nighttime (Jin 2009). The prosperity of commerce in the Song Dynasty contributed to the pattern of the city's *fang-shi* (neighborhood and market unity), and the abolition of the square and lane system changed the tradition of opening the door in the workshop of the ordinary people's residence and shifted to living facing the street, and this change made the openness of the city greatly enhanced (Wang 2016). Many spatial carriers to satisfy people's leisure and entertainment needs appeared in Bianzhou city during the Song Dynasty, such as taverns, teahouses, bathhouses, washe, goulan, etc., and even places such as temples and monasteries and gardens also became temple markets or night markets of unprecedented scale. In addition, during the same period, the city of Guangzhou expanded its development space through the construction of "Zicheng" and "Dongcheng", and the traditional "Fanfang" remained an important recreational and commercial center (Tao and Huang 2006). Between the two Song dynasties, wazi existed in large numbers as a type of square open space, and became the best place for public activities and recreation (Jin 2009). During the Qing Dynasty, there was also a large-scale boom in the construction of private gardens in the royal gardens of Beijing, the palace of Chengde, and the towns of Jiangnan and Yangzhou. The two waves of garden construction in the north and south were brought together, making the Qing Dynasty a heyday in ancient garden construction in China (Fu et al. 2009).

After the modern Opium War, the traditional street system of Chinese cities was impacted by Western forces, and the public open space of Western countries began to be introduced, and city squares, parks, cinemas, dance halls, amusement parks, and golf courses appeared in some cities of the trading ports (Deng and Wu 2017), and a set of obvious dual street public space system was formed in Chinese cities at the end of the 1940s. On the one hand, there are traditional streets with mixed uses and diverse functions; on the other hand, there are Western-style streets with clear structures and relatively homogenized functions (Murphey 1974). Shanghai during the Republican period already had European-style entertainment venues such as Paramount and Metropolis, and movie theaters also entered the city's leisure activities, and the urban leisure space showed a trend of multi-polar radiations, and the public leisure space in Shanghai during this period experienced a period of dichotomous segregation and development, agglomeration and fusion, center formation, and radial diffusion (Lin 2016; Deng and Wu 2017).

With the founding of the People's Republic of China, the leisure space in Chinese cities showed a remarkable revolutionary and idealized color. Large urban plaza gatherings became the main theme of urban activities, and China entered an era of passionate plaza leisure. Dance floors, old-fashioned bars, teahouses, playgrounds, and other leisure spaces in large cities were completely outlawed, and large urban plazas became important places for mass residents to walk, gather, yangge, and practice in the morning. In factories, workshops, and large-scale construction sites, workers' cultural palaces became important leisure spaces for people, and formed the characteristics of revolutionary literature and art, which belonged to the new mode of literature and art after the founding of new China. The drama "Red Flag Song" stirred up a red whirlwind, opera and song were honored as the people's art, and radio became an important space for the dissemination of literature and art, and listening to books and operas through the radio became a good way of leisure for many people. Eight model operas such as "Wisdom Takes Tiger Mountain", "The Red Detachment of Women", "Shajiabang" and "Cuckoo Mountain" and several old movies based on them became indelible symbols of leisure activities in the leisure and historical memory of the city dwellers of that period.

After the reform and opening up, China's urban leisure space has become more diversified, and a cultural market system has been initially established, including the market for cultural and artistic performances, the movie and television market, the audiovisual market, the cultural entertainment market, the cultural tourism market, the art training market, and the artwork market. At the beginning of the reform and opening-up period, people's leisure activities mainly included playing mahjong, watching movies, visiting exhibitions, watching cultural performances, listening to operas and concerts, listening to Peking Opera, watching TV, listening to cassettes, break-dancing, watching videos, soaking in bars, soaking in teahouses, planting flowers and raising fish, keeping pets, singing karaoke, playing billiards, and playing games in game halls, and so on. After entering the twenty-first century, with the rapid development of China's economy and society as well as the support of information technology, the way of home leisure is more abundant, the community and city-level green park system is more perfect, the socialization mode from the traditional face-to-face to the virtual network and reality combination, the development of e-commerce and the emergence of mobile Internet, more so that people can not leave their homes to complete a variety of leisure experience. VR virtual reality, live broadcasting, and

WeChat ecology have all led to qualitative changes in leisure activities and leisure modes in Chinese cities (Yu et al. 2018; Wang and Wei 2020).

8.1.4.2 Development and Evolution of Leisure Space Patterns in Chinese Cities

The evolution of urban leisure space is the result of the continuous satisfaction of the demand of urban leisure subjects and the continuous adjustment of urban supply, the medium of evolution is leisure demand and leisure supply, the evolution presents two mechanisms of spatial aggregation and dispersion, and it is influenced by four factors: socio-political, economic development, technological level, and cultural patterns (Feng 2007) and ultimately, the evolution of urban leisure space is a spiral upward process. The urban leisure space has appeared corresponding geographical differentiation along with the evolution of urban space, and the geographical differentiation of urban leisure space is formed under the comprehensive effect of social, economic and natural factors. In this process, the suburban leisure and recreational space has been gradually pushed outward with the continuous expansion of the built-up area of the city, and the urban land has been influenced by different industries, and the evolution of the urban space has appeared. Economic development force, facility construction force and leisure demand force are all forces that drive the evolution of urban leisure space (Lou and Li 2017).

In the planned economy era after the founding of the People's Republic of China, urban leisure space was dominated by city squares, supplemented by specialty parks. In the era of market economy, with the dismantling of unit walls and compounds, the opening of urban streets and public spaces formed a diversified urban leisure space system, with highly commercialized streets, city squares of different types, fashionable upscale districts, and shopping malls at all levels becoming the dominant forms of urban leisure space (Jin 2009). Shifts in economic patterns have influenced the evolution of urban recreational space. In addition, urban transportation hubs and transportation outreach also affect the aggregation of urban leisure space in transportation hub areas. As the internal and external transportation environments of cities begin to change, leisure service facilities in cities continue to improve and optimize (Lou and Li 2017).

The spatial evolution process of Chinese cities as a whole shows a trend of gradual spreading from the internal core to the outside, with leisure and recreational spots and recreation business districts within the city forming a multi-point layout, a plate pattern, and a facade through the connection of transportation systems. The adjustment of the industrial structure within the city also makes the industrial land transfer to the outside, and the leisure service industry becomes the main industry in the city center location. The booming economic development of the tourism industry has also contributed to the evolution of urban leisure and recreational space (Feng 2007). This evolution has gone through a dynamic change process, i.e., the discrete stage, the polarization stage and the diffusion stage (Tao and Huang 2006).

After the rapid economic growth of Guangzhou in the 1980s and 1990s, the urban leisure and recreation business districts began to polarize on a large scale, and this polarization still exists today, and there is a trend to increase in some areas (Xu and Chen 2018). Guangzhou has basically formed a relatively complete and reasonable recreation business district system, i.e., the form of urban recreation business districts with large-scale shopping and recreational areas such as Beijing Road, Shangxiajiu and Tianhecheng area as the first-level carriers (Tao and Huang 2006). Hong Kong's urban leisure space is formed in the process of historical development, and shows dynamic development in the continuity of time and space. As a multi-centered metropolis, Hong Kong has formed a number of recreation business districts, each of which has different development history, individual characteristics, functional structure, thematic positioning and development factors (Tao and Li 2005). The spatial expansion of open spaces in Dalian is characterized by "wide spread + agglomeration", "sports and leisure in the city and sightseeing in the suburbs", and the construction of open spaces basically conforms to the law of distance decay, while the sports and leisure open spaces in the northern hinterland appear the phenomenon of "anti-distance" (Wang et al. 2012a, b).

Socio-political centers, economic agglomeration, cultural attraction, and transportation hubs are all factors that influence urban leisure space to produce agglomeration effects (Feng 2007). Expansion of organizations, differentiation of residential space, extension of transportation facilities, and the pursuit of ecological balance are also factors that influence urban leisure space to produce a discrete effect (Feng 2007). Concentration mechanism and dispersion mechanism in the evolution of urban leisure space have only relative significance in reality, and there is no strict boundary between the two, the elements that can lead to spatial concentration can become the causative factors of spatial dispersion under certain spatial and temporal conditions (Lou and Li 2017). Similarly, elements that can cause spatial dispersion may have different spatial outcomes for different spatial objects, which may be dispersed or may become the beginning of concentration (Feng 2007; Lou and Li 2017). The evolution of urban open space occurs when the four driving forces of government, enterprises, residents and recreationists work together through the two major driving forces of recreation demand and recreation supply, and combined with the four major forces of economy, society, technology and culture, and under the joint action of centralized and decentralized mechanisms (Feng 2007; Tao and Li 2005). From the viewpoint of the behavioral pattern of recreationists, different levels of spatial structures such as residential area-level recreation space, community-level recreation space, district-level recreation space and city-level recreation space have been formed to meet the different spatial and temporal needs of residents. From the viewpoint of spatial pattern, various spatial structures have been formed, such as RBD, recreation clusters, recreation sectors, and recreation belts around metropolis from the city center to the suburbs; from the viewpoint of spatial form, various forms have been formed, such as points - sightseeing spots, surfaces - recreation centers, lines (belts) - recreation corridors and their various combinations (Feng 2007) in terms of the spatial process, it has formed a pattern of cohesion from a single primary node, gradually to the primary node radiating to secondary nodes and end nodes through the

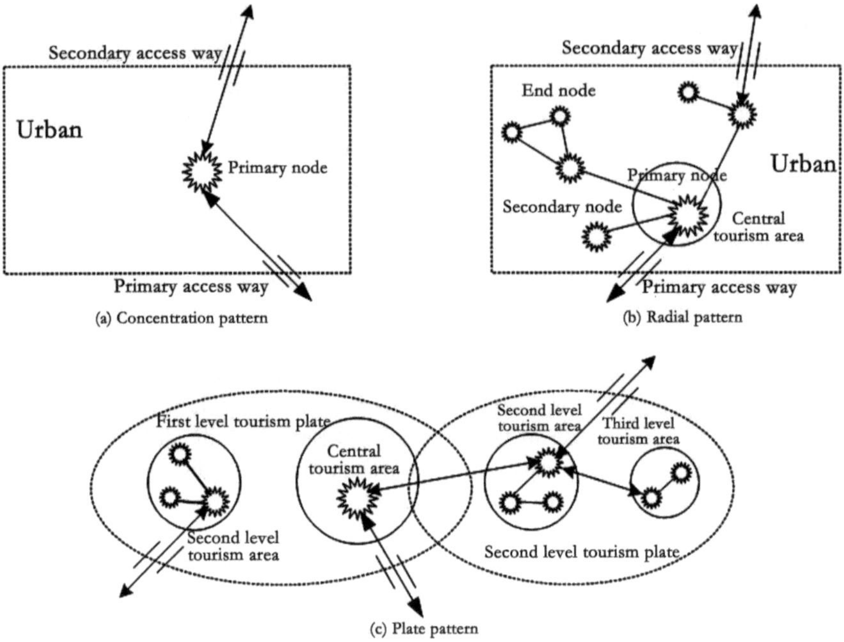

Fig. 8.2 Evolution of urban tourism space (*Source* Wang 2007)

diffusion of links to form a combination of organic plates (Fig. 8.2) (Wang 2007). The evolution of urban leisure space in China is a process of interlocking leisure demand and supply, driven by the aggregation and disaggregation of the city's economic and social development and the transportation hub system.

8.2 Structural Patterns and Spatio-Temporal Differences of Urban Leisure Space

8.2.1 Structural Patterns and Spatial Categorization of Urban Leisure Space

The different development histories of Chinese cities, the different locations of leisure spaces, and the different layouts of the resources on which the spaces are based have led to the formation of diversified patterns of leisure space structure in Chinese cities, which are mainly characterized by the core-periphery structure, the point-axis structure, the concentric-circle structure, the banding structure, and the network structure, etc. (Zhang 2008). Despite the diversity of leisure spatial structure patterns, the elements of the overall urban leisure spatial structure are relatively consistent, including supply elements, demand elements, and related corridors and routes. The urban leisure space units embodied in these elements are sightseeing and recreation points, recreation centers, basic tourist routes and tourism corridor, tourism aggregation and distribution center, theme streets, park roads, etc. (Wu 2005). Some scholars have also refined the basic urban recreation space units of "point, line and surface", i.e. point - sightseeing recreation point, line - recreation corridor and surface - recreation center. These recreational spatial units form different combinations, including monocentric structure pattern, cluster structure pattern, ribbon structure pattern, mosaic composite structure pattern, circular structure pattern, radial structure pattern, ring-radial structure pattern, beaded structure pattern, and other structural patterns (Fig. 8.3) (Feng, 2010).

Chinese scholars have summarized various models of urban leisure and recreational space (Table 8.1), including the galactic pattern (Yu 2003), the "polar nucleus-scattering points-belt" model (Xiao and Song 2008), the "field-factors" dichotomy model (Ye and Bao 2009a, b), and the "polar nucleus-cluster-sector-ring" composite spiderweb pattern (Feng 2007a, 2007b). Although these spatial structure patterns differ, the overall presentation of the circle structure pattern reflects the overall situation of the spatial structure pattern of urban leisure in China (Yu et al. 2018).

The structural patterns of urban leisure and recreational space systems in China are different, and there are certain changes in the structural patterns of leisure and recreational space systems in the same region at different times. In the early stage of urban development, the number of leisure spaces was small and the types were single, and the urban leisure and recreation centers were clearer and showed a mono-core structure pattern, such as Hangzhou in the early stage with the West Lake as the core recreation center, and nowadays, in some small cities, only one large city park is needed to satisfy the recreational needs of most people. With the social and economic development and the increase of urban residents' demand for leisure and recreation, urban leisure and recreation centers are often distributed in multiple places, which

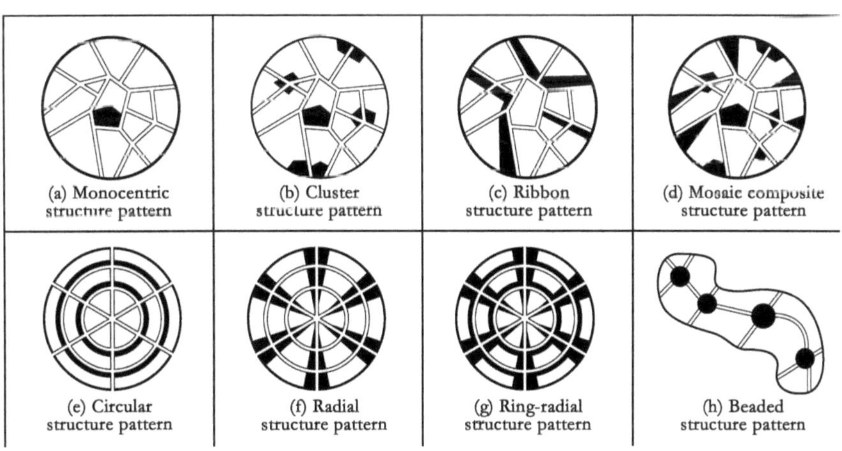

Fig. 8.3 Schematic diagram of the spatial structure pattern of urban open space (*Source* Feng 2007)

Table 8.1 Composite structural model of urban public open space system (*Source* Yu et al. 2018)

Type	Graph	Applicable area	Proposer
Galactic pattern	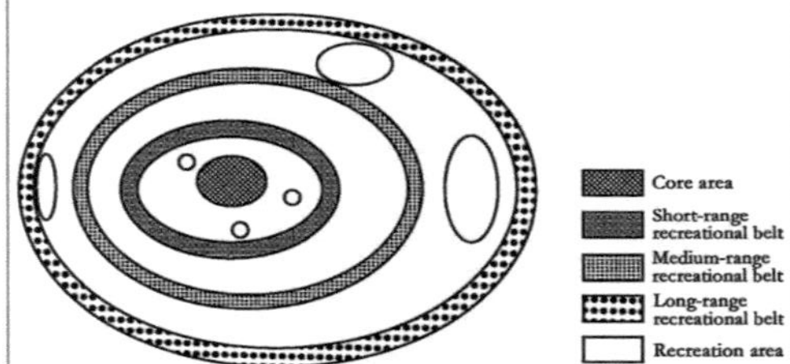	–	Yu (2003)
“Polar nucleus-scattering points-belt"pattern	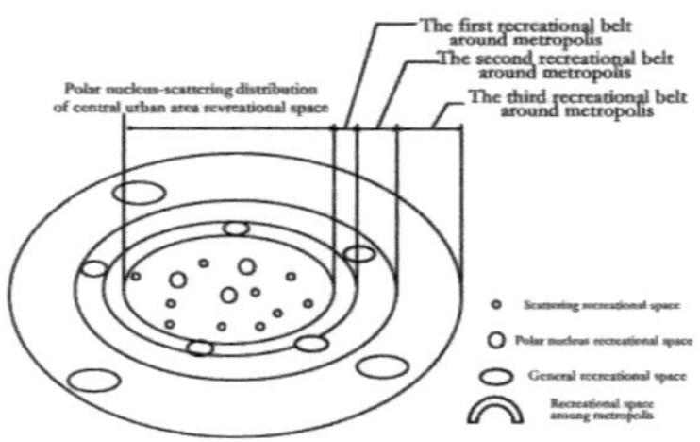	Dalian	Xiao and Song (2008)

(continued)

Table 8.1 (continued)

Type	Graph	Applicable area	Proposer
Circle model of the "field-element" dichotomy	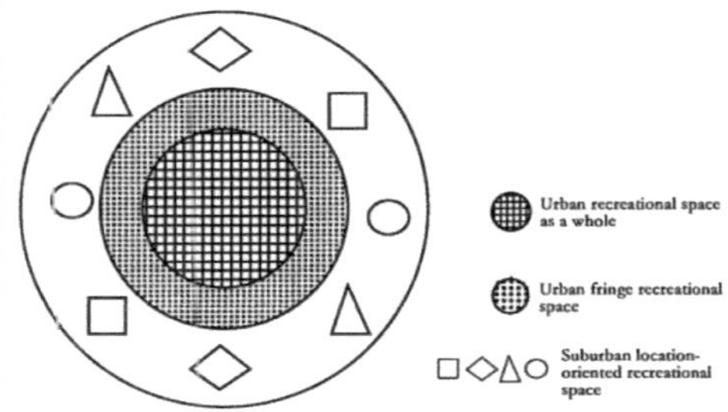	–	Ye and Bao (2009a, b)
Composite spiderweb pattern of "polar nucleus-cluster-sector-circle"	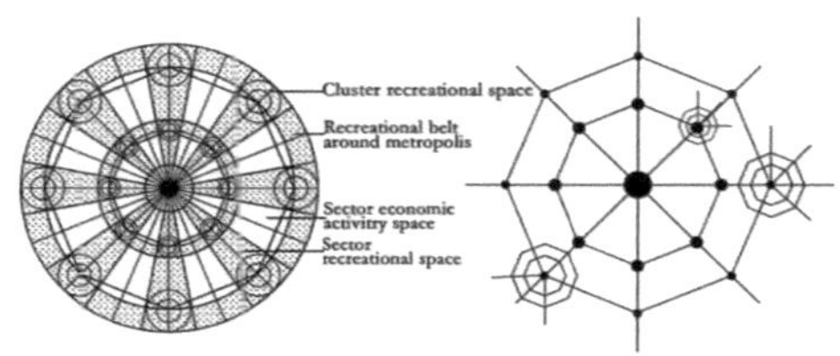	Guilin	Feng (2007a, b)

are manifested in the pattern of multi-core structure or cluster structure (Deng and Wu 2017). In the study of urban tourism space, some scholars have proposed that there are seven types of spatial layout patterns, including single node, multi-node, and chain patterns at the micro level, and four patterns at the macro level, including city center, point-axis space, urban tourism circle, and the recreation belt around metropolis (Bian 2005). It has been pointed out that the spatial expansion pattern of urban recreation business districts often develops from a single nucleus fixed structure to a multi-core structure, and the whole is often in the form of a chain, ring or net, thus forming a leisure and business district system (Tao and Li 2005). The spatial layout of recreation business districts in most cities follows the "pole-and-belt structure". In this rule, "poles" refer to RBDs of different levels and scales, as well as recreational towns; "belts" generally refer to peri-urban recreational belts (Wang et al. 2009).

Urban leisure space structure mode is largely influenced by the urban spatial pattern and natural environment, urban leisure space relying on different resource layouts, forming a diversified layout mode. For example, by the influence of rivers and waterfronts, it is easier to form characteristic leisure and recreation corridors, which are manifested in the mode of banding structure, Chongqing, Shanghai, Wuhan, relying on the water resources of the Yangtze River, all have characteristic waterfront recreation belts. Influenced by topography and landscape, cities located in plain areas tend to form a ring or circle structure pattern (Zhengzhou, Chengdu, Beijing). Comprehensively speaking, China's urban recreational space system is diverse in type, rich in function, and broad in scope, and it usually manifests itself in a composite structural pattern (Hu et al. 2015).

The above spatial structure models all serve as a reference paradigm, and each model theory has its own limitations. In addition, the urban leisure space structure model will be affected by many factors such as city scale, natural environment, social and economic development, etc. Therefore, the urban leisure and recreational space structure model needs to be analyzed in combination with the characteristics of different regions when it is applied in practice. The construction of urban recreational space structure model must fully take into account the specific conditions of the city, including the city's own natural geographic environment conditions, especially the level of urban socio-economic development. At the same time, it should also take into account the characteristics of urban residents' demand for recreation and the characteristics of urban recreation supply.

Chinese scholars have adopted different classification criteria to categorize urban leisure space according to their research objectives. They mainly include: according to the attributes of service objects, they are divided into two categories: for local residents, for foreign tourists and local residents (Wu et al. 2003); according to the function of recreation, they are divided into commercial open space, self-supporting open space, and public provisioning open space, etc. (Ma 2005); according to the degree of commercialization, they are divided into commercial open space, semi-commercial open space, and public welfare open space (Feng 2007); according to the nature of activities, it is divided into park type, fitness type, sports activity field type and other types (Wei et al. 2016); According to the spatial form, it is divided into surface, block and line open space from the material spatial form, and divided into flat space (park walking paths, citizens' recreation squares, public reading rooms in libraries, etc.) and three-dimensional space (mountain climbing, parachute jumping, scuba diving, undersea sightseeing and exploring, etc.) from the behavioral spatial form (Qin 2003); according to the spatial range of leisure activities, it is formed into a physical circle with the urban residents' living place at its center, including family leisure, urban leisure, leisure around the city, off-site leisure and virtual leisure (Lü 2011); according to the formation mechanism of recreational space, it is divided into recreational material space and recreational behavior space (Feng 2007); According to the attributes of tourism resources, it is divided into natural sightseeing open space, humanistic sightseeing open space, artificial recreation open space, sports and leisure open space, and folklore experience open space, etc. (Li et al. 2011); according to the range of services from large to small, it is divided into regional open space, city-level open space, community open space, and indoor open space (Qin 2003); according to the geographic location, it is divided into urban open space, suburban Open space is divided into urban open space, suburban open space, and urban fringe open space (Wu 2005); according to the frequency of use, it is divided into daily open space, weekend open space, and holiday open space (Feng 2007); According to the spatial function exclusivity or not, it is divided into specialized and non-specialized open space; according to the degree of human influence, it is divided into natural open space, semi-natural open space and artificial open space (Feng 2007). Most of the classification criteria for urban leisure space in China use a single attribute as the criterion for division, which is dominated by the field and level of concern of scholars' research, making it more difficult to form a unified understanding (Yu et al. 2018). Moreover, the types of urban leisure space will evolve with the development of society and changes in residents' leisure needs, which is dynamic and multidimensional, so the classification of urban leisure space will be more complicated.

The components of urban leisure space structure include leisure supply elements, leisure demand elements, and leisure corridors and leisure routes linking supply and demand. Among them, the leisure supply elements include leisure industries and extended industries with economic functions, landscape elements with ecological functions, and public leisure service elements with social functions (Lü and Huang 2014). The rational structure of open space in urban areas will promote the formation of a semi-networked structure, i.e., open spaces in residential areas constitute a relatively closed open space (Xiao and Song 2008). In existing studies, the recreational demand of urban residents is spatially characterized by a "circle" pattern, and the number of recreationists gradually decreases with the increase of distance. At the same time, the recreation and business district in the urban space is characterized by a "multi-polar nucleus" pattern, and the recreation land in the suburbs of the city is in the form of a recreation belt around the city, which also shows a circle structure (Hu et al. 2015). From the perspective of the urban area as a whole, due to the spatial separation of the residential areas, these facilities show a decentralized nature; for the municipal parks, squares and other open spaces in the urban area, their functions are mainly for the use of the residents in the area, and they generally show a decentralized characteristic (Xiao and Song 2008); while commercial open spaces, especially shopping centers and department stores, tend to show an aggregation phenomenon, which is characterized by a "unipolar nucleus" or "multipolar nucleus" spatial distribution.

8.2.2 Urban Recreation Business Districts in China

After entering the twenty-first century, China's urbanization has accelerated, and the value and significance of urban recreational business districts have been highlighted along with the transformation of the internal space of cities and the adjustment of land policies and industrial structures. Among the types of urban leisure space, there are more studies on recreational business districts, and in the last 20 years, there have been more studies on the planning, development and development countermeasures of urban RBDs, spatial structure and layout, formation mechanisms and processes, and case studies in the Chinese academic literature (Zhu et al. 2014). The research on urban leisure business in China can be divided into three phases, namely, the initial development period (1998–2003), the rapid development period (2004–2007), the consolidation development period (2008–2014), and the stable exploration period (2015–2020).

During the initial development period (1998–2003), there were few research results on urban recreational business districts (RBDs) in China. In their study of Guangzhou, Bao and Gu (1998) proposed the concept and types of RBDs, and introduced the progress of related research in foreign countries. Since then, the research on RBDs in China has officially begun, focusing on conceptual derivation, analysis of phenomena, understanding of concepts, and synthesis studies (Bao and Gu 1998; Gu and Bao 2002; Hou et al. 2002; Bian and Zhang 2004). At the same time, research on the process and driving mechanism of the formation of urban RBDs has also been carried out, and the spatial structure of urban RBDs has been analyzed, and an empirical study has been carried out on the commercial street area of Hunan Road in Nanjing (Huang and Hou 2001). In addition, the empirical study of Guangzhou's urban RBD, Tianhe City, analyzed the formation process and development characteristics of the urban RBD and summarized its formation mechanism (Bao and Gu 2002). Yu and He (2003a, b) analyzed the spatial layout characteristics of urban recreation business districts through the three aspects of the spatial manifestation of urban RBDs, the spatial layout pattern of urban RBDs, and the spatial distribution pattern of urban RBDs. On the whole, at the early stage of research, Chinese scholars lacked in-depth studies on the theoretical framework, development mechanism, and connotation of characteristics of recreation business districts (Zhu et al. 2014).

During the period of rapid development (2004–2007), the research on RBD in Chinese academia started a boom, which promoted the rapid development of RBD research. With the concept of RBD and the deepening of empirical research, the transformation of old cities and land exchanges during the transition period of cities provided the basis for RBD research, and empirical cases of RBD research appeared in many Chinese cities, including Hong Kong (Tao and Li 2005) and Shenzhen Overseas Chinese Town (Dong and Li 2006), Shangxiajiu Street and Tianhe City in Guangzhou (Bao and Gan 2005; Tao and Huang 2006), Nanjing Road and Chenghuang Temple in Shanghai (Zhang 2005; Zhang 2007, 2007a), Fuzimiao in Nanjing (Li 2006), Guanqian Street and Chengdu City in Suzhou (Liu et al. 2005; Liu 2006), Xi'an (Zhu and Zhuang 2006), Tianjin (Chen 2005), Jinan (Wang and He 2004), Jianghan Road, Wuhan (Yan and Deng 2004; Xu and Yang 2006), Changsha (Chen and He 2006), Taizhou, Zhejiang (Xie and Zhou 2005), Wantou, Ningbo (Shen and Zheng 2007), Qinhuangdao (Guo et al. 2006), Guilin (Wu et al. 2004; Jiang and Zhou 2005), etc. The research perspectives in this period involve multiple disciplines, with geography, urban planning and architecture being the most common fields (Zhu et al. 2014), focusing on the spatial structure of RBDs in Chinese cities, proposing a variety of spatial structure models, and analyzing the formation mechanisms and development patterns of RBDs. Among them, the research results on RBD development and related cases in Guangzhou, Nanjing and Shanghai are more abundant and systematic (Zhu et al. 2014). The research methodology of urban RBD has not yet formed a system, mostly using a combination of quantitative and qualitative research techniques (Zhu et al. 2014), and the relevant case study results still lack macro summary analysis, and there are certain differences in the understanding of RBD, which also leads to problems such as the characteristics of RBD, unclear definition, and unclear classification in the study.

During the period of consolidation and development (2008–2014), the objectives and depth of urban RBD research in China have changed, evolving from macro-objectives to micro-objectives, and from shallow theoretical research to deeper theoretical and practical research. Some researchers put forward the dilemma of RBD from theory to practice and the problems in planning practice, pointing out that urban RBD research should pay more attention to "people-oriented" and conduct in-depth research at the micro level (Fang and Bi 2007). Evaluating and measuring the relevance and social value of recreation and commercial districts from a human-centered perspective (Wang 2008; Li 2014), these studies applied mature theories and methods from other disciplines, such as constructing a measurement scale for the value of visitors to urban recreation and commercial districts in terms of five aspects, namely, quality, efficiency, cost, society, and hedonism, based on the theory of customer value in marketing (Li 2014). Theoretical deduction and phenomenal induction methods were used to derive a system of variables such as life patterns, recreation and leisure attitudes, recreation and leisure behaviors, and factors influencing the choice of recreation and leisure venues (Xu and Wang 2011). In this period, the research object is mostly focused on a certain characteristic or current situation of RBD cases (Song 2011), while the research scope is relatively more open, and in-depth study of traditional spatial morphology and spatial structure of the planning and design and other aspects.

During the period of stable exploration (2015–2020), the research on urban RBDs in China has been relatively stable in the past five years, with no special research hotspots. With the background of cultural and tourism integration and the policy guidance of new urbanization construction, RBD research is mostly carried out from the perspective of cultural tourism, from the use of intangible cultural heritage in recreation business district of historical and cultural districts (Zhao 2015), to the construction of urban recreation business districts guided by Tsingtao Beer culture (Yu 2015), and the optimization of historical space in ancient Kaifeng City under the perspective of cultural tourism (Shi and Wang 2015). The methodology of the research is mostly utilizing the previous research methods and lacks innovation. The content and level of the research in this period are not particularly prominent, and most of the case studies are conducted with other small and medium-sized cities as the object of empirical analysis, combined with an analysis of the factors affecting the evaluation of the satisfaction of the recreational and commercial area. Among them, Zhu et al. (2015) used Gini coefficient, spatial interpolation, kernel density analysis, geodetector and other methods, combined with ArcGIS software, to analyze the spatial and temporal distribution characteristics and causes of urban RBDs in Beijing, which is an innovation in research methods. The study of urban RBDs in China has always been accompanied by the analysis of related concepts (Zhu et al. 2014), such as CBD (Central Business District), TBD (Tourism Business District), CAZ (Central Activity Zone), CED (Central Economic Development Zone), and so on. The differences between these concepts and RBD are throughout the research process.

Chinese urban RBDs can be categorized according to their functions, characteristics and development methods. In terms of function, RBDs are classified into large shopping centers, characteristic shopping and pedestrian streets, historical and cultural renovation zones in old cities, and cultural and tourism zones in new cities (Bao and Gu 1998); and in terms of spatial manifestation, they are classified into pedestrian streets, recreation centers, natural scenic recreation zones, and recreational towns (Yu and He 2003a, b); and in terms of material and spatial characteristics, RBDs are classified into tourist city, commercial center and recreational facilities (Zhang 2005); in terms of spatial patterns, RBDs are classified into superposition, companion, growth pole and point-axis patterns, forming eight types of patterns, including CBD superposition, architectural cluster superposition, scenic area companion, historical and cultural companion, tourist property, landscape community, transportation and distribution, and theme and cultural street (Zhang 2005); and in terms of evolutionary approach, RBDs are categorized into typical urban RBDs, urban evolution of traditional urban CBDs, and modern large-scale shopping centers (Zhang and Sang 2006).

RBD is formed by the combination of leisure elements and economic elements in the process of urban development, and the geographical space in which it is located is generally the central polar nucleus area, and the land use structure is mostly an upgraded substitution mode, which attracts the gathering of leisure and consumption facilities through the clustering effect formed by economic consumption and then pulling up the value of the surrounding land. The spatial structure of RBD is formed through the neighborhood effect and the transmission effect and the self-organization effect (Hou et al. 2002; Shi 2007). Domestic research in China on the spatial structure and layout of RBDs mostly adopts a qualitative approach, and analyzes the spatial and temporal dimensions by summarizing the literature or combining with case studies. The evolution of RBD spatial structure in different cities has different characteristics, but most of them have the spatial layout evolution pattern of "point - multi-point - line - surface" distribution, as well as the three evolutionary processes of discrete stage, polarization stage and diffusion stage (Tao and Li 2003, 2005; Tao and Huang 2006). The location of urban RBDs generally relies on specific factors such as specific tourism resources, markets, convenient transportation and low land prices, etc. The spatial layout of RBDs in most cities follows the "pole-and-belt structure", and a system of evaluation indexes reflecting the appropriateness of the development of urban RBDs has been constructed in terms of the economy, society and the environment (Yu 2003). The typical city image representative area in traditional cities is the main location of RBDs, while emerging cities can also form urban RBDs through the development of commercial leisure complexes based on tourist attractions (Bian and Zhang 2004). There is no uniformity in the location of RBDs in destination cities with different images, which is closely related to the emphasis on the shaping of urban personality in urban tourism development. Urban recreation areas are mainly divided into three spatial layout modes: resource-oriented, market-oriented, and product-oriented (Wu 2005), and have generated spatial combinations of single-core, multiple-core, belt-network, and comprehensive modes.

8.2.3 Recreation Belts Around Metropolis in Chinese Cities

Driven by the surging demand for popular short-distance recreation near the city, the development of recreation belt around metropolis has become a common trend in the suburban development of China's big cities. Many large cities, such as Shanghai, Beijing, Wuhan and Jinan, have started to pay attention to the development of recreation space and facilities in suburban areas (Dang 2011), and attempted to improve the urban open space system and drive the development of urban–rural interface areas through planning and creating recreation belts around metropolis with their own characteristics (Zhao and Xu 2008). The emergence of recreation belt around metropolis in Chinese cities is also a manifestation of recreation suburbanization, and with the proposal of new urbanization construction, the process of suburbanization has entered a new stage: the uneven development of residential suburbanization, industrial suburbanization, and commercial suburbanization has made the suburban space form a highly heterogeneous social space (Chai and Feng 2014), and the overall characteristics of the suburban space are manifested in the functional zoning of the city and the multi-core agglomeration, which gradually pushes the urban spatial structure from monocentric development to polycentric development. Similarly, the suburbanization of leisure is also characterized by the development of urban recreation business districts from a single core to a multi-centre, combined with the city's three-dimensional transportation corridors, extending the suburban belt-shaped leisure space. The continuous expansion of urban space has brought about the continuous change of the recreation belt around metropolis, and the early research on this belt has some specificity and ambiguity (Zhao and Xu 2008), and researchers have used different expressions such as "suburban", "urban–rural interface", "countryside", "green belt around the city", etc. (Huang et al. 1999; Chen 1995; Sha and Wu 1997). Wu (2001) formally put forward the concept of recreation belt around metropolis on the basis of empirical research on tourism development in the suburbs of Shanghai. The spatial distribution and spatial distance of the recreation belt around metropolis are influenced by the degree of leisure demand of urban residents, the degree of access to transportation facilities, investors in recreation resources, and the government's industrial guiding policies, etc., and under the two-way force of land rent and travel costs, investors and tourists reach a compromise (Wu 2001).

The main types of recreation belt around metropolis in China are humanistic sightseeing tourism land, artificial recreation tourism land, natural sightseeing tourism land, and sports and leisure tourism land (Su et al. 2004). According to the recreational function and the degree of commerciality of its land use, all open spaces are divided into two basic types: commercial (resort and residential areas, theme parks, commercial recreation areas, and ancillary recreation areas) and public welfare (scenic spots, forest parks, historical and cultural cities, memorial sites, museums, cultural relics and monuments, wilderness, rivers, lakes and reservoirs, and other areas) (Zhang 2004). Combined with the empirical research in Changchun, the types of open spaces around the city include five categories of tourist vacation lands, rural ethnic and folklore tourism lands, cultural heritage tourism lands, industrial tourism lands and urban recreational parks from the types, grades and distribution states of the patches (Wang 2004); and the five types of vacation open spaces, folklore open spaces, cultural heritage open spaces, sightseeing open spaces and recreational parks are divided into five types according to the individual characteristics of the tourist resources, the main tourist functions, the direction of the development of the tourist areas, the operating characteristics and the demand of the source market (Zhou and Liu 2005).

Research on recreation belts around metropolis in Chinese academia lags behind the actual development of the industry (Zhao and Xu 2008). Existing research on recreation belts around metropolis mainly focuses on the relationship between the formation mechanism of recreation belts around metropolis (Wu 2001; Pan and Jin 2014; Feng et al. 2013) and urbanization (Zhou and Liu 2005), the evaluation of tourism resources in recreation belts around metropolis (Cao and Sun 2008; Tao and Tian 2009), and the spatial structure of recreation belts around metropolis (Su et al. 2004; Zhao and Wu 2009; Yang and Ma 2015) and the spatial configuration of open space (Liu and Wang 2007; Nan et al. 2012), and the behavioral characteristics of the recreation demand market and recreationists (Wu et al. 2007; Chen and Wang 2012), etc., and new perspectives on the development pattern and system composition of the recreation belt around metropolis have also emerged (Guo and Wang 2008a, b; Li and Tan 2016), but in-depth studies on the developmental changes and influencing factors of the recreation belt around metropolis are relatively scarce. The research on recreation belt around metropolis mainly explores its spatial structure, mostly taking a specific case as the research object, and less research analyzes the spatial composition and pattern of the recreation belt around metropolis from the perspectives of resources, products and markets, as well as analyzes the development of the recreation belt around metropolis from the dual perspectives of time and space.

In terms of research methods, the radius of gyration method and the spatial distribution curve are commonly used in the study of recreation belt around metropolis (Li et al. 2011; Dang et al. 2017). In the study of Beijing, 1172 samples of open spaces were selected, and ArcGIS buffer zones and spatial superposition statistics were used to calculate the radius of gyration of open space types per 10 km circle in the recreation belt around metropolis (Li et al. 2011). In the study of Shanghai, 401 recreation areas in the recreation belt around metropolis were selected, and People's Square was taken as the city center of Shanghai, and the straight-line distance between it and the recreation areas was calculated; the temporal evolution stage of the recreation belt around metropolis was analyzed according to the opening time of the recreation areas; and the methods of average city center distance, radius of gyration method, and spatial distribution curves, nearest-neighbor index, point density, center of gravity of the distribution, and standard deviation ellipse were used to study the spatial evolution characteristics of the recreation belt around metropolis in Shanghai at different stages (Dang et al. 2017).

The influencing factors of the recreation belt around metropolis in Shanghai are land occupation status, transportation connection between the city and other attractions, and tourist behavior. The recreation activity space for urban residents in Shanghai is spreading in a distance-decaying manner within the recreation belt around metropolis and extending along the transportation arteries. The attractions in the recreation belt around metropolis in Shanghai are generally connected to the central city through express arteries, while the weak connections between the attractions in the suburban counties make it a radial network system (Wu 2001). Under the combined effect of the above factors such as land use, traffic accessibility and crowd flow, the structure of the city recreation belt and its circles is formed with the city center of Shanghai as the source (Wu 2001). The spatial distribution of recreation belt around metropolis in Shanghai shows a clear distribution of circles, and the overall spatial extrapolation trend, the spatial distribution of recreation sites around metropolis shows a trend from discrete to agglomeration, with higher and higher correlation, and the spatial pattern of recreation site agglomeration shows the development trend of "point - nucleus - axis - block" (Dang et al. 2017).

The spatial and temporal evolution of China's recreation belt around metropolis is the result of various driving forces under the combination of market supply and demand mechanism, open space distribution mechanism, and government regulation mechanism. Market supply and demand is the most important mechanism for the formation and evolution of recreation belt around metropolis. From the supply point of view, the adequacy of tourism and open space supply in the peripheral areas of cities, the quality of open space resources, the level of supply of open space facilities and activities, and the accessibility of transportation infrastructures are the resource bases for the formation and development of the open space belt in the peripheral areas of cities. The former two are mainly endogenous, and both the market and the space decide the quantity, quality, type and level of the production and demand of leisure products, where to produce or where to obtain demand, and where to cluster or where to diffuse, mainly through the profit-taking choice of the open space as a micro subject. The latter, on the other hand, is mainly exogenous, regulating and guiding in order to influence the state collection of open space choices, or directly participating in and realizing the major matters of the development strategy of the recreation belt around metropolis. The three mechanisms complement each other, interact and constrain each other, and together form a comprehensive power mechanism that drives the spatial and temporal evolution of the recreation belt around metropolis.

Under the joint action of agglomeration and diffusion forces, the spatial dispersion of open space tends to a spatial equilibrium, and this spatial choice converges from the micro level of open space to the meso and macro levels, forming the spatial equilibrium of the overall distribution of the recreation belt around the metropolis, and forming the dynamic trajectory of the equilibrium evolution with the passage of time, so as to realize the spatial and temporal structure of the spatial and temporal evolution of the recreation belt around the metropolis. The formation and development of recreation belts around metropolis in China is more due to the development of urbanization, which has led to the increased ability and motivation of urban residents to travel (Zhou and Liu 2005). The development of cities and the recreation belt around metropolis interact with each other. The development of Chinese cities provides a large population base for the recreation belt around metropolis, and with the promotion of urban transportation, the short-distance travel of Chinese urban residents is more convenient, which promotes the development of the recreation belt around metropolis and the expansion of the spatial distance, and meanwhile, the government's macro-control and the related development policies have a direct role in promoting the formation and development of the recreation belt around metropolis. Centripetal urbanization and diffuse urbanization in China's urban development process have different impacts on the recreation belt around metropolis. Centripetalism is conducive to the development and enhancement of the recreation belt around metropolis, while diffuse urbanization threatens the space of the belt around metropolis, resulting in the disappearance of the local culture (Wang et al. 2010). The suburban area within the recreation belt around metropolis is a zone where urban culture and local vernacular culture are in intense conflict and collision. With the gradual expansion of urban space, the artificial landscape of the city continues to replace the natural landscape of the countryside in the suburbs, and the original local culture will gradually be influenced by urban culture. The disappearance of the "natural attributes" of peri-urban recreation sites has caused them to lose the charm of attracting tourists, and thus the phenomenon of "enclaves" has emerged in the recreation belt around metropolis (Wang et al. 2010), which has also formed the reality of sending recreation flows to other recreation nodes in a specific space-time or advantageous nodes with strong potential functions (Liu et al. 2017). China's rapid urbanization, while driving the development of tourism economy in the recreation belt around metropolis, has also brought negative impacts that cannot be ignored. Influenced by the profit-driven influence, duplicated construction, resource waste and vicious competition have appeared in the recreation belt around metropolis (Guo and Wang 2008a, b). The negative impact of the adjustment of China's urban industrial structure on the recreation belt around metropolis The continuous development of urbanization will promote the adjustment of the industrial structure of the urban economy. The rise of urban land prices and the construction of livable cities have forced some primary and secondary industries to move to the periphery of cities. The outward migration of industries will have a negative impact on the landscape structure and ecological environment of the recreation belt around metropolis. In terms of the ecological environment, as the primary and secondary industries that have moved out are mostly highly polluting industries, they have caused serious pollution and damage to the habitat and tourism environment within the recreation belt around the metropolis, which has brought great pressure and challenges to the ecological environment of the recreation area, and is not conducive to the protection and management of the environment of the recreation belt around the metropolis (Wang et al. 2010).

8.2.4 Spatio-Temporal Differences of Urban Leisure Spaces in China

The geographic scope of Chinese cities spans a wide range, and the time course of urban development and evolution varies, resulting in a wide range of differences in the structure of China's urban spatial systems, and different patterns of leisure spatial structure in different cities. Large cities such as Beijing, Shanghai, and Guangzhou have also seen a series of evolutions in their leisure spatial structure patterns at different times (Liu et al. 2000; Wu et al. 2002; Bao and Gu 2002; Zhang 2008; Lou et al. 2016; Xu and Lou 2019). The differentiation of the leisure level of Chinese cities and the different processes of leisureization have formed the differences in the spatial structure patterns of urban leisure. From the perspective of prefecture and city administrative units, the regional distribution of China's excellent tourist cities shows an imbalance of more in the east and less in the west (Hu 2013). In general, the overall level of urban leisureization in China is low, and the level of leisure reception and services is high while the supply of leisure space and environment is low among the main urban leisureization indicators (Lou et al. 2016). Leisure space and environment constrain the process of urban leisureization to a certain extent. Based on the perspective of city comparison, differences in the consumption ability of leisure life also cause differences in the level of urban leisureization (Liu and Ma 2016; Ma et al. 2019a). Among the major provincial capitals in China, the leisure level of 31 cities shows a gradient development trend, with Beijing, Shanghai and Guangzhou in the first gradient, and the five gradients represent high, high, average, low and low levels of urban leisure respectively (Lou et al. 2016).

The spatial differentiation of leisure space in Chinese cities, as a whole, roughly shows a "diagonal bar" distribution decreasing from east to west, and there are obvious differences between the cities on both sides of the "Hu Huanyong Line" (Lou et al. 2016). Taking the Yangtze River as the boundary, the recreation level of cities in the southeast of the Hu Huanyong Line also has obvious differences, with cities in the north of the Yangtze River (except Beijing) having a slightly lower level of recreation, and cities in the south of the Yangtze River showing a "circle" structure that decreases from the outside to the inside. The spatial differences in the level of sub-indicators show similar characteristics, but the spatial differences between indicators are still slightly different. The level of leisureization in cities in the east, central and west decreases in order, but some western cities also have more prominent urban leisureization development due to rich tourism and leisure resources and developed leisure-related industries (Lou et al. 2016).

The spatial difference formed by the differentiation of Chinese urban leisure economy and tourism economy presents a polarized situation (Sun and Dai 2014), forming two types of high-level convergence and low-level convergence, and this distribution pattern has not been fundamentally changed with the improvement of the overall level of development of the tourism economy, and the central and western provinces (districts and municipalities) still have not got rid of the relative lagging behind situation. Song and Ma (2010) used 2007 tourism economic data to conduct a preliminary exploration of the characteristics of spatial differences in China's provincial tourism economy in general and at the local level. However, the study utilized only one year's worth of data, which could not reflect the spatial characteristics and distributional patterns of China's regional tourism economic differences, especially the evolution of spatial interactions and correlations since 2000. Some scholars have used spatial statistical analysis methods to analyze the degree of interdependence, mutual influence and correlation patterns, as well as the spatial agglomeration pattern and evolution of China's regional tourism economic development level from 2000 ~ 2011 (Sun and Dai 2014).

Influenced by the level of regional economic development, the differentiation of urban leisure economy and tourism economy in China is more significant, and the differences in the level of leisureization among Chinese regions basically coincide with the distribution pattern of differences in the level of regional economic development (Liu and Ma 2016). The level of urban leisure in eastern China is generally higher than that in central and western China, with the traditional first-tier cities in the east having a relatively high level of leisure, and the key first-tier cities in central and western China also having a substantial increase, but there is still a large room for improvement compared to the eastern cities (Sun and Dai 2014; Liu and Ma 2016).

From the perspective of leisure consumption differences, China's urban leisure consumption capacity is also characterized by a high in the eastern region and a step-down in the central and western regions, but the gap is beginning to narrow (Ma et al. 2019a, b). The Yangtze River Delta region, the Pearl River Delta region, and the Beijing-Tianjin-Hebei region in the eastern region encompass the top cities in terms of urban leisure consumption capacity, while Wuhan, Xi'an, and Chengdu among the cities in the central and western regions are an important manifestation of the urban leisure consumption capacity, which has been raised by a large margin (Ma et al. 2019a). The Northeast region shows a multipolar development of leisure consumption (Gan et al. 2015). By analyzing the leisure consumption capacity of 31 Chinese cities from 2003 to 2015, it is reflected that regional differences in the leisure consumption capacity of Chinese cities show a trend of gradual convergence, and from the perspective of the three major sub-indicators, namely, leisure consumption potential, leisure consumption environment, and leisure consumption structure, the value of the Thiel index has declined, but the magnitude of the differences, as well as the trend and magnitude of the decline, are different. The biggest difference is in the leisure consumption environment, although the overall trend of the difference is shrinking; the regional difference in leisure consumption structure is mainly in transportation and communication as well as cultural, educational and recreational expenditures, which shows a relatively stable and then rapidly shrinking trend, and the proportion of the shrinkage is the biggest; the regional difference in the potential for leisure consumption is the smallest, and the overall trend of the shrinkage is in the middle of the oscillation. In terms of the internal differences among the three regions, the internal differences in the leisure consumption ability of urban residents in the eastern region are the largest and the speed of convergence is the fastest, and the variability of its internal development has been maintained at a high level (Gan et al. 2015; Ma et al. 2019a); the internal differences in the leisure consumption ability of urban residents in the central region are the smallest, but they have experienced the development trend of the U-shape; the internal differences in the leisure consumption ability of urban residents in the western region are in the middle and show a trend of increasing and then narrowing (Ma et al. 2019a).

Among the research methods on the spatial differences in urban leisure in China, the comprehensive index evaluation model and the Thiel index determination method are used more often (Lou et al. 2016; Liu and Ma 2016; Gan et al. 2015), and the presentation of the results is based on spatial mapping and ArcGIS to realize the graphical display of spatial differentiation characteristics (Hu 2013; Sun and Dai 2014; Gan et al. 2015; Lou et al. 2016). The coefficient of variation method and the Cobb Douglas function were used to measure the relative differences in leisure consumption, i.e., leisure consumption capacity, in each of the 31 cities. The natural breakpoint method was then used to visualize regional differences in leisure consumption capacity and the evolution of those differences. Currently, the methods of measuring regional differences include the Gini coefficient, Moran's I and Thiel index (Ma et al. 2019a). Circle analysis methods, exploratory spatial data analysis (ESDA), and principal component analysis are also used in the differentiation analysis of urban regional leisure economy and urban leisure space (Fang et al. 2013; Hu 2013; Sun and Dai 2014).

Factors affecting spatial differences in urban recreation mainly include geographic basic areas, regional economic development, transportation systems, macro policies, and government management strategies. In terms of geographic location, the geographic environment on both sides of the "Hu Huanyong Line" is very different, and the different natural geographic conditions such as climate and geomorphological features bring about differences in the urban development process, the construction of transportation facilities, and the process of economic and urban modernization, which has some explanatory power for the spatial differences in urban leisureization in China (Lou et al. 2016). In terms of regional economic development, China's reform and opening-up measures have included a focused and gradual approach to socio-economic reform, and thus China's regional economic development as well as the leisure and consumer economy are inevitably characterized by regional imbalances (Liu and Ma 2016; Ma et al. 2019a). Accompanied by the deepening of reform and opening up, the emergence of the driving role of the eastern region's economy, and the positive impact of regional economic cooperation, the rapid development of China's national economy has also gradually benefited cities in the central and western regions (Liu and Ma 2016). In terms of macro policies, China's western development strategy, new urbanization strategy, cultural and tourism integration strategy, "One Belt, One Road" initiative, leisure industry and other policies have gradually shown their effects, with the east driving the development of the central and western parts of the country, and the overall regional disparity in urban recreation has been narrowing amidst fluctuations.

Regional differences in the level of urban casualization in China are bound to exist, and moderate differentiation is acceptable (Liu and Ma 2016). The level of regional urban casualization in China should match the level of urban socio-economic development, consider the coordination between urban leisure space and urban development, and should develop urban casualization with regional characteristics according to local conditions (Li and Jin 2012). Cities in the central and western regions of China have great potential for the development of leisure space, while cities in the east are facing the new normal of slowing economic growth and the problems of supply-side structural reform and accelerated aging, and the ability to expand and enhance leisure consumption and leisure space is under some pressure (Ma et al. 2019a).

On the whole, the development of urban leisure in China has shown a continuous upward trend, and has tended to be balanced by a high degree of agglomeration, but there are still some local differences in distribution, and the urban–rural difference has contributed to the formation of a dualistic pattern of urban and rural leisure space. In terms of local space, urban leisure space, like residential space, is subject to polarization and segmentation, and the phenomenon of divergence in leisure space is becoming more and more obvious. Leisure spaces in Chinese cities have become differentiated, and have become vehicles for providing differentiated leisure activities for people of different classes, groups and social status. The diversified development of Chinese cities inevitably brings about the diversification of social classes, and the complementary interaction between urban and rural leisure modes during China's rapid urbanization process has formed the "urban complex" and "local complex" in the psychology of residents, which has led to the formation and development of recreation belt around metropolis and brought about the differentiation within the urban leisure space. This has led to the formation and development of urban recreation zones, as well as the differentiation of urban recreation spaces. By appropriately balancing the overall layout of urban leisure space, effective interaction between living space and neighborhood space based on residential space and shaping humanistic care is truly realized as the spatial foundation of a harmonious city.

8.3 Interaction Between Urban Leisure Space and Leisure Behavior

The relationship between leisure space and leisure behavior is more complex, and the study of leisure behavior is based on micro-individual-oriented research, focusing on the analysis of leisure time and space behavior in the daily life of individuals, as well as the synergistic nature of leisure behavior and the maintenance of social relationships (Zhao et al. 2016). The research with the social group of urban residents as the research object focuses on considering the leisure structural needs of urban residents and the influencing constraints on the behavior of carrying out leisure activities (Zhao et al. 2014a, b). The relationship between urban leisure space and behavior needs to be studied from the micro-individual's leisure needs and then to the social group's leisure mode from multiple perspectives, and it is necessary to pay attention to the leisure needs centered on human beings, and the urban habitat environment has also become an important perspective in the study of urban space in geography (Li and Li 2010a, b). By analyzing the background and process mechanism of the generation of leisure behavior, grasping the leisure demand of urban residents, and studying the interactive relationship between leisure space and leisure behavior, it is possible to enhance the utilization rate of urban leisure space, provide a reasonable leisure space layout, and reduce the contradiction between supply and demand of leisure demand.

Differences in leisure behaviors are influenced by multiple factors, and at the same time, such differences also affect the choice of urban leisure space. The leisure behaviors of Chinese urban residents still mainly take place in the family and community spaces, and the leisure behaviors are relatively single and the corresponding leisure spaces are relatively concentrated. In the study of daily leisure behaviors in Shenzhen and Beijing over different periods of time, it was found that the leisure space of urban residents in China is still dominated by leisure within the family and community leisure, and this characteristic has not changed fundamentally from the case study of Shenzhen city in 2001 (Liu and Chai 2001) to the case study of Beijing city in 2012 (Xu and Chai 2012). Compared to U.S. cities, Chinese urban residents travel short distances and are relatively spatially concentrated, while the leisure behavior space of Western urban residents is more diffuse, with U.S. residents traveling longer distances for leisure (Zhao et al. 2014a, b).

The application-oriented nature of domestic research in China is obvious, with research in geography focusing on spatial layout, and the field of transportation focusing on travel destinations and needs, to a certain extent ignoring the driving force and complexity of leisure behaviors (Zhao et al. 2016), and lacking research on micro-individuals' leisure needs, leisure psychology, as well as leisure behaviors. At this stage, a large number of empirical studies on urban leisure in China focus on grasping the laws at the macro level, and there is still a relative lack of micro-individual spatial and temporal analysis of leisure behavior (Zhao et al. 2016). With the shift to "humanism" (Li and Li 2010a, b), domestic research perspectives in China have begun to focus on the theoretical framework of the evolution of urban leisure from the macro to the micro level, in addition to the application of big data, 5G, social networks, the Internet of Things (IoT), and GIS technology, which provides more and richer research opportunities for the transition from the huge amount of discrete data on micro-individuals to the scale of macro-groups and the trend of urban leisure. In addition, the application of big data, 5G, social network, Internet of Things and GIS technology provides more and richer research fields from the massive discrete data of micro-individuals to the large-scale trend of macro-group. People-oriented urban leisure is the direction of modern urban development (Han and Huang 2013). The microcognitive methodology of humanism and the local theory of human-land relationship as the object of research can make up for the weakness of domestic leisure research in the study of leisure behavior and psychology, and open up a new way of thinking for research in the field of leisure.

8.3.1 Toward a Postmodern View of Leisure Space

China's urbanization process has experienced a period of rapid development after the reform and opening up, with rapid changes in the internal space of cities. With the outward migration of inner-city manufacturing space, the upgrading of residential space and the expansion and integration of transportation space, recreational space has also undergone great changes, from traditional urban green space parks and urban recreational plazas to richer recreational business districts, neighborhood "pocket parks", and community recreational green spaces pursuing the creation of gardens (Zhang 2007; Wei et al. 2018). These changes have been influenced by the post-modern new urban planning theory "Los Angeles School", forming the post-modern leisure space trend, which emphasizes more on the theory of place spirit of leisure space, stresses the diversity and personalization of leisure needs, and enriches the environment and visual presentation of leisure space, so that the city has the local characteristics of "urbanity". The local characteristics of "urbanity" are strengthened, the emotional design expression of urban leisure space is enhanced, the local factors of leisure space are created, and the standard templates and similar architectural spaces in modernized cities are reduced.

In the 1970s, human geography under the influence of "humanism" began to emphasize the transformation of cities from the modern "physical space shaping" of functional rationality to the "local community building" of socio-cultural diversity (Zhang 2012). Influenced by humanism, urban planning and construction began to emphasize the quality of micro-individuals' daily life space and the construction of urban habitat (Li and Li 2010a, b). The core ideas of the humanistic view of the city are sense of place, place spirit, and place-making, and the nature of the human–environment relationship is used to study the positive place attachment and negative sense-of-place avoidance tendencies between urban residents and the urban environment they live in (Tang 2011; Zhang 2012; Li et al. 2017), and to study the spatial cognitive process and place reconstruction process of urban residents (Li 2000; Han and Huang 2013; Yuan and Wu 2018).

Urban leisure space includes urban residents' leisure behavioral activities, leisure experiences, and individual expressions of emotional significance, and it is a dynamic space. The function of leisure and livability in the urban habitat is the residents' aspiration and pursuit, and its historical evolution process can highlight the value judgment process desired by the leisure people (Gu and Song 2001; Li and Li 2006; Zhang 2012; Li et al. 2014), and this value judgment can be displayed in the leisure space within a certain period of time, and the preferences and evasions of the leisure users continuously transform the leisure space, i.e., the structure of the urban leisure space is continuously reconstructed through the practice of localization (Zhang 2012). Imagery in leisure space has been defined as the brain schema of experiences, memories, emotions, and immediate sensations that are used to interpret information and guide leisure behaviors, and imagery provides a stable order between concrete things and concepts (Li et al. 2008). The emotional expression of urban leisure space is also the source of residents' "place sentiment", which are the labels of local groups extended from the expression of individual emotions (Tang 2007; Zhang 2008, 2009; Zhu and Liu 2011). Through the study of urban imagery, Kevin Lynch (2001) integrated the concept of urban landscape into the theory of environmental cognition and formed a system of urban structure cognition, which expresses the perceptual and sensory forms of the environment through the cognitive maps of the place, and expresses the memory of the urban space through the local images.

Toward a postmodern view of leisure space, it is advocated to explore the relationship between "people and places" between leisure subjects and cities from the perspective of humanistic microcognitive methodology, with place theory as the basic research perspective, and to construct a conceptual framework for the study of a postmodern view of place in urban leisure (Han and Huang 2013). From the viewpoint of the subject of leisure, the city has different levels of population distribution, complex structure and diversified needs, and these factors lead to the diversity and complexity of the characteristics of leisure behavior; from the viewpoint of leisure behavior, the leisure behavior has the characteristics of temporal and spatial non-fixed, change elasticity and travel randomness (Li and Li 2010a, b); from the viewpoint of the carrier of leisure realization, urban leisure behaviors require specific places and spaces for leisure activities, corresponding to the internal places and spatial environments of the city (Li et al. 2014). Leisure behavior occurs based on spatial carriers, and space and behavior constitute places; From the perspective of locality, there is diversity and historical evolution in the differences in the geographical characteristics of leisure and the expression of emotional meaning formed by social groups. The application of place theory to urban leisure research (Zhang 2012; Han and Huang 2013) will open up new ideas and expand a broader research space for the field of leisure at the psychological, behavioral, and locality levels. The postmodern view of leisure space promotes the creation of urban characteristics, pursues the manifestation of place spirit, and through the concept of "place regeneration", develops the historical and cultural heritage of the process of urban leisure development as a kind of urban leisure resources, reproduces the trajectory of urban leisure development, and allows the city to reproduce different places of leisure space to manifest place spirit and urban characteristics. Reinventing local vitality is carried out by means of local functional reorganization, mechanistic stitching, and cultural heritage (Zhang 2012). For example, Shanghai Xintiandi has become one of the city's fashionable, leisure, cultural and entertainment centers by restoring the exterior of historical buildings and remodeling the interior structure (Jiang et al. 2011); and Beijing has transformed the declining former state-run No. 798 Factory and other old factories in the electronics industry into new landmarks of Beijing's urban culture as well as places for art, leisure and consumerism (Ning and Jin 2008).

8.3.2 Formation of Urban Spatial Imagery for Leisure Behavior

Urban residents generate the perception of leisure space through their leisure behaviors, forming the residents' general perception of leisure space (Gu and Song 2001; Feng 2005). These elements of leisure space cognition and perception will accumulate to form subjective impressions in the residents' minds, that is, virtualized spatial imagery constructed from the elements that residents can recall about leisure space.

After being perceived and cognized by people, the physical space undergoes a transformation from a potential environment to an effective environment, which is then identified with by the participants of the leisure space. When the positive identification emotion accumulates, the imagery for the physical space is formed through this environmental psychological experience, which then generates the motivation for various behaviors. Spatial imagery is the re-presentation of material space in the subjective psychological environment of individuals, and is also a stable urban structure that is generally acceptable to individual or groups of urban residents. The theory of spatial imagery is used to study urban spatial layout and planning, which can effectively satisfy the satisfaction and spiritual identity of urban residents, and become an important part of the study of leisure behavior and leisure space.

Lynch first proposed the concept of urban imagery (Li and Li 2006), in which the memories and imagery of urban individuals for localized spaces have differences, while the overall memories and imagery of urban groups have universality, which can be regarded as the "placeness" or "place spirit" of the city. The superposition of individual spatial imagery forms group spatial imagery. Studying the formation process of individual spatial imagery of urban residents and summarizing the elements of urban leisure imagery can provide good planning guidance for the layout of urban leisure space (Gu and Song 2001; Feng 2004, 2005). The distribution, recognizable degree and spatial combination structure of imagery space elements in urban imagery form the characteristics of urban imagery space (Li and Li 2006). Urban leisure space imagery is the impression and meaning of urban leisure space formed in the minds of urban recreationists (including urban residents and urban tourists) through observation, experience, perception and cognition of urban leisure and recreational space or places, and after storing, understanding and reassembling the stimulation of environmental information on leisure and recreational space (Tian 2013). The imagery of urban open space originates from the perception and cognition of urban recreationists on open space, and the open space that is well perceived and cognized by recreationists usually has a high degree of imaginability (Liang and Zhang 2007), which makes recreationists feel mentally relaxed. The specific embodiment of the humanization of leisure space is to realize the participation, experience and interaction between people and space. Participation, experience and interaction, as emphasized here, have mainly physiological and psychological dimensions (Guo et al. 2008). On the physiological level, residents feel the physiological response through their actual experience. On the psychological level, residents take the initiative to enter a specific leisure space scene and personally experience the meaning expressed by the scene, so that the subject can produce the feeling of being there psychologically and resonate with the leisure space environment in terms of both visual perception and spiritual experience, thus producing a strong sense of participation.

The formation process of urban leisure imagery space is the process of people's emotional accumulation. The study of the degree of emotional accumulation cannot be analyzed through specific quantitative data, and research information is generally obtained through interviews and field life experiences in the field. The differences in the mapping process and characteristics of urban residents' recreational cognitive maps lead to the differences in the types of cognitive maps, reflecting the residents' unique spatial thinking of mapping, different spatial organization, and the psychology of different recreational environments. Empirical studies on Chinese cities show that the hierarchical level of Chinese urban recreational imagery elements such as paths, signs and nodes is high, and the imagery content is dominated by plazas, recreation and business centers and characteristic recreational neighborhoods, important scenic roads and bridges, characteristic buildings (clusters), and historical and cultural heritage and monumental sites (Wang and Hu 2007); the spatial pattern of imagery presents the characteristics of high in the middle and low in the periphery, and it has a strong tendency to coalesce in the city center (Tian 2013). Lynch's study found that residents' impressions of spatial locations can be verbalized, and that such representations are more similar to hand-drawn sketches of imagery (Lynch 2001; Wang et al. 2007). There are also many similarities in the results of different residents' representations of this spatial imagery, i.e., they are all composed of basic points, lines, and surfaces. This is related to the basic geometric and map knowledge that urban residents receive, and this generalized type of knowledge also facilitates the study of cognitive maps of urban residents (Li and Li, 2006). Urban public squares generally have a regular geometric space, centripetal is a clear feature of the regular space, which provides a basis for the creation of an enclosing environment, such as in the center of the circular space to facilitate the gathering of people. Dalian's Zhongshan Square has a standard circular spatial form, Zhongshan Square is located in the geometric center of the space, and the roads emanating from the center extend to different directions, and the spatial form creates a strong psychological cue to the residents who are relaxing in the square, and their leisure behaviors are also affected by the clear directionality of the square (Li et al. 2013).

In the imagery of leisure space, the emotion generated for the place and space is the main factor influencing the individual's leisure behavior. After the generation of leisure motivation, the decision-making stage of choosing leisure space and leisure behavior will be influenced by the emotional factors of leisure space. The process of generating spatial place sentiment is a process from cognition to awareness to motivation and creativity (Li et al. 2019). "Place cognition" is divided into direct and indirect cognition. Direct cognition, mainly in the leisure activities in the actual participation in the direct contact, sensory and experience, cognitive channels and ways to visual-based, assisted by hearing, touch, taste, smell. Indirect cognition, mainly in the leisure activities before or during the process, to get the evaluation and experience of word-of-mouth-based, assisted by media advertising, network information, exhibitions, friends and relatives recommended. Generally, indirect cognition is earlier than direct cognition, and indirect cognition affects the effect of direct cognition and has an orienting effect on direct cognition.

The process of cognitive stage is an important stage in the formation of sense of place in leisure space, and it is also the stage that determines the final development direction of place cognition. The desire to return to the place and the behavior of returning to the place generated through the cognition of place will directly affect the formation of the sense of place in the stage of consciousness, which is mainly manifested in the cumulative effect of the behavior of returning to the place to deepen the experience of recreation and form the "place identity", and at the same time, there is a risk in the process of returning to the place. At the same time, there is also a risk in the reentry process, that is, there are changes in the process or things that directly change the recreationist's cognition, which cause the difference in the cognition of the place, i.e., the emergence of "place resistance" (Li et al. 2013). Highlighting the security of spatial image elements and focusing on breaking the bottleneck of interactive development of leisure and urbanization. In the process of leisure space development and town construction, it is necessary to highlight the recognizability of leisure space, as well as the layout of spatial places with popular memories, so that more people can obtain positive emotional identity and profound spatial images from leisure space.

The relationship between urban leisure space and urban residents' leisure behaviors in Chinese cities has formed a system of place theories such as placeness, place meaning, and place attachment (Han and Huang 2013). The geographic location of Chinese cities spans a wide range, and the unique natural landscapes, folk cultures, and urban characteristics of each place create a diverse urbanity, and urban residents have emotional meanings for urban leisure spaces and physical spaces through leisure activities (Li et al. 2008); meanwhile, urban residents and foreign tourists spread such perceptions in words, pictures, and videos through social networks and word-of-mouth communication, which forming a specific city cultural card and city imagery label. In the continuous change of urban leisure space imagery, from the generation of leisure demand to the implementation of leisure behavior, all are influenced by urban place sentiment, which in turn form positive place sentiment and negative place sentiment. The image of urban leisure space will also guide the planning of leisure space for improvement, in addition, the mobility and creativity of the residents will also make the urban leisure space self-improvement.

8.3.3 A New Model of Leisure Space: Leisure Urbanization as a Way of Life

The development stage of Chinese cities is still in the transition from industrial society to post-industrial society, and cities in different geographic regions are in different stages of economic and social development, and cities also possess diverse living spaces (Yu et al. 2018). On the basis of industrial economy and service economy, urban leisure economy, as an economic form generated to meet people's leisure consumption, is essentially an experience economy (Zheng 2005). Experience economy focuses on people's perception and experience, and measures the level of service by the psychological feelings in the process of people's activities. In the era of experience economy, the economic form, governmental role, family organization and life style have changed, and the development of urban leisure activities has also shown the trend of humanization, ecology, experience and industrialization (Zhao and Jiang 2013), and leisure urbanization with the purpose of leisure has gradually become the way of life of urban residents (Li and Ni 2009). Against the background of this shift in the concept of leisure consumption, the development of urban leisure space is bound to respond to people's interests and values of leisure life in terms of function and value, and gradually develop in the direction of experiential and humanization (Guo et al. 2008).

The new industrial system represented by information technology is constantly developing, and the development of the urban economy has changed from being dominated by material products to being dominated by information production and service production. Information technology has blurred the boundaries between work and leisure, and work in urban groups is no longer a guarantee of life, but an experience of life. This has impacted the traditional work model and leisure space scale, changing the binary division between work and leisure during industrial society (Li and Jin 2012). The state of leisure existence is more advanced than the state of work existence, and belongs to the level of higher-level needs after the satisfaction of basic human survival needs (Lü and Zhang 2011; Zhao and Jiang 2013). In the period of information technology and Internet, on the basis of rich material resources, the demand of urban residents for personal skills demonstration has given rise to diversified talent performances and personalized leisure activities. From "game live broadcast" to "show live broadcast" to "carry goods live broadcast", the original leisure activities in the virtual network society are combined with the consumption activities in the real society. Playing games and performing such traditional leisure activities have become the state of work. At the same time, new types of experiential leisure are emerging (Guo et al. 2008). For example, experiential leisure theme activities, agricultural leisure experience activities, and Internet virtual leisure activities. The diffusion and penetration of information technology has reconfigured the overall activity time and leisure activity time of urban residents, and at the same time, it has also had a deconstructing and reconfiguring effect on the existing urban leisure space (Han and Huang 2013).

With the development of the economy and the smooth flow of information communication, the phenomenon of homogeneous appearance has appeared in China's cities, where the original cultural and local characteristics of the cities have been replaced by standardized concrete and steel communities, and the names of the roads are also converging. Under the model of leisure urbanization, it is necessary for urban leisure space to pay attention to the combination with the cultural characteristics of the city, form a leisure space rich in urbanity, and gradually guide residents to form positive local attachment and emotional identity. In addition, under the new model of leisure space, urban leisure space realizes seamless connection with work space, living space, activity space, etc., and guides the integration of ecological leisure and work.

8.4 Chapter Summary

Urban leisure space is a product of the process of urban development, and leisure behavior determines residents' perception of urban space to a certain extent, and "behavior" and "space" depend on and influence each other. On the one hand, this chapter analyzes the driving forces for the formation and existence of urban leisure space from the perspectives of leisure demand and the driving forces of the leisure economy, and analyzes the process of the evolution of China's urban leisure space from the perspective of the contradiction between leisure demand and the supply of leisure space; on the other hand, in view of the differences in the resources that different regions rely on and the evolution of urban leisure space, it summarizes the structure patterns of the urban leisure space in China, and classifies the categories of urban leisure space from the perspectives of different research purposes.

With the increasing demand for leisure activities and leisure space by urban residents in China, the research on urban leisure in China is gradually developing in the direction of "humanization", and the research on urban leisure has gradually shifted from the direction of leisure space, leisure system, leisure facilities, etc. to the exploration of urban residents' leisure behavior. Residents' leisure behavior and urban leisure space influence and promote each other. Residents' leisure behavior is the external manifestation of leisure motivation, and the classification of urban residents' leisure behavior varies due to different research methods and research perspectives, but all of them are characterized by subjective initiative, individual differences, diversity of forms, compatibility and correlation, and restriction. At different stages of social development, urban residents' leisure styles evolve under the joint influence of social values, residents' sense of well-being, and socio-economic factors, with the characteristics of liberalization, diversification, advancement, personalization and systematization becoming more and more obvious.

Besides, this chapter also explores the relationship between the leisure behavior of Chinese urban residents and leisure space, which is mainly divided into the following parts, namely, the influence of the construction of leisure space on leisure behavior, the spatial cognitive process of leisure behavior, and the spirit of leisure place and leisure behavior mode. Urban residents are the main body of urban leisure space, urban leisure space serves for urban residents, and only urban space with leisure behavior can really be a leisure place. As the two interdependent and interactive elements of urban leisure space, "space" and "behavior", leisure behavior profoundly influences the spatial cognition, and spatial cognition determines the degree of experience of leisure activities. Leisure space is the carrier of residents' spirit of leisure space and leisure behavior pattern; the inheritability of the spirit of leisure space is one of the important factors affecting the characteristics of leisure behavior; and residents' unique leisure behavior pattern is the symbol of regionalized leisure spirit. From the perspective of open space, place perception and return behavior will directly affect the formation of people's sense of place, and the difference in the sense of place will also cause the difference in the direction of place perception, which can be avoided through the management and planning of open space, as well as effectively change the place identity.

With the splitting and reorganization of urban functional areas, urban leisure has become a new mode of urbanization that focuses on residents' lifestyles. The urban leisure space has gradually evolved into a whole-area leisure space centered on the internal space, extending and spreading outward until it covers the recreational belt around the city. The key to the goal of creating leisure townships with urban leisure space as a way of life lies in the harmonious development of man and nature, and the perfect integration of township construction into natural and human resources requires us to take corresponding countermeasures for different categories of leisure space, such as dispersing small-volume space, enriching the mass-type leisure space, and creating a hierarchical space for consumption, and so on; and at the same time, it is also necessary to take into account the relationship of interests between leisure space and urbanization. The rapid development of urbanization is the premise for guaranteeing the wide dissemination of leisure culture and the fair improvement of leisure space facilities, and conversely, the development of leisure industry and economy drives the progress of the leisure city, and the establishment of the leisure system enriches the connotation of the leisure city.

In a word, leisure space is the leading content of urban leisure "hardware", and residents' leisure behavior, leisure cognition and leisure perception are the important factors of urban leisure "software". Whether a city can become a "leisure city" in the real sense, it is necessary to analyze the subjectivity of the city as a "place" and the perception, experience, attitude of the leisure subject as well as the relationship between them, in order to break through the elements of spatial image and the bottleneck between the development of leisure space and urbanization, and create a comprehensive leisure city with a unique flavor that truly belongs to the residents, is based on leisure life, and has a moderate concentration of leisure service industries.

References

Bao J, Gan M (2005) Comparative study on Urban recreation business district in Guangzhou old city. Landsc Arch 1:80–83 (In Chinese)

Bao J, Gu S (1998) Preliminary research on Urban RBD. Planners (4):59–65 (In Chinese)

Bao J, Gu S (2002) Formation and development of Urban Recreational Business District (RBD) in Guangzhou. Hum Geogr 17(5):1–6 (In Chinese)

Bian X (2003) Study on the spatial structure of Urban tourism. Geogr Geo-Inf Sci 19(1):105–108 (In Chinese)

Bian X (2005) Spatial structure of tourism resources in Yangtze River Delta cities. Resour Dev & Mark 21(4):354–357 (In Chinese)

Bian X, Sha R (2008) Industrial mechanism of Urban tourism spatial structure formation in the Yangtze River Delta: from the perspective of tourism enterprise location choice. Hum Geogr 23(6):106–112 (In Chinese)

Bian X, Zhang S (2004) Development of Urban recreation business district in China. Econ Geogr 24(2):206 211 (In Chinese)

Cao Y, Sun X (2008) Evaluation of tourism resources in Zhengzhou's Peri-urban recreational belt based on analytic hierarchy process. J Henan Univ (Nat Sci Ed) 38(5):497–501 (In Chinese)

Chai Y, Liu T, Ta N (2013) A framework for multi-scale Urban spatial reconstruction and planning applications based on individual behaviors. Areal Res Dev 32(4):1–14 (In Chinese)

Chai H, Feng J (2014) Social space formation in suburban large-scale residential areas based on relocation: a case study of Beijing's Huilongguan Community. Areal Res Dev 33(5):77–81 (In Chinese)

Chen Y (1995) Clarifying the concept of Urban-Rural Fringe. Geogr Territ Res 11(1):47–52 (In Chinese)

Chen J (2005) Development and construction of Urban recreation business district. City 1:52–54 (In Chinese)

Chen Y (2013) Development history and typology of Urban recreational spaces. Chin Landsc Arch 29(2):69–72 (In Chinese)

Chen Y, He Q (2006) Research on the construction of Urban recreation business district: a case study of Changsha. Tropical Geography 4:379–383 (In Chinese)

Chen H, Wang X (2012) Demand characteristics of Peri-urban recreational belts in large cities: a case study of Wuhan City. J Southeast Univ (Philos Soc Sci Ed) 14(2):107–111 (In Chinese)

Dang N (2011) Suburban recreational spaces in the leisure era: research on Recreational Belt Around Metropolis (ReBAM). Shanghai People's Publishing House Shanghai (In Chinese)

Dang N, Wu B, Yu Q (2017) Spatio-temporal evolution and dynamic mechanisms of Shanghai's Recreational Belt Around Metropolis (ReBAM) from 1970 to 2015. Tour Trib 32(11):81–94 (In Chinese)

Deng L, Wu J (2017) Evolution process and patterns of Urban public entertainment space structure in modern Shanghai (1843–1949). Urban Plan Forum 3:95–102 (In Chinese)

Dong G, Li L (2006) Study on the growth mechanism and industrial structure evolution of Urban RBD: a case study of OCT in Shenzhen. Planner (3):71–74 (In Chinese)

Fang Y, Bi D (2007) A review of domestic research on Urban Recreation business district (RBD). Contemp Econ Manag 4:13–17 (In Chinese)

Fang Q, Bu J (2003) Research on the design of Urban waterfront recreational spaces. Planners 9:46–49 (In Chinese)

Fang Y, Huang Z, Lu W, et al (2013) Spatial differences and mechanisms of tourism economy in Chinese cities. Geogr Geo-Inf Sci 29(6):100–104, 110 (In Chinese)

Feng J (2004) Spatial restructuring of Chinese cities in the transition period. Science Press Beijing (In Chinese)

Feng W (2006) Current status and key development areas of Urban recreational space research in China. Adv Earth Sci 21(6):585–592 (In Chinese)

Feng W (2010) Structural patterns of Urban recreational space systems. Arch J (S2):150–153 (In Chinese)
Feng J , Zhou Y (2008) Restructuring of social spatial differentiation in transitional Beijing. Acta Geogr Sin 63(8):829–844 (In Chinese)
Feng X, Yu J, Meng X (2013) Formation mechanism and sustainable development of Peri-urban recreational belts in inland cities: a case study of Urumqi. Ecol Econ 2:131–136 (In Chinese)
Feng J (2005) Spatial perception and image structure of Beijing residents. Sci Geogr Sin 25(2):142–154 (In Chinese)
Feng W (2007) Analysis and integration of Urban recreational spaces. Chongqing University, Chongqing (In Chinese)
Fu C, Bai C, Cao W (2009) A history of Urban development in China. Social Sciences Academic Press, Beijing (In Chinese)
Gan J, Guo F, Chen C (2015) Spatio-temporal evolution of urbanization spatial differentiation in Northeast China Since 2000. Sci Geogr Sin 35(5):565–574 (In Chinese)
Gu S, Bao J (2002) Impact of Guangzhou's Urban Recreational Business District (RBD) on Urban development. Sci Geogr Sin 22(4):489–494 (In Chinese)
Guo W, Lu Y, Sang E (2006) Reflections on establishing recreation business district in Qinhuangdao. Commer Res 2:184–186 (In Chinese)
Guo X, Guo E, Chen Y (2008) Leisure economy and the development of urban leisure spaces City. Plan Rev 12:79–86 (In Chinese)
Gu C, Song G (2001a) A study on Beijing's Urban image space and its components. Acta Geographica Sinica 56(1):64–74 (In Chinese)
Guo L, Wang W (2008a) Growth patterns and cultivation paths of recreational Belt Around Metropolis (ReBAM): a perspective of experience economy. Tour Trib 23(2):55–59. (In Chinese)
Guo L, Wang W (2008b) Literature review on recreational belt around metropolis. Ecol Econ (1):114–115, 150. (In Chinese)
Han G, Huang A (2013) Application of place theory in Urban Leisure. Hum Geogr 2:125–130 (In Chinese)
Hou G, Huang, Zhao Z (2002) Formation and spatial structure of Urban commercial recreational districts. Hum Geogr 17(5):12–16. (In Chinese)
Hu H (2013) Spatial distribution and transportation accessibility of China's outstanding tourism cities. Sci Geogr Sin 33(6):703–709 (In Chinese)
Hu J, Zhong A (2012) Aggregation and dispersion of mass cultural entertainment spaces in modern Hankou and Urban development. Wuhan Univ J (Humanities Sci Ed) 65(4):25–32 (In Chinese)
Hu W, Shen S, Qiu F (2015) Construction of Urban residents' Leisure space system: a case study of Xuzhou. Mod Urban Res 6:70–77 (In Chinese)
Huang Z, Hou G (2001) Formation mechanisms of Urban commercial recreational districts in large cities. Geogr Territ Res 17(4):44–47 (In Chinese)
Huang Z, Hou G, Xu S (1999)Sustainable development of suburban tourism and preliminary exploration of Agritourism: a case study of Nanjing. J Nanjing Norm Univ (Nat Sci Ed) 4:3–5 (In Chinese)
Jiang Y (2020) The impact of winter Olympics on host city tourism and leisure: a literature review. Tour Trib 35(4):1–3 (In Chinese)
Jiang L, Zhou Y (2005) Discussion on land use planning adjustment for Yangshuo's recreational business district. Planners 1:97–100 (In Chinese)
Jiang W, Chen K, Ma X (2011) Spatial production in China's old city renewal: a case study of Shanghai Xintiandi. Urban Dev Stud 18(10):84–89, 96. (In Chinese)
Jin S (2009) Construction and deconstruction of public recreational spaces in metropolitan areas: a case study of Shanghai. East China Normal University Shanghai (In Chinese)
Kevin L (2001) The image of the city (Fang Y, He X by trans). Huaxia Publishing House (In Chinese), Beijing
Li N (2006) Constructing Urban recreational commercial areas: A case study of the development of recreational commercial areas in Nanjing. J Nanjing Univ Fin Econ (2):30–33 (In Chinese)

Li, X Li J (2006) Analysis of Dalian's Urban spatial image. Acta Geograph Sin 61(8):809–817 (In Chinese)
Li X, Zhang Y Gao J (2014) Types and spatial patterns of Urban human settlements: a case study of Shahekou District, Dalian. Sci Geograph Sin 34(9):1033–1040 (In Chinese)
Li R Zhu H Tang L (2017) Study on negative sense of place in the context of Urban-Rural migration: a case study of Tangxia Village, Guangzhou. Hum Geogr 32(3):27–35 (In Chinese)
Liu R Ma H (2016) Regional differences in urban recreation development in China. Urban Probl (10):30–36 (In Chinese)
Li X, Jin P (2012) Characteristics and spatio-temporal differences of Urban human settlement quality in China. Sci Geogr Sin 32(5):521–529 (In Chinese)
Li X, Ni Y (2009) Coordinated development of Urbanization and Urban human settlements in China's top livable cities over the past decade. J Arid Land Resour Environ 23(3):8–14 (In Chinese)
Li J, Tan L (2016) Dynamic system of Recreational Belt Around Metropolis (ReBAM) development from an ecological civilization perspective. J Hubei Univ (Philos Soc Sci Ed) 43(6):130–134 (In Chinese)
Li X, Liu X, Ji B (2008) Differentiation characteristics of livability in Dalian's Urban communities. Prog Geogr 27(4):75–81 (In Chinese)
Li R, Guo F, An Y (2011) Types and spatial structure characteristics of Peri-urban recreational areas in Beijing over the past decade. Hum Geogr 26(1):118–122 (In Chinese)
Li X, Li H, Tian S (2013) Research on recreational systems from the perspective of human settlements. J Liaoning Norm Univ (Nat Sci Ed) 2:269–273 (In Chinese)
Li X, Guo Y, Tian S (2019) Spatio-temporal evolution and driving forces of coupling coordination in Urban human settlement systems in Liaoning Province. Sci Geogr Sin 39(8):1208–1218 (In Chinese)
Li X, Li J (2010a) Current status and prospects of human settlement research in geography. J Liaoning Norm Univ (Nat Sci Ed) 33(1):112–117 (In Chinese)
Li X, Li J (2010b) Cultural turn in physical geography. Prog Geogr 29(6):740–746 (In Chinese)
Li H (2014) Spatial pattern and optimization of Urban ecological recreational spaces in Shanghai. Econ Geogr 34(1):174–180 (In Chinese)
Li J (2014b) Comparative study on tourist value in Beijing's Urban recreation business district. Urban Probl (1):29–34 (In Chinese)
Li P (2014c) Modern public transportation and Urban lifestyles: pre-war "Capital" Nanjing. Lanzhou Acad J (9):67–74 (In Chinese)
Liang Y, Zhang J (2007) Spatial structure of Leisure image in Wuhan. J Central China Norm Univ (Nat Sci Ed) 41(4):632–635 (In Chinese)
Lin Z (2016) Spatio-temporal evolution pattern of urban tourism leisure public space in Shanghai. Tour Sci 30(2):79–94 (In Chinese)
Liu E (2001) Traditional Chinese leisure culture. Stud Dialectics Nat 5:63–64 (In Chinese)
Liu S (2006) Formation and development suitability study of Chengdu urban RBD. Commer Res 17:183–185 (In Chinese)
Liu Z, Chai Y (2001) Spatial structure of weekend leisure activities in Shenzhen. Econ Geogr 21(4):504–508 (In Chinese)
Liu R, Ma H (2016) Regional differences in urban recreation development in China. Urban Probl (10):30–36 (In Chinese)
Liu J, Wang R (2007) Analysis of influencing factors on suburban recreational land allocation. Tour Trib 22(12):18–22 (In Chinese)
Liu Z, Chai Y, Gong H (2000) Characteristics of leisure time use among Shenzhen residents. Hum Geogr 15(6):73–78 (In Chinese)
Liu Y, He S, Li Z (2005) Space-time structure of daily activities among urban poor in Nanjing. Chin J Popul Sci S1:85–93 (In Chinese)
Liu L, Xu X, Wu B (2017) Recreational Belt Around Metropolis (ReBAM): a new spatial element worth exploring. Areal Res Dev 36(2):56–60, 73 (In Chinese)

Lou J, Li L (2017) Evolution process and influencing factors of urban recreation development in Chengdu. Tour Sci 31(1):12–27 (In Chinese)
Lou J, Xu A (2011) Development stages and evolution characteristics of urban leisure functions in Shanghai. Tour Sci 25(2):16–22 (In Chinese)
Lou J, Liu S, Li L (2016) Development level and spatial differences of urban recreation in China. Urban Probl 11:29–35 (In Chinese)
Lü N (2011) Connotation and practice of leisure cities. Econ Her 2:52–53 (In Chinese)
Lü N, Huang X (2014) Functional study of urban leisure: a case of Beijing as a world tourism destination Urban. Dev Stud 21(3):99–105 (In Chinese)
Lü N, Zhang H (2011) Review of urban leisure and leisure industry research. Commer Times 13:111–112 (In Chinese)
Ma H (2002) Prospects for China's leisure tourism industry in the next decade. Qilu J 2:19–26 (In Chinese)
Ma H, Lou J, Liu R (2019a) Spatio-temporal evolution of urban residents' leisure consumption capacity in China. World Regional Stud 28(6):145–155 (In Chinese)
Ma X, Luo L, Huang H, et al (2019b) Application of WeChat-Alipay appointment registration service in obstetrics and gynecology outpatient departments of tertiary hospitals. China Digit Med 14(9):96–98 (In Chinese)
Ma H (2005) Analysis of Western urban recreational space planning and design. Qilu J (6):147–153. (In Chinese)
Mao X, Song J, Yang H (2012) Spatial pattern changes of Beijing urban parks (2000–2010). Prog Geogr 31(10):1295–1306 (In Chinese)
Miao J (2003) Central cities: spatial perspectives of the leisure economy. Stud Dialectics Nat 11:73–78 (In Chinese)
Murphey R (1974) The treaty port and china's modernization. In: Elvin M, Skinner GW (eds) The Chinese city between two worlds. Standford University Press, Standford, pp 17–72
Nan Y, Hu H, Zhu F (2012) Analysis of recreational belt around medium-small cities: a case study of Yanji City. Hum Geogr 27(2):62–66 (In Chinese)
Ning Z, Jin S (2008) 798 Art District as a cultural tourism attraction in Beijing: a perspective of spontaneous market formation. Tour Trib 23(3):57–62 (In Chinese)
Pan L, Jin X (2014) Formation mechanism and spatial structure characteristics of cross-administrative recreational belts around cities. Stat & Decis 21:105–108 (In Chinese)
Qin X (2003) Systemic analysis of urban recreational space structure: a case study of Ningbo. Econ Geogr 23(2):267–271, 288 (In Chinese)
Sha R, Wu J (1997) Preliminary study on ecological design of tourism landscapes in urban-rural fringe. Geogr Territ Res 14(3):54–57, 63 (In Chinese)
Shan N (2000) Open space landscape design. Liaoning Science and Technology Press Shenyang (In Chinese)
Shan F (2019) Research on Urban leisure sports development in China from ecological civilization perspective: cases of Hangzhou, Wuhan and Chengdu Shanghai University of Sport Shanghai (In Chinese)
Shen L, Zheng Y (2007) Exploration of new urban RBD: conceptual design of Ningbo Wantou RBD. Arch J 7 4 6 (In Chinese)
Shi H, Wang X (2015) Historical space optimization of ancient city RBD from cultural tourism perspective: case of Kaifeng. Spec Zone Econ 6:125–126 (In Chinese)
Shi C, Zhang J, Li Y (2008) Spatial differentiation of urban leisure formats China. Popul Resour Environ 18(5):90–95 (In Chinese)
Shi Y (2007) Spatial structure of RBD in Shanghai central city. Tongji University (In Chinese)
Song J (2011) Organic interaction between historical district renewal RBD and urban CBD: case study of Chengdu Daci Temple historical district renewal planning. Mod Urban Res 26(1):39–43 (In Chinese)
Song H, Ma Y (2010) Spatial analysis of provincial tourism economic disparities in China. Econ Manag 32(10):114–118 (In Chinese)

Su P, Dang N, Wu B (2004) Tourism site types and spatial structure characteristics of Beijing recreational belt around metropolis. Geogr Res 23(3):403–410 (In Chinese)
Sun P, Dai X (2014) Spatial statistical analysis of regional tourism economic disparities in China. Tour Sci 28(2):35–48 (In Chinese)
Tang W (2007) Research progress and framework of sense of place. Tourism Tribune 22(11):70–77 (In Chinese)
Tang W (2011) Characteristics of urban residents' recreational place attachment: Case study of Nanjing Confucius Temple. Sci Geograph Sin 31(10):1202–1207 (In Chinese)
Tao W, Huang R (2006) Evolutionary process and influencing factors of urban RBD spatial structure: case study of Guangzhou. Hum Geogr 21(3):10–13 (In Chinese)
Tao W, Li L (2003) Growth study of urban SRBD system: case study of Suzhou historical city. Tour Trib 18(3):43–48 (In Chinese)
Tao W, Li L (2005) Evolution model of urban RBD spatial structure in Hong Kong. City Plan Rev 29(6):69–75 (In Chinese)
Tao T, Tian J (2009) Theory and empirical research on recreational belt around megalopolis: analysis based on tourism resource value of Shanghai's new "Three Cities and Seven Towns". Financ & Trade Res 35(7):110–121 (In Chinese)
Tian F (2013) Characteristics of urban recreational space imagery under national leisure background: case study of Nanchang. Res Sci 35(5):1095–1103 (In Chinese)
Wang X (2008) Role of RBD in promoting urban development and enhancing quality of life. Planners 24(7):38–42 (In Chinese)
Wang B (2016) Comparative study on block system under Chinese and Western residential cultural backgrounds. Comp Econ Soc Syst 5:136–144 (In Chinese)
Wang J, He J (2004) Formation mechanism of urban recreational business district: case study of Jinan City. J Taishan Univ 2:87–89 (In Chinese)
Wang H, Hu S (2007) Image space and tourism planning of Zhenyuan ancient city. Areal Res Dev 26(3):61–65 (In Chinese)
Wang Q, Wei J (2020) Twenty-year changes in leisure demands of Chinese urban residents. J Harbin Inst Technol (Soc Sci Ed) 22(1):45–51 (In Chinese)
Wang M, Chai Y, Gao Y (2007) Geographical research progress on cognitive map spatial analysis. Hum Geogr 22(5):10–18 (In Chinese)
Wang Y, Feng N, Shi C (2009) Research progress on urban leisure space. Urban Probl 3:28–33 (In Chinese)
Wang H, Guo L, Song L (2010) Impact of Dalian's urbanization evolution on its recreational belt. J Beijing Int Stud Univ 32(11):32–38 (In Chinese)
Wang H, Guo L, Zhu Y, et al (2012a) Spatial expansion patterns and characteristics of recreational areas in Dalian. J Liaoning Norm Univ (Nat Sci Ed) 35(2):258–263 (In Chinese)
Wang G, Pan Z, Lu Y (2012b) Changes in regional patterns of China's interprovincial migration and influencing factors: Analysis based on 2000 and 2010 census data. Chin J Popul Sci 32(5):2–13, 111 (In Chinese)
Wang Q (2004) Spatial structure of recreational landscape in Changchun suburbs. Northeast Normal University, Changchun (In Chinese)
Wang X (2004b) Research on Urban living space structure in China. Science Press, Beijing (In Chinese)
Wang D (2007) Evolution and optimization of urban tourism spatial structure: case study of Suzhou. Urban Dev Stud 14(1):21–26, 32 (In Chinese)
Wei X, Li Y (2007) Urban leisure and leisure cities. Tour Trib 10:71–76 (In Chinese)
Wei W, Zhou J, Luo M (2018) Community park satisfaction analysis and planning strategies from "urban people" perspective: case study of Zhongnanlu Subdistrict, Wuchang District Wuhan City. Plan Rev 42(12):55–66 (In Chinese)
Wei F, Xi Y, Shu TC (2016) Development concepts and strategies of urban recreational space from spatial justice perspective: Enlightenment from U.S. experience. J Hum Settlements West China 31(5):51–56 (In Chinese)

Wu C (2005) Spatial units and structure of urban tourism. Urban Planning Forum 3:82–87 (In Chinese)

Wu B, Jia J (2002) Development of tourism and recreational functions in urban waterfront areas: case study of Wuhan. Geogr Territ Res 18(2):99 102 (In Chinese)

Wu B, Dong L, Tang Z (2003) Research on the classification and attributes of public recreational spaces. Chin Landsc Arch (5):49–51 (In Chinese)

Wu B, Dong L, Tang Z (2003) Classification and attributes of public recreational spaces. Chin Landsc Arch 5:49–51 (In Chinese)

Wu B, Wu J, Dang N (2007) Local residents' recreational preferences around cities in tourist cities: case study of Hangzhou. Hum Geogr 22(2):27–31 (In Chinese)

Wu G, Cheng D, Wu, et al (2004) Development of urban RBD in Guilin. Urban Probl (4):33–35, 39 (In Chinese)

Wu B (2001) Study on Urban Ring Recreational Belt (ReBAM): a case study of Shanghai. Geograph Sci 21(4):354–359 (In Chinese)

Wu L (2001a) Introduction to human settlement environment science. China Architecture & Building Press, Beijing (In Chinese)

Wu L (2001b) Exploration of human settlement environment science. Planners 17(6):5–8 (In Chinese)

Wu Q (2001c) Theory and practice of residential spatial differentiation in large cities. Science Press, Beijing (In Chinese)

Xia J, Wang Y (2010) From reconstruction to regeneration: protecting residential authenticity in historic and cultural district Urban. Dev Stud 17(2):134–139 (In Chinese)

Xiao G, Song W (2008) Optimization of urban recreational space structure: case study of Dalian China. Popul Resour Environ 18(2):86–92 (In Chinese)

Xie L, Zhou J (2005) Development of urban recreational business district (RBD): case study of Taizhou. Tech Econ Manag Res 5:126–127 (In Chinese)

Xu X, Chen Z (2012) Service hierarchy and spatial layout characteristics of urban parks based on leisure demand: case study of central Guangzhou. Trop Geogr 32(3):293–299, 320 (In Chinese)

Xu J, Wang L (2011) RBD development direction based on consumers' leisure behavior: case study of Changsha. Econ Geogr 31(7):1213–1218, 1225 (In Chinese)

Xu X, Chai Y (2012) Gender differences in daily leisure behavior of Beijing residents. Hum Geogr 27(1):22–28 (In Chinese)

Xu X, Chen Z (2018) Evolution and influencing factors of public leisure service level in Guangzhou. Areal Res Dev 37(6):58–63 (In Chinese)

Xu A, Lou J (2019) Regional differences and causes of urban leisure development in China. World Regional Stud 28(6):98–108 (In Chinese)

Xu F, Yang K (2006) Construction of modern urban RBD system: case study of Wuhan. Tour Trib 21(3):24–29 (In Chinese)

Yan H, Deng H (2004) Environment and functions of urban recreation business district: case study of Jianghan Road Wuhan. Urban Probl 1:40–43 (In Chinese)

Yang L, Ma X (2015) Spatial structure characteristics of peri-urban recreational belt in Changsha. Econ Geogr 35(10):218–224 (In Chinese)

Yang Z, Zhou K (2008) Revisiting leisure cities and urban leisure. Tour Trib 12:51–57 (In Chinese)

Ye S (2009) Conceptual differentiation of leisure, recreation and tourism from means-end perspective. J Guangxi Univ Natl (Philos Soc Sci Ed) 31(S1):26–30 (In Chinese)

Ye S, Bao J (2009a) ROP-ENCS: a typological framework for urban recreational space morphology studies. Trop Geogr 29(3):295–300 (In Chinese)

Ye S, Bao J (2009b) Foundation for characterizing urban recreational space morphology: field model vs. element model. Geogr Geo-Inf Sci 25(3):99–102 (In Chinese)

Ye S, Ye T, Wu X (2015) Exploring integrated management institutions for urban recreational spaces. J South China Univ Technol (Soc Sci Ed) 17(3):49–54, 113 (In Chinese)

Yu S (2003) Urban tourism and Urban recreation. East China Normal University Press Shanghai (In Chinese)

Yu C (2015) Tourism development research of Qingdao Beer Culture Leisure Business District. China Ocean University Qingdao (In Chinese)
Yu G, Ma H (2006) Dialogue on the concepts of 'leisure' and 'recreation'. Stud Dialectics Nat 22(9):86–91 (In Chinese)
Yu L, Liu J, Li T (2018) Research progress on urban public recreational space in China. Acta Geogr Sin 73(10):1923–1941 (In Chinese)
Yu S, He S (2003a) Sociological analysis of urban recreation. J East China Norm Univ Nat Sci (2):54–61 (In Chinese)
Yu S, He S (2003b) Layout study of urban recreation business district (RBD). Hum Geogr 18(4):10–15 (In Chinese)
Yuan J, Wu Y (2018) Loss and reconstruction of locality in urbanization. J Huazhong Univ Sci Technol (Soc Sci Ed) 32(1):41–46 (In Chinese)
Zhang J (2008) International trends in leisure studies and preliminary discussion on China's research propositions. Tour Trib 23(5):68–73 (In Chinese)
Zhang Z (2012) Place theory: generalized thinking on constructing human-environment settlement science system. Dev Res 7:47–55 (In Chinese)
Zhang H (2004) Tourism development and land use in metropolitan Peri-Urban recreational belts: case study of Xi'an. Shaanxi Normal University, Xi'an (In Chinese)
Zhang J (2005) Morphological patterns of recreation business district in metropolitan Shanghai. Areal Res Dev 24(3):63–67 (In Chinese)
Zhang F (2007) Living conditions of migrant farmers in Beijing's semi-urbanized areas. Institute of Geographic Sciences and Natural Resources Research, Chinese Academy of Sciences, Beijing (In Chinese)
Zhang L (2007) Theoretical study on spatial structure of urban RBD: case studies of Shanghai City God Temple and Suzhou Guanqian Street. Areal Research and Development, 26(5): 65–69, 84 (In Chinese)
Zhang J (2009) Study on the relationship between metropolitan life and city space. J Tongji Univ (Soc Sci Sect) 20(4):51–58 (In Chinese)
Zhang W (2007) Pocket parks: oases escaping urban hustle. Chin Landsc Arch 23(4):47–53 (In Chinese)
Zhan G Sang Z (2006) Conceptual differentiation between CBD and RBD and their functional extensions. Tourism Tribune 21(12):77–80 (In Chinese)
Zhao H (2015) Utilization of intangible cultural heritage in HRBDs. J Shanxi Norm Univ (Nat Sci Ed) 29(1):109–112 (In Chinese)
Zhao X (2020) From education to leisure: Shifts in museum audience research perspectives. Museum Manag 1:75–81 (In Chinese)
Zhao C, Chen M (2016) Breakthrough and reconstruction: conceptual trends in urban leisure supply optimization. Acad Forum 39(4):73–77 (In Chinese)
Zhao X, Jiang Q (2013) Diversification trends of leisure tourism in the experience economy era. Res Financ Econ Issues 6:140–145 (In Chinese)
Zhao Y, Xu W (2008) Research progress on ReBAM in China over the past decade. Econ Geogr 28(3):492–496 (In Chinese)
Zhao Y, Chai Y, Gui J (2016) Frontiers of urban leisure space-time behavior research in China. Tour Trib 31(9):30–40 (In Chinese)
Zhao M, Wu B (2009) Spatial structure evolution of resort areas around Beijing. World Regional Stud 18(4):134–140, 133. (In Chinese)
Zhao Y, Chai Y, Kwan M (2014a) Comparative study of travel behaviors between Chinese and American urban residents: Beijing vs Chicago. Geogr Res 33(12):2275–2285 (In Chinese)
Zhao Y, Chai Y, Martin D (2014b) Socio-cultural effects on behavioral companionship choices: Beijing vs Utrecht. Sci Geogr Sin 34(8):946–954 (In Chinese)
Zhao C (2013) Localization and modernization: two dimensions of urban leisure development. Acad Forum 36(3):130–133, 141 (In Chinese)

Zheng S (2005) Urban leisure development based on integration theory. Econ Geogr 25(2):228–231, 251 (In Chinese)
Zhou L, Liu J (2005) Interaction between ReBAM and urbanization: a case study of Changchun. Hum Geogr 20(6):98–101 (In Chinese)
Zhu Y, Zhuang J (2006) Formation mechanisms of urban recreational business district (RBD) in Xi'an, an ancient capital. Mod Urban Res 4:53–58 (In Chinese)
Zhu H Liu B (2011) Concepts of sense of place, place attachment, and place identity: Distinctions and research implications. J South China Normal Univ (Nat Sci Ed) (1):1–8 (In Chinese)
Zhu H, Liu J, Li L (2014) Research progress on urban leisure commercial districts in China. Prog Geogr 33(11):1474–1485 (In Chinese)
Zhu H, Liu J, Tao H (2015) Spatio-temporal distribution characteristics and formation mechanisms of urban leisure business districts in Beijing. Acta Geogr Sin 70(8):1215–1228 (In Chinese)

Chapter 9
Living Space of Different Groups: A Diverse Mosaic

9.1 Foreigners in China and Foreign Enclaves

9.1.1 Changes of the Foreign Population in China

With the economic slowdown in developed countries and the tightening of immigration policies, emerging and economically dynamic developing countries are gradually becoming major hubs for foreign activity. As the world's leading trading nation and second largest economy, China has seen a rapid increase in inbound foreign tourism. Between 1990 and 2013, the number of foreign visitors to mainland China increased from 1.7473 million to 26.2903 million. Since the 1990s, the number of long-term foreign residents in China has also increased, with a tendency to concentrate in major metropolitan areas. However, as a traditionally non-immigrant country, China has imposed strict regulations on foreigners' residence, employment and access to public services. For example, foreigners with residence permits are prohibited from working in China without permission from the relevant authorities. Despite these restrictions, the flow of foreigners into China has continued to grow. According to the sixth national census (which for the first time included people from outside China as census takers), the number of permanent foreign residents living in China for more than three months reached 600,000 in October 2010, a fourfold increase over the previous decade.

According to census data (National Population Census Office of the State Council 2012), of the 593,832 foreign residents in China in 2010, 336,245 were male and 257,587 were female. By nationality, the top ten foreign groups were from South Korea (120,750), the United States (71,493), Japan (66,159), Myanmar (39,776), Vietnam (36,205), Canada (19,990), France (15,087), India (15,051), Germany (14,446) and Australia (13,286). In terms of place of residence, the top ten regions hosting foreign residents were Shanghai (143,496), Beijing (91,102), Guangdong (74,011), Yunnan (45,801), Jiangsu (30,928), Shandong (30,172), Fujian (27,386), Zhejiang (26,765), Liaoning (22,723) and Guangxi (21,465). About 20% of these

J. Feng et al., *Urban Living Space in China*, Urban Sustainability,
https://doi.org/10.1007/978-981-95-0595-1_9

foreign residents had been in China for more than one year, while more than 40% had been in China for more than two years. The main reasons for entering China were study (25.9%), employment (22.7%) and business (18.3%). The educational level of these foreign residents showed a polarised trend, with more than half having at least a bachelor's degree, while around 20% had only a primary school education or less, consisting mainly of Southeast Asian nationals living in Yunnan and Guangxi.

With the continuous influx of foreigners into major Chinese cities, foreigner enclaves (also known as transnational migrant communities) have emerged in major cities based on nationality. For example, there is a Korean enclave in Beijing's Wangjing district, a Japanese enclave in Shanghai's Gubei district, "Koreatown" in Qingdao's Chengyang district and Shenyang's Xita street, and a "Middle Eastern street" in Yiwu. These newly emerged foreigner enclaves in the context of globalization have not only created a new international social space in the cities of China, but also brought multiple culture and new vitality to the city, and promoted the connection between foreigners and the local community in economic activities such as investment and consumption, stimulating the formation and further development of an international society and city. As foreign groups become more visible to the public, how these foreigners survive and develop in cities of China and how they construct their living space deserve attention.

9.1.2 Foreign Enclaves in China

9.1.2.1 The Korean Enclave in Wangjing, Beijing

Since the establishment of diplomatic relations between China and South Korea in 1992 and the signing of the China-South Korea Free Trade Agreement in 2015, there has been a marked increase in economic cooperation between the two countries. China has emerged as South Korea's preeminent trading partner, export market, import source, and recipient of foreign investment. In turn, South Korea has become China's third largest trading partner and a significant source of imports. The deepening of economic and trade ties has also led to a flourishing of personnel and social and cultural exchanges between the two countries. In 2013, South Korea ranked first among foreign countries with 3.97 million visitors to China, contributing to the nation's total of 26.09 million foreign visitors that year. According to statistics on overseas Koreans, in 2013, there were already more than 350,000 Koreans living in China for an extended period of time, and China has become the third largest place of immigration for Koreans after the United States (2.091 million) and Japan (893,000). According to data from the sixth national census, among the 593,832 foreigners who have resided in China for more than three months and were surveyed, Koreans, with 120,750 individuals, constituted the largest foreign group living in China on a permanent basis. In terms of the basic attributes of Koreans in China, it is evident that they generally possess a high level of education, with more than 60% having received a bachelor's degree or higher. The majority of these individuals have resided in China for less than two years (50%) or two to five years (25%), with the predominant reasons for entry being academic (35%), employment (30%), and business (18%). The majority of the Korean population is concentrated in major cities in the eastern and coastal regions, including Beijing, Guangdong, Shandong, Tianjin, Liaoning, and Shanghai, with a notable presence in Beijing specifically.

Looking at the development and changes of the Korean population in Beijing, the Korean population in Beijing maintained a strong growth trend before the financial crisis broke out in 2008. In the five years from 2003 to 2007, the population doubled, and by 2007 it had already exceeded 100,000 people. After the outbreak of the financial crisis, there were large fluctuations in the size of the population, and in 2009 the population shrank by about 30% and has since recovered, with a population of 74,025 in 2013 (Table 9.1). According to incomplete statistics, there is a Korean enclave in Wangjing, Beijing, with a population of about 60,000. Together with areas such as Wudaokou, the population is close to 100,000, making it the largest enclave of foreigners living permanently in China.

Located in Beijing's Chaoyang District, Wangjing is an area with a high concentration of foreign-related resources, such as foreign embassies in China, international organizations and representative offices in Beijing, foreign chambers of commerce, prominent foreign news media in Beijing, international schools and foreign doctors. Wangjing is strategically located in a pentagonal area formed by the intersection of four major roads: the Fourth Ring Road to the northeast, Jingmi Road, the Fifth Ring Road to the northeast and the Jingcheng Expressway. It is close to the Asian Games Village and the Olympic Village to the west, the CBD business district and the Korean Consulate to the south, and is one of the six high-end functional areas planned and built by Beijing since the 11th Five-Year Plan. Wangjing mainly consists of three residential areas: Huajiadi, Nanhuqu and Wangjing, with more than 20 residential communities.

According to Zhou et al. (2016), the Wangjing area has successfully transformed from a rural–urban fringe into a Korean enclave since the 1990s. According to their research, the Korean enclave has undergone four main stages of change, as shown in

Table 9.1 Changes in the size of the Korean population in Beijing (2003–2013)

Year	Korean population		
	General residents	Students	Total population
2003	30,000	13,000	43,000
2005	50,000	10,000	60,000
2007	70,000	30,000	100,000
2009	46,064	21,109	67,173
2011	57,600	20,000	77,600
2013	54,652	19,373	74,025

Source Ministry of Science and Technology of South Korea. Annual report of statistics on Korean nationals overseas 2013: Northeast Asia, South Asia Pacific, North America, South America, Europe, Africa, Middle East. http://www.mofa.go.kr, 2015–11-02

Fig. 9.1. In the first stage, from 1992 to 1998, Huajiadi, with its geographical advantages of being close to the Korean consulate and the capital airport, as well as good educational facilities such as the Korean International School, Beijing University of Chinese Medicine, Beijing Youth Politics College, and the School of Architecture of the Central Academy of Fine Arts, attracted some Korean students and their parents to live in Beili, Nanli, Xili, etc. in Huajiadi, forming the prototype of the Wangjing Korean residential area. In the second phase, from 1999 to 2002, ordinary Koreans began to flock to China, triggering a wave of large-scale Korean migration to Beijing. At the same time, Wangjing New Town (comprising Wangjing New Town District 1, Wangjing New Town District 2, Wangjing West Garden District 3 and Wangjing West Garden District 4) was completed in 1999, officially introducing the concept of 'Wangjing' to the public. Among them, Wangjing West Garden District 3 and District 4 have adopted a floor heating design scheme based on Korean living habits, and have the infrastructure to attract Koreans to move in. Compared with the diplomatic apartments in Yansha and the Asian Games Village, Wangjing New Town is not only more modern in construction, but also cheaper to rent, attracting many Korean investors, entrepreneurs and their families. In the third phase, from 2003 to 2008, Wangjing underwent a second development after 2003, with more high-end and high-rise buildings being built and a large number of commercial facilities beginning to move in. In addition, LG Electronics' research center, the largest outside South Korea, was established in the Wangjing Science Park, and later major companies such as Motorola and Siemens successively set up their Chinese headquarters in Wangjing. As a result, the number of Koreans choosing to live in Wangjing grew rapidly. In particular, it attracted many Korean consulates, expatriates from large Korean-funded companies, and Koreans doing business or starting businesses in Beijing, creating a spatial pattern centered on Wangjing New Town and its surrounding areas. In the fourth stage, after 2009, the Korean residential area of Wangjing experienced great fluctuations and changes. The outbreak of the financial crisis in 2008 had a significant impact, jokingly referred to in the media as the "Korean won coughing and Wangjing catching a cold" phenomenon. As the Korean won depreciated sharply, prices in Beijing rose and employment pressures increased. From 2007 to 2009, the Korean population in Beijing shrank by about 30% and has since recovered. However, the changing environment has also meant that many Korean investors and entrepreneurs have found themselves in financial difficulties or even bankruptcy, and have had to find new jobs or investment opportunities. These Koreans, constrained by economic conditions, tend to choose communities such as Nanhu Middle Garden and Nanhu East Garden, where rents are cheap. As a result, the spatial pattern of Korean settlements has shown a clear trend of outward diffusion based on the previous stage.

Overall, using 1998 as the dividing point, the spatial center of the Korean residential area of Wangjing moved northwards from Huajiadi to Wangjing New Town. After 2003, the central position of Wangjing New Town as the Korean residential area was further strengthened, and it gradually expanded into the surrounding areas, presenting a spatial pattern of "large agglomerations and small dispersals". The tendency of the Korean population to move from concentrated to dispersed settlement patterns has contributed to some extent to the expansion of the spatial boundaries

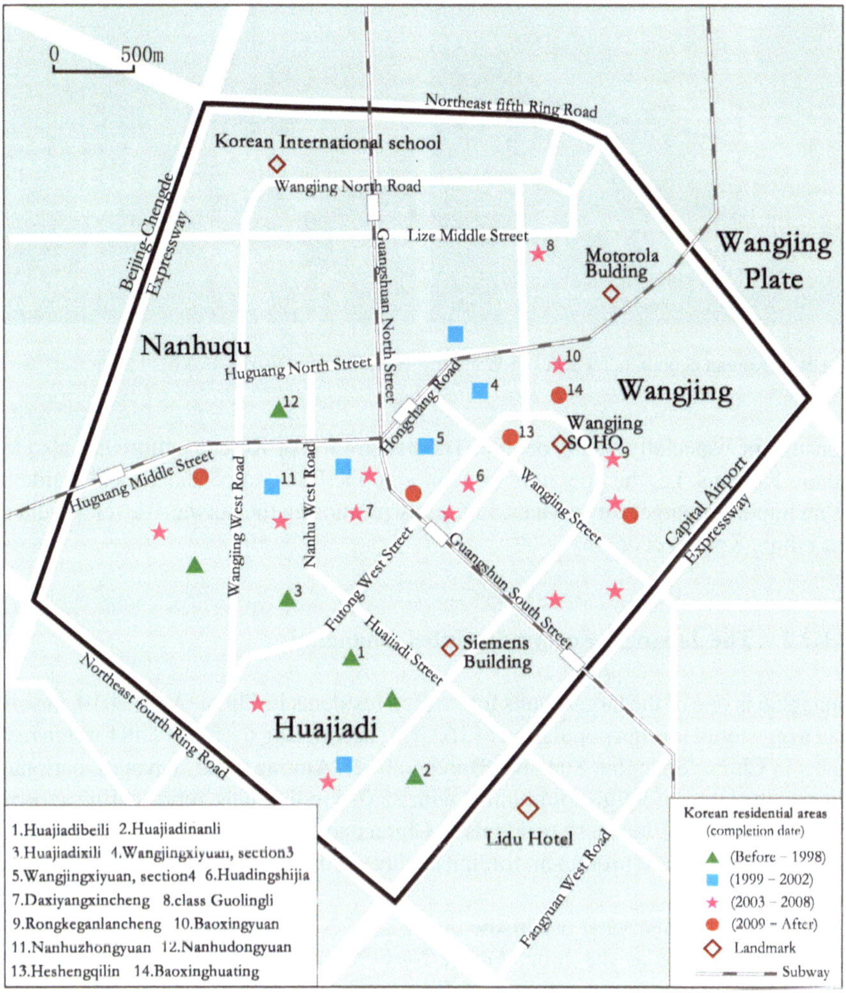

Fig. 9.1 Spatial distribution of Korean residential areas in Wangjing

of the Korean residential area. At the same time, ethnic economic facilities that mainly serve Koreans are also booming. In particular, in the 3rd and 4th districts of Wangjing West Garden, there are many shops with ethnic characteristics. These shops are mainly concentrated in catering, retail industries such as Korean food, clothing and cosmetics, and service industries such as convenience stores, Korean foot baths/massages and beauty salons. Even IP cards for international long-distance calls are sold for the exclusive use of Koreans (Fig. 9.2). In other words, all the goods and services that Koreans need in their daily lives can be found in the residential area. And under the influence of the "Korean wave", Korean entertainment, culture, popular clothing, cosmetics and other products are also very popular with Chinese

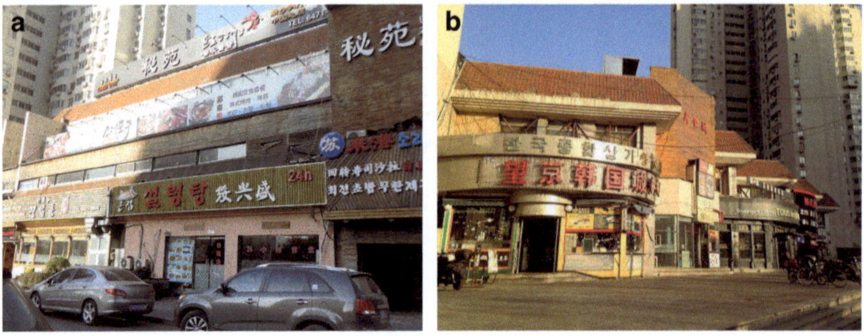

Fig. 9.2 Korean economic facilities in Wangjing West Garden District 4(a)(b)

consumers, especially young people. The penetration of Korean culture has also led ethnic Koreans, i.e. the operators of economic facilities, to consider local residents as an important target for services, indirectly promoting the outward development of the ethnic Korean economy.

9.1.2.2 The Japanese Enclave in Gubei, Shanghai

Shanghai is one of the largest hubs for foreign residents in China. As of 2010, the city had a registered foreign population of 162,481, accounting for 27% of all foreign residents in China (Shanghai Statistics Bureau 2012). Among these, Japanese nationals formed the largest foreign community, with 35,075 individuals, representing approximately 53% of all Japanese residents in China and 22% of Shanghai's total foreign population. The most prominent foreign enclave within the city is the Gubei Japanese residential area.

In the 1990s, Shanghai established the Gubei New Area, which was the largest foreign settlement at the time. Gubei New Area was Shanghai's first foreign-oriented business district (Hongqiao Economic and Technological Development Zone). Construction of support facilities began in 1986 and the first phase was completed in 1993. The total planned land area is 1,366 million m^2 and the total construction area is 3 million m^2. The aim of the plan was to solve the housing problems of foreign experts and compatriots from Hong Kong, Macau and Taiwan, and to promote investment. Therefore, Gubei New Area is equipped with appropriate office, commercial, cultural and entertainment, and educational facilities, and is a foreign-related residential area with both commercial and foreign trade functions. According to China's immigration policy at the time, foreigners were required to live in foreign-related residential areas. As a result, the vast majority of foreigners in Shanghai live in Gubei New Area, creating a foreign settlement area with a mix of nationalities.

With the rapid growth of Shanghai's foreign population in the twenty-first century, the city lifted restrictions on foreigners' residential choices in 2003. This shift marked a transition from mixed-nationality residential patterns to more segregated, nationality-based ethnic enclaves. Given the pre-existing concentration of Japanese residents in Gubei, the area gradually evolved into a distinct Japanese enclave. By 2003, both the spatial distribution and population scale of Japanese residents had undergone significant changes. By 2010, Changning District had emerged as the most concentrated area for Japanese residents, housing approximately 43% of all Japanese individuals in Shanghai, followed by Pudong New Area, Minhang District, and Xuhui District.[1] In reality, however, most Japanese residents in Changning District were concentrated in and around Gubei New Area. In this study, the "Gubei area" is defined to include Gubei New Area, Hongqiao, and Xianxia. While Gubei New Area was originally established as an international residential district, Hongqiao is home to Japanese schools and other international housing developments, making it an attractive residential choice for Japanese expatriates. Xianxia, on the other hand, hosts the Japanese Consulate and numerous Japanese corporate offices, leading to the expansion of ethnic economic infrastructure catering to the Japanese community. Over time, this development extended the reach of the Japanese enclave beyond Gubei.

In 2002, there were 24 residential communities with a high concentration of Japanese residents, 19 of which were located in Gubei New Area and 5 in Hongmei Road, Hongqiao District. In 2012, the number of residential communities with a high concentration of Japanese residents increased to 39, 34 of which were located in Gubei New Area, with no significant change in Hongmei District (Zhou and Liu 2015). The 15 newly built communities are all part of the second phase of development projects in Gubei New Area, located to the east of the first phase of Gubei New Area, as shown in Fig. 9.3.

The formation of the Japanese enclave in Gubei is primarily due to two key factors. On one hand, Gubei New Area benefits from its advantageous location near the Hongqiao Economic and Technological Development Zone and Shanghai Hongqiao International Airport. On the other hand, the area offers a high-quality living environment with well-established supporting infrastructure, making it well-suited to meet the residential preferences of Japanese nationals. Additionally, the relatively high rental affordability of the area supports their housing choices. According to a field survey conducted in 2012, there were over 300 ethnic Japanese economic facilities distributed throughout Gubei, which essentially meet the daily life needs of Japanese residents. These ethnic economic facilities include: 22 educational institutions (1 Japanese school, 2 international schools, 7 nurseries/kindergartens, and 12 tutoring centers), 16 Japanese food-related retail establishments (6 Japanese specialty food stores, 9 Japanese-affiliated supermarkets, and 1 large supermarket), more than 200 food service establishments, and 53 other facilities (8 medical institutions, 6 real estate agencies, 13 massage parlors, 21 beauty salons, and 5 audio-visual shops). In terms of spatial distribution, these Japanese ethnic economic facilities are spread throughout the Gubei area (Figs. 9.4, 9.5, and 9.6), with the highest concentration

[1] Shanghai Statistics Bureau. Shanghai 2010 Population Census Data. Beijing: China Statistics Press, 2012.

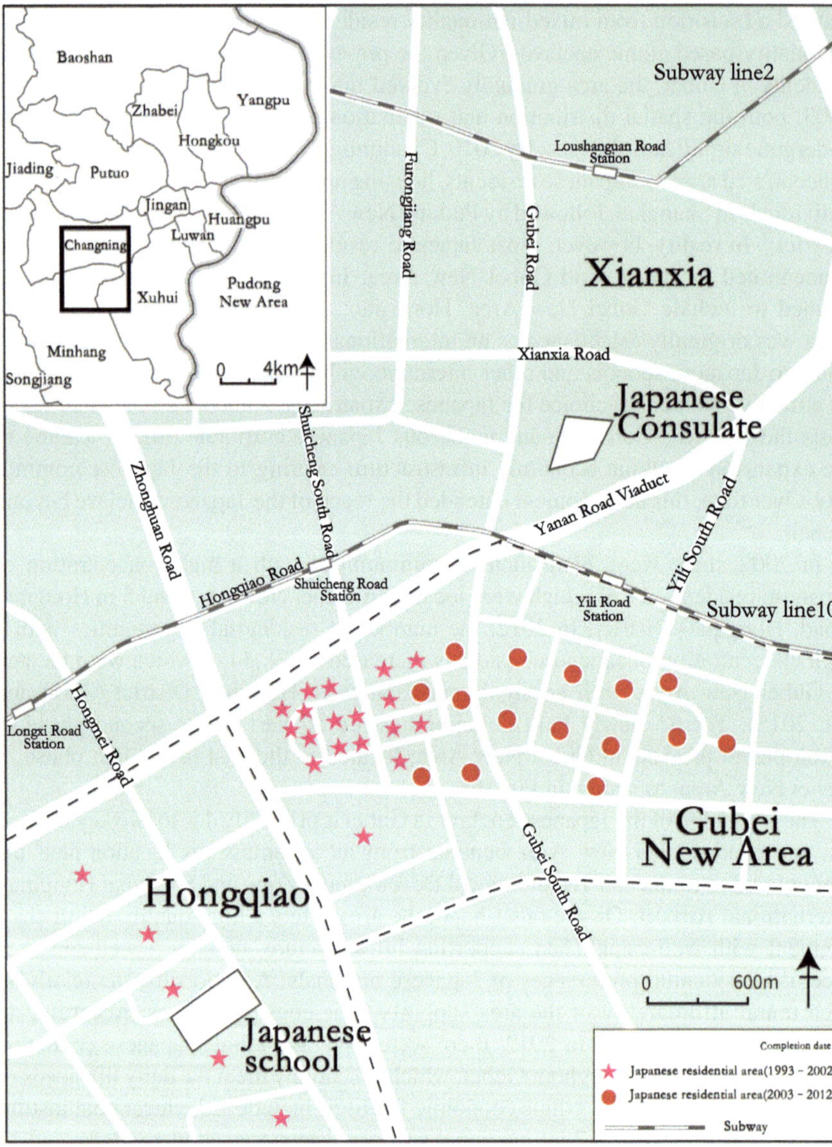

Fig. 9.3 Spatial change of Japenese residential area in Gubei New Area

found in the first phase of Gubei New Area. This concentration can be explained by two main factors. First, the first phase of Gubei New Area was the earliest site for the development of Japanese ethnic economic activities, creating agglomeration economies. Second, with the completion of the second phase of Gubei New Area, rents for residential and commercial properties in the first phase decreased, lowering the investment threshold for ethnic businesses.

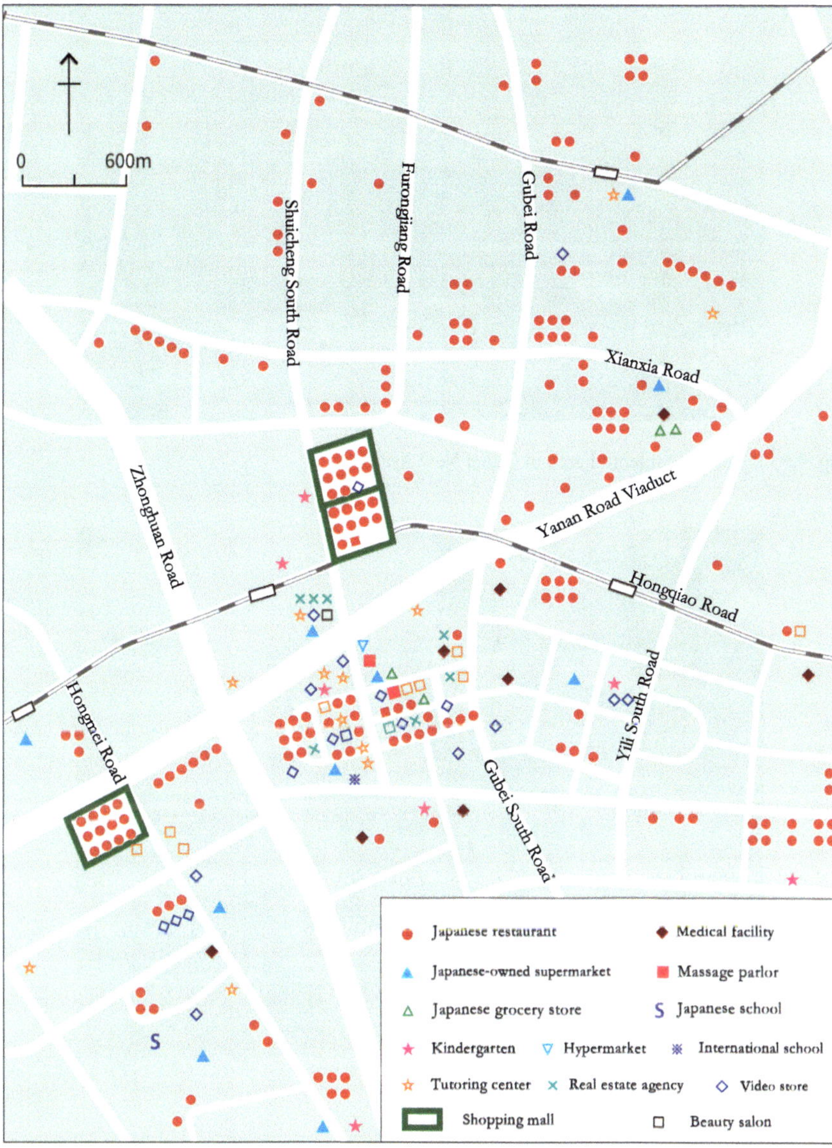

Fig. 9.4 Spatial distribution of Japanese economic facilities in Gubei New Area (2012)

Fig. 9.5 Japanese residential area in Gubei New Area

Fig. 9.6 Japanese kindergarten in Gubei New Area

9.1.2.3 The African Enclave in Xiaobei Road, Guangzhou

Guangzhou, as the starting point of the Maritime Silk Road, has been a major port city for China's foreign trade for over 2,000 years, from the Qin and Han dynasties to the Ming and Qing periods. The establishment of the "Thirteen Factories" during the Qing dynasty further solidified Guangzhou's role as the central hub for China's export trade. After the founding of the People's Republic of China, Guangzhou led the country in terms of openness to foreign trade and cultural exchange. Additionally, its proximity to Hong Kong and Macau, along with its position in the Pearl River Delta, known as the "world's factory", and its role as the host of the biannual "China Import and Export Fair" (Canton Fair), attracted a significant influx of foreign enterprises and businesspeople. Since the 1990s, the export of clothing and footwear has become the core of Guangzhou's export economy, bringing a large number of African migrants, who now form the largest foreign group in the city.

Since the 1990s, Guangzhou has gradually developed five areas with relatively concentrated foreign populations: Sanyuanli, Huanshidong, Tianhebei, Ersha Island, and Panyu. Unlike foreign enclaves in Beijing and Shanghai, Guangzhou's foreign communities are predominantly made up of African traders. The Sanyuanli and Tianhebei areas are the most notable, particularly the Xiaobei area in the Hongqiao Subdistrict of Yuexiu District. According to a study by Li et al. (2008), the total African population in Guangzhou ranges from 15,000 to 20,000, with a highly diverse mix of languages, including French, English, Arabic, and Portuguese. The majority of these African migrants come from West Africa, including former French colonies such as Guinea, Benin, Mali, Senegal, Côte d'Ivoire, and former British colonies like Nigeria and Ghana. Additionally, a significant number come from Central Africa, including the Congo, Angola, Tanzania, and Kenya. Most African migrants come to Guangzhou primarily for trade, exporting industrial products from China to Africa or engaging in other related commercial activities. A smaller portion deals in small commodities, while the majority export clothing, personal goods, and household items, with a particularly large portion involved in building materials trade, given Guangzhou's proximity to Foshan, China's largest ceramics and construction materials base.

Since the 1990s, the number of Africans residing in the Xiaobei Road area has increased significantly. Located in the Hongqiao Subdistrict of Yuexiu District, Xiaobei Road is surrounded by numerous mid- to high-rise commercial and residential buildings, such as Tianxiu Mansion, Xiushan Building, Ceramics Mansion, and Guolong Mansion. For instance, Xiushan Building alone houses nearly 200 units, with shops run by Africans and Arabs making up more than half of the commercial space. The majority of Africans living in Xiaobei Road are Muslim. In these three buildings—Xiushan Mansion, Tianxiu Mansion, and Guolong Mansion—over 400 Africans from 52 countries have settled, with Tianxiu Mansion being the first choice for newly arrived African migrants to Guangzhou (Li and Du 2012). This can be attributed to the area's favorable location, which attracts African traders. Xiaobei Road is close to key transportation hubs, including the entrance to the Airport Road, Guangzhou Railway Station, metro stations, and several bus stations. Moreover, the area is home to several major wholesale markets, such as "West Station Watch City," "Baima Clothing Market," and "Liuhua Clothing Wholesale Market." Tianxiu Mansion is only 3.5 km from Guangzhou Railway Station and less than 5 km from the old venue of the China Import and Export Fair. The rental prices in this area are generally in line with the overall rental market in Guangzhou, placing it in the mid-range category. These factors have contributed to Xiaobei Road becoming an important destination for the African community.

9.1.3 Living Spaces of Foreigners in China

After 2000, the number of foreign population in China surged, leading to the emergence of foreign enclaves of varying sizes and characteristics in major cities such as Beijing, Shanghai, and Guangzhou. This trend is expected to continue strengthening in the future. The previous sections have focused on the foreign communities with larger concentrations in Beijing, Shanghai, and Guangzhou, introducing the living spaces of significant foreign groups, including Koreans, Japanese, and Africans.

Most of these foreigners are transnational migrants who temporarily reside in China for economic reasons. They typically stay in China for a few years before moving on to another country or returning to their homeland, with no intention of settling long-term in China. As a result, the living spaces of foreigners in China tend to exhibit significant levels of replication and closure. Replication refers to the foreign nationals' tendency to quickly and effectively adapt to local life by largely maintaining the lifestyles of their home countries, especially in areas such as food culture and education. In cities like Beijing, Shanghai, and Guangzhou, ethnic economic facilities with characteristics of the migrants' home countries have been established to cater to these communities. For instance, in Gubei, many advertisements and shop signs are prominently displayed in Japanese, reflecting the integration of Japanese culture into the area. This not only introduces Japanese culture into Gubei but also helps ease the challenges Japanese people face in adjusting to life abroad, promoting peaceful coexistence between them and local residents, despite cultural, social, and lifestyle differences. Similarly, in Wangjing, numerous Korean schools and after-school tutoring centers are present, allowing Korean children to receive education similar to what they would have experienced in Korea, even while living in Beijing. Closure, on the other hand, refers to the spatial segregation of foreign communities from local residents. The closure of foreign enclaves is a complex issue, involving both voluntary and involuntary forms of separation. For example, the African community often exhibits voluntary segregation and passive isolation. Africans tend to cluster together in specific areas to minimize development costs, mitigate unfamiliarity with the local environment, strengthen social ties, and share information. However, passive isolation occurs due to the attitudes of local residents. In contrast, the Japanese community tends to practice both voluntary and active segregation. This can be attributed to the complex historical and political dynamics between China and Japan, leading the Japanese to seek residential areas that offer a sense of safety and psychological protection through a more isolated living arrangement (Zhou and Liu 2015).

In the era of globalization, as urban economic globalization is vigorously promoted, it is important to consider whether efforts should be made to encourage foreigners to integrate more actively into local society, thereby enhancing the internationalization of the city. This question warrants further research and reflection.

9.2 Overseas Chinese Farms and Living Space of the Returned Overseas Refugees in China

9.2.1 Origins and Resettlement of Returned Overseas Refugees

After the establishment of the People's Republic of China, some overseas Chinese voluntarily returned to their homeland to participate in the construction of the new China, inspired by patriotic fervor. After World War II, with Southeast Asian nations gaining independence and nationalist sentiments and international ideologies rising, some countries launched anti-Chinese campaigns, forcing a group of overseas Chinese to return. China distinguished between voluntary returnees and those returning due to oppression abroad, referring to them as guiqiao (overseas Chinese who voluntarily returned) and nánqiao (refugee overseas Chinese) (Yao 2009). Over the thirty years following the founding of the People's Republic of China, there were three major waves of overseas Chinese returnees. The first wave occurred from 1949 to 1959, when around 300,000 overseas Chinese returned, mainly scientists (mainly those returning from Europe and the U.S.), students pursuing further studies, and those choosing to settle in China. In the 1950s, many returnees were former refugees returning to participate in socialist construction. The second wave, occurring in the 1960s (mainly before 1965), saw about 300,000 returned overseas refugees from countries like Indonesia, India, and Myanmar due to the anti-Chinese movements, the Sino-Indian border conflict, and the Sino-Myanmar tensions, with Indonesia contributing the largest group (around 136,000) (Huang 2005). The third wave of returnees happened in the late 1970s after deteriorating Sino-Vietnamese relations in 1978, which led to a massive influx of overseas Chinese and Vietnamese refugees into Yunnan. In the early 1980s, with tensions between China and North Korea, as well as China and Mongolia, some overseas Chinese from North Korea and Mongolia also returned (You 2010).

From the late 1950s onward, China formulated a general policy to accept and resettle overseas Chinese, focusing on resettling them in rural and mountainous areas, and only accepting those with special skills for employment. On February 2, 1960, the State Council issued the Guidelines on the Reception and Resettlement of Returnee Overseas Chinese, establishing the "Committee for the Reception and Resettlement of Returnee Overseas Chinese" to oversee these activities.[2]

---

[2] Overseas Chinese Affairs Office of the State Council. 2007. A compilation of documents on the reform and development of overseas Chinese farms.

In 1955 and 1975, large-scale anti-Chinese incidents in Indonesia and Vietnam prompted the Chinese government to quickly settle large groups of returnee by expropriating land at the outskirts of towns and villages to establish overseas Chinese farms. Between the 1950s and 1970s, 84 state-owned overseas Chinese farms were established across seven provinces (Fujian, Guangdong, Guangxi, Jiangxi, Yunnan, Hainan, and Jilin), which collectively resettled 240,000 returned overseas refugees. Additionally, nearly 200 state-owned farms in Guangdong, Guangxi, Fujian, Yunnan, and Hainan were established by the agricultural reclamation and forestry systems, settling over 90,000 returnees.

9.2.2 *Resettlement Methods and the Evolution of Living Space*

Guangdong is the province with the largest number of overseas Chinese farms and the highest number of returnees resettled. The province has 23 overseas Chinese farms, covering an area of 1,148 km^2, which resettled 89,669 returnees from 24 countries and regions, including Indonesia and Vietnam. The number of overseas Chinese farms, total population, and the number of resettled returnees and refugees in Guangdong account for 27, 49, and 40% of the national totals, respectively (Yao 2009), making it highly representative. In May and June 2016, a research team from Sun Yat-sen University conducted a survey of 196 returnee refugees from Vietnam and Indonesia in the Huadu Overseas Chinese Farm in Guangzhou, focusing on their basic information, time of return, links with their countries of origin, living conditions, identity, and sense of belonging, along with detailed structured interviews about their experiences and changes in their living conditions.

The Huadu Overseas Chinese Farm, located in the northeastern outskirts of Guangzhou, was founded in 1955. It resettled over 5,000 returnee refugees from Indonesia, Malaysia, Singapore, Vietnam, Cambodia, Thailand, India, Laos, New Zealand, Madagascar, and Japan. The farm, which was government-invested and operated, assigned returnees of working age to different production teams for agricultural work. In 1999, Huachao Town was established, retaining the original branding of the Huadu Overseas Chinese Farm. In 2005, the farm was incorporated into Huadong Town after an administrative restructuring. Currently, the farm covers an area of 14.5 km^2, with three community committees, and has a resident population of 5,123 people, including 2,176 returnee refugees and 2,177 family members.

9.2.2.1 Work Zone: Living Difficulties and Social Isolation

In the early stages of establishing overseas Chinese farms, the government hastily designated work zones to resettle returnees. On one hand, returnees faced enormous difficulties adapting to their new lives. The majority of these returnees had been born and raised in foreign countries, and although they were Chinese nationals, their language, culture, customs, and even thought processes were heavily influenced by their countries of residence, making them very different from local residents (Table 9.2). The "rural orientation and concentrated resettlement" method of settling these returnees caused many to feel uncomfortable, as most had not been involved in farming in their countries of residence (Zhang 2013).

On the other hand, the work zone was both a production space for agricultural activities and a living space for. Since the land for the farm had been expropriated from various villages over time, the Huadu Overseas Chinese Farm was divided into seven work zone (Fig. 9.7), including Zhuhu, Wangongtang, Luochaigang, Gangtou, Xintan, Yanghe, and Banbianshan. Each of these areas is adjacent to surrounding rural areas, with the farm headquarters located in Luochaigang. Other areas such as Gangtou, Banbianshan, and Yanghe are located to the northeast, Xintan to the southeast, and Wangongtang to the north, with Zhuhu in the northwest. Except for the continuous areas of Zhuhu/Wangongtang/Luochaigang, the other work zones are scattered and not connected. This spatial fragmentation made the farm seem more isolated compared to the surrounding rural areas of Huadu District and hindered the integration of the returnees into the local community.

During the special period of 1967, many overseas Chinese returnees were compelled to sever or reduce their ties with their countries of origin due to various factors. At the same time, these returnees were also forced to forge stronger connections with the local community in order to facilitate their integration into the host society. Due to the challenging living conditions on the agricultural farms, combined with other factors, some returnees chose to leave the farms, while those who remained voluntarily underwent an extended period of adaptation, ultimately establishing a more stable lifestyle and sense of spatial belonging.

Table 9.2 Resettled overseas Chinese: challenges in adapting to daily life

	Language	Climate	Living conditions	Diet	Clothing	Work Style
Southeast Asia (Overseas Chinese Communities)	Mandarin, local languages	Tropical climate	Relatively spacious	Southeast Asian food, preference for sour flavors	Women prefer wearing skirts	Mainly engaged in business and technical labor
China (Huadu Overseas Chinese Farm)	Cantonese	Subtropical climate	five bedrooms and one living room (relatively small)	Sweet potatoes, etc.	Strict regulations on wearing skirts	Mainly agricultural work

Fig. 9.7 Work zones of Huadu overseas Chinese farm

9.2.2.2 Resettlement Area: External and Inward Integration

After 1991, with economic reforms in the Huadu Overseas Chinese Farm, the living conditions of returnees gradually improved, further stabilizing their production and living arrangements. In 1993, the farm, under the direct leadership of the Overseas Chinese Investment Corporation, no longer assigned jobs to the children of returnees, prompting them to find work outside the farm, thus initiating external integration into the local community. As a result, their living space expanded beyond the farm to surrounding rural areas, and even to Guangzhou's urban districts.

In 2009, in an effort to improve the living conditions of the returnees, the Huadu Overseas Chinese Farm underwent a housing renovation project, merging the seven work zones into three residential communities (Figs. 9.8, Table 9.3). The original work zones were repurposed for commercial development, with only a few buildings where the returnees had previously lived left behind. In this transition from scattered work zones to concentrated residential areas, the living spaces of the returnees became more compact, and the shortening of spatial distance fostered internal integration, especially for the second and third generations of returnees.

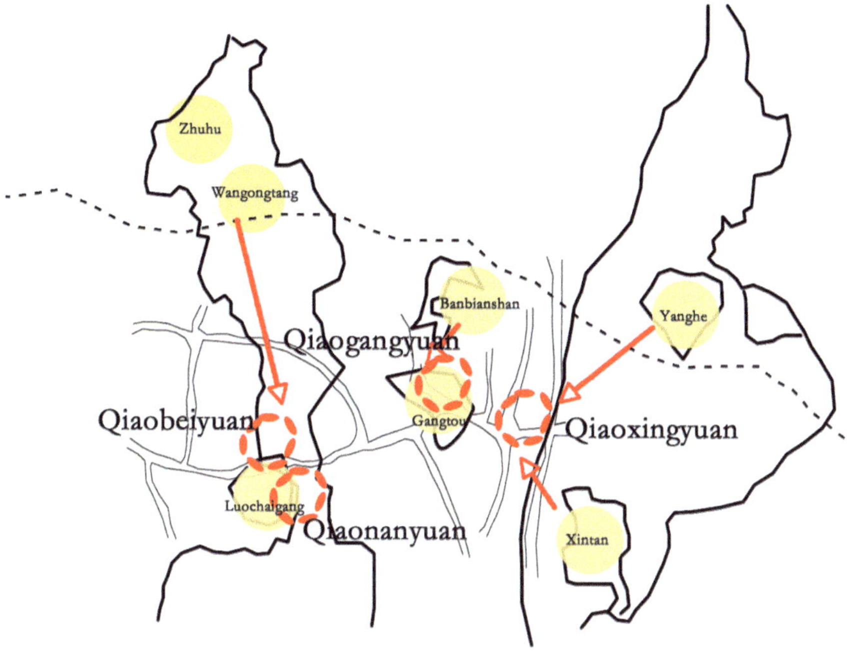

Fig. 9.8 Inward integration of returned overseas refugee

Table 9.3 Inward integration of returnees

Original work zone	Current residential area	Population details	Community committee
Luochaigang (Main Farm)	Qiaonanyuan, Qiaobeiyuan	Total population: 1578 Total households: 698 Indonesian returnees: 165 Vietnamese returnees: Data unavailable	Luochaigang Neighborhood Committee
Wangongtang			
Zhuhu			
Gangtou	Qiaogangyuan	Total population: 680 Total households: 300 Indonesian returnees: 153 Vietnamese returnees: 354	Gangtou Neighborhood Committee
Xintang	Qiaoxingyuan	Total population: 1453 Total households: 643 Indonesian returnees: 99 Vietnamese returnees: 324	Beixing Neighborhood Committee
Yanghe			
Banbianshan			

9.2.3 Generational Change and Evolution of Living Space

Since the establishment of the Huadu Overseas Chinese Farm in 1955, the children of the returnees have mostly grown up and started families of their own, with some even having a fourth generation born in the farm. Over the course of three generations, the returnees have experienced varying degrees of assimilation, passing through generational changes in language, profession, and psychology, which have also influenced the evolution of their living spaces.

9.2.3.1 Generational Differences

The first generation of returnees faces the greatest barriers in terms of language and cultural differences, with their almost insular living circles resulting in limited assimilation and a stronger retention of the culture of their countries of origin. As a result, their living spaces tend to be more enclosed and independent. In contrast, the first generation maintained close overseas ties, initially through frequent correspondence with relatives abroad. Although these communications were interrupted between 1966 and 1976, they resumed after 1976 and have continued to the present day. Nowadays, phone calls and WeChat are the primary means of communication with overseas family members, making this generation the most connected with overseas relatives, as well as the main force maintaining these relationships within the entire family. In terms of resettlement policies, prior to the large-scale anti-Chinese movement, the first generation of returnees was generally resettled according to the principle of "placement according to nationality, mainly on farms, with skilled individuals assigned according to their abilities." However, after the anti-Chinese movement, returnees were properly arranged to work on state-run overseas Chinese farms based on national needs and individual preferences. On one hand, they enjoyed preferential treatment as employees of state-owned farms; on the other hand, there were significant restrictions on their placement and personal skills.

After 1991, significant economic reforms began at the Huadu Overseas Chinese Farm, and in 1993, the farm's direct management shifted to an Overseas Chinese Investment Corporation. From this point on, second-generation returnees were no longer assigned jobs by the farm and had to seek employment outside the farm. The overseas Chinese affairs office encouraged strong returnees to work in state-owned enterprises, lifting previous employment restrictions and facilitating the second generation's migration out of the farm. As a result, second-generation returnees increasingly integrated into society, and the new interpersonal networks they formed—interwoven with the local community's living spaces—further promoted the growth of the "overseas Chinese" identity, leading to a dual integration of both space and emotion. However, despite the support provided by the overseas Chinese network to help the second generation leave the farms, many still faced significant challenges in securing decent jobs in the cities due to a lack of social background and lower educational levels. The fear of failure and the subsequent gossip also deterred many from attempting to leave the farm. In terms of overseas relations, the connection between the second generation and their overseas relatives gradually weakened, with both the frequency of visits abroad and communication via mobile phones significantly decreasing. Nevertheless, influenced by the first generation, the second generation still strongly recognized the value of overseas Chinese culture and continued to place importance on their overseas connections.

In contrast to the first and second generations, the third generation of returnees exhibits more diverse characteristics. According to survey results, third-generation returnees describe themselves as "rooted in China," emphasizing the "patriotism of their grandparents when they returned to China." Due to the assimilation effects of their environment, the "overseas Chinese" identity has become a cultural and symbolic image for third-generation returnees, occasionally appearing in their minds. Some have managed to leave the farm through personal efforts, gaining university admissions or seeking better opportunities, and have settled in cities outside the farm. In their work and daily lives, they no longer emphasize their identity as returnees to the same extent as their parents did, and their living spaces have largely merged with those of regular urban residents.

9.2.3.2 Influence of Intermarriage

Language barriers, differences in clothing, difficulties in reconciling dietary habits, and varying beliefs have all created an invisible boundary between the overseas Chinese farms and the surrounding rural areas. The construction of the overseas Chinese farms involved the free requisition of rural land, and in the early stages of the farms' establishment, the preferential treatment for returnee employees under the state-run system also led to conflicts between returnees and the surrounding villagers during production and daily life.

Over time, the relationship between returnees and villagers has become more integrated as the generations changed. On one hand, at the central market of the overseas Chinese farm, three dance teams (group dances, pair dances, and aerobics) perform every evening, drawing large numbers of returnees and villagers to join in. Additionally, movies shown on major holidays in rural areas attract returnees, further enriching collective activities that symbolize the integration of the living spaces of returnees and villagers. On the other hand, regarding intermarriage, before 1978, the marriage rate between the overseas Chinese farms and surrounding villages was low, about 1.7%, mostly internal marriages. However, by 2007, the intermarriage rate had risen to 13.2% (Sun 2009).

9.2.4 Living Space of Overseas Chinese Returnees

9.2.4.1 Identity and Sense of Belonging

The majority of returnees resettled on the overseas Chinese farms had ancestors who had migrated long before, meaning that the overseas Chinese identity was one of the multiple identities they carried during the cultural adaptation process. Although many returnees lack knowledge of modern Chinese society, they still retain a strong sense of "roots." Having lived on the overseas Chinese farms for a long time and enjoyed the benefits of being state-owned enterprise employees, they identify not only as overseas Chinese but also as employees (or retirees) of the overseas Chinese farms.

In the cultural adaptation process, the memory of each generation varies, leading to significant intergenerational differences in their sense of identity and belonging. Survey results show that the vast majority of returnees strongly identify with the labels "overseas Chinese," "Chinese," and "employee (retiree) of the overseas Chinese farm." However, the first generation generally does not consider themselves "Chinese citizens," but rather citizens of their countries of origin, and they regard their original homeland as "home." With each successive generation, 80% of the second generation identifies as "Chinese citizens" with their home being their "current residence," while all members of the third generation identify themselves as "Chinese citizens" with their home being their "current residence."

Although the third generation recognizes the overseas Chinese farm as their "home," most of them rarely participate in community activities and do not have a strong willingness to do so, showing little recognition of the importance of these activities. This reflects that they do not truly feel a sense of belonging or identity with the overseas Chinese farm, and their willingness to contribute to its development is low.

9.2.4.2 The Impossibility of Replicating a Multicultural Living Space

When returned overseas refugees first came back to China, although the government provided welfare and assistance, their "difficult" status as a disadvantaged group has been difficult to erase over the years. The former "refugees" and their descendants still face challenges in the rapid urbanization of China, particularly the urban–rural divide. Research into the current living conditions of several returnee families shows that the vast majority of returnee households are at a middle-to-low income level. On average, a typical four-person family, in the most ideal conditions, has a total monthly income of about 13,000 yuan, not accounting for additional costs after children marry.

In terms of living space, the Overseas Chinese Farm currently has a fragmented Southeast Asian-style architectural landscape, but it has not formed a certain scale system. The local government is working to restore a multicultural living space, create a "Southeast Asian-style street", and develop the tourism economy. However, the returning overseas Chinese refugees do not fully agree with this transformation.

The older generation of returnees generally believes that the government's development of the cultural street does not truly contribute to cultural development but rather exploits the "returnee" element for commercial purposes, which cannot be sustained in the long run. As the vice president of the Indonesian Returnee Association lamented about the "cultural street" and cultural development, he noted with regret that, as the older returnees pass away, membership in the association has declined, and the younger generation has little awareness or participation. The third generation now has minimal contact with overseas relatives, and there is limited investment from overseas Chinese in the local development. The identity of "overseas Chinese" has gradually faded with the passing of generations.

9.2.4.3 Summary

Under the influence of multiculturalism, the changes in the living space of the overseas Chinese farm can be explained according to John W. Berry's "Intercultural Adaptation Model". The result of cultural adaptation is a change from "marginalization" to "assimilation", and the living space changes from closed and isolated to open and integrated.

In this process of cultural adaptation, returnees have been assimilated into mainstream culture, and their distinctiveness has gradually diminished. Their living and cultural spaces have gradually integrated with those of mainstream culture, a process influenced both by the returnees' own weak connections to their origins and by the lack of a multicultural environment at their place of residence.

9.3 The Living Space of Migrant Population in Cities

9.3.1 Areas of Migrant Settlement in Cities

With the rapid development of urbanization in China, the trend of rural population moving to cities and small-town populations moving to large and mega cities is becoming increasingly obvious. The resulting settlements of migrant populations can be considered a natural product of the urbanization process. According to the statistics of the sixth national census, at the end of 2010, the floating population in China reached 242 million, of which about 150 million were migrants moving from rural to urban. According to the current urban population statistics, migrants moving from rural to urban are the main force of China's new urban population and belong to the urban population in the statistical sense. However, China has long practiced urban–rural segregation, and the migrants are outside the urban system. They cannot enjoy the same social security and social services as ordinary citizens, and encounter various institutional or non-institutional rejections in the process of flowing into the city. The study of the migrant population by geography mainly focuses on the causes, process, mechanism and spatial distribution of the rural-to-urban flow of the migrant population (Yao et al. 2008). As early as the 1980s, with the increase in the number of migrant populations, Beijing first saw the emergence of a migrant population settlement area—"Zhejiang Village". Research on the formation process and basic situation of "Zhejiang Village" was the earliest research on migrant population settlement areas (Wang et al. 1997). Since then, there has been a gradual increase in academic research on the types of urban spaces where migrant populations congregate, such as a series of empirical studies on migrant population agglomeration areas in large and medium-sized cities such as Beijing, Shanghai, and Nanjing (Liu 2000; Qian and Chen 2004; Guo 2006). In recent years, geographers have also begun to explore this research field from the perspective of living space.

In the context of reform and opening up, the social space of cities in China has gradually become more complex. The emergence of a variety of social spaces has provided the migrant population with a diverse range of living space options. Economic globalization has led to the continuous differentiation of urban social space. In some of the cities, "transnational social spaces" have emerged, i.e., social spaces for foreigners. Driven by the reform of the household registration system, new non-nationalized social spaces such as "Zhejiang villages" and "Henan villages" have emerged (Ma and Xiang 1998). Under the guidance of urban expansion policies, social spaces such as the "village community" type of urban village have been created. The government has directly built social housing in line with the positive trend of poverty alleviation, forming "nationalized" spaces, which are also the only "non-marketization" social spaces (Ma et al. 2012).

Currently, the migrant population in cities is still dominated by the working population, and "focusing on work but neglecting life" is a true reflection of their situation. From the perspective of the migrant population's living space, we can clearly understand and learn about their living patterns, living environment and living needs, which is more in line with the core idea of "people-oriented" in new urbanization. This chapter introduces the living areas of migrant people according to their living space, and divides them into four main types: the first type is urban affordable housing areas represented by low-rent housing and public rental housing; the second type is urban villages; the third type is suburban communities at the interface of urban–rural fringe; and fourth type is urban agricultural areas around the city dominated by migrant farmers.

9.3.2 Low-Rent and Public Rental Housing

Social housing is an important institutional measure implemented by the government to address the housing problems of low-income and impoverished groups. Since a significant portion of the migrant population in cities has low-income levels and poor purchasing power for commercial housing, they are eligible for social housing. There are two types of social housing related to migrants: the first type is low-rent housing, which is non-ownership, government-subsidized rental housing for low-income urban residents, mainly targeting migrant workers and other low-income groups; the second type is public rental housing, which is owned by the government or public institutions, and is rented to newly employed workers at below-market rates, including recent graduates and people who have migrated from other areas for work. This type of housing addresses the housing difficulties faced by newly employed workers and other middle-income groups.

However, in reality, social housing often fails to meet the living needs of the migrant population, and even leads to social problems due to over-concentration. Most of the social housing projects of governments are located in the suburbs, where land costs are low but the location is remote. This model has provided a large number of effective housings for low-income families in urban areas with housing difficulties, but the results have been minimal in major cities. The reason for this is that because social housing is far from the city center, the corresponding supporting service facilities such as culture, education, health, and sports are not perfect, which increases the cost of living and working for the people who move in. For example, it is inconvenient for children from outside the city to go to school, and so on. This has caused many low-income people in cities to give up moving into social housing. It can be seen that without complete supporting facilities, social housing developments that are built further and further away from the city center may well become isolated islands, resulting in the marginalization of the people who are supposed to be protected. This in turn leads to a series of new social problems, such as the formation of slums where large numbers of low-income people live, the difficulty for the migrant population to feel a sense of belonging to the city, and a lack of opportunities for social interaction (Xu 2009). In addition, the current imperfect financing model for social housing has resulted in the quality of some low-rent or public rental housing, which is supposed to be a people-oriented project, not being fundamentally guaranteed. Coupled with the relatively low level of property management services, this has a negative impact on the living environment of social housing and has also inadvertently lowered the standard of social housing in providing basic living needs for low-income families. This section uses the Jinshazhou community as an example to interpret the daily living space of residents in Guangzhou's social housing communities, in order to reveal the real living conditions of the migrant population in Guangzhou's social housing communities (Li et al. 2014).

Case study: living space of the social housing community's residents in Jinshazhou, Guangzhou

The Jinshazhou community is a large-scale social housing district built by the Guangzhou municipal government to address housing issues for severely impoverished households. It was planned and constructed by the Guangzhou city government and was once heralded as the largest public rental housing community in China. The community is located at the westernmost part of Guangzhou, in the northern part of Baiyun District, occupying a small island at the border between Guangzhou's western district and Foshan's Nanhai district. Its area includes the Pingle and Fengling communities. The Jinshazhou social housing project was completed in October 2007 and was delivered for use in February of the following year. The community consists of 64 residential buildings, with the main apartment types being one-bedroom apartments, two-bedroom apartments, and three-bedroom apartments. More than 80% of the units are smaller than 90 m^2 in floor area. The Jinshazhou community is well-equipped with supporting facilities, including secondary schools, primary schools, kindergartens, a comprehensive meat and vegetable market, a post office, a police station, a health clinic, a cultural activity station, and a senior service station. Overall,

Jinshazhou residents mainly belong to the middle and low-income groups. In recent years, due to the increase in residential population, the community has encountered issues such as inconvenience in shopping, chaotic neighborhood management, inadequate amenities, and traffic congestion. However, residents still strive to seek personal living space through continuous community optimization within the existing space and regulations.

Residents are highly motivated to improve their own housing. Housing in Jinshazhou is dominated by small units of 60–80 m^2, most of which are shared by two or three generations of a family. Although housing conditions have improved relatively compared to the past, the average per capita living space is still generally small. When it comes to maximising the use of space, the residents of Jinshazhou have shown their unique wisdom. Auntie Zhang's family now lives in a two-bedroom, one-living room apartment of about 60 m^2. She says: "We just have to find our own way. My son and daughter-in-law live in one room, and my husband, grandson and I live in the other. We use bunk beds: my husband and I sleep on the bottom bunk, my grandson on the top bunk… The dining table is foldable to save space". Due to the lack of activity facilities within the Jinshazhou community, residents often go to nearby squares and parks with better environments to exercise. In addition, markets have spontaneously sprung up around the community, and these vendors supplement the basic daily life of the community's residents. This shows that residents have further developed social spaces outside the community to meet their needs for leisure, socialising, employment, consumption, etc.

Informal employment is common among residents. As Jinshazhou residents generally have low levels of education and low monthly household incomes, many of them are doubly disadvantaged households, low-income households, people with disabilities, and elderly people living alone, so they rely on government subsidies for income. However, rather than passively relying on fixed subsidies, Jinshazhou residents are more proactive in improving their situation through informal employment. As most residents are not formally employed, they mostly work as vendors, day labourers, hawkers, taxi drivers, craftsmen and so on. For example, to meet the transport needs of the surrounding residents, there are many disabled vehicles in the community that have been converted into taxis or delivery vans, and the income is considerable. Informal employment often has irregular working hours and income, and is characterised by mobility, secrecy and discontinuity, making it easier to evade the supervision of urban management and tax-related departments, thereby maximising and providing the most flexible supplement to family income.

9.3.3 Urban Village

Urban villages are generally found in megacities, with Guangzhou and Shenzhen in the Pearl River Delta region being typical examples of a spatial 'mosaic' structure. So why have urban villages, as an organic part of the city, become places for foreigners to live in the built-up areas of the city? This is because urban villages lack the legal status of urban land and the public services that should be provided by the government, making their rental costs lower than other spaces. Thus, it is precisely because of their low cost that urban villages have adapted to the needs of new urban migrants, especially low-income migrant workers, and have attracted a large number of foreigners to live there, thus becoming known as "private security housing".

Urban villages are low-income communities or areas with a high concentration of migrant workers, and their most obvious spatial characteristics are high-density buildings and disordered spaces. Due to the peculiarities of urban village land under the dual urban–rural land structure, villagers rely on rent from rental housing as their main source of income. In order to maximize personal profits, villagers often use the ownership of homesteads to make their houses as tall and large as possible, creating the common urban village spatial landscapes of "kissing buildings," "holding hands buildings," and "one-line sky". In addition, the influx of migrant populations has caused excessive population density, which has put a heavy burden on environmental health management and fire safety management. Another characteristic of urban villages is the establishment of a special urban ecosystem of self-service through the informal economy. Due to the relaxed institutional environment and good location conditions, urban villages and their surrounding areas have become a gathering place for the informal sector, which provides a large number of jobs for low-income migrant populations. Therefore, most migrants choose to live in urban villages and spend their daily lives there. They pay great attention to various consumer, commercial, entertainment and other commercial facilities in urban villages (Xue and Huang 2008). Although the migrant population living in urban villages is clearly non-agricultural in terms of the social division of labor, social relations such as differences in household registration, differences in systems, differences in social production relations, and reorganization of the division of labor still make them different from full-fledged "urban residents" in many respects, and they have become a marginalized group. In this section, we take Guangzhou Ruibao Village as an example to analyze the living space characteristics of migrant settlements in urban villages (Wu et al. 2013).

Case study: the living space of migrants in Ruibao Village, Guangzhou

Ruibao Village is located in Haizhu District, Guangzhou. The village is adjacent to the Zhongda Cloth Market in the north and the Pazhou Exhibition Center in the northeast. It is conveniently accessible from the main service area of Shahe Street and the Shisanhang Clothing Market. Nearly 80% of the floating population living in Ruibao Village come from rural areas in Zhanjiang, Guangdong Province, making it a relatively typical homogeneous floating population settlement.

The social space of the agglomeration is formed through the village link. In the mid-1990s, the residents of Ruibao village were mainly migrants, most of whom came from other provinces and regions such as Jiangxi and Xinjiang, with only a small number of Zhanjiang people; in the late 1990s, as the profitability of belt production increased, the Zhanjiang people engaged in belt production recruited helpers from Zhanjiang to expand the scale of their production, at the same time absorbing many of the Zhanjiang people who were previously engaged in other jobs around the village to work in belt production. Belt production work. In the twenty-first century, as the scale and benefits of belt production in Ruibao village increased, more and more Zhanjiang people began to come to the region through the links of the township to work in the belt production industry. The Zhanjiang people in Ruibao Village have also formed a stable and close social relationship network through the identity of their fellow villagers, helping each other in life and cooperating in production. At the same time, the ethnic economic system has been rapidly formed, and various restaurants, repair shops and pharmacies with Zhanjiang characteristics have been opened. These characteristic self-service systems make Ruibao Village gradually take on the characteristics of a "second hometown" for Zhanjiang people. With the continuous arrival of people from the same hometown, the non-Zhanjiang population began to leave, and Ruibao Village gradually became a settlement area mainly for Zhanjiang people.

The production and living spaces of the migrant population overlap. There are a total of 1,127 rental units in Ruibao Village, and the number of houses has remained unchanged for a long time. However, the number of rental units increased from 4,142 in 2008 to 5,894 in 2012. Most of the buildings in the village have been added to or altered, resulting in a high building density and an outdated appearance. This has led to an increasingly dense living space in Ruibao Village (Fig. 9.9), while the overall architectural landscape of the space has become increasingly messy due to continuous alterations and expansions. In order to save costs, the production and living spaces of Zhanjiang people are becoming integrated. The raw materials can be processed at home, and most of the living space is occupied by the raw materials and the just-completed finished products. Here, living space is greatly reduced. In order to make full use of space, bed-over-bed and attic spaces are very common. Therefore, the floating population maintains low rents at the expense of housing area, while having a low-cost living and production space.

The spatial characteristics of the original urban–rural fringe are utilized. As an urban village, Ruibao Village has relatively low rents for rental housing and lax garbage and sanitation management, which maximizes savings in the living and production costs of the floating population. Ruibao Village is conveniently located, with a subway and multiple bus lines nearby. Over the long term, a supply market for belt production materials has formed in the north of Ruibao Village, and it is close to the Zhongda Cloth Market. It is very convenient for obtaining orders, purchasing raw materials, shipping finished products, and commuting to work. The floating population of Zhanjiang in Ruibao Village has taken full advantage of these spatial characteristics of Ruibao Village, choosing Ruibao Village as a place to live and as a production base, occupying and reshaping the space.

9.3.4 Urban–Rural Fringe Communities

The urban–rural fringe is a transitional area between city and countryside. The future development direction of its rural part has not been accurately positioned, while the urban part lacks development planning and a clear division of functional areas. The

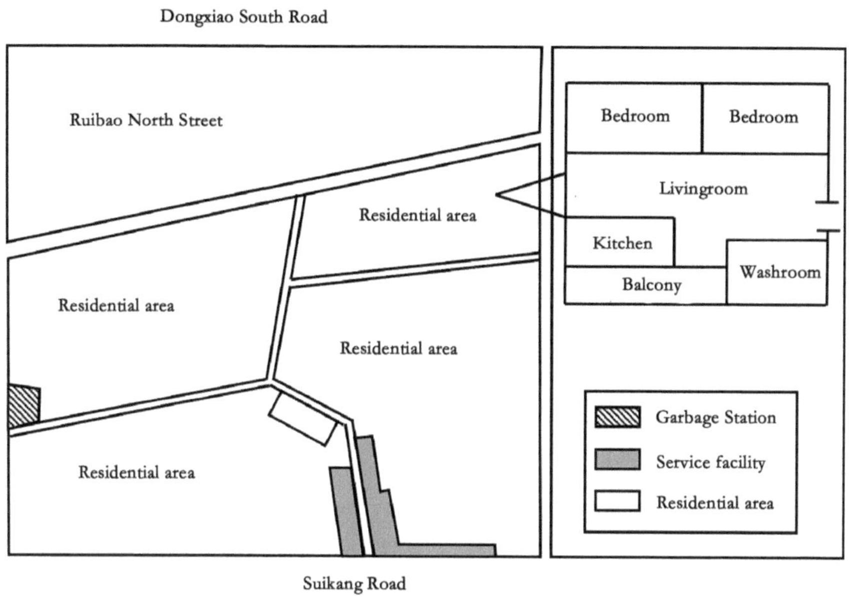

Fig. 9.9 Residential space of floating population in Ruibao village

unique geographical, administrative and demographic intersections of the urban–rural fringe have formed a unique socio-economic structure.

Why has the urban–rural fringe become a settlement area for the urban migrant population? First, the near-field diffusion of urban industries has created more employment opportunities in the urban–rural fringe, and the construction of residential buildings in the suburbs has also made the urban–rural fringe a focal area for the employment of migrant construction workers. Second, a large number of vacant rural homes on the rural side of the urban–rural fringe provide living space for the migrant population. Finally, driven by rural and industrial ties, the settlement of the migrant population in the urban–rural fringe will further attract more migrants to make a living there (Song and Wu 1997).

The migrant population is concentrated in the urban–rural fringe areas, which are characterized by closely related production and living spaces that form an internal self-circulating economic space. At present, the secondary industry in the urban–rural fringe areas is dominated by township enterprises and small and medium-sized enterprises, while the tertiary industry is dominated by housing rental, small catering and retail. A large amount of unused land has become a haven for illegal construction. A large number of floating population and registered farmers in the urban–rural fringe areas form a special interdependent and mutually beneficial relationship with land and rented housing as the medium. Related Internet cafes, small workshops, garbage collection stations, and low-cost social service industries have gathered around the urban–rural fringe, forming a self-serving and self-circulating economic and social cycle. This area not only includes rural areas, village-turned-communities, and various non-public enterprises, but also borders and intersects with urban communities, Party and government organs, schools, hospitals, etc., forming a complex and diverse social spatial pattern. Existing research mainly focuses on the spatial production of urban–rural fringe areas, exploring how cities govern such spaces and manage migrant communities, such as the study of the "black cluster" of M waste pig farms in Guangzhou (Liu and Wang 2011). This section uses the example of the garbage village on the urban–rural fringe of Beijing to highlight the characteristics of the "gray space" where migrant populations congregate in the suburbs.

Case study: the living space of migrants in Waste Recycling Village at Beijing's urban–rural fringe

Located in the north of Beijing, Bajia Village and Dongxiaokou Village are the two largest waste distribution centers among the hundreds of waste villages in the suburbs of Beijing. Bajia Village is located between the Fourth and Fifth Ring Roads in the north of Beijing. It is close to Zhongguancun and Haidian University District, and has gradually developed into a large waste village with electronic waste as the main focus and other types of waste as a supplement. Dongxiaokou Village is located 5 km north of the Fifth Ring Road. On this 500 mu land,[3] it once handled a quarter of Beijing's centralized waste recycling, and more than 30,000 waste recyclers made their living here. The first company to enter Dongxiaokou was a waste recycling company called Fuyou Xinyuan. The owner of the company leased 60 acres of wasteland from the villagers and divided the area according to different types of renewable resources. Dongxiaokou has developed a management model similar to that of a farmers' market, in which the recycling company cooperates with businesses to share the work: the former rents out stalls and is responsible for management, while the latter recycles, sorts, and sells waste. Since then, a number of large waste recycling companies, such as Xinfa Waste Material Recycling Center and Beijing Zhonghai Rongtong Material Recycling Center, have gradually moved into Dongxiaokou, building a waste city.

The production space is relatively organized, while the living space is mixed. The spatial texture within the waste village is different from that of an urban village with rental housing. Instead, it is similar to an early Zhejiang village, which is a spatial form that mixes production activities with living. Since waste management requires relatively open space for storage and sorting, and waste requires relatively spacious access routes for entry and exit, the waste village has adopted a courtyard organization centered on the waste market in the local space at the urban–rural interface. Scattered individual waste recyclers rent housing in the periphery, resulting in mountains of waste and a dirty and untidy living environment. Over time, the spatial morphology and internal functional organization of the waste village has also undergone continuous self-improving evolution, guided by upstream and downstream links in the industrial chain.

---

[3] 1 mu $\approx 666.67m^2$.

Kinship networks shape the social space of waste villages. More than 90% of Beijing's waste collectors originate from Gushi County in Henan Province and its surrounding areas. These geographic ties serve as the primary channel for migrants entering Beijing's waste recycling industry and also function as an informal network for disseminating market information. The most common organizational structure in waste villages is family-based entrepreneurship, which connects various stages of the waste recycling supply chain, from collection and sorting to baling, transportation, and reuse. Processed waste materials are shipped to processing hubs in Handan, Baoding, Tangshan, and even Hangzhou and Guangzhou for further refinement. Henan dialect is the dominant language spoken in the village, and a long-distance bus route, the "Beijing-Gushi Line," serves as a direct link between the village and the migrants' hometowns. Many migrants have built their livelihoods by working in waste recycling and have subsequently contributed financially to infrastructure projects in their hometowns, such as building roads, bridges, and schools. In the evenings, an informal Henan-style night market emerges along the village's periphery, further reinforcing the cultural and economic ties that migrants maintain with their home regions. In this way, Henan migrants have reconstructed a sense of belonging within these waste villages, turning them into hybrid urban–rural spaces.

In 2011, the Beijing Municipal Government announced the Comprehensive Land Use Plan for Dongxiaokou Township (2006–2020), which reclassified the village's land for urban development, initiating a three-year demolition process. As electricity and water supplies were cut off and entire sections of shantytown housing were razed, tens of thousands of villagers were forced to relocate. Fuyou Xinyuan, the first waste recycling company to establish itself in Dongxiaokou, was among the businesses slated for demolition. Following its closure, the owner moved his operations further north to Nanqijiazhuang Village, while other waste business owners sought alternative locations beyond the Sixth Ring Road. Some entrepreneurs chose to return to their hometowns or exited the recycling industry entirely. This process of displacement and relocation reflects a unique cycle of "retreat" and "resettlement"—as urban expansion continually encroaches on peri-urban spaces, migrant communities are pushed further outward, forcing them to re-establish themselves at the shifting edges of the city (Tao et al. 2014).

9.3.5 Peri-Urban Agricultural Zones

In the process of rapid urbanization in China, due to the shortage of agricultural production skills among the local agricultural workforce in cities, an aging population, and part-time work, migrant workers have gradually replaced the local agricultural workforce in the field of agricultural production. However, this type of migrant population does not easily attract the attention of society and academia. This is because, in terms of industry type, most migrant workers in cities are engaged in manufacturing or the tertiary industry, while the number of migrant farmers engaged in agricultural activities is relatively small, and their role in urbanization is relatively secondary. Although they are a minority, they cultivate a large area of land and bear the important responsibility of urban environmental regulation and food security. Therefore, migrant farmers in peri-urban agricultural areas should not be ignored, and research on their living space is an important way to supplement our understanding of the living space of the urban migrant population.

Migrant farmers represent a mode of rural labor migration that is "leaving one's hometown but not changing one's professional attributes". The main group is tenant farmers, and the nature of their labor determines that they are more stable than mobile. From the perspective of basic means of subsistence, migrant farmers generally have poorer living conditions than urban residents and local villagers. However, migrant farmers themselves do not have high expectations for their living environment. They don't care whether they are satisfied or not, and they don't consider working away from home to be a pleasure. This reflects the fact that migrant farmers do not regard the city as a place to settle permanently, and the idea of returning to the countryside when their working lives are over still plays a decisive role (Zhang 2007). In terms of the social integration of the floating agricultural population, most migrant farmers live only in their own circles, generally interacting with fellow villagers, and lack interaction with local people, so they cannot integrate well into the local community (Ma et al. 2012). In addition, the city's neglect of the migrant farmer community, the imperfect security systems such as medical insurance and social security, and the difficulty in solving the problem of children's education have all contributed to the difficult living conditions of this socially disadvantaged group of migrant farmers, making it difficult for these rural laborers to stay in the city for a long time. "The city that they cannot integrate into, and the countryside that they cannot return to" is becoming an inevitable experience for migrant farmers. This section uses the four villages of Huanggang, Qianluoma, Dongjiangzhouying and Xiaodian in the suburbs of Beijing as examples to analyze and interpret the real-life situation of migrant farmers in semi-urbanized areas of Beijing (Zhang 2007).

Case study: living space of migrant farmers in the suburbs of Beijing.

Huanggang Village, Qianluma Village, Dongjiangzhouying Village and Xiaodian Village (Table 9.4), all located in the suburbs of Beijing, have a certain number of migrant populations. Among them, the migrant population in Huanggang Village has already exceeded twice the local population. In Xiaodian Village and Huanggang Village, only a very small proportion of the migrant population is engaged in agriculture, and most of the migrant population is engaged in secondary and tertiary industries in the village enterprises or in urban areas. Most of the migrant farmers come from poor areas in Chengde, Hebei, Henan or the mountainous areas of Chongqing.

The living conditions of migrant farmers are generally poor. A field survey conducted in Xiaodian Village, Chaoyang District, revealed that many local villagers had converted their entire courtyards into rental housing for migrant farmers. However, most migrant farmers opted to construct simple single-story houses near the farmland they rented, while only a small portion rented houses from local farmers.

Table 9.4 Number and distribution of farming households in the surveyed areas

Village	District/ county	Local population	Agricultural labor force	Migrant population	Migrant farmers
Dongjiangzhouying village	Shunyi	257	130	140	10
Qianluma village	Shunyi	651	262	32	24
Xiaodian village	Chaoyang	1108	384	636	60
Huanggang village	Chaoyang	1303	68	2800	30

In Qianluma Village, migrant farmers predominantly resided in the solar greenhouse district in the southern part of the village, maintaining a noticeable distance from the residences of local villagers. Since these houses were self-built, they were extremely rudimentary in both design and construction, often replicating the living conditions of their hometowns. The majority of migrant farmer households occupied one or two rooms, ranging from 20 to 40 m^2 in size. These houses lacked heating during winter, and kitchens and toilets were even more basic—often assembled using prefabricated boards and placed in a corner of the courtyard. Many lacked courtyards altogether, with houses directly built adjacent to the ends of greenhouses or polytunnels.

Migrant farmers largely retain their traditional rural lifestyles. Although they have begun using liquefied gas and electricity for cooking, the use of firewood and coal remains widespread. The adoption of time-saving and less-polluting energy sources such as liquefied gas and electricity indicates a gradual shift in their living habits after migration, where increased income has led to a consideration of time costs. Given that most migrant farmers engage in vegetable cultivation, their agricultural waste disposal methods remain rudimentary. Approximately 70% of surveyed farming households discard or compost decayed vegetables, a practice that is not only slow to decompose but also negatively impacts air quality and complicates village sanitation efforts. Migrant farmers tend to reside in remote areas far from village centers, where local administrative authorities have limited incentives to enforce environmental improvements, further delaying changes to unsanitary living practices.

Migration necessitates a reconstruction of social networks within living spaces. Upon migrating, migrant farmers must navigate new social relations with local farmers, land-leasing households, and local administrative bodies such as the village committee. These relationships vary depending on regional background, personality differences among migrants, and the organizational structure of land leasing. For instance, three migrant farming households in Dongjiangzhouying Village arrived in Beijing relatively early and settled next to local villagers, fostering close and harmonious relationships with them. In contrast, Qianluma Village hosts a larger migrant population, with over ten migrant households having built houses adjacent to their leased greenhouses, forming a self-contained community with minimal interaction with local villagers, beyond basic acquaintance. Furthermore, as each migrant household directly contracts with the village committee, their relationship with the committee is closer than that of local villagers. Besides paying annual rent, migrant farmers frequently report infrastructure issues within their residential areas, such as maintenance of power substations and irrigation wells. Consequently, the village committee takes migrant farmers' interests into account when formulating policies.

9.4 Yuppies, Urban Subcultures, and Their Spatial Landscapes

9.4.1 Urban Graffiti

9.4.1.1 Art Without a Place

Graffiti originated in the United States in the 1960s, often emerging in urban slums as a means for marginalized communities to express emotions through vibrant colors and bold styles. It serves as a dynamic interaction between urban residents and their living spaces. However, graffiti has historically been exploited by political radicals and street gangs as a means of territorial marking or ideological expression, leading governments to categorize it as a public nuisance requiring eradication. Over time, as societal perceptions evolved, graffiti's association with lawlessness and violence gradually diminished. Many countries have since introduced formal regulations for managing graffiti, such as designating legal graffiti walls and establishing procedural guidelines for its governance. Today, graffiti has gained recognition as an artistic expression of urban identity, contributing to the cultural landscape of cities worldwide.

China's graffiti scene emerged relatively late, developing in the 1980s. Guangzhou, as a frontier of economic reform and a hub of cultural exchange with Hong Kong and Taiwan, became a center for graffiti's growth (Figs. 9.10 and 9.11). Consequently, many contemporary graffiti artists regard Guangzhou as the birthplace of Chinese graffiti art. Unlike in Western countries, where urban regulations differentiate between vandalism and artistic graffiti, China's municipal management policies primarily classify all unauthorized street painting as public defacement, leading to an ongoing "cat-and-mouse" game between graffiti artists and urban authorities. Many graffiti practitioners in China are art students with higher education backgrounds, who are proactively exploring ways to integrate graffiti into urban culture while promoting its legitimate development. However, due to deep-seated administrative biases against graffiti, there remains a lack of dialogue between graffiti artists and city officials. As a result, this subcultural art form struggles to find a legitimate foothold in urban spaces.

Fig. 9.10 Urban graffiti: Lingnan Institute of Technology

Fig. 9.11 Urban graffiti: Jiangnan South Road

9.4.1.2 Spoilers of City's Order or Leaders of Individuality?

Globally, graffiti has become an emblem of individuality and creativity in major cities. For example, Spring Street in New York, East London, the Left Bank of Paris, and the remnants of the Berlin Wall all feature renowned graffiti walls, attracting numerous visitors.[4] Despite its growing artistic recognition, municipal authorities worldwide remain hesitant to embrace graffiti, frequently citing concerns over urban cleanliness and aesthetic order. In China, resistance to graffiti is evident in incidents such as the sudden removal of a large-scale graffiti wall created by both domestic and international artists under Changsheng Bridge in Zhuhai on September 23, 2015, which was replaced with a stark "NO GRAFFITI" sign. Similarly, an elaborate graffiti installation on a demolition site in Jing'an District, Shanghai, was erased overnight, sparking debate within Shanghai's municipal government.[5] The fundamental question remains: Are graffiti artists pioneers of urban individuality or disruptors of public order? In recent years, the discourse surrounding graffiti has escalated to a broader urban planning issue.

Field research in Guangzhou suggests that despite the existence of a few designated graffiti walls, local graffiti artists have developed a strong online and offline social network to sustain their community. Platforms such as "Tutu," a social network dedicated to mapping graffiti locations, TAG (Talk About Graffiti) salons, which facilitate discussions on graffiti culture, and "Maidi" workshops, which focus on teaching graffiti design, have become increasingly active within the subcultural art scene. Guangzhou's graffiti artists are leveraging these social networks to cultivate new talent and advocate for the recognition of graffiti as a legitimate form of artistic expression. As a result, graffiti culture is gradually being integrated into various urban spaces, taking root, flourishing, and evolving within the city's social fabric.

9.4.2 Tianguang Market

9.4.2.1 The Hundred-Year "Ghost Market"

Late at night, the streets of Guangzhou are filled with dark shadows. People walk around with flashlights, selecting items they like in the grocery stores. This mysterious market, which opens in the early morning and closes at dawn, is called the "Ghost Market" or "Tianguang Market" (the Cantonese term "dawn" also means "daybreak"). It is a folk market that mainly sells second-hand cheap goods such as old furniture, utensils, clothes, shelves, as well as antiques, calligraphy and paintings, ancient books, potted plants, etc.

Tianguang market first appeared in Guangzhou and later spread to Hong Kong, Nanjing, and other cities. Its origins can be traced back to the late Qing Dynasty and early Republican era. Historical records suggest that its prototype emerged from an outdoor meat and vegetable market located on the western outskirts of Guangzhou during the Ming Dynasty. Due to the region's hot climate, buyers and sellers preferred

[4] Wang Linsheng. Street art has nowhere to go, so what's the future of urban graffiti walls? http://news.99ys.com/news/2015/0716/9_194702_1.shtml. 2015-07-16.

[5] Ma Xiangbin. On the ruins of graffiti, the city's attitude was "eradicated overnight". http://money.163.com/15/0126/14/AGT4HF2400253B0H.html, 2015–01-26.

to conduct transactions before sunrise, leading to the early hours of operation. The modern Tianguang market officially took shape in the late Qing and early Republican periods when wars and social upheaval resulted in a flood of privately owned antiques and valuables entering underground markets. Unable to trade through formal channels, merchants smuggled these high-value items into Guangzhou, where they were sold in unregulated street markets under the cover of darkness. By the mid-Republican era, Tianguang market had reached its peak, evolving from a specialized antique market into a comprehensive secondhand marketplace, centered around Lanmalu (modern-day Zhongshan Seventh Road). Following the founding of the People's Republic of China, Guangzhou authorities implemented stricter urban management policies, renovating Lanmalu and enforcing regulations that led to the decline of Tianguang market. However, in the 1980s, the market resurged to accommodate the needs of Guangzhou's expanding migrant and low-income populations, reemerging in various locations, each specializing in different types of goods.[6]

Among these, Tianguang market at Haizhu Middle Road specializes in books, stamps, and antique coins, operating on Saturdays from 5:30 AM to 9:00 AM. Tianguang market at Guangta Road, known for books and antiques, attracts literary enthusiasts and operates daily from 6:00 AM to 8:30 AM. Wenchang North Road Tianguang market is renowned for antiques and, during the early 1980s, became a hub for smuggled relics and artifacts from tomb raiders across the country.

9.4.2.2 The Threefold Marginality of Tianguang Market

Tianguang market exhibits three distinct aspects of marginality, which contribute to its unique role in urban informal economies. First, the marginality of goods: Most items sold in Tianguang market originate from scrap collection, waste stations, and scavenging activities. Some also come from expired inventory of regular retail stores (Table 9.5). These products are diverse and inexpensive but vary greatly in quality, making the market a space where second-hand and discarded goods are repurposed for economic survival. Second, the marginality of time: The market's nocturnal trading hours are a historical legacy stemming from the illicit nature of its commerce. A century ago, the market was notorious for selling smuggled or stolen goods, shaping its clandestine identity. While its modern form has largely detached from this past, it still carries the label of "cheap and second-hand" rather than entirely shedding its former illicit reputation. The persistence of its unconventional trading hours is not only a necessity for the urban underclass but also a way to maintain a sense of dignity in their economic activities. Third, the marginality of participants: Vendors at Tianguang market are primarily scavengers, waste recyclers, and second-hand goods resellers. These individuals spend their daytime hours collecting discarded items or working low-wage jobs, and at night, they dedicate a few hours to

[6] Liu et al. (2013). The Market of the Cracks: Resting at Dawn, Market Research on the Ground Floor of the Tian Guang Market at Ximenkou, Guangzhou. Third Prize for the 2014 National Social Practice Research Course Assignment for Undergraduate Students of Urban and Rural Planning.

Table 9.5 Marginality of Tianguang market

Category	Goods item	Price at Tianguang market	Price at supermarket	Wholesale price
Food	Meat Rice Box	¥2/box	¥13.5/box	¥8.9/box
	Garden Swiss Roll	¥1/pack	¥3.8/pack	¥2.8/pack
	Instant Noodles	¥1/pack	¥4.5/pack	¥2.46/pack
Medicine	Medicine	¥5/box	¥20/box	¥15.5/box
Clothing (Non-branded)	Jeans	¥5–10/pair	¥30–50/pair	¥8–20/pair
	Slippers	¥5/pair	¥12/pair	¥8/pair
	Tops	¥1–5/piece	¥30–50/piece	¥20–35/piece
Electronics	Nokia N96	¥30/unit	¥500/unit	¥298/unit

Source Research report by Liu Hui et al.

Table 9.6 Behavioral characteristics of street vendors in Tianguang market

Scavengers	Arrive at Tianguang market around 2:30 AM and leave around 8:00 AM. During the day, they primarily engage in scavenging and often rely on free temple meals and temporary shelters for survival. At night, they sell collected items, earning meager income barely sufficient for subsistence. Their earnings from Tianguang market constitute nearly their entire financial support
Waste recyclers	Mostly migrant laborers performing physically demanding and unstable jobs with low wages during the day. They use a few nighttime hours at Tianguang market to sell collected recyclable items, supplementing their income. Their earnings from Tianguang market represent one of their key financial sources
Second-hand clothing resellers	Primarily engaged in reselling used clothing, either by piece or by weight. During the day, they source and sort second-hand clothes at Oujia Used Goods Market, then sell them at Tianguang market at night. Most of these vendors come from Henan Province

Source Summary of the research report by Liu Hui et al.

selling their findings, supplementing their meager income (Table 9.6). The economic stratification resulting from urban diversification has created a group of people who exist on the periphery of formal economic structures. Their need for alternative spaces and flexible trading models is a crucial factor behind the century-long survival of Tianguang market in the face of urban modernization.

9.4.2.3 Survival in the Margins: What Lies Ahead?

Guangzhou's Tianguang market, a marketplace deeply embedded in Cantonese culture, has persisted in the urban margins despite modernization. It offers a diverse range of goods, from expired canned food priced at just a few yuan per box to Ming and Qing dynasty antiques valued at tens of thousands. It caters to both necessity and passion—some seek bargains for survival, while others hunt for treasures in the darkness before dawn. A seasoned dealer remarked that the "fairness" of lies in its unique economic logic—one could either spend millions on counterfeit antiques or, with patience over a decade, curate an entire antiquarian bookstore.

Tianguang market operates on a low-margin, high-volume business model. While prices are strikingly low, most goods are second-hand, heavily worn, or seemingly beyond further use—yet they still find buyers. However, its prominence within Guangzhou is fading. The Hualin Jade Tianguang market has been redeveloped into a formal commercial complex; the electronic goods market on Liwan Road was shut down due to complaints about public disturbance; and on Wenchang Road, many street vendors, known colloquially as "ghost sellers," transitioned from informal street stalls to renting legitimate storefronts, marking a form of successful transformation.[7] As Guangzhou's urban landscape continues to evolve, Tianguangxu's survival raises critical questions: Is it a testament to the city's inclusivity towards the lower classes, or is it a symptom of urban disorder that must be eliminated? Its fate remains uncertain, and only time will reveal the answer.

9.4.3 *The Bar Streets*

9.4.3.1 Space Beyond Consumption

Since their introduction to China in the late 1980s, Western-style bars have proliferated across major cities such as Beijing, Shanghai, and Guangzhou, giving rise to bar-centric districts like Sanlitun (Beijing), Xintiandi (Shanghai), and Huanshi Road (Guangzhou). In recent years, bar streets have come to symbolize urban fashion and vibrancy, attracting crowds of young urbanites seeking nightlife experiences. Behind this nocturnal economy, significant financial gains have emerged, prompting cities beyond China's first tier metropolises to invest in developing bar streets. The pursuit of "nighttime GDP" is often accompanied by efforts to brand these cities as "modern international metropolises."

In fact, consumption today has transcended its economic connotations and has a complex social construction function. Bar Street is not just a place for consumption, but also a space that encompasses various social relationships. Bar Street has become a stage that carries the nightlife of a special group in the city, blending complex connotations such as individuality, restlessness, and ambiguity. It either embraces or rejects outsiders, and may also be deconstructed and reconstructed due to the "local nature" of the local society and culture.

[7] Zhou Wei. Infatuated with the Tian Guang Market. http://news.ifeng.com/a/20160222/47533011_0.shtml. 2016-02-22.

9.4.3.2 From Enclaves to Place

Judging from the development of bars, they have gradually evolved from a marginalized urban public space rejected by the middle class to a consumer space suitable for the middle class occupying the central city area. With the deepening development of globalization, bars are spreading all over the world (Bromley and Nelson 2002; Kraack and Kenway 2002; Mack 2006). This section takes Guangzhou as an example to explore the development process of bars influenced by globalization since the reform and opening up from two main lines: economy and culture.

From the perspective of economic globalisation, the first bars in Guangzhou were mainly located near high-end foreign hotels, high-end residential areas and consulate areas in the eastern part of the city. These areas are relatively hidden, monotonous and closed, and they are living spaces established by foreigners and businessmen to maintain their original lifestyles. At the same time, they have become 'enclaves' in the urban space due to differences in consumption patterns and subjects (Lin and Wang 2011). However, as Guangzhou has become more open, these bars have gradually been transformed into consumption spaces open to the general public. At the same time, Guangzhou's bars have gradually integrated the characteristics of Guangdong's late-night snacks, with menus featuring a variety of tasty snacks. The clustering of bars has led to the formation of several distinctive bar streets, such as Fangcun Style Bar Street, Padi Bar Street, Yanjiang Road Bar Street and Huanshi East Bar Street. This is a localised choice made by the bar streets to adapt to Guangzhou's transformation into an international metropolis and to meet the living needs of Guangzhou's diverse urban communities.

From the perspective of cultural globalization, the emergence of certain bars in Guangzhou can be traced back to the music tea houses rooted in local urban culture (Cao 2010). Historical records suggest that Guangzhou's bar culture spans over a century. During the late Qing Dynasty, with the expansion of commercial activities and the increasing influence of Western culture, some hotels began to incorporate ballrooms, and singing halls emerged in tea houses and taverns. These singing halls—later referred to as music tea houses—became the prototype of bars in China. In the 1980s and 1990s, as Hong Kong and Taiwanese pop culture spread and ideological constraints loosened, karaoke dance halls flourished in Guangzhou. In response, hybrid bars with strong local characteristics and diverse entertainment functions emerged. These venues not only featured live music performances but also offered a variety of shows, including acrobatics and comedy acts, akin to modern Hello Performing Arts Bar in Huanghuagang and the scenic bars along White Goose Pool.

On the one hand, Guangzhou Bar Street has transformed from a Western cultural enclave into a space with local consumer characteristics in a multicultural context. On the other hand, it has combined Guangzhou's karaoke culture with the endogenous music bar as a local landscape, which not only reflects the homogenized spatial production of bars and other slow cultures with global consumer response in Guangzhou, but also connects the night culture with Lingnan's local consumer colors to present local heterogeneity (Lin and Wang 2011).

9.4.3.3 The Production and Consumption of bar Streets

In addition to the aforementioned local characteristics, bars also have the functions of space production and space consumption. Lin and Wang (2011) used the Huanshi East Bar Street as an example, selecting "Little Mountain", "Chinabox" and "Lu Dian" as research objects. Through qualitative research methods such as non-participant observation and in-depth interviews, they analyzed the understanding of bar consumption places from the discourse of the actors, analyzed the meaning hidden in the language, and then explained the construction of the bar consumption space. Lin and Wang (2011) found that, since the beginning of the twenty-first century, bars have gradually become symbols of fashionable leisure, with an identity in social space that gives social significance to consumer identity. In addition, the government has further emphasized through urban planning the special social functions that the Huanshi East Bar Street assumes in a specific urban location, giving the Huanshi East Bar Street its social significance and making it a symbol of local consumption in Guangzhou.

The spatial imagination of bars is collectively shaped by multiple actors, including bar operators' planning imagination, consumers' consumption imagination, managers' service imagination, and performers' cultural imagination. These elements together construct a reproduced space that emerges from the interactions among diverse stakeholders. The spatial imagination of bars is intricately linked to various social relations and is established through spatial conceptualization and symbolization. Essentially, it represents a layered assemblage of imagined spatial narratives, all centered around the conceptual framework of "bar space."

Finally, within the dimension of reproduced space, the material (physical space) and human mental imagination, consumption experience, and consumer relations are interwoven and mutually influential. This interplay achieves a synchronous integration of material and mental space, ultimately leading to the formation of a "new space." The social significance and place attachment of the Huan Shi East bar consumption area reflect an integrated human–environment relationship, contributing to the construction of a new spatial order that transcends mere perceptual interactions between people and space.

9.4.4 *Artists' Enclave*

Artists, as a niche social group in urban society, have been among the earliest to exhibit a cultural motivation for migration from cities to rural areas. They bring urban lifestyles and humanistic sensibilities to the countryside, distinguishing themselves to some extent from capital investors who drive rural urbanization through manufacturing industries. In navigating the tension between sustaining their livelihood and adhering to artistic ideals, they continuously seek a balance.

The rural areas of the Pearl River Delta exhibit diverse development models due to the interplay of policies, location advantages, and economic capital, making them a focal point of academic discussion. This section examines two artistic clusters—Dafen Oil Painting Village in Shenzhen and Xiaozhou Village in Guangzhou—to explore how artists, as the primary agents of spatial intervention, have driven bottom-up urbanization. The analysis aims to shed light on the diverse forces and unique trajectories shaping China's rural urbanization.

9.4.4.1 Industry-Driven: Dafen Village for Oil Painting

The formation of Dafen Oil Painting Village originated from Hong Kong art dealer Huang Jiang. It is close to the second line of the Shenzhen Special Economic Zone, outside the customs area, which makes it convenient to return to Shenzhen and Hong Kong. In addition, cheap rent and low labor costs provide a unique geographical and social environment. The flexible international market information and large market demand in Shenzhen and Hong Kong provide an extremely advantageous market environment for the formation of Dafen Oil Painting Village. Therefore, after 1989, the development of the oil painting industry, similar to the processing and foreign trade model, caused a large number of painters to gather in Dafen Village, the rental housing economy to rise, and the rural form to begin to change. If it is said that before 1995 Hong Kong painters played a leading role in the distribution network of Dafen oil paintings, then after 1995, self-employed painters became the dominant group in Dafen Oil Painting Village. These entrepreneurs used the China Import and Export Fair as a stepping stone, and relied on the special geographical relationship between Guangzhou and Shenzhen to drive the rapid development of the Dafen oil painting industry. Since then, with strong government guidance, the embryonic oil painting industry has taken shape. The development of the industry has promoted population agglomeration, and coupled with the urban development and expansion of Buji Street, Dafen Village has transformed into an urban village. In 2001, after the completion of the first oil painting market in the village, Oil Painting First Street, although the art community was still dominated by painters and supplemented by painters, the class of original painters sprouted and developed rapidly. After the financial crisis in 2008, the oil painting industry market shifted to a "two-pronged" approach of originals and copies, and exports and domestic sales (Qian et al. 2006).

The living space of artists in Dafen Oil Painting Village is deeply interwoven with their production space and social networks. Artists construct highly interdependent networks that facilitate both collaboration and innovation, forming a "Dafen" production agglomeration model.

Collaboration occurs at multiple levels, including cooperation between oil painting producers and between oil painting workshops and auxiliary industries. This collaborative production model reduces costs, enhances flexibility in handling orders, and improves both efficiency and quality.

Innovation is driven by the openness of production knowledge and artistic techniques within the cluster. New techniques evolve and disseminate rapidly through the network. Additionally, innovations in subject matter and the integration of oil painting with other art forms continue to develop and refine within this collective environment (Cai and Tong 2014).

9.4.4.2 Culture-Driven: Xiaozhou Village in Guangzhou

Unlike Dafen Oil Painting Village, where industry drives the gathering of artists, Xiaozhou Village's transformation from an ordinary ancient village in the Lingnan water town to a famous art village was initially driven by the culture of individual grassroots artists. In addition to being attracted by Xiaozhou Village's beautiful natural environment, artists generally believe that China's rapid urbanization process has eroded the human relationships of traditional Chinese society, and that commercial thinking has dominated interpersonal interactions. This "lost" traditional human kindness can be found in traditional rural communities like Xiaozhou Village. Moreover, they agree that the rise of modernity in China and the high degree of social modernization have severed the relationship with traditional Chinese history and culture. Xiaozhou Village can realize their cultural consumption of history with its unique traditional rural architectural landscape (He et al. 2012). As a result, artists have moved into Xiaozhou Village, and by renovating and updating the community's old buildings and other physical environments to create work and living spaces that meet their needs, they have showcased and realized their own cultural characteristics to the outside world. The gathering of artists has created their own unique artistic and cultural atmosphere in Xiaozhou Village.

Subsequently, the low housing rents in Xiaozhou Village, combined with its unique artistic atmosphere, attracted a large number of art training institutions, art college students and art exam students to rent accommodation in the village. The students who moved in have become the dominant force and main participants in the village. Their rapid arrival has created a large demand for clothing, food, housing, transportation, etc., and the huge profits have become the fundamental reason for the community's large-scale physical renewal and spatial reconstruction. The rental of houses has brought a stable and relatively high rate of return, which has stimulated the investment enthusiasm of local villagers. Since the village stopped approving new homesteads as early as 1997, in order to have more houses for rent, many villagers began to demolish old houses and build new ones. This wave of demolishing old buildings and building new ones has resulted in the disappearance of a large number of old buildings in Xiaozhou Village. The unique traditional water town atmosphere that was once characteristic of Xiaozhou Village is dying out. Xiaozhou Village is gradually becoming less attractive to artists from an aesthetic point of view. Artists, as the original builders of Xiaozhou Village's artistic atmosphere, are being crowded out, divided and surrounded by student-like spaces. Recently, some urban elite artists and art dealers have successively moved into Xiaozhou Village, which has further accelerated the replacement process of the original grassroots artists. Xiaozhou Village is slowly becoming a so-called "art" space under consumerism, losing its original individualistic grassroots artistic flavor, which is quite saddening.

9.5 Chapter Summary

China is undergoing the world's largest and most complex urbanization process and socioeconomic transformation. While the urban economy is growing rapidly, residents' lifestyles have also undergone unprecedented changes, leading to a drastic restructuring of urban living spaces as the carrier of residents' lives. Residents' living spaces are an important part of the urban system and play an important role in urban development and residents' daily lives.

So how should we understand living space? With regard to the definition of living space, previous studies have tended to understand living space in two ways. The first is to consider living space as the spatial manifestation of the concept of "dwelling", i.e. living space as dwelling, and living space as the human living environment, a "static" dwelling space. For example, Zhang (2007) believes that living space is the space occupied by people's daily lives. People's daily lives involve constant movement between home, work, consumption places and non-consumption public places. Therefore, living space is the place and space where various activities of daily life take place. Wang and Wang (2011) believe that living space refers to the places and spaces where people carry out various activities to meet different needs in life. The second group, which tends to understand living space as the space of "activity" concepts, i.e. living activity space, believes that it is the space occupied by human activities and tends to understand living space as the "dynamic" space of daily activities. For example, Feng (2010) believes that living space is the spatial projection of urban residents' activities; Wang (2011) believes that living space is a concrete and real experiential space of daily life, and is the sum of places that accommodate various daily activities that occur or take place; Zhang and Chai (2013) use the term "living activity space to represent living space, and believe that living activity space includes not only the living space formed by long-term behaviours such as moving, but also the space for daily activities formed by work, housework, shopping, leisure, socialising and moving between them. Zeng (2015) suggested that the definition of living space can be expanded to include the space that accommodates both the concept of 'residence' and the concept of 'activity'.

As cities serve as essential living spaces for residents, the deepening process of urbanization has heightened public concern regarding the quality of urban living spaces and the need for their spatial optimization. Consequently, urban living spaces have become a key research focus in human geography, particularly regarding the evaluation of urban spatial quality, planning strategies, and the underlying factors that shape urban living conditions. Despite the growing body of research in this area, significant gaps remain. Chai and Gong (2001) pointed out that while studies on living spaces across various temporal and spatial contexts help in understanding general patterns of human activities, they often overlook the individual variability in daily life practices. Recent studies have begun addressing specific social groups and locations, focusing on the elderly, people with disabilities, low-income populations, women, and urban public spaces such as squares, parks, and shopping centers. In contrast, international research on urban living spaces has long adopted a more socialized perspective, emphasizing issues of spatial equity, residential segregation, and social exclusion of marginalized groups. These studies prioritize the living conditions of disadvantaged populations and the socio-spatial dynamics within large cities, reflecting an increasing emphasis on the social determinants of urban quality of life. This highlights two critical research imperatives: first, a deeper exploration of the living spaces of diverse social groups in specific urban contexts to enhance our understanding of urban complexity and spatial differentiation; second, an increased focus on the social dimensions of urban space, integrating historical, social, and cultural perspectives to interpret urban spaces not merely as physical entities but as historically and socially constructed phenomena (Lü 2008).

In this context, this chapter focuses on the more representative and diverse groups within Chinese cities, including foreigners living in China, returned overseas Chinese, migrant populations, graffiti artists, street vendors, artists and other diverse groups (these diverse groups have begun to attract attention in the context of the transformation of urban space, but there is still very little research on them). It sorts out and explores the current living conditions of these groups and the diverse urban living space landscape formed around their lives, revealing the historical, social, spatial and cultural nature of the city, as well as the intensification of urban spatial differentiation and the accumulation of conflicts among social groups. For example, the spatial formation of the international social landscape represented by ethnic enclaves is the result of the combined effects of factors such as politics, the economy, society and personal choice, but it also reveals the segregation of foreigners in China and local residents in terms of living space. The changes in the living space of the returnee diaspora in China although it has helped the returned overseas Chinese integrate into mainstream society, it has also caused the "overseas Chinese nature" of the returned overseas Chinese themselves to gradually disappear; the living space where groups such as the foreign population live together reveals that their lifestyle is still "production-oriented and life-neglecting"; and the living spaces of niche groups are all unique, with legal, formal spaces such as bar spaces that have become slow-culture spaces with local consumer characteristics in a multicultural context, and art spaces that have become a balance for artists between maintaining a realistic livelihood and adhering to artistic ideals.

As can be seen from the discussion in this chapter, urban space is the embodiment and result of the spatial production of different social groups. This space is also a product of history, society and culture. With the structural differentiation of urban subjects, the urban spatial structure has shown a tendency towards fragmentation. The living spaces of different social groups are very different, and their combination has resulted in a collage-style mosaic of urban living spaces. The marginalized and disadvantaged groups in the city often find themselves in "islands" in the urban space, and are subject to unequal and unfair treatment by society. Because of this, the existence of a fragmented urban living space in the form of a mosaic not only calls for us to understand the diversity and complexity of urban space from different perspectives, and to further explore the mechanisms behind some of the social problems currently faced by Chinese cities, but also calls for us to pay more attention to the development of different social groups and to social equality, progress and diversity. To this end, we need to consider what kind of urban space should be created for the future development of Chinese cities. This also requires us to consider how to promote social justice, social equity, and diverse social value in urban planning through the allocation of spatial resources and the rational layout of space, so as to promote fairness in urban social space and the social integration of marginalized and vulnerable groups.

References

Bromley RDF, Nelson AL (2002) Alcohol-related crime and disorder across urban space and time: evidence from a British city. Geoforum 33(2):239–254
Cai Y, Tong X (2014) Cultural Industry upgrading in global value chains: a case study of Dafen village. Hum Geogr 29(3):115–120 (in Chinese)
Cao L (2010) Investigation and Research on Bar Music and Its Cultural Context in Guangzhou. Music Art 4:44–53 (In Chinese)
Chai Y, Gong H (2001) Time-geographic research on urban society. J Peking Univ (Philos Soc Sci Ed) 38(5):17–24 (in Chinese)
Feng J (2010) Spatial Perspectives on Urban Society. China Architecture & Building Press, Beijing (in Chinese)
Guo Y (2006) Spatial aggregation and restructuring of migrant populations in metropolitan fringes: a case study of Minhang district, Shanghai. AR Res Dev 5:34–38 (in Chinese)
He S, Qian J, Xu Y et al (2012) Spatio-temporal evolution of rural gentrification under rapid urbanization. Acta Geogr Sin 67(8):1044–1056 (in Chinese)
Huang X (2005) History and current status of returned overseas Chinese. Hong Kong Social Sciences Press Ltd, Hong Kong (in Chinese)
Kraack A, Kenway J (2002) Place, time and stigmatized youth identities: bad boys in paradise. J Rural Aff 18(2):145–155
Li Z, Xue D, Michael L et al (2008) Social spatial analysis of African enclaves in Xiaobei road, Guangzhou. Acta Geogr Sin 63(2):207–218 (in Chinese)
Li J, Zhang W, Sun T et al (2014) Agglomeration characteristics and economic performance of urban clusters in China. Acta Geogr Sin 69(4):474–484 (In Chinese)
Li Z, Du F (2012) Production of new urban social spaces under "transnational commercialism": an empirical study of the African economic zone in Guangzhou. City Plan Rev 36(8):25–31 (in Chinese)
Lin G, Wang L (2011) Locality and spatiality of bars under globalization: a case study of Guangzhou. Scientia Geographica Sinica 31(7):794–801 (In Chinese)
Liu M (2000) Study on Shipai floating population settlement: a comparison with Beijing "Zhejiang Village". Mark & Demogr Anal (5):42–47. (in Chinese)
Liu Y, Wang F (2011) Space production and black cluster in urban-rural fringe: case study of M garbage pig farm in Guangzhou. Scientia Geographica Sinica 31(5):563–569 (in Chinese)
Lü L (2008) Urban spatial turn" and new urban geography. World RegNal Stud 17(1):32–38 (in Chinese)
Ma J, Huang Z, An M et al (2012) Impact of agricultural population mobility on rural development under urbanization: a case study of the Pearl River Delta. J Anhui Agric Sci 40(2):1089–1091 (in Chinese)
Ma LJC, Xiang B (1998) Native place, migration and the emergence of peasant enclaves in Beijing. China Q 155:546–581
Mack P (2006) A social and cultural history alcohol. Oxford University Press, London

Population Census Office of the State Council (2012) Tabulation on the 2010 population census of China. China Statistics Press, Beijing (in Chinese)
Qian Q, Chen Y (2004) Preliminary study on floating population settlements in Chinese metropolises: cases of Beijing "Zhejiang Village" and Guangzhou Shipai area. City Plan Rev 27(11):60–64 (in Chinese)
Qian Z, Yan X, Wang A (2006) Urban cultural industry agglomeration: Shenzhen Dafen Oil Painting Village. Trop Geogr 26(3):269–274 (in Chinese)
Shanghai Municipal Bureau of Statistics (2012) Shanghai 2010 population census data. China Statistics Press, Beijing (in Chinese)
Song Y, Wu W (1997) Spatial agglomeration characteristics, formation mechanism and regulation strategies of migrant population in Beijing. Econ Geogr 17(4):71–75 (in Chinese)
Sun Y (2009) Field survey of marriage circles in Guangdong Huadu Overseas Chinese Farm. Ba Gui Qiao Kan 1:74–77 (in Chinese)
Tao D, Tong X, Feng K (2014) Grey space production in urban-rural fringe through "scrap material village" perspective. Urban Plan Int 29(5):8–14 (in Chinese)
Wang H, Liu S, Sun L (1997) "Zhejiang Village": a unique way for Chinese peasants to enter cities. Sociol Res 1:58–69 (in Chinese)
Wang K (2011) Review on urban living space research. Prog Geogr 30(6):691–698 (in Chinese)
Wang L, Wang X (2011) Deconstruction and quality reconstruction of urban community living space structure. Scientia Geographica Sinica 31(1):22–28 (in Chinese)
Wu Q, Huang S, Cui W (2013) Development trends of urban consumption space from retail format evolution perspective. Mod Urban Res 5:114–120 (in Chinese)
Xu P (2009) Social exclusion theory and housing issues of migrant workers. J High Corresp Educ (Philos Soc Sci) 12:34–36 (in Chinese)
Xue D, Huang G (2008) Regulation beyond formal governance: Spatial agglomeration and survival of informal sectors in urban villages: Case study of Xiadu Village, Guanzhou. Geogr Res 27(6):1390–1398 (in Chinese)
Yao H, Xu X, Xue D et al (2008) Research progress on floating population in China. Urban Problems 6:69–76 (in Chinese)
Yao J (2009) From refugees to citizens: identity transition of Vietnamese returnees in Huadu overseas Chinese farm. Sun Yat-sen University, Guangzhou (in Chinese)
You Y (2010) Analysis of returned overseas Chinese resettlement policies (1949–1979) based on Guangdong archives. Bagui Qiaokan 4:36–40 (in Chinese)
Zeng W (2015) Urban residents' living spaces during transitional period: a case study of Nanjing. Nanjing Normal University, Nanjing (in Chinese)
Zhang F (2007) Living conditions of migrant farmers in Beijing's semi-urbanized areas. Institute of Geographic Sciences and Natural Resources Research, Chinese Academy of Sciences, Beijing (in Chinese)
Zhang J (2013) Identity confusion of returned overseas Chinese in farms and government resettlement policies. J Huaqiao Univ (Philos Soc Sci) 1:31–36 (in Chinese)
Zhang Y, Chai Y (2013) Suburbanization of living activity space. Prog Geogr 32(12):1723–1731 (in Chinese)
Zhou W, Liu Y (2015) Formation characteristics of ethnic economy in Japanese enclaves: a case study of Shanghai's Gubei area. Geogr Res 34(11):1–16 (in Chinese)
Zhou W, Liu Y, Quan Z (2016) Formation and evolution of ethnic enclaves of Koreans in China under globalization: a case study of Beijing's Wangjing. Acta Geogr Sin 71(4):651–667 (in Chinese)

Chapter 10
"New" Living Space

10.1 Network Technology, the Information Society, and the Development of Virtual Living Spaces

10.1.1 Virtual Living Space in the Context of Information and Communication Technology

10.1.1.1 The Deepening of E-commerce Services

At the inception of the e-commerce platform, its biggest advantage over the physical shopping space was its low cost. The advantages of low costs are primarily evident in three dimensions: temporal costs, as users are able to efficiently search for products; spatial costs, as consumers are exempt from transportation expenses and can make purchases from the comfort of their homes; and economic costs, as e-commerce platforms minimize overheads such as rental expenses for physical stores, leading to comparatively lower prices for products. However, as the variety of products and the number of users continue to grow, the virtual shopping space supported by e-commerce platforms is further evolving. First, with the integration of big data technology, virtual shopping spaces are no longer homogeneous and uniform; instead, they are "tailored to each individual," providing personalized recommendations for every consumer. Secondly, after the explosive phase from 2008 to 2012, e-commerce platformization and platform departmentalization have become the trends; after 2016, omni-channel retailing, personalization, and convenience emerged as new characteristics of virtual shopping spaces (Wang B 2017). Third, the virtual shopping space is also no longer simply providing physical goods, but has shifted to a full range of consumer offerings such as services and experiences. For example, product categories shift from traditional items like clothing, beauty products, and books to virtual services such as video memberships and mobile top-ups, and even to services such

J. Feng et al., *Urban Living Space in China*, Urban Sustainability,
https://doi.org/10.1007/978-981-95-0595-1_10

as housekeeping and emotional counseling. The categories available in virtual shopping spaces are becoming broader and deeper into the physical shopping space and even the living space of residents. From this perspective, virtual shopping spaces, in terms of breadth, are expanding their boundaries, enhancing the range of categories and the quality of services; in terms of depth, they are rooted in real life, constructing a comprehensive and integrated "shopping space."

10.1.1.2 The Balanced Restructuring of Virtual Shopping Spaces

In the early development of e-commerce platforms, the level of information infrastructure had a significant impact, leading to the virtual shopping space in first-tier and new first-tier cities far surpassing that of cities at other levels. However, as a developing country, China has a large population concentrated in third- and fourth-tier cities, which means there is a vast market. Some emerging e-commerce platforms focus more on lower-tier markets, creating a shopping space at the bottom level with low prices and simple "bargaining" methods. Early e-commerce platforms entered the market with a "three lows" strategy (low cost, low price, low profit margin) and gradually evolved into platforms with "high cost, high price, high profit margin." However, they are also facing the impact of a new batch of "three lows" enterprises (Jiang and Ren 2017). The various levels of virtual shopping spaces in China are gradually being built, improved, and iterated based on this foundation.

At the same time, from another perspective, virtual shopping spaces are gradually expanding their spatial components. Online platforms, offline entities, and warehousing logistics have all become important components of the virtual shopping space (Du and Jiang 2017); consumer finance, supply chain finance, and payment services have become crucial factors supporting the operation of the virtual shopping space (Wang B 2017); at the same time, traffic distribution, content recommendation, and e-commerce socialization are also important factors in maintaining the vitality of the virtual shopping space. From this perspective, virtual shopping spaces are gradually forming a full-link space of "traffic - shopping - after-sales."

10.1.1.3 The Interaction Between Virtual Shopping Spaces and Physical Shopping Spaces

Virtual shopping spaces are always interacting in complex ways with physical shopping spaces, exhibiting different interaction patterns in various scenarios. For example, Zhang et al. (2016) summarized that the role of online shopping versus traditional shopping mainly include four types: substitution, complementarity, neutrality, and correction. Additionally, the development of online shopping has led to two hypotheses: the innovation diffusion hypothesis and the efficiency hypothesis. In terms of shopping spaces in Chinese cities, the significant expansion of early virtual shopping spaces had a noticeable substitution effect on traditional physical shopping spaces. However, with the development of models such as "new retail" and "O2O" (online-to-offline) in recent years, virtual shopping spaces have instead promoted the development of physical shopping spaces (such as "Tmall Small Stores"). At the same time, physical shopping spaces demonstrate strong irreplaceability at the community scale (addressing the last mile issue), while virtual shopping spaces are also attempting to solve the convenience of community shopping through devices that combine "virtual and reality," such as unmanned vending machines. At the same time, there is a trend of mutual promotion and joint development between virtual shopping spaces and physical shopping spaces. Under the influence of virtual shopping spaces, physical shopping pays more attention to the improvement of service quality; many emerging models, such as "online booking - in-store pickup" for physical stores and the establishment of physical stores to enhance the image of Taobao shops, reflect the interaction between virtual shopping spaces and physical shopping spaces, based on the classic four effects, showing more complex, mixed, and interactive patterns, and continuously shaping the spatial morphologies of both.

10.1.2 Virtual Living Service Space in the Context of Information and Communication Technology

The development of information and communication technology, especially the advancement of mobile payment, has created a virtual living service space (Lin et al. 2018), establishing a bridge between government management and residents' lives. In July 2015, the "Guiding Opinions of the State Council on Actively Promoting the 'Internet+' Action" was released, which mentioned "fully leveraging the efficient and convenient advantages of the internet to improve resource utilization efficiency and reduce service consumption costs. Vigorously develop emerging consumption using the Internet as a carrier and interacting online and offline. Accelerate the development of emerging services based on the internet, such as healthcare, health, elderly care, education, tourism, and social security. Innovate government service models and enhance the government's scientific decision-making capabilities and management levels." With the gradual implementation of the "Internet +" initiative, cell phone recharge, life payment, ETC charging, hospital electronic registration system, etc. have collectively created a convenient, efficient and transparent virtual living service space. The significance of virtual living service spaces lies not only in efficiency and cost reduction, but more importantly, in enhancing the management level of relevant departments and improving the management service system.

10.1.2.1 Virtual Medical Service Space

In recent years, the development of virtual medical services in China has mainly come from two aspects: online medical services and hospital electronic appointment systems. On one hand, online medical services essentially provide online healthcare services, offering efficient and convenient answers for common diseases faced by patients. As relevant departments further regulate and rectify virtual medical services such as online medical services, the scope of online medical services has been clearly defined, and the relevant behaviors have been further standardized. On the other hand, the hospital electronic appointment system reduces waiting time, limits the operations of scalpers, and improves consultation efficiency through the way of "online booking-mobile payment-direct waiting for consultation". For example, the online appointment system launched by Alipay in collaboration with certain hospitals allows patients to directly visit the outpatient department after making an online payment, significantly reducing the time patients spend seeking medical care (Ma et al. 2019). Compared to traditional medical service spaces, virtual medical service spaces reduce the waiting time before and after "registration" and "number queuing," making the spatial path more streamlined and the utilization more efficient. An important development in virtual medical services in recent years is telemedicine service. With the support of communication technology and related devices and software, it is possible to achieve remote connections between local hospitals and renowned hospitals, remote consultations between attending specialists and patients, as well as collective diagnoses by experts from different hospitals in conjunction with patients remotely. It is even possible to conduct remote surgeries using relevant specialized equipment (Ma et al. 2014). Telemedicine is an emerging medical enterprise encouraged by the country, which is of great significance in alleviating the tightness of China's high-quality medical resources, reducing the economic burden on patients, and realizing timely treatment for patients.

10.1.2.2 Virtual Public Service Business Hall

The virtual public service business hall undertakes various functions such as service processing, service payment, and customer feedback. From certain perspectives, its functionality is already close to that of a physical public service business hall. This type of virtual public service business hall includes mobile service halls and utility payment services (such as water and electricity bills), etc. This type of virtual public service business hall includes mobile business halls and utility payment services (such as water and electricity bills). For example, the virtual electricity business hall completely virtualizes the user-side utility payment process through the way of "mobile business halls-fully automated meter reading-mobile payment", significantly reducing time and space costs (Liu 2015). Similarly, the virtual mobile business hall establishes a complete virtual telecommunications service space through the way of "virtual user accounts-online service processing-online service feedback", eliminating queue time costs and improving service processing efficiency.

10.1.2.3 Virtual Transportation Service Space

The virtual transportation service space in the current Chinese cities mainly contains two forms: smart travel service based on smart city and ETC charging system based on traffic management, which together constitute the current virtual transportation service space in China. Based on smart phones and the Internet of Things, smart travel services have constructed a virtual transportation space primarily characterized by the flow of traffic information, allowing for real-time interaction with users' travel (Chai et al. 2014). At the same time, users' travel information and navigation data are fed back into the traffic database, further promoting the evolution of the virtual transportation service space. From this perspective, the virtual transportation service space constructed by smart travel services focuses on the flow of traffic information, enhancing spatial efficiency and utilization levels (Zhai 2015). The ETC toll collection system, which was vigorously promoted in 2019, has not only reduced costs but also gradually improved departmental management levels and traffic safety conditions. From the perspective of virtual space, the ETC system effectively constructs a virtual service space that integrates local toll collection, user payments, mobile payments, and traffic safety. On one hand, ETC significantly reduces travel time costs; on the other hand, with the promotion and in-depth development of ETC, the input of highway toll information and vehicle traffic information will greatly enhance the quality of public services and management levels of relevant government departments (Qian 2013). From this perspective, the virtual living service space is essentially based on mobile payment technology, connecting service providers and users. By building service systems such as mobile business halls, it creates a virtual living service space that encompasses user payment, service usage, and departmental management. This not only significantly reduces costs and improves efficiency but also enhances the quality of public services, promotes the reform of the public service system, and drives the development of the physical public service system. In summary, the significance of virtual living service spaces lies not only in reducing residents' time and financial costs, but more importantly, in the fact that these spaces are inherently more transparent, efficient, and streamlined. They can effectively lower management costs and enhance management efficiency. Virtual living service spaces are bound to play an increasingly important role in residents' living spaces.

10.1.3 Virtual Social Space in the Context of the Information Society

Since the rise of information and communication technology, virtual social spaces have always been a focal point of interest across various sectors of society. Overall, virtual social spaces are characterized by decentralization, dynamism, fragmentation, dispersal, immediacy, and cyberization (Tao 2014); virtual social spaces can generally be divided into two types: strong ties dominated socialization and weak ties dominated socialization (Chang 2014). The academic community has a relatively clear understanding of the basic characteristics, types, and patterns of virtual social spaces. However, in recent years, with the transformation of China's social and economic structure and the deepening development of the internet, new forms of virtual social spaces have emerged that are worth noting.

10.1.3.1 Otaku Culture

The concept of otaku culture actually originated in Japan, referring to "a collective term for a group of people who are immersed in subcultures and deeply connected to manga, anime, video games, personal computers, science fiction, feature films, and handmade models" (Azuma 2001); However, in the context of China, with the development of the internet, especially represented by live streaming social media like Bilibili, otaku culture has gained more spatial attributes. It is often associated with "spending long periods indoors," "rarely socializing or exercising," and "indulging in the internet and virtual worlds" (He 2018). From this perspective, in the context of urban living spaces in China, otaku culture actually reflects the phenomenon of extreme expansion of virtual social interactions and extreme shrinkage of physical social spaces. On this point, many scholars have explained accordingly. For example, Jiang (2014) proposed the mechanism of "the more WeChat, the more anxiety, the more indifference," that is, the virtual social space gradually squeezes the physical social space. Moreover, the virtual social space has a self-reinforcing characteristic, where the path of "dependence - anxiety - greater dependence" causes many users to become trapped in the virtual social space, unable to extricate themselves, and sinking deeper and deeper (Du 2018). The virtual social space is a realm that is vastly different from the real world for many users. By displaying dynamics, users can realize the function of shaping their own image, which fully satisfies the need for "self-presentation" as Goffman puts it (Goffman 1989), and this kind of social pleasure is an important reason why the virtual social space can continue to expand and proliferate itself (Nong 2014).

In addition, the virtual social space represented by otaku culture forms a tight closed loop with other virtual spaces, further excluding the entry of physical social spaces. For example, the products offered by virtual shopping spaces, the public services provided by virtual living service spaces, and the online-offline services such as takeout provided by O2O platforms. The virtual social space for individual users can be almost infinitely extended and proliferated at a single point in geographic coordinates, isolated from but closely related to the physical world.

10.1.3.2 Subculture and "Information Cocoon"

The concept of subculture emerged in the mid-twentieth century, originating from the Chicago School. From the perspective of cultural studies, subculture is a subordinate culture that relies on style and alternative symbols to challenge mainstream culture, thereby establishing identity recognition. Subculture has three main characteristics. First, subculture has a "resistant" feature, meaning it resists mainstream culture. Second, subculture has a "stylized" feature, which means it has its own distinct style. Third, subculture has a "marginal" feature, indicating that it exists on the fringes of the social and cultural hierarchy (Zhang 2014). The emergence of the internet era has further accelerated the development of subcultures. Through various media resources, technological means, and rapid dissemination methods, a group of "interest groups" represented by subcultures has formed (Chen 2016). From this perspective, the virtual social space behind subcultures is actually somewhat distanced from the mainstream virtual social space, exhibiting a certain rebelliousness, using specific symbolic and ritualistic stylistic symbols, and forming relatively clear boundaries in virtual space (Huang 2011).

Furthermore, in recent years, the emergence of big data technology and intelligent recommendation algorithms has accelerated the expansion of such subcultural virtual spaces. Human acceptance of information is guided by interest, which gradually confines and closes individuals within virtual social spaces. Personalized recommendation algorithms exacerbate this situation—this is what Cass Sunstein referred to in 2008 as the "information cocoon." In the internet, public attention is limited, often only able to see the content they want to see and hear the voices they want to hear, ultimately leading to a self-imposed isolation and a narrowing of information. The emergence of the "information cocoon" is a phenomenon where the sub-social spaces formed by the fragmentation of virtual social spaces gradually distance themselves from the public and reinforce their own boundaries. In other words, the limited nature of public attention and intelligent recommendation algorithms have led to the formation of echo chambers in virtual social spaces, each with clear boundaries and "distances." The emergence of these echo chambers has made the virtual social space even more insular, resulting in a more polarized and uncontrollable development of secondary virtual social spaces (Liu 2017).

10.1.3.3 Internet Traffic and Hotspots

Similar to physical spaces, there are also many central and hotspot areas in virtual social spaces. Although virtual social spaces are decentralized and dispersed, they are still quite uneven. For example, based on agenda-setting theory, the more emphasis the media places on an issue, the higher the public's level of concern about that issue; and this agenda is often determined jointly by the media and the audience (McCombs and Shaw 1972). From this perspective, there will be imbalanced hotspot areas in virtual social spaces, and these hotspots are influenced by external forces (capital and power) and the collective participation of the public, rather than being a completely public, decentralized virtual space. Furthermore, as social media continues to evolve, information is no longer a scarce resource; instead, users' attention has become the scarce resource. These centralized hotspot areas effectively compensate for the shortcomings of people's brief attention spans and short memory times in the fragmented era (Zhang 2018).

In recent years, a typical representative of the hotspots in virtual social spaces is the "internet celebrity." Internet celebrities build a scenario-based living space through social media, creating concrete ideal life scenarios that establish emotional connections with their audience. By resonating emotionally, they gain the favor and attention of their fans (Guo 2019). In other words, the hotspot areas in virtual social spaces are no longer "instantaneous," but rather "sustained" and sticky. In the current virtual social space, "attention" has become the most important resource. Areas where "attention" is concentrated are the hotspots of the virtual social space, holding significant discourse power and having a substantial influence on the direction of public opinion in the virtual social space (Yuan 2010).

10.1.3.4 Information and Communication Technology and the Characteristic of Evolution of Virtual Living Spaces

The first is fragmentation. Under the influence of information and communication technology, behaviors in virtual living spaces often exhibit characteristics of temporal and spatial fragmentation (Zhao et al. 2016). With the support of mobile internet and smartphones, virtual living spaces are often discontinuous in terms of time scale, yet the level of participation in residents' lives is very deep. Moreover, this fragmentation contains two levels of meaning: the fragmentation of virtual-reality behavior and the fragmentation of virtual-virtual behavior. Actions in the virtual living space can be interrupted at any time by events occurring in the physical space, while actions in the virtual living space can also be interrupted by other virtual activities, such as suddenly needing to reply to QQ or WeChat messages while online shopping. The Spatio-temporal fragmentation across multiple scenarios and scales is a key characteristic of the virtual living space.

The second is blurred boundaries. This ambiguity is mainly reflected in two aspects: the blurred boundaries between virtual space and physical space, and the blurred boundaries between different types of virtual spaces. Online-offline products like O2O make the boundary between the virtual and the physical increasingly indistinct, while transformations represented by new retail further blur the boundaries of virtual spaces such as shopping, socializing, and entertainment (Wang B 2017).

The third is technology dependency. The development of virtual living spaces has a strong reliance on technological advancement. From the perspective of technological products, websites like Taobao and Tmall are important supports for virtual shopping spaces, while social products like WeChat are essential components of virtual social spaces. From a specific technological standpoint, mobile payment technology has made the construction of virtual shopping spaces possible; 4G technology has directly given rise to virtual social spaces represented by live streaming. The upcoming 5G also holds great promise for the development of virtual living spaces. Therefore, the development of virtual living spaces has a strong dependence on the progress of information and communication technology.

10.2 Interaction Between Paravirtual Communities and Residents' Living Space

A paravirtual community is a type of community space that exists between real communities and virtual communities, based on a certain geographical premise. Its characteristic is that residents of real communities engage in interpersonal interactions in virtual spaces through real-name or semi-anonymous methods. This interaction combines the trustworthiness of real communities with the anonymity of virtual communities, as well as the convenience and cross-temporal characteristics that arise from the physical absence (Li 2019). A paravirtual community is essentially an online geographical community. For individuals, it is not a completely new space, but rather a partial extension of real-life communities into cyberspace. There is a fairly close interaction between the "real" and the "virtual" in paravirtual communities. The "real" serves as the foundation for the "virtual," while the "virtual" also profoundly influences the "real." Common examples of paravirtual communities include community websites, BBS (bulletin board systems), real-name QQ groups, and WeChat groups, which often have functions such as information sharing, social interaction, community self-governance, and community services. Paravirtual communities play an important role in strengthening community belonging, enhancing community identity, promoting community integration, and improving community development through effective interaction with residents' real-life spaces.

10.2.1 Information Resource Sharing and Meeting Residents' Daily Life Needs

Information searching is often the initial motivation for people to use online platforms. Paravirtual communities rely on the efficient dissemination capabilities of the internet and low communication costs to become information distribution platforms for community residents, thereby fulfilling their essential information and news service functions. Their uniqueness lies in the fact that they are based on real communities, consisting of residents who share similar discourse systems and common concerns (Li 2019). Therefore, the shared information resources in paravirtual communities are often more effective and credible, providing more direct benefits to residents' lives. Moreover, the seeking and obtaining of external information is an important channel for individual socialization. Therefore, participation and interaction in paravirtual communities can significantly influence an individual's sense of belonging and integration in real communities (Cui 2017). The role of community residents gradually transforms from information receivers to information contributors. Through the process of sharing and mutual assistance, residents establish a common foundation of interests, and the bond between residents and the community continues to strengthen.

The Huilongguan cultural and residential area in Beijing mentioned earlier has a planned residential population of about 300,000, making it one of the largest relatively independent residential communities in the country and even in Asia (Zhang 2002). Among this population, the proportion of migrants is nearly 80% (Feng et al. 2017). The Huilongguan Community Network is a typical paravirtual community, where the "community forum" serves as a free communication area covering topics such as shopping, emotions, workplace issues, home decoration, and real estate. Registered community residents can share information, exchange experiences, and help each other. The "information center" acts as the community's electronic bulletin board, primarily publishing and recording recent news, work, and activities within the community. The "trading market" contains information related to second-hand buying and selling, job recruitment, and rental listings (Wang 2010). The community network showcases a wide variety of information and resources, significantly reducing the cost of information search for residents while meeting their diverse daily needs and greatly enhancing the convenience of life. This is especially true for "new immigrants," for whom the community serves as a starting point for settling in a new city. The exchange of information and mutual assistance among community residents helps them quickly integrate into the community and fosters a sense of identity and belonging.

Piz Chun is an urban village located outside the Fifth Ring Road in Chaoyang District, Beijing. It has a local population of over 2000 and more than 40,000 migrant workers. The "Workers' Home" in Pi Chun is a grassroots social organization, officially known as the Beijing Workers' Home Cultural Development Center. Its main service target is the migrant worker population in Pi Chun, and it has developed into a new important platform and key player in cultural exchange and community building for workers nationwide.

During the research process, it was found that the construction of a paravirtual community is of great significance for the self-organization and self-service of the "Workers' Home" in Pi Chun, with WeChat group chats being the main method. The members of the union WeChat group are almost all workers from Pi Chun, and the content of communication within the group is mixed, mainly consisting of job recruitment information, event announcements, and casual chatting. Compared to one-on-one private chats, WeChat group chats have the characteristics of low privacy and openness, meaning that even people who are not familiar with each other in real life can chat in the group without it feeling awkward. This is crucial for migrant workers who lack familiarity and trust with one another. Especially in WeChat groups, short-term job advertisements are diverse and come from a wide range of sources, often being forwarded and recommended by fellow workers. Some even receive confirmation from opinion leaders among the workers, such as "the information is genuine, and there are no wage arrears." Since being scammed and having wages withheld are the most common issues workers face in the city, the information shared within worker groups is considered more trustworthy than other unfamiliar online channels, making it better suited to meet the needs of the workers. Overall, migrant workers, as marginalized and vulnerable groups in the city, find it difficult to access public resources such as education, healthcare, and entertainment due to the barriers of household registration and the pressures of heavy work. In unfamiliar geographical spaces and fragmented social relationships, they struggle to seek support and help. However, paravirtual communities, such as WeChat groups established among migrant workers, can alleviate feelings of loneliness and powerlessness by emphasizing collective organization. Interpersonal interactions in paravirtual communities help enhance the connection between the population and the community. At the same time, through the sharing of space, resources, information, and emotions, the lack of trust and high levels of vigilance among migrant workers will also improve, ultimately promoting community integration and cohesion.

It is worth noting that at the beginning of 2020, during the pandemic prevention and control efforts for COVID-19, communities, as the basic units of urban governance, served as an important fortress for preventing external input and internal spread. Channeling pandemic prevention and control resources and efforts down to the community level became one of the most critical measures in our country's response. In this special public health event, to block the spread of the virus, the movement of people has been greatly restricted, which naturally has brought difficulties and troubles to people's daily lives. For example, during the lockdown in Wuhan, communities implemented 24-h closed management. To reduce contact between individuals, epidemic prevention work and residents' daily needs largely relied on online paravirtual communities, with WeChat groups playing the most significant role. Community workers promote epidemic prevention knowledge, collect demands for food and daily necessities, and arrange for home delivery in WeChat groups. Residents can learn about the epidemic situation around them, offer mutual comfort and assistance, and build a sense of solidarity through group chats. These demonstrate the significant role that paravirtual communities can play in ensuring the normal operation and harmonious stability of communities during a crisis. When physical community spaces are "closed," residents can still obtain information and support in paravirtual community spaces and meet their basic daily needs.

10.2.2 Expansion of Interaction Space and Restructuring of Residents' Social Relation Network

The community is the basic spatial unit of urban society. A community without neighborly relationships and social interactions is a material shell devoid of social content (Feng et al. 2017). During the planned economy period, the organizational form of Danwei had multiple functions, including production, living, and entertainment. The community met residents' needs for social interaction, emotional belonging, and resource acquisition, thus forming a strong sense of community connection (Li 2019; Zhang and Yi 2019). However, the interaction dilemma and emotional crisis of urban community residents have become a prominent and common problem nowadays, characterized by social estrangement and indifference among neighbors (Hu 2015). The reason for this change is not that modern communities lack spaces and places for residents to engage in social activities compared to traditional communities; on the contrary, modern communities have more specialized activity spaces and facilities, such as squares, green areas, and fitness equipment. The fundamental reason lies in the fact that with the social transformation and deepening marketization in our country, commodified housing has dismantled the network of familiar relationships built on kinship and community ties in traditional neighborhoods, leaving only geographical relationships intact, which has promoted the heterogeneity and looseness of community populations. In addition, the independent housing model emphasizes privacy and security, which has intensified the estrangement among neighbors in the community (Feng et al. 2017). The decline of neighborly communication and community interactions in physical spaces not only weakens the social relationship networks among residents but also disrupts the strong attachment between individuals and the community, leading to serious issues of individual "atomization" and community "hollowing out" (Gan 2019; Li 2019). As a result, the community exists only in form and lacks substance. Apart from meeting residents' housing and living facility needs, it cannot provide any emotional support, resources, or assistance. The enthusiasm for community development is lacking, and it has even harmed the quality of life for residents to some extent.

However, based on internet technology, paravirtual communities formed by local residents, due to their diversity and convenience, can help residents break through traditional neighborhood communication methods and establish new social networks in virtual spaces. This greatly expands the interaction space of the community and increases the vibrancy of community interactions. At the same time, the communication in paravirtual communities promotes people's gathering offline, for example, many activities organized online rely on physical spaces to be realized. This gathering inevitably strengthens the bonds between people (Sessions 2010). Therefore, the new social connections built in paravirtual communities will ultimately manifest in real life and promote the restructuring of social relationship networks among residents in real communities.

For example, in the Huilongguan community network in Beijing, residents have spontaneously established various community groups based on different interests and hobbies, such as sports clubs, public welfare organizations, and community volunteer medical service organizations. They also actively organize community activities online and promote them to attract more participation (Hu 2015; Yin 2016). To some extent, the social lives of community residents are closely embedded in the community network, and the significance of this semi-virtual space even surpasses that of the physical space (Zheng 2010). Residents can always achieve precise connections with corresponding community groups in different sections of the community network based on their interests, hobbies, and needs, which is difficult to accomplish efficiently in physical space. Through online communication, residents can meet new friends and establish new social connections. Additionally, since many activities still rely on physical spaces, such as sports, the social interactions in paravirtual communities can promote the full utilization of community public service facilities and enhance the vitality of public spaces in the community. Even when the existing infrastructure and public activity spaces in the community cannot meet the demand, the frequent community activities initiated by the community network can encourage the establishment of new activity spaces, further improving community development and enriching residents' lives (Yin 2016). In addition, through the development of offline activities, the new social connections established by residents in virtual spaces will be tested in real spaces, some of which will be strengthened and become an important part of the residents' community relationship network. Research shows that some internet users in community networks significantly know more people in actual community life, and these relationships were initially established through the community network (Zheng 2010). It is evident that social interactions in paravirtual communities have become an important aspect of the way Chinese urban community residents interact at the moment, and even a prerequisite for conducting social activities in physical spaces.

It is worth noting that, in addition to the community network that covers the entire community, there are also other paravirtual communities that functionally refine according to different neighborhood scales. For example, in the Peking University Xiaojiahe residential area, there are WeChat groups for residents based on different zones (such as Zone 1, Zone 2, Zone 3, Zone 4) and different building numbers (such as a specific building within a certain zone), and there are even WeChat groups for residents of different units within the same building. In these groups, individuals are required to use their doorplate number along with their real name as their nickname. The smaller the scale of a real space corresponding to a paravirtual community, the more relevant the interests of the people involved, and the more opportunities there are for them to meet in real life, the closer the virtual community is to a real community. For example, a WeChat group for residents of a specific unit in a certain building within a neighborhood may involve a dozen households, all using their house numbers and real names as personal identity markers, leaving almost no "virtual" relationships. At this point, the WeChat group is merely a communication space for people with real identities occurring in a virtual space. In a sense, small-scale paravirtual communities can better promote member participation and interaction, significantly improving the indifferent neighborly relationships in modern cities.

The "Literary Group" WeChat group of the "Workers' Home" in Beijing's Pi Chun is a paravirtual community established based on shared interests. Individuals who actively participate in interactions often use their real names and are familiar with each other in real life, while onlookers or "lurkers" mostly choose to remain anonymous, observing the group's dynamics and obtaining useful information. The members of this WeChat group include workers, university teachers, university students, and some members of the public. The content of the exchanges revolves around literary interests, primarily focusing on sharing original works, commenting, and forwarding good articles, which reflects a certain level of professionalism. The

"Literature Group" WeChat group interacts with the weekly lecture activities organized by "Workers' Home," allowing workers to meet more people, especially from other social classes, such as university teachers and students who come to give lectures. This interaction helps to break down the barriers of household registration, occupation, education, and income that limit the social circles of workers, preventing the "isolation" of their social networks.

10.2.3 Community Supervision and Co-governance Under Public Affairs Participation

Paravirtual communities such as community websites, online homeowners' forums, WeChat groups, and QQ groups provide residents with open channels for expressing their interests (Gao 2017). Residents can expose issues related to community construction, business services, property management, and more. Even when faced with infringement incidents, driven by common interests, community residents can form temporary rights protection organizations within these paravirtual communities and create a collective voice, thereby helping to cultivate a shared community will and strengthen community unity (Wu 2014). Activities such as suggestions, complaints, and reports in a paravirtual community, due to their wide openness, can put pressure on the relevant stakeholders, making it easier to receive timely responses and handling. This effectively protects the rights and interests of community residents and is conducive to the development of the community towards a more just and improved direction. Gao (2017) uses the case of homeowners in the Landi community opposing the construction of a substation in their "backyard" and ultimately securing economic compensation as an example of a NIMBY (Not In My Backyard) movement. This illustrates that paravirtual communities, as platforms for information disclosure and discussion, can unite the common will and strength of community residents, enabling them to achieve identity recognition and value convergence. Notably, during this NIMBY movement, the homeowners in the group had not yet moved into the community, but they established connections online under their real names, shared experiences, discussed strategies, organized mobilization, and extended their online social relationships into offline activities. In addition, paravirtual spaces like community networks often establish government workstations. On one hand, they publish the work and achievements of government agencies, allowing community residents to participate in supervising the government's performance of its duties, thus promoting top-down community management. On the other hand, they help higher authorities understand public sentiment, ensuring that the needs of the masses are communicated and addressed. Through a bottom-up feedback mechanism, effective communication between both parties is strengthened (Zhou et al. 2013; Li 2019). During the COVID-19 pandemic prevention and control period, community residents were fully mobilized to form a community epidemic prevention and control collective where everyone has a responsibility. The paravirtual community thus took on the functions of community supervision and joint governance by residents. For example, in the "Wuhan Sister-in-Law's Scolding" incident, a property owner criticized the supermarket in a WeChat group for issuing "AB Yin-Yang packages" targeting different neighborhoods and complained about the supermarket's bundled sales. This prompted community workers to promptly express their apologies, begin to understand the detailed needs of the property owners, cancel the supermarket's group purchase packages, change suppliers, and increase personalized supplies.

The paravirtual community is an important platform for residents to participate in community public affairs. To some extent, it helps enhance the community's self-governance capability, which is an essential force that modern communities must integrate and maintain, but which was lacking in the traditional Danweilization communities. In Danweilization communities, the Danwei coordinates and manages the work and life of community members, which is a top-down administrative management model rather than a horizontal self-organizing mechanism (Zheng 2010). In modern urban communities in our country, the administrative power represented by the community committees still holds a dominant position, resulting in generally low participation among community residents and a lack of public spirit (Zhang 2005; Jiao and Wang 2019). In this context, a paravirtual community formed by self-organized residents serves as an outlet for public expression and a discussion platform. It can promote residents' expression of opinions, protect their interests, assume community responsibilities, and share community achievements. This fully mobilizes residents' subjectivity and initiative, allowing the bottom-up operational mechanism to function effectively and helping to build a strongly aggregated community.

10.3 "Micro Life": The Impact of WeChat on Residents' Living Space

10.3.1 The Functions of WeChat and the Restructuring of Residents' Living Space

WeChat is a real-time communication software launched by Tencent in 2011, which is free to download and install, and supports sending text, images, and conducting voice and video chats. Over nearly a decade of development, WeChat has surpassed and almost replaced the original marketing goal of mobile SMS services. It has profoundly influenced various aspects of urban residents' lives with its increasingly powerful and diversified features and continuously innovative technology, as defined by WeChat itself—"WeChat is a lifestyle." As of September 2018, WeChat had over one billion active users each month. Its existing features include online chatting, mobile payments, sharing on Moments, operating public accounts, and city services. As a brand new media platform, WeChat has brought a new lifestyle to people, and its rich functionalities have greatly changed people's habits and social behaviors. In fact, it has created a new human living environment or domain—"Micro Life" (Kuang 2014; Tang and Wang 2016; Tang and Tang 2016), which redefines the relationship between urban residents and urban life, and reconstructs the living space of urban residents.

Firstly, WeChat, as a mobile technology, can break the boundaries of various functional spaces. The division of living spaces, working spaces, leisure spaces, etc., based on functionality has become severely ineffective in the WeChat era; the boundaries between these spaces have become blurred. As long as WeChat is open on the phone, one can receive work tasks, social resources, and entertainment information at home. Even when in a specific functional space, one can simultaneously complete multiple tasks or plans. With the support of WeChat's powerful features, working from home is no longer a novel phenomenon, social interactions do not require stepping outside, and shopping can be done anytime and anywhere through online malls and online payments, which means that physical functional spaces are being weakened.

Secondly, WeChat has embedded new communication technologies and mobile payment functions into all aspects of urban residents' lives. As a new lifestyle technology that supports instant communication, builds social networks, integrates public information, and provides practical means for daily living (Tang and Wang 2016), WeChat has infiltrated nearly all life scenarios, making online-offline interactions a quickly accepted and relied-upon new way of life for urban residents. In restaurants and physical stores, WeChat Pay has become a very common payment method, allowing consumers to complete every dining and shopping transaction without carrying cash. New payment methods such as QR code payments and facial recognition payments have entered the consumption scenarios of urban residents. The WeChat Moments feature has also become an important aspect of social interaction, where users share their personality traits and life situations in a self-presenting manner, forming a complete social context alongside offline interactions. Additionally, the sharing of information on WeChat has integrated into the scene of knowledge creation, enabling users to access knowledge and information without leaving their homes. While becoming recipients of information, they have also become producers of big data and leaders in self-media (Gou et al. 2016). It is worth noting that WeChat, as a location-based social networking service (LBSNS), can guide the flow of people by matching users' interests with their locations, thereby reconstructing the meaning of real-world locations (Wu et al. 2016). WeChat emphasizes a peer-to-peer structure from user to user, and by embedding geographic location information into online social processes, it not only reflects users' emotions, lifestyles, and tastes (Wang 2013), but also guides real-life social outings, leisure activities, and shopping trips.

With the continuous upgrades of the WeChat version, WeChat can "one-click access" various city services and public transportation, bringing convenience to urban residents' lives. Through WeChat, urban residents can achieve various goals such as mobile phone top-ups, utility bill payments, medical appointments, administrative service handling, and public transportation via buses and subways. This convenient online service not only reduces the cost for urban residents to access public services but also alleviates the functions of offline public service spaces, leading some physical public service spaces to gradually withdraw from the daily lives of urban residents.

Overall, "Micro Life" is a brand new domain, and the impact of WeChat on the living spaces of urban residents is multifaceted and layered: the popularity of WeChat has broken down the boundaries of various functional spaces in urban residents' lives; WeChat's functions are fully integrated into people's life scenarios; at the same time, WeChat has realized location enhancement based on geographic information sharing; it has also incorporated some functions of urban public service spaces, altering the structure of urban residents' living spaces. In the WeChat era, the living spaces of urban residents have been restructured. The following analysis will focus on the significant impact of WeChat on residents' living spaces, represented by workspaces, consumption and leisure spaces, and social spaces.

10.3.2 Impact of WeChat on Residents' Workspace

Since the launch of WeChat in 2011, it has gained favor among professional groups due to its simple interface and instant communication capabilities, becoming increasingly used in work scenarios. The continuous connectivity and instant accessibility of WeChat have enhanced the flexibility of work time and location, meeting the demand for quick responses and efficient communication in the workplace (Huang and Yang 2018). Additionally, the availability of mobile communication technology ensures that professional groups can engage in knowledge processing and collaborative knowledge production on WeChat. A survey indicated that in 2016, nearly 60% of new friends added by WeChat users came from work relationships, and over 80% of users used WeChat for work purposes, with first-tier cities being the main force for WeChat office use.[1] While urban residents use WeChat for mobile office work, they have also created an around-the-clock workspace that breaks down functional boundaries: WeChat has increased the freedom of work locations and the flexibility of work hours, shifting traditional office spaces to new types of workspaces. The constant and intensive flow of information has invaded urban residents' other living spaces, constructing an omnipresent work scene (Meng 2018).

Freelancing is no longer a novel topic today. With the support of WeChat for office work, freelancers can choose to work from home or in public places like cafes, as long as there is a continuous wireless internet connection, the workspace can be anywhere. With information technology meeting daily work needs, traditional offices and office buildings have become less important; rented, temporary office space, represented by maker spaces, can then become a workspace for workers. For entrepreneurial teams that cannot afford rent, communication based on WeChat, which is not limited by time and space, is the most important. In this new type of workspace, the boundaries between superiors and subordinates, as well as between departments, become less distinct as the physical space boundaries dissolve, and informal discussions have

[1] Data source: https://tech.qq.com/a/20170424/004233.htm#p=1.

become an important part of the work chain (He and Shen 2015). Another type of new office space is the interaction-oriented office area, which weakens the layout of personal offices and cubicles, incorporates informal communication spaces and micro-relaxation areas, emphasizes innovation and communication, and relies on the instant transmission of information brought by communication tools like the internet and WeChat. Shenzhen Overseas Chinese Town and Beijing Baidu Tower are typical representatives of interaction-oriented office areas (He and Shen 2015). Urban residents can enjoy a better working environment and communication atmosphere in these new workspaces that break traditional office limitations. On one hand, WeChat enhances people's work efficiency; on the other hand, it serves as a tool for "connecting" people across different times and spaces, promoting collaboration among workers.

However, from another perspective, this all-weather, information-dense communication actually results in the invasion of workspaces into other areas of life. Relevant surveys show that WeChat has become the primary communication tool for users at work, with over 40% of users joining WeChat groups with more than 100 people for internal company communication, and more than 30% of users joining such groups to expand their professional networks.[2] These phenomena indicate that users are likely required to handle work at any time and place, while also dealing with the professional social interactions that come with work. The urgency and anxiety of work are increasingly projected onto communication tools like WeChat (Meng 2018). Continuous fragmented work tasks invade the other living spaces of urban residents, squeezing their family life and leisure time. The public service advertisement "Put down your phone and spend quality time with Dad and Mom" highlights this conflict and contradiction in urban life.

10.3.3 Impact of WeChat on Residents' Space of Leisure Consumption

WeChat, as a new lifestyle technology, is also transforming the leisure spaces of urban residents. On one hand, it changes traditional business models and urban commercial spaces through mobile payment methods; on the other hand, it provides urban residents with more autonomy in their leisure activities and consumption patterns.

Traditional business models have been greatly impacted by WeChat, whose vast user base has allowed it to firmly establish itself in the information age and become a perfect marketing channel and e-commerce entry point (Wang 2014). The combination of WeChat Pay and third-party interfaces has significantly met the daily consumption needs of urban residents—various e-commerce platforms such as Didi Chuxing, Dianping, and JD.com have gained access to WeChat's user resources, and it is the development of e-commerce that has led to the transformation of traditional business models. Traditional commercial squares and department stores facing

[2] Data source: https://tech.qq.com/a/20170424/004233.htm#p=15.

losses have taken two measures: first, they have launched online shopping platforms to gain support from WeChat user groups; second, they have adjusted the types of businesses in commercial areas to attract offline consumers, with experiential consumption spaces being one of the strategies. In the context of the WeChat era, commercial centers place greater emphasis on differentiated competition, focusing on the development of experiential consumption and reducing the prevalence of highly homogeneous department stores and retail. For example, according to research by He and Shen (2015), the proportion of leisure, entertainment, and dining in some shopping centers in Guangzhou has reached nearly 50%.

In addition to the increase in experiential commercial spaces, the autonomy of urban residents in their daily consumption and leisure activities has also been enhanced through the connection of WeChat Pay. The "2019 WeChat Pay Small Business Operation Big Data Report" shows that over 50 million individual merchants and businesses are now linked to WeChat Pay,[3] with WeChat QR code payments covering a wide range of industries. Many restaurants in the city (including McDonald's, KFC, etc.) offer self-service ordering and self-checkout functions, while convenience stores and supermarkets have also implemented self-selection and self-payment options, allowing consumers to complete their purchases without any interaction with staff. Additionally, museums, art galleries, and parks in the city have introduced WeChat guided tours, where scanning a QR code provides information about items of interest, making these leisure spaces more autonomous and personalized public spaces.

WeChat, as a location-based social networking service, also plays a role in reconstructing real-world locations. Urban residents can obtain information about virtual commercial spaces through WeChat, allowing merchants in traditionally disadvantaged locations to increase the number of offline consumers (Chen et al. 2018) and achieve new development. The sharing of geographic location information alongside other content in WeChat social interactions can effectively enhance the visibility and accessibility of leisure and consumption spaces. More and more urban residents are discovering niche, novel, and uniquely representative leisure and consumption spaces based on this, and some consumption and leisure spaces located in traditionally disadvantaged geographic areas have become popular "check-in spots" worth experiencing, leading to a "revival" of these spaces.

10.3.4 Impact of WeChat on Social Interaction Space of Residents

According to Tönnies' definition, a community can be understood as a closely-knit living community characterized by locality, consciousness, behavior, and interests, composed of a homogeneous population with shared values (Tönnies 1999). The emergence of the internet has brought new connotations to "community," where

[3] Data source: con://baijiahao.baidu.com/s?id=1646806111371773926&wfr=spider&for=pc.

online members can delineate boundaries based on a sense of belonging and collective identity, thus forming virtual communities. A typical example of this is the online communities established through WeChat. The communication features and Moments sharing function of WeChat have made the formation of virtual communities possible, maintaining and expanding users' social relationships. Through online-offline interactions, they deepen the connotation and dimensions of social interaction spaces. The WeChat virtual community is not an isolated social phenomenon; rather, it serves as one of the social methods for urban residents. When online relationships continue to develop, they transition into face-to-face interactions offline (Nie et al. 2013). Therefore, virtual communities effectively supplement and extend the real social interaction spaces of urban residents.

For contemporary urban residents, WeChat social interaction is a very important component of daily social life. Users engage in self-presentation and impression management through private chats, group chats, and posting on Moments (Wang and Lan 2017). This serves as a form of self-expression in virtual communities, setting the context for social interactions between users. On one hand, WeChat reflects and maintains real-life social relationships; on the other hand, it expands new social connections, integrating virtual social circles with offline social circles through the connections of strong and weak ties, creating a virtual community primarily based on strong ties, supplemented by weak ties. While likes and comments on Moments maintain weak ties in the social network at a low cost, WeChat social interactions among acquaintances also enhance "in-group identity" and "small circle culture" (Jiang 2014). In the decentralized, bidirectional communication where the roles of communicator and audience instantly switch (Wang J 2017), WeChat helps users cultivate and accumulate social capital.

In the virtual community established on WeChat, strong relationships provide members with a sense of belonging and security, while weak relationships always facilitate the exchange and sharing of diverse information. Based on this, the virtual community also organizes collective actions. These functions of WeChat's virtual community complement the shortcomings of offline social interactions for urban residents. The interaction between urban residents online and offline has transformed the meaning of social interaction space and increased its dimensions—residents' social interaction space should no longer be understood as merely a physical social interaction space, but rather as a composite space of online and offline social interactions, as WeChat has already embedded itself in social scenarios and become a direct source of people's social experiences. In fact, with the increase of commercial housing communities and the decline of Danwei, realistic community interactions have become increasingly rare. Online communities, represented by owners WeChat groups, can compensate for this isolation. Additionally, Moments records and shares the offline social behaviors of urban residents, linking real social interaction spaces, events, and users' social relationships, thus becoming the context for the next social behavior and a basis for personal information judgment. WeChat also helps users achieve multi-line social interactions within the same physical space, breaking the limitations of time and space, which traditional face-to-face socializing cannot accomplish. Overall, the virtual community built by WeChat has become an important component of the social interaction space for urban residents, intertwining online and offline social interactions to collectively define the residents' social interaction space.

10.4 Fashionable Living Spaces: The Impact of Fashion Concepts and Styles on the Changes of Living Spaces

The concept of fashion first emerged in the field of clothing. In the late Middle Ages, the forms and functions of clothing began to change: from draped styles to tailored designs, shifting from concealing the body to expressing it (Yang 2012). In the nineteenth century, the Industrial Revolution made large-scale industrial production possible (Svendsen 2010), and fashion transitioned from the lives of the upper class to those of ordinary people, gradually permeating various aspects of social life.

In 1908, American sociologist Roosevelt first defined fashion, and since then, different scholars have interpreted the concept of fashion from various perspectives. In fact, fashion is both a popular behavioral pattern and represents a set of values with specific cultural connotations (Zhou 1995). Today, fashion has become a "widespread social phenomenon" (Simmel 2001), and the integration of these two aspects can roughly outline the full picture of fashion. Japanese scholar Minami Hiroshi categorized fashion into three main types: objects, behaviors, and ideas in his work "System: Social Psychology" (Zhou 1995). Specifically, fashion can manifest as a concept or value, such as environmental protection, minimalism, or a certain aesthetic style, which guides our daily behaviors, especially consumption behaviors; fashion can also manifest as a way of acting, such as going to bars or backpacking; more commonly, fashion is expressed through the pursuit of specific material forms, such as the craze for luxury goods or trendy brands. It can be said that the pursuit of the goal of "fashion" dominates contemporary social life.

A deep understanding of the essential characteristics of fashion will help us more accurately define the lifestyle and living space associated with fashion. Ma (2005) summarized that fashion has 12 characteristics: novelty, differentiation, imitation, conformity, transience, triviality, scalability, periodicity, extremity, luxury, class, and the contradiction between innovation and conservatism. In fact, the novelty of fashion, the coexistence of differentiation and conformity, and its class nature are the essential characteristics of fashion as a lifestyle. Firstly, regarding novelty, it is the inherent characteristic of fashionable things. A fashionable life means taking the novelty of things as the basic orientation for action, constantly pursuing the new. Secondly, concerning the contradiction between differentiation and conformity, German sociologist Simmel revealed the inherent contradiction of fashion: the original intention of fashion is to pursue differentiation from others; however, in the process of seeking difference, people inevitably tend to imitate others (Yang and Zeng 2006). Therefore, the distinction and conformity are the fundamental contradictions of fashion. Ultimately, in terms of class, the distinctiveness of fashion is initially reflected in the differences between classes. People from higher classes are often the creators of fashion, while those from lower classes can only be followers and imitators of fashion. Thus, fashion serves as an important symbol for distinguishing people's social status (Sun 2008).

10.4.1 Practicing Fashion Concepts Through Lifestyle

The trend towards a fashionable lifestyle is closely linked to the formation of modern society. In modern society, traditional structures have collapsed, and people can no longer rely solely on single standards such as economic status to achieve self-identity. The decentralization and diversification of standards have led to a crisis of identity. In this context, fashion, as a means of constructing identity, offers possibilities for self-realization (Sun 2007). Whether through the possession of material items like clothing or by practicing a sophisticated and advanced lifestyle philosophy, fashion distinguishes the self from others, achieving self-identity while showcasing uniqueness. Additionally, the class nature of fashion gives it specific symbolic meaning, representing a higher status and value. The pursuit of fashion by individuals becomes a process of seeking social recognition from certain specific groups (Cao 2011). In summary, for individuals, fashion is an important means of self-realization; it helps us achieve a unique self and seek broader social recognition. The relationship between fashion and identity is a significant reason why we tirelessly engage with fashion.

When fashion is so closely related to self-realization, it itself becomes a life ideal. The pursuit of this ideal is reflected in a comprehensive change in lifestyle, which permeates all aspects of daily life. New lifestyles defined by fashion continuously emerge with the changes of the times, not only altering people's daily behavior patterns but also, unknowingly, creating new living spaces.

10.4.2 Lifestyle Changes in Pursuit of Fashion

Novelty is the essential characteristic of fashion. The novelty of things encompasses different types, which can refer to the generation of new ideas from nothing, such as the new taste orientations of the urban middle class, the rise of consumer trends, and new lifestyles. It can also refer to the renewal of old things, such as the improvements brought by new technologies like internet information technology to existing activities. Correspondingly, the fashionable lifestyles that people eagerly pursue can be classified into the following three categories.

10.4.2.1 Fashionable Lifestyles Based on New Technologies

Every technological advancement made by humanity has brought about revolutionary changes in human life. In the past decade or so, the rapid development of new technologies represented by the internet and the Internet of Things (IoT) has greatly disrupted people's existing lifestyles. The lifestyle supported by new technologies has made life more convenient and intelligent. At the same time, this lifestyle also has a certain consumption threshold, which naturally carries symbolic meanings of social status and wealth. Therefore, this lifestyle has naturally become a fashion that everyone pursues.

In terms of home life, the combined use of the Internet and the IoT has made today's home living increasingly intelligent. When no one is at home, people can use mobile phones and other terminal devices to remotely control various appliances in the house. For example, they can turn on the air conditioning to adjust the temperature before arriving home, remotely control the light Switches, start the rice cooker from a distance, and monitor home security remotely. The intelligent home life allows people to arrange their time and spatial activities more efficiently, providing a more comfortable living environment and a higher quality of life.

In terms of shopping, the internet has removed the limitations of time and space, allowing people to select and purchase the products they want online anytime and anywhere. Additionally, with the help of IoT technology, they can monitor the delivery progress of their products in real-time. In terms of travel, smart transportation based on IoT technology can monitor real-time road conditions, enabling drivers to plan routes and adjust their travel to avoid congested areas. Smart transportation also helps travelers stay informed about public transport routes and arrival times, thus avoiding wasted time. Furthermore, with the support of new technology transportation tools, such as electric vehicles and electric scooters, people are beginning to pursue new modes of transportation, often branding them as fashionable. In the realm of social interaction, the existence of the internet has completely broken the time and space constraints of socializing. People can quickly find like-minded strangers through social apps. Various public social platforms also provide space for self-expression, becoming trendsetting fashion platforms that spark waves of pursuit among people.

10.4.2.2 Fashion Lifestyles as Symbolic Consumption

Fashion and consumption are always inseparable; people's pursuit of a certain fashion concept ultimately culminates in consumer behavior. Fashion consumption is not aimed at satisfying basic survival needs, but rather is a mix of both practical and enjoyable purposes. Baudrillard's theory of symbolic consumption points out that contemporary consumption is symbolic consumption, where the possession of material goods actually carries significant symbolic meaning (Huang 2007) and is key to personal identity recognition. The pursuit of a fashionable lifestyle is also a consumption of the fashion symbol, primarily manifested through leisure activities and fashion consumption.

Some leisure and entertainment activities that align with the tastes of the urban middle class have become sought-after fashionable activities. Cinemas, bars, and cafes have become the preferred trendy social venues for people. Various fitness programs focused on body management are regarded as fashionable by urban white-collar workers. The enthusiasm for certain niche or subcultures, such as rock culture, has also become a fashionable entertainment activity for urban youth. Through their consumption of these leisure and entertainment activities, people showcase their unique tastes, fully experience a fashionable lifestyle, and complete the construction of their identity as urban middle-class individuals.

Actively consuming image-related fashion is also a practice of fashion concepts. People enthusiastically join the waves of consumer trends surrounding clothing, jewelry, bags, cars, shoes, and skincare products, integrating the pursuit of material trends into their lifestyles.

10.4.2.3 Fashionable Life Philosophies and New Lifestyles

Life philosophy can be seen as a program for one's lifestyle. In recent years, some young people in cities have begun to adopt "XX-ism" as their guiding philosophy to regulate their daily behaviors. The most typical examples include low-carbonism, vegetarianism, minimalism, and philanthropy. Since most practitioners of these philosophies are well-educated and possess a certain social status and resources, these life philosophies naturally carry symbolic significance and thus become a trend. Under the influence of these fashionable life philosophies, practitioners need to make corresponding adjustments in their daily behaviors. For instance, low-carbon advocates tend to use environmentally friendly transportation, avoid plastic bags and disposable utensils, and purchase fewer unnecessary and non-eco-friendly items. Fashionable life philosophies have created a new fashionable lifestyle.

10.4.3 Formation of Fashionable Living Space

Living space is the place where daily life activities take place. On one hand, it is shaped by daily behaviors guided by lifestyle; on the other hand, it is deliberately transformed by the concepts represented by that lifestyle. The fashionable living space corresponding to a fashionable lifestyle is no different. A fashionable lifestyle represents a certain philosophy and ideal of living, which imposes specific requirements on the functionality and aesthetic expression of space, dominating people's imagination and design of space, and profoundly influencing the changes in urban landscapes.

10.4.3.1 Fashionable Design of Urban Public Environments

The fashion that is sought after is reflected in various material forms, and urban spaces are no exception. From architecture to neighborhoods, from residential buildings to public spaces, their external manifestations are influenced by fashion, with design language and overall style trending towards fashionability.

From the perspective of urban environment design, the emergence of public environment design companies has led to increasingly free expressions in the design of urban public environments, and the aesthetic value has become increasingly important in urban public environment design (Liu 2017). People's endless pursuit of a fashionable lifestyle naturally permeates the field of urban public environment design, where the design style of urban public facilities and the overall urban public environment are based on the concept of fashion. The idea of fashion profoundly influences the design of public environment facilities, such as urban public bicycle rental systems, smart bus stops, electronic traffic guidance systems, public restrooms, and regionally themed sculptures. In these urban public environment designs, designers consider incorporating fashionable elements to create a space that better aligns with people's fashionable lifestyle concepts. In addition, the design concepts and styles of public environments also aim for fashion, giving rise to design styles such as green and environmentally friendly, intelligent, ethnic, and post-industrial.

10.4.3.2 Fashionable Consumption Space

Consumption is an important way to practice a fashionable lifestyle, and urban consumption spaces have changed due to the rise of this lifestyle. On one hand, fashion consumption spaces that carry symbolic significance have emerged and occupy an important position in urban spaces. As more and more new leisure and entertainment activities are embraced as fashionable, a number of new consumption spaces have also arisen, such as coffee shops, bookstores, galleries, and bars that can host professional music performances. These spaces not only need to accommodate fashionable leisure activities but also need to align with fashionable aesthetic styles. The rise and transformation of fashion consumption spaces have also changed the functional zoning of cities, leading to a reconstruction of the internal spatial structure of urban areas under the influence of fashionable lifestyles. On the other hand, existing consumption spaces are continuously updated to meet the ever-changing demands of fashion, which impose new requirements on architectural or decorative styles. From the perspective of external spaces in commercial districts in our country, there has been a shift from the originally economical and practical design of exterior walls to the adoption of large glass decorations, the emergence of numerous advertising banners, and the integration of urban architecture with modern lighting technology (Guo 2012). Fashion elements are constantly being added to architectural design. Providing fashionable scene experiences, enhancing spatial flow design, and emphasizing spatial interaction are also important techniques for the fashionization of commercial spaces (Li and Liu 2007). In addition, historic and cultural districts have also become emerging spaces for cultural symbol consumption. The combination of historical culture and nostalgia has attracted considerable attention from the urban middle class. The renovation of these districts retains historic and cultural characteristics while incorporating many fashionable elements through methods such as physical darning and functional replacement (Liu and Chu 2010). Represented by Shanghai's Xintiandi, Chengdu's Kuanzhai Alley, and Beijing's Nanluoguxiang, a series of historic and cultural districts that integrate leisure, entertainment, and trendy shopping have become popular consumer locations and new fashionable attractions in the city.

10.4.3.3 The Transformation of Urban Living Spaces Guided by Fashion Lifestyle Concepts

Practitioners of fashionable lifestyle concepts choose a unique way of living, and their daily behaviors are regulated by these fashionable lifestyle concepts. To achieve a specific fashionable lifestyle, practitioners will make self-selective choices regarding their living spaces. When this fashionable lifestyle becomes more widely disseminated, urban living spaces will undergo autonomous transformations to meet these needs.

Low-carbon advocates typically choose green transportation methods such as walking and cycling. To achieve this lifestyle, they actively select walkable neighborhoods and areas with good green landscapes. The popularity of such a fashionable lifestyle will inevitably influence the decisions of urban managers, leading to adjustments and changes in residential area design, road traffic organization patterns, and urban green space landscape design. The lifestyle of vegetarians will also give rise to the emergence of vegetarian consumption spaces in cities, and their personal living spaces will be confined to areas that meet the needs of their lifestyle, having a subtle yet profound impact on the urban spatial structure.

10.5 Chapter Summary

The study of human geography and urban geography requires a process of in-depth exploration and innovation (Gu 2009). Attention to new technologies such as ICT is itself an important way for urban research to achieve innovation. The impact of ICT has permeated the daily lives of urban residents in China, creating new forms and types of living spaces. These "new" living spaces effectively complement traditional urban living spaces and are worth exploring. In the context of information and communication technology, virtual living spaces have developed, including the deepening of e-commerce services, the balanced reconstruction of virtual shopping spaces, and the interaction between virtual shopping spaces and physical shopping spaces. These developments have expanded the breadth and depth of urban living spaces, resulting in a more complex character of living spaces in an environment where physical and virtual interactions are evolving. The advancement of information and communication technology has also led to the emergence of virtual living service spaces, such as virtual medical service spaces, virtual public service halls, and virtual transportation service spaces. These virtual living service spaces play an increasingly important role in residents' living spaces, with significance not only in terms of efficiency and cost reduction but also in enhancing the management level of relevant departments and improving the management service system. In the context of an information-based society, virtual social spaces also exhibit many new forms and characteristics. Otaku culture, subcultures, "information cocoons," traffic, and trending topics are all important vocabulary in virtual social spaces. Under the influence of information and communication technology, the evolution of virtual living spaces shows features such as fragmentation, blurred boundaries, and technological dependence.

In Chapter 6, a certain length is devoted to discussing the positive role that community networks play in facilitating social interactions and enhancing community cohesion in the large residential area of Huilongguan in Beijing. In fact, aside from community networks, BBS, real-name QQ groups, and WeChat groups are also common forms of paravirtual communities. They often serve functions such as information sharing, social interaction, community self-governance, and community services. Paravirtual communities enable the sharing of information resources to meet the increasingly complex daily needs of residents, expand social interaction spaces, and reconstruct residents' social relationship networks. They also facilitate community supervision and joint governance in public affairs participation, allowing bottom-up operational mechanisms to function effectively, which helps in building a strongly cohesive community. Paravirtual communities, through effective interaction with residents' real-life spaces, are significant for strengthening community belonging, enhancing community identity, promoting community integration, and improving community development.

If we were to investigate the software or technology that has had the greatest impact on the lives of Chinese people over the past decade, there is no doubt that WeChat would be at the forefront. As a mobile technology, WeChat has broken the boundaries of various functional spaces, embedding new communication technologies and mobile payment features into all aspects of urban residents' lives. By providing "one-click access" to various urban services and public transportation, it has brought convenience to urban residents' lives and has led to the gradual withdrawal of many physical public service spaces from daily lives. WeChat has impacted residents' workspaces through its all-weather working style that breaks down functional boundaries. It has influenced residents' leisure spaces by creating experiential and autonomous consumption and leisure activities. Through "WeChat socialization," it has become an important component of residents' social spaces, thereby affecting those social spaces. "Micro Life" has become a lifestyle, and urban residents in China have developed a significant dependence on WeChat. It is hard to imagine what life spaces would look like without WeChat.

The evolution of lifestyle towards fashion is closely linked to the development of modern society. The combined use of the Internet and the Internet of Things has made today's home life increasingly intelligent. Fashion and consumption are always inseparable; the pursuit of a certain fashion concept always culminates in consumer behavior, and fashion consumption blends practical and enjoyable purposes in consumer behavior. Fashionable life philosophies have created new fashionable lifestyles, with some young people in cities starting to adopt "XX-ism" as a guiding philosophy for their daily lives. A fashionable lifestyle represents a philosophy and ideal of living, with specific requirements for the functionality and aesthetics of space, dominating people's imagination and design of space, guiding the transformation of urban living spaces, and profoundly influencing the evolution of urban landscapes.

References

Azuma H (2001) Zoological postmodern era: How the Otaku have influenced Japanese society. Chu Xuan first translation. Dahong Art Co., Lt., Taipei (in Chinese)

Cao H (2011) Analysis of individual differences and social universal identity in Simmel's fashion theory. J Shanxi Norm Univ (Soc Sci Ed) 38(S3):125–127 (in Chinese)

Chai Y, Shen Y, Ta N (2014) Smart travel applications based on spatio-temporal behavior research. City Plan Rev 38(3):85–91 (in Chinese)

Chang X (2014) An empirical study on social relationships and interactive spatial characteristics in virtual cyberspace. Wuhan University Press, Wuhan (in Chinese)

Chen L (2016) New media spaces and the dissemination of youth subcultures. Jiangsu Soc Sci 4:199–205 (in Chinese)

Chen D, Yang Y, Li E et al (2018) Impacts of mobile smart device usage on multi-task shopping behaviors and commercial micro-locations of beijing residents. Sci China Earth Sci 3:353–365 (in Chinese)

Cui Y (2017) An empirical study on the impact of online community participation on individual sense of community belonging. SE Commun 4:44–47 (in Chinese)

Du Q (2018) Transformation of social interaction concepts in the context of virtual social networks. J Chin Multimedia Network Teach 7:160–161 (in Chinese)

Du R, Jiang K (2017) New retail: connotation, development drivers, and key issues. Price: Theor Pract (2):139–141 (in Chinese)

Feng J, Wu F, Zhou P (2017) Neighborhood relations and social space regeneration in suburban large-scale residential areas: a case study of Beijing's Huilongguan. Prog Geogr 36(3):367–377 (in Chinese)

Gan M (2019) Research on the characteristics of residents' relational networks based on community identity: a case study of Tangshan. Hebei University of Economics and Business, Shijiazhuang (in Chinese)

Gao X (2017) Production and action of virtual resistance space in NIMBY movements: a case study of blue land community in City B. J Nanjing Tech Univ (Soc Sci Ed) 16(4):60–71 (in Chinese)

Goffman E (1989) The presentation of self in everyday life. Zhejiang People's Publishing House, Hangzhou (in Chinese)

Gou S, Li G, Zhang K et al (2016) Spatio-temporal behavior of "tourists" on social media platforms: a case study of professor W's WeChat "moments." Tourism Tribune 31(8):71–80 (in Chinese)

Gu C (2009) The transformation of Chinese human geography. Acta Geogr Sinica 64(10):1175–1183 (in Chinese)

Guo D (2012) Design of external spaces in urban commercial districts under the context of symbolic consumption. Southwest Jiaotong University, Chengdu (in Chinese)

Guo S (2019) Research on the daily life of internet celebrities under bourdieu's field theory. South China University of Technology, Guangzhou (in Chinese)
He W (2018) From Otaku to two-dimensional: academic perspectives and knowledge archaeology of a youth subculture. Journalism Commun Res 25(10):40–59, 127 (in Chinese)
He L, Shen C (2015) Transformation and restructuring of urban space in the "internet+" era. In: China Urban Planning Annual Conference Abstract Collection (in Chinese)
Hu Y (2015) Research on community interaction spaces in the information age. Beijing University of Civil Engineering and Architecture, Beijing (in Chinese)
Huang B (2007) Review of baudrillard's theory of symbolic consumption. J Qinghai Norm Univ (Philos Soc Sci Ed) 3:1–4 (in Chinese)
Huang Y (2011) Cultural dissemination in SNS networks. Hunan Normal University, Changsha (in Chinese)
Huang Y, Yang L (2018) Factors influencing social use of WeChat among occupational groups: an exploratory study based on grounded theory. Journalism Commun 11:66–74 (in Chinese)
Jiang J (2014) WeChat addiction: social illusion and self-loss. Soc Sci Nanjing 11:96–102 (in Chinese)
Jiang Y, Ren X (2017) Causes and development strategies of new retail from the perspective of the "wheel of retailing" theory. Econ Forum 1:99–101 (in Chinese)
Jiao R, Wang K (2019) Discovering subjectivity: dilemmas and solutions in urban community participation. Res Dev 5:147–154 (in Chinese)
Kuang W (2014) A quantitative study on the development of WeChat in China. Chin J Journalism Commun 36(5):147–156 (in Chinese)
Li J (2019) Reconstruction of social relationship networks in urban communities under the "internet+" background: a case study of District B in City H. J Chongqing Univ Educ 32(5):5–9 (in Chinese)
Li C, Liu J (2007) Super-experiential fashion: non-architectural expressions in pan-commercial and entertainment buildings. New Architect 3:30–33 (in Chinese)
Lin W, Feng J, Li Y (2018) Impact of ICT on spatial changes of residence and employment in urban villages: a survey of five urban villages in Beijing. Prog Geogr 37(2):276–286 (in Chinese)
Liu S (2015) Thoughts on development and reform of electric charge payment system in power supply enterprises. Autom Instrum (11):241–242, 245 (in Chinese)
Liu S (2017) Research on artistic design of urban public environmental facilities incorporating fashion elements. Huazhong University of Science and Technology, Wuhan (in Chinese)
Liu F, Chu D (2010) Reconstructing "urban cultural capital": preliminary study on "fashionization" trend and development strategies of historical blocks. Chin Overseas Architect 3:76–78 (in Chinese)
Ma H (2005) Analysis of western urban recreational space planning and design. Qilu J 6:147–153 (in Chinese)
Ma H, Chen Q, Qin P et al (2014) Development status and related issues of telemedicine technology at home and abroad. J Med Inform 35(12):35–39 (in Chinese)
Ma H, Lou J, Liu R (2019) Spatio-temporal evolution of urban residents' leisure consumption capacity in China. World Reg Stud 28(6):145–155 (in Chinese)
Mccombs ME, Shaw DL (1972) The agenda-setting function of mass media. Public Opin Q 36(2):176–187
Meng W (2018) WeChat as a mirror of the Chinese spiritual landscape. People's Tribune 2:116–117 (in Chinese)
Nie L, Fu C, Cheng D (2013) WeChat moments: virtual communities from a social network perspective. Journalism Rev 5:73–77 (in Chinese)
Nong Y (2014) Mobile internet in the micro-era: light acquaintance sociality, communicative pleasure, and ethical anxiety in new stranger society: a case study of WeChat. Lit Cult 3:91–99 (in Chinese)
Qian C (2013) Research and application of expressway ETC data mining. Chang'an University, Xi'an (in Chinese)

Sessions LF (2010) How offline gatherings affect online communities: when virtual community members "meetup." Inf Commun Soc 13(3):375–395
Simmel G (2001) The philosophy of fashion. Culture and Art Publishing House, Beijing (in Chinese)
Sun P (2007) Sociological review of dressing fashion. J NW Norm Univ (Soc Sci) 4:26–32 (in Chinese)
Sun P (2008) On Simmel's view of fashion. J NW Norm Univ (Soc Sci) 6:95–99 (in Chinese)
Svendsen L (2010) Fashion: a philosophy (trans: Li M) Peking University Press, Beijing (in Chinese)
Tang K, Tang J (2016) Analysis of interpersonal interaction in WeChat moments: concurrently discussing the rise and governance of micro-lifestyle. J Jiangsu Adm Inst 1:79–87 (in Chinese)
Tang K, Wang D (2016) WeChat as a lifestyle: concurrently discussing the concept of micro-life and its media-social orientation. J Harbin Inst Technol (Soc Sci Ed) 18(5):46–51 (in Chinese)
Tao D (2014) Understanding micro-culture in the micro-era. China B Rev 3:4–5 (in Chinese)
Tönnies F (1999) Community and society (trans: Lin Rongyuan). The Commercial Press, Beijing (in Chinese)
Wang Y (2010) Community democracy and the creation of starfish-type community networks: analysis of Beijing Huilongguan community network operation. J Party Sch Hangzhou Municipal Committee CPC 3:86–89 (in Chinese)
Wang SS (2013) I share, therefore I am: personality traits, life satisfaction, and facebook check-ins. Cyberpsychology Behav Soc Networking 16(12):870–877
Wang Q (2014) Business model innovation research on WeChat platform. J Zhengzhou Univ (Philos Soc Sci Ed) 6:89–93 (in Chinese)
Wang B (2017) The essence, causes and practical trends of "new retail". China Bus Mark 31(7):3–11 (in Chinese)
Wang J (2017) "WeChat people" and life styles in networked era. Tianjin Soc Sci 4:89–93 (in Chinese)
Wang L, Lan J (2017) Self-presentation behavior in WeChat moments among youth groups: a virtual ethnography study. J Jinan Univ (Philos Soc Sci Ed) (12):115–125, 128 (in Chinese)
Wu L (2014) Mechanism analysis of community websites' impact on residents' community participation. Sci Technol Inf (3):12, 10 (in Chinese)
Wu S, Li X, and Wu X et al (2016) Guidance effect and spatial impact of location-based social network service information flow. World Reg Stud 25(1):159–165 (in Chinese)
Yang D (2012) The course of fashion: a detailed account of its history, leaders, and theories. Peking University Press, Beijing (in Chinese)
Yang X, Zeng Y (2006) Aesthetic redemption in modern life: Georg Simmel's fashion theory. J Sichuan Int Stud Univ 5:80–85 (in Chinese)
Yin Z (2016) Research on urban residential space in network society context. Tsinghua University, Beijing (in Chinese)
Yuan J (2010) Ideal and reality of microblogging: three challenges in constructing public space via social media. J Zhejiang Norm Univ (Soc Sci Ed) 35(6):20–25 (in Chinese)
Zhai Q (2015) Research and evaluation on urban virtual-real space correlation based on residents' activities. Nanjing University, Nanjing (in Chinese)
Zhang W (2002) Sales boom in Beijing Huilongguan cultural residential area: the largest affordable housing project in China. China Real Estate Inf 7:13 (in Chinese)
Zhang B (2005) Causes of motivation deficiency in urban community participation. Henan Soc Sci 7:22–25 (in Chinese)
Zhang S (2014) Youth subcultural groups in social networking sites and their cultural reconstruction. Southwest University, Chongqing (in Chinese)
Zhang Y (2018) Rationality analysis of Weibo hot search list from communication perspective. J News Res 9(16):29, 31 (in Chinese)
Zhang X, Yi Y (2019) Structural changes of neighborhood relations in China from historical perspective. J Yibin Univ 19(10):75–84 (in Chinese)
Zhang Y, Zhen F, Zhang Y (2016) Online shopping behavior characteristics of Nanjing residents: a case study of books and clothing. Prog Geogr 35(4):476–486 (in Chinese)

Zhao Y, Chai Y, Gui J (2016) Frontiers of urban leisure space-time behavior research in China. Tourism Tribune 31(9):30–40 (in Chinese)
Zheng Z (2010) Community diversity and integration in post-danwei era: community building with a case study of community network practice. Lanzhou Acad J 11:116–119 (in Chinese)
Zhou X (1995) Theoretical exploration of social fashion. Zhejiang Acad J 3:62–65 (in Chinese)
Zhou C, Jiang H, Gao J (2013) Formation mechanisms of socio-spatial differentiation in urban public services: a case study of parks in Guangzhou. City Plann Rev 37(10):84–89 (in Chinese)

Chapter 11
Conclusions and Perspectives

11.1 Conclusions

Urban living space refers to the association of the social life of urban residents with space or its reflection in space, and it is the trajectory left by the social activities and behaviors of the residents on the urban space. Since the reform and opening up, especially since the transition of China's economic system to the market, China's rapid urbanization has gained development and triggered a drastic reconfiguration of the space of large cities; at the same time, the technological changes in the new era have also led to a great change in the connotation of urban living space in China, and urban living space has become an important carrier for interpreting China's rapid urbanization at the present time. This is the main background facing the study of Chinese urban living space in the new era, and of course, the research methodology and perspective of urban living space need to be diversified and further explored. Taking Beijing, Guangzhou, Dalian and other cities as the main empirical research areas, this book starts from analyzing the concepts related to living space, firstly, analyzing the urban "lifestyle" and "quality of urban life", and then analyzing the urban living space represented by the residential space, interaction space, activity space, leisure space, and so on. On this basis, it extends the scope of urban living space in terms of living space of different groups and "new" living space, and shows the latest forms of urban living space in the context of technological change and the cultural evolution of the times.

The traditional living, working and public service spaces of Chinese urban residents are all centered around the unit, and with the disintegration of the unit system, the traditional living spaces of Chinese urban residents have undergone a tremendous process of reconstruction. Focusing on the period after the founding of the People's Republic of China, the evolution of Chinese urban residents' living space can be divided into three phases, i.e., the phase of living space dominated by the unit system (1949–1987), the phase of decline of the unit space (1987–2000), and the phase of diversity of living space dominated by the market economy (2000–Present).

J. Feng et al., *Urban Living Space in China*, Urban Sustainability,
https://doi.org/10.1007/978-981-95-0595-1_11

Academics have conducted many studies on the structure of urban living space and its changes, urban community living space and its planning, daily activity space of urban resident, and the quality of urban living space, etc. However, the establishment of a framework for the study of urban living space in China as well as a systematic empirical study have yet to be further developed.

Lifestyle is an effective entry point for revealing the way society operates, and it is a traditional research topic in sociology. Academic research on it has gone through the process of gradually developing from a marginal concept of subsidiarity to an independent research field, as well as shifting from the traditional grand generalization theory to the return of daily life, which is a change of great significance. With the rapid increase in China's level of economic development and the acceleration of social transformation, the lifestyles of Chinese urban dwellers have undergone radical changes, and understanding this change is the prerequisite for recognizing China's urban living space. The most prominent features of the traditional lifestyles of Chinese urban residents are: daily interactions based on neighborhood relations and a sense of local community, consumption behaviors based on the development and utilization of physical commercial spaces, collective recreation and entertainment based on public venues, and convenient and frequent trips based on public transportation systems. Economic, social and cultural changes have had a profound impact on the lifestyles of urban residents, with institutional reforms, the power of capital, technological innovation, cultural development and social change having a particularly significant impact. Against the above background, the new lifestyles of Chinese urban dwellers exhibit the following characteristics, including: flexible mobility relying on private cars and public transportation systems; convenient shopping relying on the Internet and efficient logistics and distribution; rich cultural and leisure activities relying on symbolic consumption; a new mode of social interaction relying on information technology; and low-carbon lifestyles adhering to the concept of sustainable development. It is worth emphasizing that the evolution of urban lifestyles has shaken and rebuilt the original Chinese urban living space. The new urban living space after the disintegration of the unitary system, the new leisure life in the context of mass culture, the new lifestyle in the Internet era, and the new lifestyle in the era of flexible transportation all provide us with insights into the tremendous impact of lifestyle evolution on the living space of Chinese urban residents.

As China's level of urbanization increases, urban construction has begun to focus on improving "quality". In this context, the quality of urban life has attracted the attention of more and more scholars. The evaluation of quality of urban life and the living environment is the basis for grasping the urban living space. Although academia has not yet formed a unified concept of quality of urban life, the urban quality evaluation index system has attracted the attention of scholars, and subjective indicators of residents' feelings have been added to the previous objective external conditions. The quality of urban life has characteristics such as spatial differences and social hierarchy. The quality of life and living environment in different residential areas of Chinese cities are clearly differentiated. The closer a community is to the city center, the better the transportation convenience and public facilities and services.

The higher the grade of the residential community, the higher the requirements for the quality of urban life. The spatio-temporal differences in the quality of the living environment in Chinese cities are significant. With the development of urbanization and the improvement of social productivity, the overall level of the urban living environment has been greatly improved. Cities with high-quality living environments are distributed in a “cluster” along the southeastern coast of China. From the perspective of the living environment, the reconstruction of quality of urban life is explored, combining the subjective and objective with the soft and hard living environments, and analyzing the connection and influence between the subjective and objective. The construction of quality of urban life is an important part of improving the urban living environment and is closely related to the daily lives of urban residents.

Residential space is an important part of urban living space. From the perspective of residential function, the fundamental unit of urban residential space refers to the smallest spatial unit in urban residential space that is indivisible at the urban scale and has a certain degree of homogeneity and integrity in terms of spatial structure, social function, and administrative planning and management. From the perspective of spatial statistics, the fundamental unit of residential space is the street or town scale, which is the most feasible for obtaining data. Empirical studies in cities such as Beijing based on census data at the street and township level show that China’s urban industrial restructuring and industrial spatial evolution have led to corresponding changes in the structure of urban residential space. There is a clear differentiation between urban occupations and occupational spaces, and the white-collar population has a significant impact on the formation of new residential spaces. The spatial distribution of the residential population engaged in the primary and secondary industries and the residential population engaged in production and transportation is more concentrated. The spatial distribution of the elderly population, migrant population, the spatial distribution of the Uyghur population, the illiterate population, and the rural population is becoming less consistent with the spatial distribution of the residential population, reflecting more prominent urban problems such as the separation of employment and residence of the employed population, the commuting of the migrant population, and the left-behind elderly population. The distribution of the migrant population is becoming more widespread and showing a trend of spreading towards the suburbs. The trend of intellectuals being concentrated is becoming more and more pronounced. The housing system is becoming more complex, and commercial housing is becoming more widespread, but its spatial differentiation is also increasing. In terms of the mechanism of evolution of the urban social spatial structure, individual differences, family differences and regional differences are the three levels that constitute the process of differentiation of urban social space. Behind this are the driving forces of administrative power, market power and social power. The interaction of these forces promotes the continuous evolution of the urban social spatial structure. The livability of urban communities is also an important indicator reflecting the condition of urban residential space. The evaluation of community livability requires the establishment of a special indicator system for evaluation from the following modules: the rationality of the physical space, the completeness of service facilities, the supporting conditions of the infrastructure, the beauty of the environment and landscape, the care and harmony in the neighborhood, and the degree of efficiency of governance services.

Social interaction space is an important part of urban living space. The empirical analysis of a large residential area represented by the Huilongguan community in Beijing reflects the general characteristics of the current social interaction space of residents in Chinese urban communities based on social relationship networks and neighborhood relations. Overall, the pattern of differentiated sequences of social relations networks of urban residents is more complex, and the separation of geographic space and the territoriality of social facilities have an important impact on individual social interaction networks. Neighborhood relations in cities show the characteristics of more unfamiliar and shallow interaction. The social life facilitated by urban virtual and semi-virtual spaces represented by the community Internet has played a positive role in the integration and development of communities, effectively fostered a sense of belonging and identity in the community, and exerted a spatial and temporal reorganization effect of "re-localization" to make up for the shortcomings of traditional neighborhoods and form a new kind of neighborhood relationship. It can be said that modern neighborhoods and communities, under the influence of new technologies such as the Internet, are interpreting a new way of social interaction and forming a new community social space. In the process of their development, the formation of a common territory, the formation of social relations, the emergence of a common social life, the establishment of a community culture and the formation of a common consciousness constitute the mechanism for the establishment and maintenance of the "new community". In addition, there is a dialectical and unified relationship between the network of interpersonal social relations and space, which is a continuous process of interaction between them, with social relations being limited by space and social space mediating social relations.

Activity space is also an important part of urban living space. In the life path of families and individual residents, as the most important form of space for daily activities, the interaction between employment and residential space shows the process of change in residents' Long-term behavior space, and the constraints on residents' long-term behavioral activities are highly differentiated and distinctly different from short-term behavioral activities. Individuals' employment and migration behaviors interact closely, are jointly influenced by environmental, personal and family factors, and contribute to changes in individuals' daily activity patterns. When examining daily activity space of urban resident from life paths and life logs, it is necessary to combine important life events such as residential space, work space, economic and social roles, and changes in family structure in the life course of families and individual residents as research clues to explore their impact on individuals' daily life paths. Changes in residential space and work space bring about changes and constraints in the job-housing relationship and geographic context of each individual in the household, affecting the opportunities for minimizing coupling constraints in the performance of different daily activity spaces. The rise in social status affects the mobility of the family and the way of leisure activities by improving the economic conditions of the family. Changes in family structure and changes in the social roles of individuals within the household, in turn, directly affect the patterns of division of domestic labor within the household and the ways in which resources within the household are utilized. Life in urban settlements widens gender differences in activity and mobility, but the household acts as a coordinating mechanism that influences and constrains individuals' daily activity patterns through different strategic choices in the household, helping them to better cope with the pressures they face in their own lives, thus constituting a more complex nesting of the local order of activity. Fixed resource utilization, spatial and temporal configurations in the family form relatively fixed norms and order of behavioral activities of family members, which in turn promote the occurrence of daily activity, and the process of forming this order is also a process of continuous adaptation and interaction between human beings and space.

With the increasing demand for leisure activities and leisure space by urban residents in China, the research on urban leisure in China is gradually developing in the direction of "humanization", and the research on urban leisure has gradually shifted from the direction of leisure space, leisure system, and leisure facilities to the exploration of urban residents' leisure behavior. There are many ways to classify the leisure behaviors of Chinese urban residents, but all of them are characterized by subjective initiative, individual differences, diversity of forms, compatibility and relevance, and restriction, etc. At different stages of development, the leisure styles of Chinese urban residents have evolved under the joint influence of social values, residents' sense of well-being, and socio-economic factors, and the characteristics of liberalization, diversification, advancement, personalization and systematization have become more and more visible. Chinese urban leisure space has undergone socio-spatial differentiation, becoming a vehicle for providing differentiated leisure activities for different classes, groups and people of different social status. The diversified development of Chinese cities inevitably brings about the diversification of social classes. In the process of China's rapid urbanization, the complementary interaction between urban and rural leisure forms the "urban complex" and "native complex" in the psychology of the residents, which prompts the formation and development of the leisure belt around the city. This has led to the formation and development of urban recreation zones, as well as the differentiation of urban recreation spaces. The leisure behavior of Chinese urban residents and leisure space are mutually influential and promote each other. Leisure behavior profoundly influences spatial cognition, and spatial cognition determines the degree of experience of leisure activities. Leisure space is the carrier of residents' spirit of leisure place and leisure behavior mode, the inheritability of the spirit of leisure place is one of the important factors affecting the characteristics of leisure behavior, and residents' unique leisure behavior mode is the symbol of regionalized leisure spirit. Leisure space is the dominant content that constitutes the "hardware" of urban leisure, and residents' leisure behavior, leisure cognition and leisure perception are the important factors that constitute the "software" of urban leisure. Whether a city can become a "leisure city" in the real sense, it is necessary to analyze the subjectivity of the city as a "place" and the perception, experience, attitude of the leisure subject as well as the relationship between them, in order to break through the elements of spatial image and the development bottleneck between leisure space and urbanization. Only then can we break through the element of spatial imagery, break the development bottleneck between leisure space and urbanization, and create a comprehensive leisure city with a unique flavor that truly belongs to the residents and is based on leisure life, with an appropriate concentration of leisure service industries.

Focusing on the more representative and multi-cultural groups in urban China, and sorting out the living status of different types of groups and the multi-cultural living space landscapes formed around their lives, can reveal the historical, social, spatial, and cultural nature of the city and their interrelationships, and at the same time, help to grasp the mechanism of the occurrence of some of the current prominent social problems, especially the deep-seated mechanism of the intensification of social spatial differentiation and the accumulation of conflicts among social groups. The mechanisms. Ethnic ghettos, the living and activity spaces of groups of returnees in China, and the living spaces of disadvantaged and niche groups all show that urban space is the result of the spatial production of different social groups, and that this space is at the same time a historical, social and cultural product. It is worth pointing out that these distinctive living spaces of different groups in the city include both legal and formal spaces, such as the bar space which has become a slow cultural space with local consumption characteristics in the context of multiculturalism, and the art space which has become a balance point for artists to balance the contradiction between maintaining a realistic livelihood and adhering to the ideal of art; as well as informal spaces, such as graffiti spaces. The existence of these different types of living spaces for different social groups has led to the urban living space jigsaw puzzle with the characteristics of a collage mosaic, further leading to the diversification, fragmentation and further complication of urban spaces, which urgently needs to draw the attention of urban planning and management.

The impact of information and communication technology has permeated the daily lives of urban residents in China, generating new forms and types of living space, and these "new" living spaces are an effective complement to traditional urban living spaces. Under the background of information and communication technology, the development of virtual living space, including the deepening of e-commerce services, the spatial restruction of virtual shopping space, and the interaction between virtual shopping space and physical shopping space, has expanded the breadth and depth of urban living space, making the living space show more complex characteristics under the environment of physical and virtual interaction. Under the influence of information and communication technology, the evolution of virtual living space is characterized by fragmentation, boundary blurring and technology dependence. A paravirtual community realizes the sharing of information resources, expands the space for interaction and creates effective interaction with the real living space of residents; it also realizes community supervision and common governance through participation in public affairs, so that the bottom-up operational mechanism can play an effective role and help to build a strong aggregation-type community. Currently, "micro-life" has become a way of life, and WeChat has had an impact on residents' workspace by breaking down functional boundaries and working throughout the day. WeChat has an impact on residents' leisure space by creating experiential and autonomous consumption and leisure activities, and it has an impact on residents' social communication space by becoming an important part of residents' social interaction space through "WeChat socialization". The gradual fashionization of lifestyle is inseparable from the evolution of modern society. The combination of two major technologies, the Internet and the Internet of Things (IoT), has made today's home life more and more intelligent. Residents' fashionable consumption is mixed with the dual purpose of practicality and enjoyment, and the concept of fashionable life has created a new fashionable lifestyle. Fashionable lifestyle represents a kind of living concept and ideal of life, which has specific requirements for the functional and aesthetic performance of space, dominates people's imagination and design of space, guides the change of urban living space and profoundly influences the evolution of urban landscape.

11.2 Research Perspectives

Urban living space in China is a dynamic and fascinating topic of eternal research. With the development of information and communication technology, technological innovations have had an increasing impact on the social life of urban residents, and in addition, their lifestyles and living spaces have been influenced by cultural and consumer concepts. In this context, the urban living space in China has been constantly renewed, evolved and developed, and new topics have emerged in the research field of urban living space.

This book attempts to conduct a systematic theoretical discussion on the important aspects of urban living space in China by using a research approach based on empirical research, with a view to summarizing the characteristics and new development trends of urban living space in China. As the research idea focuses on establishing a research framework for urban living space in China and conducting empirical research on the major aspects of urban living space covered by the framework, it is inevitable that there are some omissions, and some contents have not been analyzed due to the limitation of space, which are to be improved by further research in the future.

The following is an outlook on the future academic directions of urban living space in China that are worthy of further research

(1) Comparative study of living space in Chinese and foreign cities

Researchers can select representative cities in Europe, the United States, Japan, South Korea and other countries and regions, as well as representative cities in China, and design questionnaires and interview surveys specifically targeting urban living space, and at the same time, combine the population census, housing census and other data of relevant cities in China and abroad, and carry out a comparative study on the characteristics, evolution and formation mechanism of urban living space in China and abroad.

(2) Research on regional differences in urban living space in China and their formation mechanisms

To conduct empirical research on representative cities in the eastern coastal region, central region, western region, and northeastern region of China, to explore the differences in urban living space in major regions of the country and the economic and human geographic basis of their formation, to explore the mechanism of the formation and development of the representative urban living space in each region, and to explore the driving mechanism of the formation of regional differences in urban living space in China. This is a macro-level study at the regional level.

(3) Comparative study of the living space of Chinese urban and rural residents in the context of rapid urbanization

This is an interesting topic and it is a pity that we have not yet seen any systematic research in this area. The study of traditional communities in sociology began with rural communities and expanded to urban communities. Rapid urbanization in China has led to a large number of rural populations moving into cities and towns, and quite a number of rural communities in central and western China have evolved into "hollow villages" due to population exodus and the decay of housing compounds, while villages like those in southern Jiangsu Province and northern Zhejiang Province continue to maintain their "vitality" due to the transformation of the rural economy. On the outskirts of big cities, there are also a large number of urban villages, which are inhabited by the floating population. The floating population living in the urban villages, on the one hand, lives in the towns, and on the other hand, maintains inextricable links with the villages where they used to live. Therefore, it is of great significance to study the living space of different actors in these urban and rural settlements, and to explore the impact of China's rapid urbanization after the reform and opening up on their living space, in order to decipher China's urban–rural spatial transformation and the direction of the development of new urbanization.

(4) Urban living space in China based on the perspective of consumer behavior

Researchers can focus on the current characteristics, historical evolution and development mechanism of urban living space in China from the perspectives of consumption behavior, especially shopping behavior and new consumption modes and behaviors. The study of the interaction between the consumption behavior and living space of urban residents in China, as well as the relationship between residents' consumption behavior and the use of facility space (including cyberspace) under the new consumption mode, will undoubtedly be of great significance in interpreting the latest evolutionary characteristics of urban living space in China in the new era.

(5) The living space for urban residents in China based on information and communication technology

Although this book has touched on this topic, it has not been expanded. In fact, this topic requires specialized, systematic and in-depth research to explore the characteristics and development and evolution mechanism of the lifestyles and living spaces of Chinese urban residents in the context of information and communication technology, and to study the interaction between the development of information and communication technology and the change of urban residents' lifestyles and the evolution of their living spaces, as well as the dynamic mechanism of the interaction.

(6) "Micro" living spaces for urban residents in China

With WeChat having become a way of life for residents, systematic research on the relationship between WeChat and the lifestyles and living spaces of urban residents in China, and exploring the urban living spatial structure and its evolution based on WeChat and its evolution, are of great significance for understanding urban spatial transformation in China.

(7) Research on urban living spatial structure and its evolution in China during the transition period

Based on the empirical research, according to the characteristics of urban living space in China in different periods and the background of its formation and development, researchers can explore the model of urban living space structure and the driving mechanism of its evolution in transition and post-transitional period from the theoretical perspective, and summarize the model of urban spatial structure and its evolution based on the social life in China from the theoretical level.

(8) Subcultural groups' living space in China's cities

Subcultural groups themselves are also part of urban residents, and rather than exploring the commonalities of urban living space as a whole, the thesis of this study focuses more on the differences, zooming in on various types of subcultural groups, and specializing in surveys and empirical analyses of the characteristics of their living spaces. With the development of fragmentation of urban space and the so-called post-city, post-suburbia and even post-modernism concepts, urban niche cultural groups or various subcultural groups are getting more and more attention, and their living space is an interesting research topic as it plays an enriched and complementary role in recognizing the general sense of urban living space.

(9) Research on virtual living space in urban China

The topic of virtual communities and paravirtual communities has been covered many times in this book, but not in depth due to the limitation of space as well as by the research topic of this book, which is undoubtedly a topic worthy of systematic research. Virtual living space has been integrated into the lives of ordinary residents in Chinese cities, and has become an important and inseparable part of daily living space. It is important to focus on the composition, types and characteristics of virtual living spaces in urban China, their development and evolution, and their relationship with technological evolution.

(10) Research on fashionable living space in Chinese cities in the new era

This book has only made a beginning in exploring the impact of fashionable life on urban living space; in fact, this topic requires specialized and in-depth research.

Fashionable life is a reflection or a living state of economic, social and cultural evolution, and different groups of people have different perceptions and acceptance of fashion, which requires a systematic study of the perceptions, states, characteristics and trends of development and evolution of fashionable living space of different types of people.

(11) Study on the living space of Chinese urban residents in the 5G era

The development of 5G, the fifth generation of mobile communication technology, is bound to cause a huge revolution in the field of mobile communication. China's Huawei is a global leader in the research and development of 5G-related technologies, and it is foreseeable that Chinese city dwellers will be the first to enjoy the impact of 5G on their daily lives because of China's Huawei's proximity to the city. In this context, the study of Chinese urban residents' living space in the 5G era is undoubtedly a brand-new topic of great significance. Through the relationship between the change of mobile communication technology and the evolution of urban living space, we will further explore the regularity of the relationship between technological change and urban living space from the theoretical level.

(12) Study on planning for living space of Chinese urban residents

In recent years, academics have carried out research on residents' living space planning based on residents' travel characteristics and use of public service facilities. In fact, residents' living space planning research should also be put on the agenda, especially in the context of intelligent transportation and smart city development, urban residents' living space is facing major changes in comparison with traditional living space. The physical facilities and public service facilities needed in the living space of urban residents need to be re-planned and laid out, and the spatial and temporal utilization of the living space needs to be rationally decided, all of which need to be researched first.

Index

J. Feng et al., *Urban Living Space in China*, Urban Sustainability,
https://doi.org/10.1007/978-981-95-0595-1

Q

R

S

MIX
Papier aus verantwortungsvollen Quellen
Paper from responsible sources
FSC® C105338

If you have any concerns about our products,
you can contact us on
ProductSafety@springernature.com

In case Publisher is established outside the EU,
the EU authorized representative is:
Springer Nature Customer Service Center GmbH
Europaplatz 3, 69115 Heidelberg, Germany

Printed by Libri Plureos GmbH
in Hamburg, Germany